AF183463

Landscape Bionomics
Biological-Integrated Landscape Ecology

Vittorio Ingegnoli

Landscape Bionomics Biological-Integrated Landscape Ecology

Springer

Vittorio Ingegnoli
Università degli Studi di Milano
Milan, Italy

ISBN 978-88-470-3926-1 ISBN 978-88-470-5226-0 (eBook)
DOI 10.1007/978-88-470-5226-0
Springer Milan Heidelberg New York Dordrecht London

Softcover reprint of the hardcover 1st edition 2015

Printed on acid-free paper

Springer is part of Springer Science+Business Media (www.springer.com)

"The beautiful is the root of science and the goal of art, the highest possibility that humanity can ever hope to see"
Rothenberg David (2011)
Survival of the Beautiful: Art, Science and Evolution. Bloomsbury.

"quasi novitas nos magis quam magnitudo rerum debeat ad exquirendas causas excitare"
Marcus Tullius Cicero (44 BC)
De Natura Deorum

The biperspectivable systems view of landscapes, functioning simultaneously as natural and cognitive systems, and therefore as a tangible bridge between nature and mind, open the way for close cooperation among landscape researchers and scientists from all other relevant disciplines and professions
Zev Naveh (2001)
Landscape & Urban Planning 57

Foreword

The Authors of Ancient Greece and Rome never explicitly treat the concept of landscape, even if they knew the theoretical principles to measure the territorial surface and to represent the space in perspective, e.g. through the Euclidean geometry. For instance, Herodotus, one of the first famous historians (and geographer) of the Mediterranean Regions, does not present in his works a true perception of the environment. The same we can affirm for the books reporting the history of countries not yet well known, as Gallia for Julius Cesar or Germany for Tacitus. On the contrary, a good concept of environment is shown in the Bible, representing the base of the message centred on the "Promise Land". No doubt that the contiguity between the desert and the green areas, rich of fruits and pastures, together with the religious message, leads to a clear intuition of the different life conditions offered by the environment.

A long evolution is necessary in our culture to arrive to an exhaustive conception of "landscape". Only during the final Medieval Period a new cultural relation between man and his environment begins to emerge, as we can see in the Tuscany paintings of the late fourteenth century. A first attempt to the prospective representation may be recognised in the Scrovegni Chapel (Padua, 1305), painted by Giotto; but it is limited to the inner architectures. Anyway, as cleverly highlighted by Jessup, only after few years we can find a significant turning point, given by a letter of Francis Petrarca. The famous Poet, at that time living in Provence, writes to his master in Italy describing the rise on Mount Ventoux in 1334. Arrived on the top, he gives a look towards the valley and then far away to the Alps, hence to the sea: thus, three spatial coordinates are established from the point where the Poet was placed. This is the first written document referring a prospective vision, so Jessup identifies the beginning of a new phase of human history: the Modern Eve.

In this frame, the note painter Paolo Uccello assumes a prominent position, studying the problem of perspective both on a mathematical basis and on practical experiences: he arrives to final results in his works around 1450. The perspective vision spreads rapidly in the Tuscan paint becoming the main way to represent nature. But, in general, it is a vision of nature from afar, without a true relation with the subject of the picture. An accurate representation of nature is reached only thanks to the genius of Dürer, from the landscapes of his juvenile works to the detailed Rasenstueck, a true scientific representation of a vegetation community, frequently found near the villages.

The term "landscape" is used firstly by Tiziano, in a letter to Philip II of Spain. This famous painter reproduces, on the background of his pictures, typical dolomitic landscapes which he knows very well, being born in Cadore. In Renaissance paintings the landscape is generally interpreted and represented as a perspective view of a natural or agricultural environment containing human figures, which constitutes the principal subject of the painting and the main attraction of the observer. An interesting evolution, the inversion of the goals, emerges with the landscape painters at the end of sixteenth century, not many years after Titian: the environment becomes the principal subject of the paintings, while the human figure is in a marginal position.

The cultural interpretation of the landscape in the Renaissance and near after is mainly limited to an aesthetic significance, as a component of a picture or of a story. We may frequently find this aspect in the works of the landscape painters. Usually, the nature of Central Europe is represented in these paintings, what can be understandable for the French Poussin or the Dutch painters, less for Salvator Rosa, working in Naples, or in Lorrain, working in Rome: within their pictures we can see normally the deciduous oaks, while are uncommon the Mediterranean plants, even the most typical as the holm (*Quercus ilicis*). Therefore, these representations not always result truly natural. Similarly the aesthetic sense dominates in the landscape descriptions of Canaletto, Guardi or Piranesi or, passing to literature, of Goethe or Manzoni.

The landscape assumes also a scientific significance, but only more recently: at the beginning of nineteenth century Von Humboldt realises his famous exploration of Latin America, frequently describing the landscapes, mainly characterised by the components of the tropical forest. His representation of the vegetation on the slopes of Chimborazo acquires an historical relevance as the first intuition of the vegetation belts. Von Humboldt starts from a painting vision, useful to define the landscape as an ecological unit characterised by the presence of indicator species. In his reasoning this pictorial view is analysed into its components parts, mainly the tree species which characterise the landscape as a visual element, but in other cases he may add the animals, e.g. the chamois on the Alps, the guanaco on the Andes or the kangaroo in Australia. In this way we pass from a global systemic vision of the landscape as a pictorial observation to the precise definition of its individual components, which can be considered as indicators. This method was widely developed in Europe, on the Alps by Unger and later by Kerner for the Danube area and Rikli (1942–1948) and Beguinot (1913) in the Mediterranean environment.[1]

[1] Unger F. 1836—Über den Einfluss des Bodens auf die Veitheilung der Gewächse, nachgewiesen in der Vegetation des nordöstlichen Tirol's. Wien, 366 pp. Kerner A., 1863—Das Pflanzenleben der Donauländer, Innsbruck. Béguinot A., 1913—La vita delle piante superiori nella laguna di Venezia e nei territori ad essa circostanti. Studio biologico e fitogeografico, pubblicazione n. 54 dell'Ufficio idrografico del R. Magistrato alle Acque, Venezia. Rikli M., 1942–1948—Das Pflanzenkleid der Mittelmeerländer, 3 Bde., Bern.

In the cited scientific works, as in many other of the first twentieth century, plant species with an indicator value for the landscape aspects are limited only to few very indicative ones, mainly trees (except in arid areas). Anyway, in the same period, a new paradigm spreads: in natural conditions plant species tend to group themselves according to combinations that can be repeated in similar environmental conditions: these combinations are indicated with the name of community, association, Pflanzengesellschaft and their study as phytosociology. This stimulating idea is developed by Braun-Blanquet and, following him, during the first half of the twentieth century many researchers explore with a uniform method the vegetation of Central and Mediterranean Europe. This first synthesis regards the North-Western Germany (Tuexen, 1937) and indicates the importance of understanding the plant associations: through their analysis it is possible to deepen both the ecology of individual species and the definition of the characters of a territory, in a new synthetic vision. In a series of annual conferences in Germany (prior to Stolzenau and then Rinteln) scholars from all over Europe come together and discuss methods and results: I was fortunate to have participated and I keep a very vivid memory of them. An important argument is the vegetation cartography, which in that time is mainly carried out by prospecting on the ground, having its proper centre within the laboratory directed by Tuexen.

The cartographic representation of the plant communities directly leads to the identification of reiterated patterns, corresponding to different landscape types. It is merit of Tuexen to have in-depth investigated into the individual components observable in the landscape, through the census of the presence of both individual species and distinct plant communities. In this way, we achieve a connection with the results of phytosociological analysis, which in the meantime had developed in Europe. The landscape is interpreted as consisting of a plurality of plant communities, which tend to reach a degree of complexity gradually increasing up to a maximum compatible with the available physical resources: the concept of potential natural vegetation is a consequence, which at first had a wide application. The definition of landscape with the floristic-sociologic method allows us to analyse and describe in terms of quantity and quality with uniform (and comparable) criteria the set of vegetation cover of the European and Mediterranean countries and to propose a classification for different types of landscape as a result. At the basis of this interpretation is the assumption that vegetation develops in a deterministic way, but this is a first approximation without confirmation in the real world and therefore today it seems that the concept of potential natural vegetation should be subjected to a severe criticism and review.

The phytosociological approach spreads in the first half of the twentieth century in Central Europe, to become quite prevalent here. However, it can be considered questionable for the conception of the species as a fixed entity, corresponding to the Linnaean concept. Consequently, the community also appears a sort of mosaic, divisible into a collection of small stones: a simplistic approach limited to the descriptive level. A different path is traced by scholars especially in the USA, who interpret the presence of the species as a transient state subject to change and focus the interest on the changes themselves. This represents a shift, from the simple record of the species present, to the turnover of the same due to ecological factors. In the 60s and 70s, Whittaker deepened the study of the vegetation on the basis of gradients analysis, based both on changes in physico-chemical factors than on abstract data, particularly biodiversity, for which different states (alpha-beta-etc.) were distinct and quantitative evaluation methods proposed.

Whittaker too participates at the workshops organised by Tuexen in the 60s, with very open and tolerant discussions: many of us (then young participants) remind sympathy with his work and his ideas are widely implemented mainly by Dutch researchers, but also by Italian and German one. Over the years they penetrate deeper into the Central European scientific culture leading to a gradual overcoming of a strictly deterministic concept. At the end of the 70s Whittaker deepens his experience with the study of Mediterranean-type ecosystems in Israel, in collaboration with Naveh, which even then has a wide experience of research at the regional scale. The Mediterranean vegetation here has an extraordinary diversity of species and groupings, with reports in rapid change as a result of grazing, fire and other forms of impact. A meeting lasted a few years, but that has a profound influence on Naveh, almost re-greening the biblical message mentioned at the beginning.

In his synthesis of 1984 and 1993 Naveh[2] presents a holistic view, based on general systems theory, on the landscape, which is interpreted as a complex system: it must be considered a whole (holon) with its own appearance (Gestalt), which is the result of the relationship between natural and anthropogenic factors. The prevailing view at the time, based on the results of the study of natural sciences and ecology, is overcome by an interpretation of human-centred, with openness to applications and must be supported by proper policy decisions: a goal toward which we should strive for, but that is seldom achieved. In this Naveh, starting from entirely different premises, approaches to the concept of cultural landscape (Settis).[3]

In this foreword I start from an overview of the long history of the development of the concept of landscape in Western culture, first in the artistic vision and then as a subject of scientific inquiry. I then detail the more recent developments of which I have witnessed (and in part also co-actor), in an attempt to outline the state of the art on this issue. These developments are the subject of many discussions with Vittorio, and also of a reflection together, which was published recently.[4] But still many insights are possible, which could also develop important application aspects. So I leave it to Vittorio the difficult task to continue and deepen the discussion in this new book, and I wish the reader to derive pleasure and benefit.

Rome
January 2014

Sandro Pignatti

---

[2] Naveh Z., Lieberman A.S., 1984—Landscape Ecology. Theory and Applications. 2nd Edition 1993. Springer, New York.

[3] Settis S., 2010—Paesaggio Costituzione cemento. La battaglia per l'ambiente contro il degrado civile, Torino, Einaudi.

[4] Ingegnoli V. e Pignatti S., 2007—The impact of the widened landscape ecology on vegetation science: towards the new paradigm. Rendiconti Lincei 18: 89–122.

Preface

A very short history of this discipline and of its name has to be premised. At the beginning of my studies on the environment, I, as a post-graduate scholarship student in Territorial Planning at the Polytechnic of Milan, presented a synthetic but crucial work at the International Congress on "Environment and Engineering", held at the Science Museum of Milan in 1971: the paper was titled "Ecologia Territoriale e Progettazione" (Territorial Ecology and Design). The main principles of my studies contained in that work constituted the basis of a new discipline, the development of which put me in touch with different researchers as S. Langé (Polytechnic of Milan), O. Ravera (University of Milan), K. Buchwald (University of Hannover), S. Pignatti (University of Rome), R.G.H. Bunce (University of Lancaster), F. DiCastri (UNESCO, Paris), RTT. Forman (University of Harvard) and Z. Naveh (University of Haifa).

At the beginning of the Eighties, DiCastri, as the International Coordinator of Ecological Sciences, asked me to refer my studies to Landscape Ecology, recently recognised as a new important branch of Ecology at the World Congress of Veldhoven (1981). After a meeting with R. Forman in Harvard and some workshops with Sandro Pignatti and Zev Naveh in Central Italy, I decided (1990–1991) to reach a second degree in Natural Sciences to properly develop this field of studies.

After it (1994–1995) I had the possibility to deepen my researches and to participate to significant landscape studies, e.g. on vegetation in the Sila Piccola Park, CNR (Italian National Council of Research) and on landscape structure in South Africa with the University of Natal (together with M.J. Samways and J. Ott). I was also invited to write the chapter on Landscape Ecology in "Frontiers of Life", edited by Baltimore, Dulbecco, Jacob & Levi-Montalcini.

In the early 2002, Richard Forman was asked by Springer to give a title to the new book of mine. Forman, who wrote the foreword, suggested "Landscape Ecology: a Widening Foundation". The reason is clear: the discipline re-founded by me had taken a new direction, becoming "the biological integrated landscape ecology".

This fact is inserted in the acquired observation that, today, we have two main Schools of Landscape Ecology: the American and the European-Mediterranean one. The first is mainly devoted to the study of the reasons of the spatial distribution of species and communities on the territory; the second is more holistic and concerns the landscape functions and the human presence on territory too, as pointed out by Peter Weisberg (Reno University) in a recent seminar at the University of Ancona. The American School of landscape ecology is important, because it opened general ecology to scale problems and spatial structure of ecosystems. The importance of the holistic School results in allowing the landscape to be recognised as a living entity. If this concept is properly developed within a biological-integrated landscape ecology, it brings inevitably to a more complete discipline, which I proposed to name "landscape bionomics" (2010).

The recognition of the landscape as a specific level of organisation of life on Earth leads to very significant changes both in the definition and the assessment of a landscape. At a territorial scale, in a given geographical area, the "landscape" is defined as the "biological integration of a set of plant, animal and human communities and of their system of natural, semi-natural and human-cultural ecosystems in a certain spatial configuration". The meaning of this hyper-complex system is to be a living entity, not an inconsistent set of separate issues and themes (water, air, soil, species, pollution) in which some interrelations can be found!

Inevitably, the changes in how to assess and manage the environment follow. Indeed, we recognise the structures and functions of each landscape, which is the peculiar behaviours that go beyond the traditional relationships among the components, due to systemic laws. Thus, one can speak of a "health" and a number of syndromes (or disease) of the landscape. This fact is very important, because it has been demonstrated that the pathological changes of a landscape, or part thereof, can affect human health too, even in the absence of pollution! Moreover, we must note that a discipline like this follows criteria similar to medicine thus considering an ecologist as an "ecoiatra".

This new direction especially forces to a deep change in the applications of ecology. The landscape units (LU) need a diagnosis of their healthy state and need ecological models and indices able to evaluate these hyper-complex systems in their intrinsic integrated components. First of all, vegetation science must correspond to this vision, because mainly vegetation (composed by autotrophic species), both natural and anthropic, has the task to structure a landscape as living entity. This correspondence changed many concepts of vegetation science, as exposed by Ingegnoli and Pignatti in "The impact of widened landscape ecology on vegetation science: towards the new paradigm" (Rend. Fis. Acc. Lincei, Ingegnoli and Pignatti 2007).

This new direction enhances also the cultural and ethic implications, because men's responsibility on Nature and Society, thus on the whole Creation, grows drastically, passing from a geographic structure to a biologic one! Even the concept of sustainability should change, because it is no more sufficient that economy has to consider "also" the ecological aspects: on the contrary, economy should be recognised as a chapter of ecology!

Pope John Paul II, making Saint Francis the Saint Patron of Ecology to underline his care for creation, said: "*not to behave like dissident predators where nature is concerned, but to assume responsibility for it, taking all care so that everything stays healthy and integrated, so as to offer a welcoming and friendly environment even to those who succeed us*". Remember that Francis of Assisi arrived to affirm:

Be praised, my Lord, through our lady Mother Earth, who feeds us and rules us, and produces various fruits with coloured flowers and herbs.

Therefore, our Mother Earth is able not only to feed us but also to "govern us". This observation precedes and reinforces what science recently understood, recognising that a landscape is a specific level of organisation of life, able to rule even the most high creature, man, indicating him to assume responsibility for it, otherwise the "ecosystem services" can be switched in *peril* for men and the entire life. To assume responsibility we have to know how; without knowledge it is properly difficult to follow the laws of Nature managing our environment and our landscapes. Pay attention to the fact that you can intervene on the environment with the best intentions but causing damage!

This is the main focus of the book: the need to study the "landscape units" through methods following "clinical diagnosis", considering ecologists as "doctors" of ecological systems, i.e. "*ecoiatri*", and the need to implement the knowledge of the laws concerning our environment, facilitating our responsibility towards it: these are aspects contained in the title of this book, from the Greek, then Latin, "Bionomia", meaning "doctrine of the laws of life".

As this discipline is *like medicine*, the concepts of "man's impact on nature" and "strategic environmental assessment" change their significance becoming the ways to verify landscape pathologies, to suggest therapy and to check the capability of the involved subject to survive to the operations without damage. A comparison with the normal state of bionomic systems is needed:

- Man's actions on nature bring to an impact only if they are out of scale; otherwise
- We are in presence of Coevolution.

Thus even "mitigation and compensation" concepts must turn into "strategic interventions of rehabilitation and therapy", which has to be performed on a living entity!

To reach these challenging aims, we divided the book in chapters, which can be grouped into the following (conceptual) sections:

1. *Theory,* Chaps. 1, 2, 3, 4 —The definition of landscape as a living entity needs an hyper-complex model and both physical and biological basis. We must follow a correct epistemology to create the new discipline of Landscape Bionomics. After a synthesis of System Theory, we have to explain landscape structure (anatomy) and functions (physiology). The landscape evolution, alteration and pathology are the remaining arguments.
2. *Analysis,* Chaps. 5, 6, 7—A more consistent pool of ecological-bionomic methods, indicators and indices are presented on vegetation, fauna, human population and communities and on the entire context. These analyses bring new important contributions especially in vegetation science and urban ecology.

3. *Evaluation and diagnosis,* Chaps. 8, 9—Non-equilibrium thermodynamics absolutely leads to an historical evaluation of the landscape, an indispensable tool for a good diagnosis. Specific criteria and methods for landscape diagnostic evaluation have been added.
4. *Therapy,* Chaps. 10, 11—Ecological therapy and environmental design (and planning) must be expressed in a wider methodology, which brings to the figure of the physician of the landscape (ecoiatra).
5. *Applications,* Chaps. 12, 13, 14—Experiences derived from European researches in which the Author applied his studies: e.g. CONECOFOR (Forest Focus, with the EU Forest State Services and Gottingen University), rural landscapes (with Leuven University), urban parks (comparison among Milan, Wien, Berlin), urban landscapes (Berlin, Milan). These experiences are applied in eight case studies: on Alpine landscapes, on plain landscapes and on the comparison between suburban-rural landscapes near Milan and Brussels.
6. *Ethics,* Chap. 15—Being the landscape a living entity, we must underline the bio-ethical principle, linked also with epistemological and economical aspects of sustainability.
7. *Main abbreviations & definitions*—A sort of brief glossary, edited by Elena Giglio: very useful.

We initiate in these last years to collaborate with the university of Leuven (Belgium) to study the suburban-rural landscapes. So, I have to underline that this book, written by Ingegnoli, add some other valid researchers in Chap. 14.

Milan
January 2014

Vittorio Ingegnoli

Acknowledgements

The recent evolution of my thinking has been influenced by discussion with Virginio Bettini, Fiorenza De Bernardi, Richard T.T. Forman, Hubert Gulink, Martin Hermy, Giuseppe Carlo Lozzia, Giuseppe Magro, Ernesto Marcheggiani, Sandro Pignatti. Good discussion also took place with many students of my graduate classes at the university of Milan (included the Mountain University of Edolo) and with some post-graduate students at the universities of Leuven and Venezia. I thank each of them warmly.

A particular mention to Hubert Gulink, Frederik Lerouge and Ernesto Marcheggiani for the valid collaboration in the research on rural-suburban landscapes near Brussels and Milan and to Sandro Pignatti for the discussion on the research on Alpine forests and the concise bionomic state of vegetation (CBSt).

Encouragements to persist in my studies, notwithstanding the difficulties to survive as researcher in my country, came from many colleagues and friends, especially Domenico Albanese, Anna Barbati, Andrea Castellotti, Stefano Bocchi, Francesco Lombardo, Almo Farina, Marco Ferraguti, Andrea Galli, Anna Giorgi, Tino Langé, Juergen Ott, Bruno Petriccione, Nicola Saino, Giuseppe Scarascia-Mugnozza, Francesco Sassi, Enzo Siligardi, Claudio Smiraglia, Carlo Soave, Alessandro Toccolini, Peter J. Weisberg.

Many sincere thanks to Antonella Cerri and Andrea Ridolfi, editors at Springer Milan, and to Fiorenza De Bernardi, Alessandro Ingegnoli, Ernesto Marcheggiani, Sandro Pignatti, for reviewing portion of this book. A special appreciation to Elena Giglio Ingegnoli, who reviewed the entire manuscript and edited a short glossary, with an exceptional competence.

Contents

1 Landscape Bionomics and the Theory of Living Systems

1.1 Life and Its Organisation on Earth

1.1.1 The Basilar Biologic Characters of Life

> "...Discoursing in my last letter on the different practice of the Italian and Dutch Painters, I observed that 'the Italian Painter attends only to the invariable, the great, and general ideas, which are fixed and inherent in universal nature'. I was led into the subject of this letter by endeavoring to fix the original cause of this conduct of the Italian Masters. If it can be proved that by this choice they selected the most beautiful part of the creation, it will show how much their principles are founded on reason, and, at the same time, discover the origin of our ideas of beauty...". "...This it is our business to discover and to express..." as "...the art which we profess has beauty for its object...".

What is the significance of the previous sentences of Sir Joshua Reynolds [1], dating back to November 10, 1759, in relation to landscapes and life? What's the thread among Italian Masters (Fig. 1.1), beauty and landscapes? Both art and science focus on the discovery of unity among variety and diversity of our experience and of Nature: as Pythagoras and his scholars have been asserting since sixth to fifth century BC. "...the beauty of a system consists of the proportion among the components parts..." or "...harmony is the unification of diversity...". More recently, the Emerging Properties Principle asserts that "...an organic whole is more complex than the sum of its parts..." and that "...characters of the system are the consequence of the way in which the elements organise themselves".

The harmony of a territory with the laws of Nature breeds the beauty of a landscape, which has not only a visual value that can be perceived (semiology), but a deeper ecological and medical valence[1] too: so the capacity of "measuring", better quantitatively esteeming, this beauty becomes a crucial point. Landscape bionomics permits it through its capacity of evaluation of the ecological state of a landscape (Fig. 1.2).

As it has been observed by many authors, from Ashby [2] to Prigogine [3], "...landscapes, as well as parts of them, are created by Nature as its most sensitive, fine and complex structures, through non reversible processes which are time oriented. "...non-equilibrium thermodynamics becomes the most important physical discipline when complex adaptive systems, such as landscapes, are involved with life processes..." [4]. Here again, harmony and beauty related to vegetation, landscapes and life on Earth come

[1] The heavy importance of having visual access to and of being within quality landscapes in developing cognitive, emotional and behavioural connections, encouraging imagination and creativity, cognitive and intellectual development and social relationships, improving job and school performance, encourages learning, inquisitiveness and alertness, alleviating mental stress and illness, restoring the mind's ability to focus, is by now a sound and reliable scientific point.

V. Ingegnoli, *Landscape Bionomics Biological-Integrated Landscape Ecology*,
DOI 10.1007/978-88-470-5226-0_1,

Fig. 1.1 *The Ecstasy of St. Francis* (or *St. Francis in the Desert*), oil on panel, painted by the Italian Renaissance master Giovanni Bellini between 1475 and 1480. It is now housed in the Frick Collection in New York City

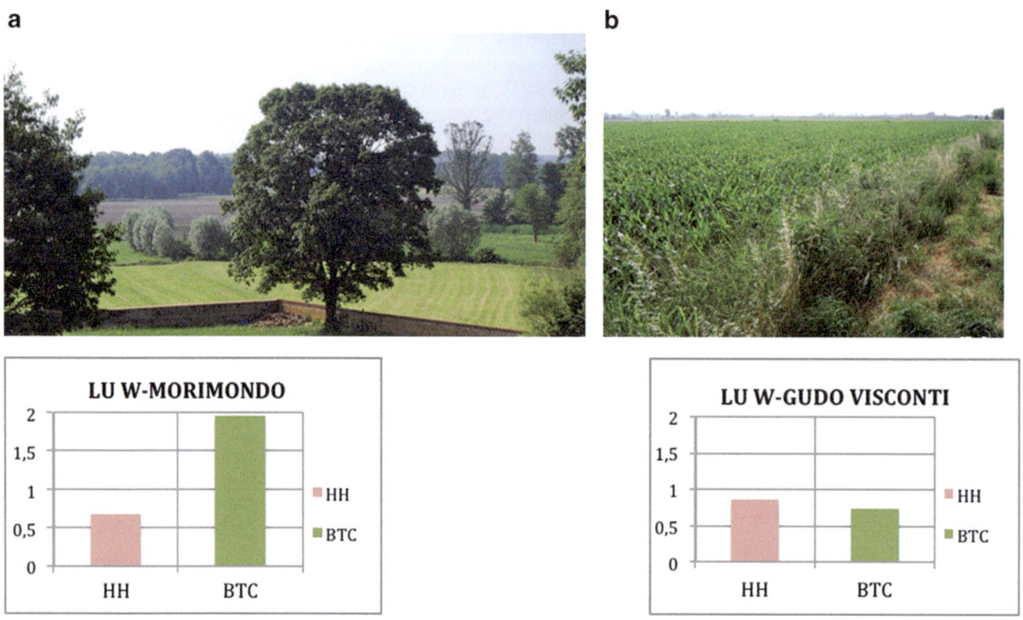

Fig. 1.2 (**a**) A portion of an agricultural *protective* landscape near Morimondo Abbey (Lombardy) with an acceptable beauty value, as confirmed by the good level of its Biological Territorial Capacity; (**b**) a portion of an agricultural *productive* landscape near Gudo Visconti (Lombardy) with a scarce beauty value due to its banalization, as confirmed by the worst level of its Biological Territorial Capacity (agricultural landscape average BTC value in Lombardy is around 1,3 Mcal/m^2/year; HH = human habitat % of landscape unit (LU))

Fig. 1.3 An alpine farm in Tirol, near Nauders (Austria), about 1,500–1,600 m a.s.l. This portion of landscape, constituted by different types of ecocoenotopes, surrounded by a *Spruce forest*, confirms that life at any scale cannot exist without its environment

from non-equilibrium thermodynamic and indeterminacy.

So, are we able to define the term "Landscape" in a way comprising the whole concept? The European Landscape Convention (ELC) held in Brussels on March 1st, 2004, tried to define the concept of landscape with a compromise between the visual and the scientific meanings, in order to "promote landscape protection, management and planning, and to organise European co-operation on landscape issues". The adopted ELC definition of landscape is the following: The landscape means an area, *as perceived by people*, whose character is the result of the action and interaction of natural and/or human factors".

This compromise forced the ELC to ignore the advanced definitions of landscape, whereas the text of Naveh and Lieberman has been underlining, at least since 1984 [5], the necessity to study the landscape through the System Theory , and the book "Frontiers of Life", edited by four Nobels in 1999, in the chapter on Landscape Ecology, written by Ingegnoli [6, 7], defined the landscape in a way still more innovative: "an adaptive complex system of biogeocoenosis as a specific level of biological spectrum. System of ecocoenotopes repeated over the land".

Trying to better understand the reported definition of landscape, we have to start observing that Life is a complex self-organising system, operating with continuous exchange of matter and energy with the outside; the system is able to perceive, to process and transfer information, to follow rules of correspondence among independent worlds (coding), to reach a target, to reproduce itself, to have an history and to participate in the process of evolution. Moreover, we observe that, in an evolutionary view, structure and function become complementary aspects of the same evolving whole.

Consequently *life cannot exist without its environment*: both are the necessary components of the system, because life depends on exchange of matter and energy and information between a concrete entity, like an organism or a community, and its environment [8, 11]. That is the reason why the concept of life is not limited to a single organism or to a group of species and therefore life organisation can be described in hierarchic levels (i.e. the so-called "biological spectrum" *sensu* E.P. Odum [12, 13]). The world around life is made also by life itself (Fig. 1.3); so the integration reaches again new levels. This is the reason why biological levels cannot be limited to cell, organism, population, communities and their life support systems: life also includes ecological systems such as ecocoenotopes [10], *landscapes*, ecoregions and the entire ecosphere (ecobiogeosphere). As all remember, the Gaia hypothesis [14] has already asserted that the Earth itself is a living entity.

SCALE	BIOTIC Viewpoint*	FUNCTIONAL Viewpoint**	SPATIAL CONFIGUR. Viewpoint***	HUMAN CULTURAL Viewpoint°	**LIVING ENTITIES °°**
Singular	Organism	Organism niche	Living space	Cultural agent	Meta-organism
Stationary	Population	Population niche	Habitat	Cultural site	Meta-population
Local	Community	Ecosystem	Micro-chore	Historic-cultural district	Ecocoenotope
Territorial	Set of communities	Set of ecosystems	Chore	Historic-cultural landscape	**Landscape**
Regional	Biome	Biogeographic system	Macro-chore	Historic-cultural region	Ecoregion
Global	Biosphere	Ecosphere	Geosphere	Noosphere	Ecobiogeosphere§

* biological and general-ecological criterium; ** traditional ecological criterium; ***not only a topographic criterium, but also a systemic one (Crf. Emergent Property Principle);
°cultural, intended as a synthesis of anthropic signs and elements; °° types of living entities really existing on the Earth as spatio-temporal- information proper levels; §remember the "Gaia Hypothesis".

Fig. 1.4 Hierarchical levels of biological organisation on the Earth

The landscape as specific "system of ecosystems" has been defined by many scientists: Giacomini [15], Buchwald and Engelhart [16], Ingegnoli [17, 18], Walter [19], Lorenz [20], Naveh & Liebermann [5], Forman & Godron [21], Odum [13, 22], Pignatti [23], Leser [24], Meffe and Carroll [25].

A few of these scientists, however, fully realised that the landscape is a life entity. In fact, remembering the concept of "noogenesis" [26]—available both for life and thought—and considering the definition of life as "the attitude to order the disorder", if we recognise some intrinsic behaviours in a complex adaptive system of ecocoenotopes (i.e. landscape), we have to define this entity as a living one. Let's go deeply.

1.1.2 The Landscape as a Peculiar Biological Level

Investigating the *environment* at different scales, it is easy to note (Fig.1.4) the present existence of four parallel hierarchies, respectively based on the biotic viewpoint, on the functional viewpoint, on the spatial (configuration) viewpoint and on the cultural (human) viewpoint, all recognising only five levels (the "white rows" in the table).

But the real environment is constituted of six levels. In fact, it is necessary to consider two hierarchic levels in the middle "biological spectrum": (1) the ecobiota, composed of the community, the ecosystem and the microchore (i.e. the spatial contiguity characters, *sensu* Zonneveld [27]), which we will name *ecocoenotope*, and (2) the *landscape*, formed by a system of interacting ecocoenotopes (the "green row").

No doubt that some characters of community and ecosystem are available also at landscape level and even the inverse is true: it is only reductionism which pretends to separate all the characters related to each level. For example, processes allowing the definition of life are *exportable* characters: each specific biological level expresses a process in a *proper* way, depending on its scale, structure, functions, amount of information and semiology. But we can note that each biological level presents *exportable* characters and *proper* ones [10]: Table 1.1 shows a synthesis of them.

Observe that each system which owns *proper* characters is an *entity*, and we can find *emergent* properties characterising cell, organism, population, ecocoenotope, landscape, ecoregion and ecosphere. That is why these levels are six types of concrete living entities, whose

Table 1.1 Schematic representation of the main characters of the life systems having a strong ecological interest

Biological levels and range of scales	Main proper characters (Ecological chapters)	Exportable characters
Organism S = 10^{-2}–10^{6} m^2 T = 10^{-3}–10^{3} years B = multicellular E = vital space	Genetic integrity, phenotypic growth, discrete bodily form, physiological autonomy, metabolism, ethology, etc. (Auto-ecology)	Basic bio-systemic: (structure, dynamics, reproduction, maintenance, etc.)
Population S = 10^{0}–10^{9} m^2 T = 10^{-1}–10^{3} years B = organisms E = minimal habitat	Genetic similarity, ecological density, age distribution, birth/death ratio, logistic growth, social behaviour, etc. (Population ecology)	Carrying capacity, habitat, etc.
Ecocoenotope S = 10^{2}–10^{8} m^2 T = 10^{0}–10^{4} years B = species/ environment E = site	Dominant/rare species, niche, succession, trophic web, speciation, competition, foraging, etc. (Community and ecosystem ecology)	Energy flux, biodiversity, disturbances incorporation, etc.
Landscape S = 10^{6}–10^{10} m^2 T = 10^{2}–10^{5} years B = ecocoenotopes E = land	Permeant populations, source-sink dynamics, ecotope role, landscape apparatuses, transformation control, etc. (Landscape ecology)	Spatial contiguity characters, context conditioning, etc.
Ecoregion S = 10^{10}–10^{12} m^2 T = 10^{3}–10^{6} years B = landscapes E = region	Biogeographical processes, fluvial basin ecology, regional geomorphic processes, zonal climate characters, etc. (Eco-geography)	Land compensation, biome characterisation, etc.
Ecosphere S = 10^{13}–10^{14} m^2 T = 10^{7}–10^{9} years B = ecoregions E = world	Atmospheric and oceanic bio-equilibrium, thermal balance and vegetation, organic limestone and plate tectonics, etc. (Global ecology)	Biogeochemical cycles, climatic cycles, etc.

Main dimensions: *S* space, *T* time, *B* biotic components, *E* environmental components

investigation needs these criteria to be integrated, better to be reconceived, remembering that *any ecological system* must include both a biological element and its environment, plus its cultural/ information contents.

The first consequence is that we *cannot describe the behaviour of a landscape* scaling up an ecological system of communities. Therefore, the use of computer clustering landscape indicators has to be very controlled and *strictly limited.*

Similarly, even if the use of the exportable characters of a landscape, the chorological ones, is useful in different branches of ecology, because the spatial aspect may aid in studying many levels of biological organisation, note that landscape ecology focused the attention on scale in ecology, but the landscape itself is not a concept valid at any scale!

As underlined, the landscape presents diverse intrinsic/proper characters, which cannot be found and studied in any other biological levels, e.g.

- Ecotissue structure of the landscape
- Peculiar landscape element structure
- Urban regions multiple structures
- Context role of ecotopes
- Landscape apparatuses
- Landscape efficiency of vegetation
- Permeant animal populations
- Source–sink dynamics
- Context control of transformation processes

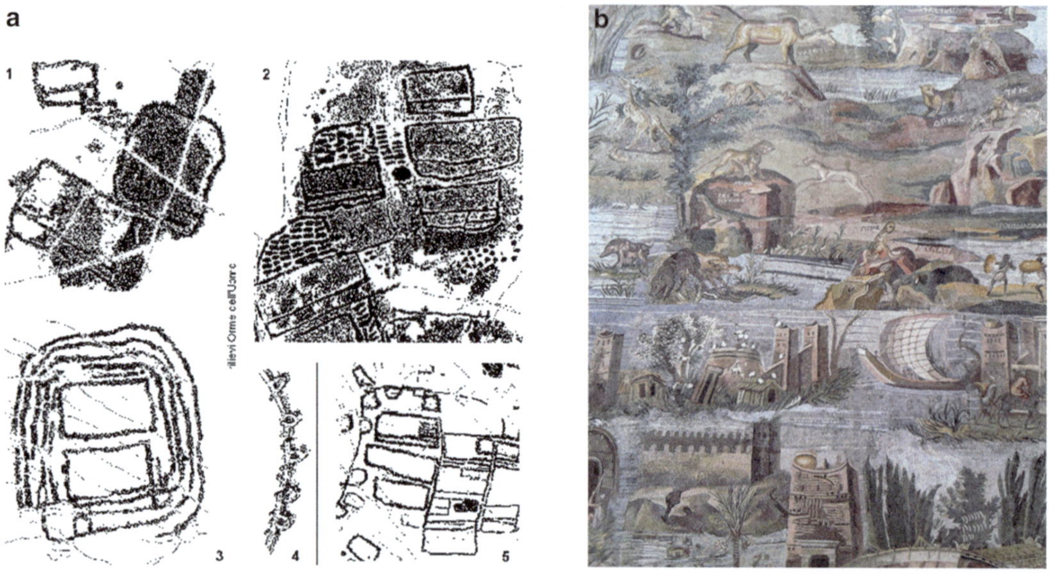

Fig. 1.5 (**a**) Paspardo, Vite, topographic compositions Camonica Valley (Lombardy) about three millennia BC (from Arcà, A. In: Tracce 9, *2nd International Congress of Rupestrian Archaeology,* 1997 Darfo Boario Terme) and (**b**) a detail of the roman mosaic of the landscape of the Nile river, about II BC, (from Palestrina Archeological Museum, (Latium), photographed by Ingegnoli) (A deep discussion regards this mosaic, concerning the question if it could or couldn't be a reflection of a type of topographical painting typical of Alexandria (Egypt) in the late third or early second century BC, related to the beginning of geographical studies and to the figure of Demetrios the topographos, the earliest recorded landscape painter (topographos: Diodorus Siculus: History XXXI.18.2)

- Human habitat/biological territorial capacity of vegetation function correlation
- Ecotope reproduction process
- Bionomic imprinting on culture
- Complex biodiversity metastability
- Landscape influence on genetics
- Landscape pathologies and their influence on man, etc.

Obviously, these characters cannot be explained in this paragraph: this book is written just to study and understand all these concepts. Let's begin taking a step backward.

1.2 The Concept of Landscape

1.2.1 The Evolution of the Concept of Landscape in the Western World

As pointed out in the book "Landscape Ecology, a Widening Foundation" [10], in Europe the first representations of topographic maps of a territory date back to the Chalcolithic Age, at the beginning of the sub-Boreal period, about five millennia by present, characterised by a cooler climate and dominated by oak and alder trees (*Quercus robur, Alnus glutinosa,* etc.). This fact is very important, because it demonstrates a high level of conceptualisation of the elements of the landscape. Well-known prehistoric rock carvings are present in the Alps, from Mount Bego (France) to Val Camonica (Italy). The map of Paspardo (Fig. 1.5a) (Val Camonica, Lombardy [28]), for instance, shows the representation of cultivated fields, paths, rivers, canals and houses as topographic drawings on the smooth local rocks. Therefore, it is not insignificant that a map of a landscape appeared at least three millennia before Christ, before its denomination through a written word. The parallel evolution of the sub-alpine peoples with the local landscapes after the last glaciation (Dryas period, 9000–8000 BC) led to

Fig. 1.6 (**a**) Frontispiece of a reprinting of De re rustica, got in Paris, by R. Estienne on 1543. (**b**) Landscape drawn by Leonardo da Vinci in 1473, now in the Uffizi Museum, Gabinetto delle Stampe, Florence

sub-alpine or mountain agricultural landscapes, in which the natural and the human patches were so well integrated and defined to be able to remain extremely stable during the following four or five millennia.

In the Roman world, the term landscape was *regio-regionis*, later *pagus*, which emphasised a geographic aspect, while the term for scenery was *prospectus* [29]. A distinction between the visual-artistic and the geographic-ecological meaning appeared in the definition of a landscape painter: "*pictor topiarius, qui regiones formas pingit*" that is "country painter, who paints the shapes of a landscape" (Fig. 1.5b).

Moreover, at the beginning of the Christian Age, some prodromes of ecology appeared in the scientific field of agronomy. Well-known treatises, like *De re rustica* by Columella [30], were not only practical but also theoretical, as shown by the description of the process of soil fertility, etc. They presented some ecological notes concerning the man-landscape relationships, too, such as the assertion that "no field is tilled without profit if the owner, through much experimentation, causes it to be fitted for the use which it can best serve" and that it "is important to plan the right proportion between the farmstead and the territory" and were reprinted many many times (Fig. 1.6a).

The emergence of the Renaissance in the fourteenth century in Italy reintroduced the concept of landscape with new interest: a very important role was played by the discovery of perspective rules. During Roman times, perspective was known, but without all its complex geometrical rules, while in the Medieval age the sense, even intuitive, of perspective seems to have been lost. The artistic world of the Renaissance was the source of many fields of science and many artists, especially painters, studied and drew the landscape in a perspective representation (e.g. Piero della Francesca). Perspective rules are important, because they cause a deep consciousness of the point of view and of the scale, and this is crucial, even for science. While drawing, Leonardo da Vinci (Fig. 1.6b) began asking why the elements of the landscape had been shaped in that particular way and what had been the reasons for their development [31]. Leonardo compared the elements of a landscape with the components of an animal body, recognising not only a structural sense, but also a functional one. Through those

Fig. 1.7 Landscape of the wide valley of Adige near Lavis (Trento), view from Fai della Paganella

studies (Fig. 1.6b), Leonardo founded the discipline of geomorphology and was conscious of the concept of vegetation.

1.2.2 The Landscape Between Visual and Scientific Meaning

In the Modern Age, the "Grand Tour", which all cultured Englishmen and northern Europeans made through the Alps to Italy in the seventeenth, eighteenth and nineteenth centuries, brought them in contact with rugged, picturesque scenery. Many painters portrayed these landscapes (e.g. Salvator Rosa, Claude Lorrain), and all those influences led to English landscape architecture. The landscape garden was a product of the Romantic movement: its forms were based on direct observation of nature and the principles of painting.

The biogeographer Alexander von Humboldt, in Germany, was the first to give a scientific definition to the concept of landscape (*landschaft*) [32] as *Der Totalcharakter einer Erdgegend*, that is *the total character of a given territory*, meaning both the perceptive and the natural aspects (Fig. 1.7).

The non-classical principle of wilderness, particularly strong in the United States, was opposed against the French idea of a deterministic (Cartesian) nature. Thoreau wrote in 1854 [33]: "In wilderness lies the preservation of the world". The concept of landscape was strictly linked to the concept of wilderness. Like the poet Wordsworth and the painter Constable in England, and like Emerson and Thoreau in the United States, Frederick Law Olmsted [34, 35] felt the great moral appeal in natural landscape beauty and wilderness and the necessity to preserve natural areas. The first preservation area of natural wilderness and scenic landscapes was created in Yosemite Valley, in California (1864), a few years later designated as the first National Park in the world by the government of the United States (1872).

Nonetheless, founding the first School of Landscape Architecture at Harvard since the end of nineteenth century, Olmsted was not able to develop the concept of landscape in a scientific way. During the past century the landscape was defined mainly as a cultural and visual subject, studied by architects and geographers. The etymology of the English name, *landscape*, that is "everything you can see when you look across an area of land", prevailed on the Latin etymology of "*pagus*", mainly geographic and more suitable for scientific purposes. The English meaning was applied even to the term "paesaggio" (Italian) or "paysage" (French), confusing their meaning with "panorama" (the same word in English and Italian). Only after the preliminary European studies (1950–1978) of Troll [36, 37], Zonneveld [38], Buchwald & Engelhart [16], Ingegnoli [17], Finke [39], Leser [40] and the international Congress of Veldhoven (The Nederland's, 1981), the scientific concept of the term landscape began to arise. After this congress, in 1982, the IALE (International Association for Landscape Ecology) was founded and a series of new, specific books were published: Ingegnoli [41, 42], Naveh and Lieberman [5], Forman and Godron [21], Farina [43].

1.3 Scientific Criteria at the Base of the Study of a Landscape

1.3.1 The Physical Basis of Life: Space-Time and Complexity

In summary, between life and its environment there are strict relationships, exchange of energy, matter and information and a priori knowledge.

Following the Theory of Relativity (Einstein), let us remember that not only energy and mass are transmutable, but even space and time. Energy can be organised as matter or information, depending on different codifications of the chronotope. In facts, the Chronotope shows four dimensions. When energy is transformed in matter it assumes three spatial dimensions (x, y, z) plus one temporal dimension (t); while if energy is transformed in information it assumes two spatial dimensions (e.g. plane wave) and two temporal dimensions (t_1, t_2).

As stated by Manzelli [44], when the visible light frequencies cross a transparent medium, the associated plane wave remains dimensioned as information (two spatial and two temporal dimensions); on the contrary, when the wave encounters the retina, the photochemical reaction is done through the conversion in a particle of the plane wave, which assumes a form available to interact with the three-dimensional structure of the matter. It is important to underline these facts, because any transformation between energy and matter needs a sort of *catalysis* through an information system, to increase the neg-entropy and to proceed toward ordered forms.

We know that the exchanges energy-matter-information, which permits the emergence of life on Earth, are of the greatest importance and changed completely the evolution of the entire Planet. A mutual interaction and an information exchange are present even between life and his environment: a sort of "a priori" knowledge.

Konrad Lorenz [20] defined life as "a process of knowledge" and Karl Popper [45] wrote: "From the beginning, life must have been equipped with a general knowledge, which we usually name 'knowledge of the natural laws'". Note that the current definition of adaptation is mainly a Darwinian one and must be changed, because it is not seen as a form of *a priori* knowledge. In facts, the definition of life contains both biological systems *and* their environment: therefore every living system follows life processes and exhibits systemic attributes.

The need of a sort of catalytic process to introduce information in a system and the definition of life as a process of knowledge within so complex hierarchic levels of biological organisation reveal the difficulties in studying the biological systems. These systems are inevitably complex or even hyper-complex once.

The complexity of ecological systems is difficult to investigate, especially if we consider landscape ecology in the same way as medicine, because clinical analysis, diagnosis and similar methods need a rich theory to be applied. Complexity may be defined as the attribute of a system (natural or cultural) which contains information that is hard to understand [46]. Complexity does not depend on the number of components of a system, but especially on the type of interaction among them. It is possible to classify at least four types of complexity:

1. *Organised simplicity:* systems formed by few components with simple interactions, which mathematics can formalise (e.g. Lotka-Volterra equation).
2. *Unorganised complexity*: systems formed by a very high number of components, with casual interactions. They may be formalised by statistical analysis and deterministic chaos (e.g. turbulence theory).
3. *Organised complexity*: systems formed by a medium number of components, with organic interactions. They cannot be completely formalised by mathematics (e.g. biological-ecological systems).
4. *Irreducible complexity:* a single system which is composed of several interacting parts that contribute to the basic function, the removal of any one of which causes the system to effectively cease operating.

Biology, thus ecology, is interested in all four types of complexity, but crucial problems generally deal with the third and fourth type. In biology, experimental data cannot be very precise, even if collected over a long period; moreover, it is practically impossible to get equations of the temporal evolution of an ecological system. Actually, it must be remembered that a biological system is able to learn (it is an adaptive system), since it is continuously changing.

On the other hand, the mathematical theory shows even the theoretical impossibility of dealing with a perfect formalisation of the third and fourth type of complexity: once some inference rules and a number of axioms are fixed, we may find exact assertions which are impossible to demonstrate as being true or false. This is a basic theorem of Gödel [47].

1.3.2 Epistemological References

So, the landscape is a living entity, represented by a hyper-complex system [6, 7], very difficult to be studied. Therefore, we need to follow the best epistemological[2] criteria in order to create the most appropriate discipline.

Perhaps the biggest limitation of the modern scientific method depends on the principles adopted by the science of motion (mechanics) since Galilei—followed by Newton, Leibnitz, Lagrange and Laplace—which became the model for every possible form of science (according to the hierarchy: mathematics, physics, chemistry, biology, psychology and sociology). Even those sciences which were concerned with aspects of reality quite dissimilar, as the phenomena of life, they had to adapt. The method, called *reductionism*, which consist of reducing all phenomena to simpler structure below, in the belief that the simplest of all is the physical, was applied to all fields of science (a remnant of neo-positivism).

This epistemological approach, however, in the world of biology is highly controversial and limited because it follows a great relief, as well explained by Konrad Lorenz [20] and Giorgio Israel [48]: the destruction of any notion of subjectivity, design and finality, the key to biological thought. Even Jacques Monod [49] -still Israel remembers- had in mind the epistemological contradiction between scientific objectivism, which implies the rejection of any kind of project, and the teleonomic character of living beings.

In fact, the only method we have for understanding remains the one "try and error", guided by the principle of approximation to the truth, considered as absolute. We note, with Popper, that if the truth were not absolute and objective we could not err, because our errors would be as good as our truth, undermining our ability of understanding. Broadly speaking, one may say, also, that without the concept of truth it would be impossible

[2] European sensu, as philosophy of science.

regardless of the accidental and abstract the essential, basic of perceptual knowledge through the well-known process of pattern-matching. Moreover, if the truth were not absolute, it would lead to a degradation of ethics, dangerous because, in the name of the most sacred things, it would lead to allow the most heinous crimes "for the good of the cause" and, as demonstrated by Alexander Solzhenitsyn [50], lie would have the upper hand.

Einstein wrote in his "Scientific Autobiography' [51] that even scholars of relief (such as Mach) could be hampered by prejudice in the interpretation of the facts: "Prejudice, which still is not gone, is the belief that the facts can and should result in scientific knowledge by itself, without free conceptual construction". The same great scientist [52] wrote that independence from the prejudices is determined by philosophical analysis and it is "the mark of distinction between a mere artisan or specialist and a real seeker after truth".

The epistemology of Einstein begins from the experience (E) and proposes a structured set of theoretical assumptions (A, axioms), intended to explain E. The theory identified by A allows to logically derive a series of consequences (T) which will then be compared with E. The result of the comparison will allow to assess the adequacy of the theoretical hypotheses A. Note that the only passage from A to T is a logic inference, while in the other two passages (E → A and T → E) the inference is predominantly intuitive, i. e. it is a free conceptual construction (Fig. 1.8).

Even the English-Polish epistemologist Jacob Bronowski [53] says that the reasons which will allow humanity to survive and scholars to continue in scientific discoveries, "will not be just or unjust rules of conduct, but more substantial lighting, at the light of which good and evil, means and ends, justice and injustice will be seen in a terrible clarity of boundaries". These epistemological positions are fully converging with the above: these lights will be produced by the light of Truth. The logical process alone is not enough, neither science nor ethics. Again Einstein used to underline that he learned more from Dostoevsky than from his colleagues of theoretical physics.

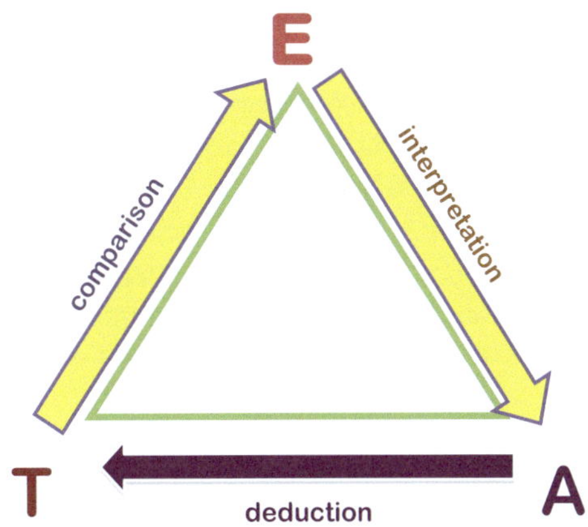

Fig. 1.8 Scheme of the epistemological relationships of Einstein, confirmed even by Popper. *E* experience (reality), *A* axioms (theoretical assumptions), *T* theoretical consequences, *violet arrow* logical inferences, *yellow arrows* inferences intuitive (free and creative conceptual construction)

1.3.3 Landscape Bionomics: A New Ecological Discipline

At present, the discipline of landscape ecology needs a revision, according to the new scientific paradigms we enhanced before. That is why Ingegnoli tried to better focalize landscape ecological elements and processes, in order to widen the foundation of landscape ecology to enhance its capacity of answering new ecological questions and planning problems, as expressed through his Biological Integrated School [10], relocating in a deeper biological vision the different approaches, first of all those by Naveh [5] and Forman [21]. The term "ecology" is today both inflated and degraded. So, the discipline of Biological Integrated Landscape Ecology has been recently named "Landscape Bionomics" sensu Ingegnoli [54, 55].

In summary, we have to underline that:

(a) The landscape, as a level of hierarchical organisation of life on Earth, *is a living entity*, so *a proper biological system*;

(b) Thus, *the landscape is a complex, adaptive, dynamic, self-organising, hierarchical system.*
(c) Its complex structural model can be based on the concept of tissue, thus being named *ecotissue* (Cf. 2.3);
(d) *Landscape bionomics* must be considered as *a discipline like medicine*, biologically based and transdisciplinary. We can properly compare a landscape scientist, which we call "ecoiatra", with a physician of a more wide and complex level of life.
(e) *Landscape pathologies*, but also their *influence on human health*, which may be dangerous *even in absence of pollution* (the environmental stress brings to higher 24 h mean cortisol excretion and to partial inhibition of feedback mechanisms) *must be diagnosed and healed.*
(f) *Territorial planning* has to be considered *as a project for surgical operations*, even in the case of "aesthetic surgery", and the *process of strategic environmental assessment as the related indispensable check-up*, necessary to verify contingent therapy and the capability of the involved subject to survive to the operations without damage.
(g) Even *culture does not implicate the subjection of nature to the dominance of man*; we may demonstrate that in many cases cultural changes of landscapes express natural needs.

Being the landscape a biological level, *it is the physiology (ecology) / pathology ratio which permits a clinical diagnosis of the landscape*, after a good analysis and anamnesis. No doubt that landscape bionomics has its own predictive theory; nevertheless, it is necessary to develop this discipline not as a simple predictive science, but also as a *prescriptive* one—again just like medicine.

The main chapters of Landscape Bionomics (see also the Preface) may be grouped in the following disciplinary components:

1. Theory, Chaps. 1, 2, 3, 4
2. Analysis, Chaps. 5, 6
3. Evaluation and diagnosis, Chaps. 8, 9
4. Therapy, Chaps. 10, 11
5. Applications, Chaps. 12, 13, 14
6. Ethics, Chap. 15
7. Short glossary

1.3.4 A Synthesis of System Theory to Study Living Systems

The concept of system was introduced (end of nineteenth century) into scientific fields by thermodynamics, the theory dealing with entropy and irreversibility at a macroscopic level. A set of elements closely interacting forms a system. The totality of relations among the components and their states constitutes the structure of the system. Because of its relations a system is a whole.

Thus, a system is always more than the sum of its elements. This fact is very important, being the basis of the holistic axiom. Holism is the view that the entirety of a complex system, such as an organism or a landscape, is functionally greater than the sum of its parts.

The Emerging Properties Principle affirms that some of the characters of a whole (i.e. a system) are determined by the properties of its elements, but other characters of the system are the consequence of "the way in which the elements organise themselves". That is: the whole is greater than its components, as the epistemological school of *Gestalt* (perception of the form) proclaimed in the first half of the twentieth century.

The first part of the mentioned properties is quite easy to understand. For example, when fungi (i.e. mycorrhizae) colonise the roots of trees, the fungi-root system is able to extract mineral nutrients from the soil more efficiently than roots alone; or when a fluvial ecosystem is combined with vegetated corridors, forming a riparian landscape, the cleaning capacity of the system is greatly enhanced. Similar mutual relationships are common in nature, and also in a well-ordered human society.

The second part is less intuitive and needs to be illustrated: when, in the development of the mammalian nervous system, there are some mutant factors changing the disposition of the

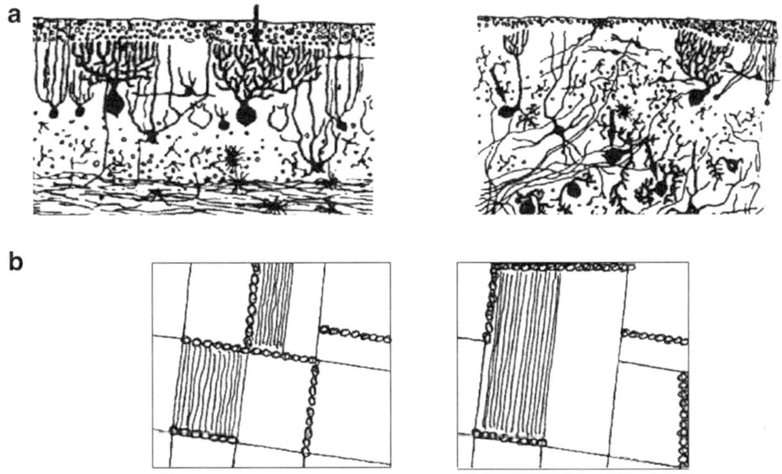

Fig. 1.9 Some examples of the Emerging Properties Principle. (**a**) Cerebellar cortex disease causing severe damage to its functions by changing only the disposition of its cells. (**b**) Changing the disposition of the same elements in an agricultural landscape means changing its ecological parameters (wind effects, humidity, animal home range, etc.)

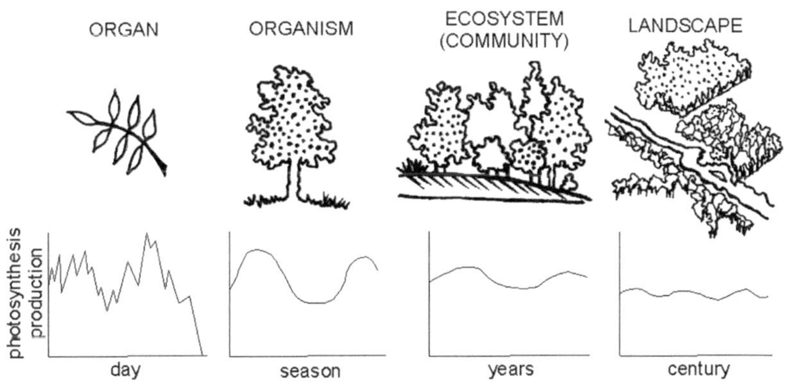

Fig. 1.10 Hierarchical levels of ecological systems. Note that a low process speed corresponds to the behaviour of a high level in a system, and vice versa, e.g. photosynthesis in a forest patch and in a single tree

same elements, the resulting system changes its functions dramatically (e.g. in the cerebral cortex); when, in the management of an agricultural landscape, we change the disposition of the same elements (e.g. fields and hedgerows), the functioning (and productivity) of the ecological mosaic modifies drastically (Fig. 1.9) (note that, if repeated at a larger scale, this process could cause desertification).

1.3.5 Living Systems are Hierarchic

Studying a patch of trees, a researcher has to inquire at a more synthetic scale if he wants to know the significance and the constraints of this patch, e.g. in what kind of vegetational landscape it is growing, what are the climatic constraints, etc.; then he has to investigate on even a more detailed scale, e.g. single trees, if he wants to know the components of the plant association and the reason of their existence.

The central concept of the Hierarchical Theory is that the organisation of a system results from differences in process rates, which change with the scale. Levels in the hierarchy are isolated from each other because they operate at distinctly different rates (Fig. 1.10). Boundaries, which are not only the physical ones, separate the set of processes from components in the rest of the system.

Thus, the inferior level components explain the origin of the level of interest (i.e. allow an inner description), while a superior level system explains the significance of the level of interest (i.e. allow the characters derived from transferred conditions to be understood).

One of the most important consequences of the hierarchical structure of a system is the concept of *constraint*. It is more correct than the non-systemic concept of “limiting factor” and it shows the behaviour of an ecological system as limited: (a) by the behaviours of its components and (b) by environmental bonds imposed by superior levels of organisation. Remember, there is a linkage between constraint and information.

1.3.6 Living Systems are Dynamic

Following the System Theory, a system is defined as *dynamic* if it is able to evolve over time. So, the output (y) depends on the *history* of the system, not directly on the input (*a*): we must hypothesise the existence of a third element, the *state*, which includes information on the past, present and potential evolution of the system. The value $x(t)$ assumed by the state at the instant t must be sufficient to determine the value of the output in the same instant: knowing the values of $x(t_1)$ and $a(t_1, t_2)$, the state (then the output) in the instant t_2 can be calculated.

A dynamic system may be defined by six sets of variables (Fig. 1.11), correlated by two functions:

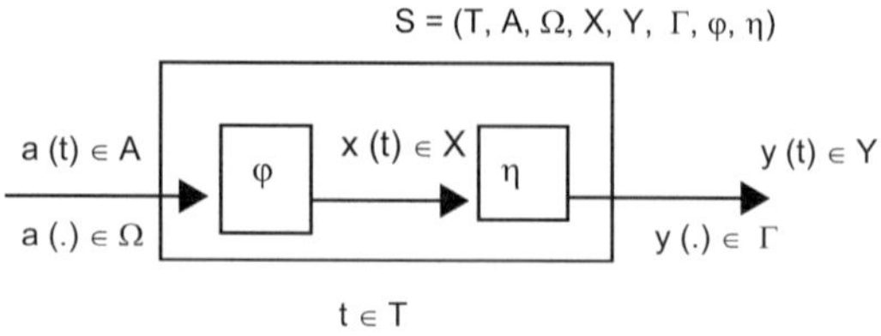

Fig. 1.11 Theoretical representation of a dynamic system, defined with six sets of variables

Input parametric variables $a(t) \in A$ and input functions $a(.) \in \Omega$

Output parametric variables $y(t) \in Y$ and output functions $y(.) \in \Gamma$

State variables $x(t) \in X$

Time variables $t \in T$

Function of the state transformation $x(t) = \varphi[t, t_0, a(.)]$

Function of the output transformation $y(t) = \eta[t, x(t)]$

The couple state-time (x, t) has great importance because the set X, T is the set of events, the *history* of the system. Once an instant t, an initial state $x(t_0)$, an input function $a(.)$ are fixed, the transition function $\phi[., t, x(t), a(.)]$ is univocally determined, and named “movement” of the system (Fig. 1.12). Dynamic changes use energy, therefore the photosynthesis (or chemiosynthesis in primeval systems) becomes necessary.

The study of the landscape of the Venice Lagoon is a fine example to show how to control the movement of this complex system (Fig. 1.13). The field of the states of the system (state of the phases) was done by a structural variable (barene/tidal area) and a functional one (biological territorial capacity of vegetation,

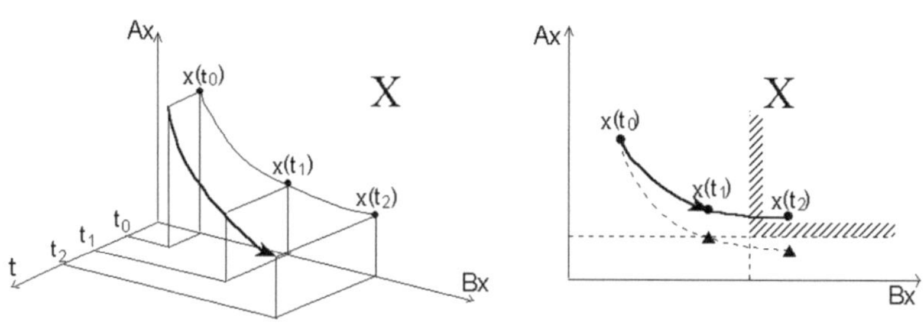

Fig. 1.12 Movement and trajectory of a dynamic system. *X* represents the field of the states of the system (state of the phases). *AX* axis and *BX* axis are, for example, a set of structural variables and a set of functional variables, respectively. The trajectory is the projection of the movement of that X field. Note the drawing of a delimited set (*right*), useful to control the trajectory

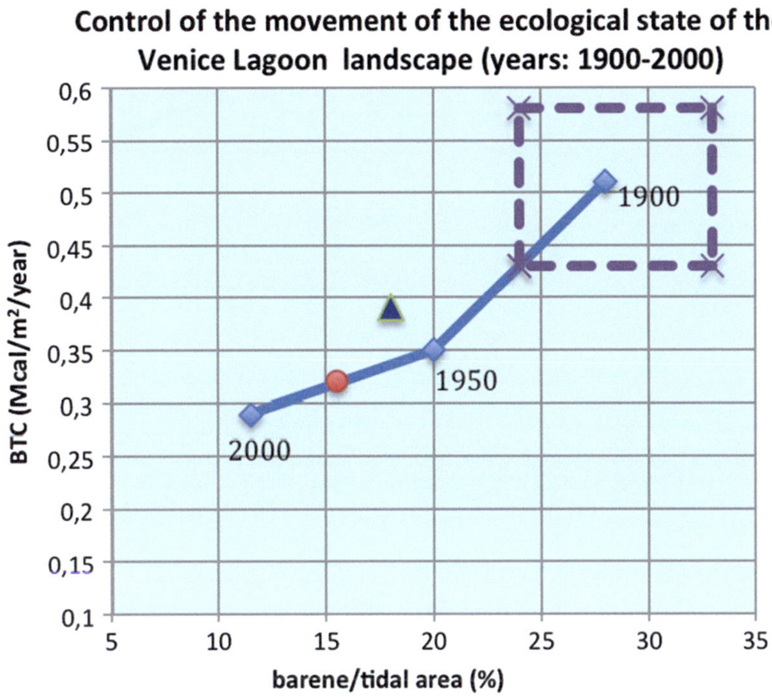

Fig. 1.13 Example of control of the movement of the landscape of Venice. The field of the states of the system (state of the phases) is done by a structural variable (barene/tidal area) and a functional one (biological territorial capacity of vegetation, BTC). Two recovery plans (year 2020) are compared: the blue scenery (through the *blue triangle*) is the best even if it does not follow back the pattern of transformation, as the *red circle*

BTC). The barene are flat islands covered by halophytic vegetation. Two recovery plans were compared: the blue scenery (triangle) is the best even if it does not follow the pattern of transformation.

1.3.7 Living Systems are Dissipative

Photosynthetic processes have the main responsibility of energy transfer in biological systems. How is it possible? Living systems must be open, otherwise, the free energy F would not be available. In an open system we have to consider two fluxes of entropy: d_eS, that is the entropy flux due to the exchanges with the environment, and d_iS, which is the entropy flux due to the irreversible processes within the system. The second term has a clear positive sign, but the first term does not have a definite sign. So the inequality of Clausius[3]-Carnot becomes: $dS = d_eS + d_iS$ (*being* $d_iS > 0$). In a period in which the system is stationary ($dS = 0$), thus

$$d_eS + d_iS = 0 \text{ and } d_eS < 0 \,(\text{being } d_eS = -d_iS)$$

So, in an evolutionary process, when the system reaches a new stationary state of lower entropy $S\ (t_1) < S\ (t_0)$, it is able to maintain it in balance by "pumping out" the disorder. But this is possible only in non-equilibrium conditions, in so-called *dissipative* systems: a dissipation of energy into heat is necessary to maintain the system far from equilibrium and to create order. The amount of entropy "pumped out" is named negentropy.

[3] The Clausius theorem states that for a system undergoing a cyclic process: $\int \delta Q/T \leq 0$, where δQ is the amount of heat absorbed by the system. The inequality holds in the irreversible case.

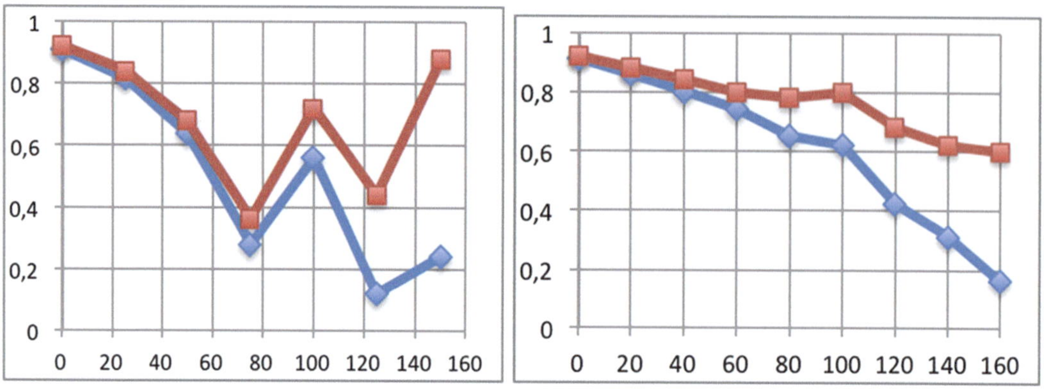

Fig. 1.14 An example of deterministic chaos. In the *left plot*, starting from two very similar initial conditions, the Bernoulli equation shows very different trajectories, after time 70. The *right plot* shows a real case: two municipalities near Milan having 91 and 92 % of their territory as agrarian in 1851, after 160 years show heavy differences, the first having only 16% while the second maintaining 60 % of it

An energy dissipation, which allows work to be done, has to be coupled, e. g. with the transformation of a system from state A_0 to state A_1. The process able to perform this transformation is an example of an *operator* (*Op*), a *rule* of action on a given function. If we express it in the form $A_1 = (Op)\,A_0$, the complete transformation process is

$$A_1 = [(Op)A_0 \cup (e_w \rightarrow e_d)]$$

where: e_w = available energy, e_d = dissipated energy. If the state of the system becomes an auto-function for a certain operator, the system does not undergo further changes. This state is called a *fixed point* of the system, and it may represent a stationary state or an *attractor*.

1.3.8 Self-organisation and Deterministic Chaos

Complex interacting systems in which cycling, structuring and auto-regulation emerge from the inside may be called *self-organising* systems. In living systems the ability to maintain a dynamic equilibrium as a whole is called *homeostasis*. It is ensured by a large number of closely interrelating cybernetic feedback mechanisms, hierarchically ordered. These biological and ecological processes of auto-regulation can be active also at the vegetation level. Auto-regulation needs information, deriving from biological and ecological processes, which can be carried out both in energetic and/or in material way: that is, energy structures itself with the help of information. Positive and negative feedbacks functions are fundamental, too. Their dynamics can be synthetically expressed by:

$$X_t = f(X_0, t, \lambda),$$

where X_t is the state of the system at time t, X_0 is the state of the system at time 0, λ is a specific parameter for the examined system indicating the acquisition of energy and matter from outside. Depending on the parameter λ and its values [9], X may tend toward a temporary stationary state (*metastable* state) or a chaotic one.

Note that the *uncertainty* given by *chaos* does not depend on complexity: in fact, even a simple deterministic system can be chaotic. As shown by Prigogine [3], if we consider the Bernoulli equation:

$$x_{n+1} = 2\,x_n(\text{Mod } 1)$$

where: Mod 1 = numbers between 0 and 1, it is easy to see that very short differences of the initial conditions can be brought to very different trajectories, as in Fig. 1.14.

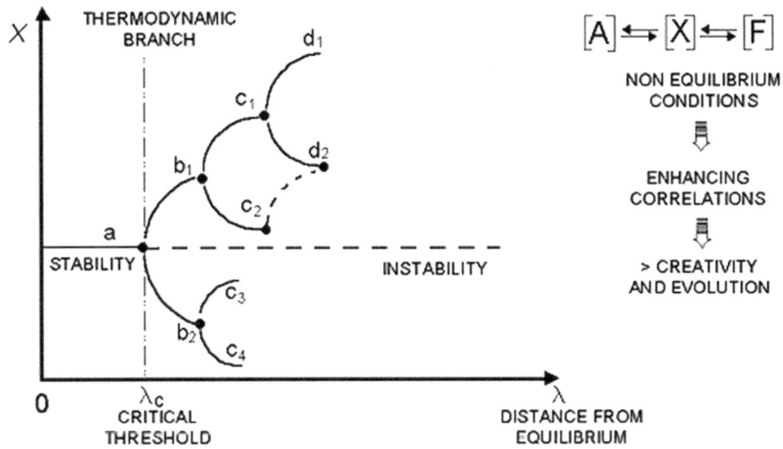

Fig. 1.15 Successive bifurcations in a non-equilibrium system: going further on the stable thermodynamic branch, the intermediate product enters a field of instability with the appearance of subsequent bifurcations. Note that the point d_2 can be reached through the path a-b_1-c_1-d_2 but also a-b_1-c_2-d_2. An historical behaviour can be observed in this process

A system is deterministic-chaotic when magnifying small differences it amplifies *initial* conditions, for instance between two trajectories. It is impossible to shorten the description of a chaotic system because of its unpredictable behaviour due to branching possibilities of evolution, thus to a manifold of attractors.

Highly chaotic webs are so disordered that the control of complex behaviours is impossible, while highly ordered webs are so rigid that it is not possible to express a complex behaviour. But if "frozen" components begin to melt, it is possible to have more complex dynamic behaviours leading to a complex co-ordination of activities within the system. Thus, the maximum complexity is reached in a "liquid" transition between solid and gaseous states, where the best capacity of evolution is expressed. For instance, it is possible to see a similar situation in DNA due to its capacity to maintain an ordered structure besides to change by mutations.

The threshold between order and deterministic chaos seems to be the most important need for complex adaptive self-organising systems (order at the edge of chaos). As these systems are dissipative, an order through fluctuations is effective in working between the above-mentioned conditions.

1.3.9 Non-equilibrium Thermodynamic and Metastability in Living Systems

A self-organised living system is able to capture intense energy fluxes and to utilise its negentropic input to produce new structures. Prigogine and Stengers showed [56] that even simple physical systems present processes of order.

Figure 1.15 shows the concentration of the intermediate product *X* in a chemical reaction: going further on the stable thermodynamic branch, the intermediate product enters a field of instability with the appearance of subsequent bifurcations. Thus, the result cannot be deterministic: when a system reaches a branching point, disturbances, like fluctuations, become important, constraining the system to choose one of the two branches of new relative stability. Therefore, the evolution of this kind of system contains an *historic* criterion in itself. The fluctuation-dissipation sequence may be viewed as a feedback process. A macro-fluctuation, due to a change of disturbances, produces instabilities leading to increased dissipation of energy and the system becomes more difficult to maintain: when a threshold is reached, characterised by the prevailing of new structures over the previous

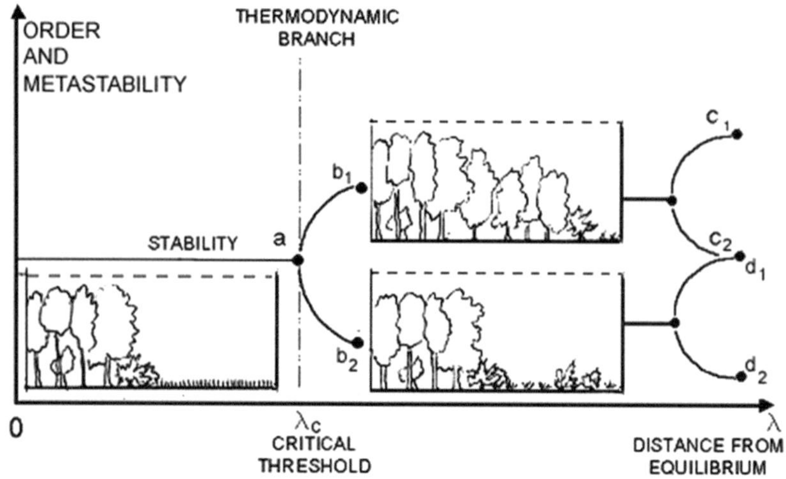

Fig. 1.16 A process of macro-fluctuation (e.g. land abandonment) lead to a bifurcation point with two possible ecological transformations: b1 and b2, which have different metastabilities. Note that after another macro-fluctuation the transformation d1 may coincide with c2 or not (c1 or d2). Examples like these show the need to revise the concept of succession, leading to significative changes in vegetation science

ones, a new organisational state emerges. That is why the Prigogine statement is "order through fluctuations". Environmental conditions are important for an ecological system at a branching point, enabling it to choose one of the two branches of new relative stability (*metastability*) (Fig. 1.16).

Under these conditions, large range mutual relations occur among the components. The matter acquires new properties, a new sensitivity of matter to itself and to its environment takes place, associated with dissipative and not reversible processes. A far from equilibrium system is able to self-organise through intrinsic probabilities, exploring its structure and realising one among the possible structures, but not a random one. This process is available from cell proteins formation to vegetation transformation.

When a system is oscillating around a steady attractor, but may even move toward another attractor, then we have *metastability* (Godron [57]; Naveh and Lieberman [5]; Forman and Godron [21]). Note that the concept of metastability is not a compromise between a form of stability and one of instability: higher or lower metastability depends on the distance from the position of maximum stability and on the height of the thresholds of local (far from equilibrium) stability. Ecological systems with low metastability have a low resistance, but a high resilience to disturbances. By contrast, high metastability systems have high resistance to disturbances. For example, a prairie patch has a higher resilience than a forest one.

The concept of metastability allows the traditional concept of ecological equilibrium to be updated: "equilibrium" does not stay around 0, but it identifies various stationary or equilibrium states far from 0. A system reaches a new organisation after instabilities and the passage to a new metastable level (see Fig. 1.15).

Remembering the theory of hierarchic systems, we know that some limitations on the dynamics of an ecological system come from inferior levels of scale and are due to the biological potential of its components. Other limits are imposed by superior levels as environmental constraints. A range of conditions emerges for every kind of ecological system, for instance a vegetation complex in a landscape, and can be expressed as the *constraints field* or optimum set of existence. Note that in many cases the majority of disturbances can be *incorporated* within ecological systems. The mentioned constraint field of an ecological system is based on a resistance strategy to a current regime of perturbations. Therefore, we can speak of disturbance incorporation when the system organisation exerts control over some environmental aspects that is impossible to control at a lower level of organisation. This process may limit possible alterations to its stationary state; meanwhile it may utilise perturbations as structuring forces.

1.4 Different Types and Levels of "Order"

1.4.1 Self-Transcendent Systems

To characterise living systems simply as hypercomplex, open, adaptive, non-equilibrium and learning systems is not enough. They are even self-transcendent, [5] "capable not only of representing and realising themselves, but also of transforming themselves. They are evolution's vehicle for qualitative change and ensure continuity. For them Being fall together with Becoming".

Thus, in the study of these systems it is not generally possible to follow simple and univocal cause–effect connections and information--structure relationships. A good example is the unsolvable linkage between physiology and pathology, as the removal of a disturbance needs the understanding of the normal process, which is rather understood through a disturbance [58]. It is not possible to analyse global entities (i.e. an organic whole) studying only their parts. As suggested by the new scientific paradigms, it is necessary to study ecological systems with the help of system theory and its corollaries on irreversibility, information, etc. That is why the previous paragraph is dealing on a synthesis of system theory. But this is not enough.

As underlined by Teilhard de Chardin [59] the evolution is going towards an ever more increasing complexity. In this process, the recent (Holocene) large mutualism between natural systems and human population brought to many types of human and semi-natural landscapes and enhanced the amount and the exchanges of information. The actions directed to landscape transformation imply the emergence of rules of correspondence between the complex structure of natural landscapes and the formation of new anthropic structures. For instance, during these transformations, the high complexity of the landscape expresses relationships allowing the maintenance or the increase of a good level of metastability through processes that produce sub-systems of interface elements.

As expressed in Sect. 1.3.9, it is thermodynamically possible that the energy introduced into a system may locally decrease entropy, i.e. the disorder. The neo-Darwinians, excluding the teleology, are forced to resort to energy. Since the possible decrease of entropy generates order, applying a quantitative and reductionist conception of science, they claim that this process can be achieved, albeit gradually and very long, very complex levels of order, as is the case of living systems. But this is an illusion.

The "principle of emergent properties" (Cfr. 1.3.4) in natural systems demonstrates that each hierarchical level acquires greater complexity and order than the previous one, but we forget that, in fact, there are different types and levels of "order", incomparable with each other.

1.4.2 Energy and Information

In the Sect. 1.3.1, we observed that energy can be organised as matter or information, depending on different codifications of the chronotope. This assumption is very interesting, but it must be deeply understood, to avoid confusion. For instance, to avoid sentences as this one: organisms receive continuously energy from sun, which produces order and information.

Here is confusion between the carrier or *support* of information and information itself, which are two very different things. The example we reported of the light wave (plane wave) should be evident, because light is indispensable for information, but it is not information by itself. Another example: a telegraphic line can only work thanks to the electric power but the telegrapher must introduce the message to be transmitted over the line. The telegrapher introduces the message and the electric power carries out the function of its transport.

Moreover, we must distinguish between the concept of *energetic order* and the concept of *information order*. We know that for thermodynamic entropy, energy degrades irreversibly in quality, converting from more ordered forms to less ordered forms. Heat is the most degraded energy since molecular agitation is the more chaotic thing one can imagine. An electrical current flowing in a conductor is a more valuable form of energy: it consists of electrons going all in the same direction.

But this physical order that we have called "energetic order" is different respect information order. Information order is always, if we analyse it, referable to a rigorous order of bits.[4] Think about the above flow of telegraphic information: the message written by the telegrapher is expressible as a sequence of bits travelling as a train on the telegraphic line. Another example, at the macroscopic level: illuminating and heating a room is not enough to put it in order or organize it, the room always remains disordered if we do not tidy it up.

1.4.3 Levels of Biological Organisation

A first class of order is intrinsic of matter, i.e. it does not require the intervention of external information, what decreases the uncertainty. Enough energy-dependent conditions, such as

- The arrangement of atoms in crystals, or
- The order in low-entropy systems far from equilibrium (dissipative)

After these two orders of organisation level, another class, incomparably superior to the previous one, concerns:

- The order of organic macromolecules such as proteins and nucleic acids (III order)
- The order of the processors[5] of the above-mentioned macromolecules like ribosomes (IV order) etc
- The order of the cell structure (DNA) with its functions for self-survival (V order)

Therefore, it is evident that information contained in DNA cannot be read by means of energy alone. But there are many other hierarchical levels of biological order, concerning:

- To the cell structure
- To the organism (e.g. human body)
- To an ecological community
- To a human ecosystem
- To the complex adaptive system of natural and human communities (landscape)

It is recalled that already proteins (III order level) contain codes that subsequent levels (e.g. IV and V) must process.

Since a processor can interpret messages only if it shares encoding, a functional protein must get along with his interpreter, hence it cannot be generated randomly. If not, in system with high complexity, can only lead to useless or destructive changes.

The coding also has to do with the meaning of information: to juxtapose a series of terms or dial randomly some sentences (if that were possible) using the word in a dictionary is not enough if you want to write a poem.

The use of the term "information" might be a bit misleading, as it depends upon the concept of compressibility. Informally, from the point of view of algorithmic information theory, the information content of a string is equivalent to the length of the shortest possible self-contained representation of that string. A self-contained representation is essentially a program. Following the Algorithmic Information Theory [60] the complex specified information (CSI) is the "*incompressible*" information, which cannot be synthesised in a more simple form or rule. An

[4] Unit of information in information theory, one bit is typically defined as the uncertainty of a binary random variable that is 0 or 1 with equal probability.

[5] The processor intended in the sense of a system able to carry out the instruction from a program performing logical operations.

example of incompressible information is the DNA, so it cannot be the output of natural laws. Being it incompressible, natural laws should contain it in a complete form, but we know it is not possible.

For example, let us consider a DNA sequence like this: CTAGGCATCATGAAATAGGAA-CAAATCATTTAG. No chemical or physical law contains this sequence or the description of the other organic macro-molecules. Being these sequences incompressible they cannot be generated by natural laws which are simple algorithms. No algorithm shorter than a sequence may generate it.

We cannot avoid to observe that all these considerations lead towards the hypothesis of the presence of intelligence in our universe. Intelligent design is a scientific theory which holds that certain features of the universe and living things are best explained by an intelligent cause and are not the result of an undirected, chance-based process such as Darwinian evolution.

Intelligent design begins with observations about the types of information produced by intelligent agents. Even the atheist zoologist Richard Dawkins [61] says that intuitively, "biology is the study of complicated things that give the appearance of having been designed for a pur pose". Darwinists believe natural selection did the "designing" but intelligent design theorist Stephen Meyer [62] notes, "in all cases where we know the causal origin of "high information content", experience has shown that intelligent design played a causal role".

Intelligent design is thus heavily dependent upon "information theory". One of its fundamental premises is that "information" which is complex (highly ordered) and specified (fits a pre-existing pattern) is not produced by naturally occurring events (chance or law-governed processes), but rather this sort of observable information and complexity is best explained as the product of intelligent action.

Intelligent design implies that life is here as a result of the purposeful action of an intelligent designer, standing in contrast to Darwinian evolution, which postulates that life exists due to the chance, purposeless, blind forces of nature.

References

1. Reynolds J (1759) The true idea of beauty. In: Johnson S (1811) The idler. W. Durrell, New York
2. Ashby WR (1962) Principles of the self-organisation system. In: Von Foerster H, Zopf GW (eds) Principles of self-organization. Pergamon, Oxford
3. Prigogine I (1996) La fin dès certitudes: temps, chaos et les lois de la nature. Editions Odile Jacob, Paris
4. Ingegnoli V (2011) Non-equilibrium thermodynamics, landscape ecology and vegetation science. In: Moreno-Pirajàn JC (ed) Thermodynamics. Systems in equilibrium and non-equilibrium. In Tech, Rijeka, Croazia, pp 139–172
5. Naveh Z, Lieberman A (1984) Landscape ecology: theory and application. Springer, New York
6. Ingegnoli V (1999) Ecologia del Paesaggio. In: Baltimore D, Dulbecco R, Jacob F, Levi-Montalcini R (eds) Frontiere della Vita, vol IV. Istituto per l'Enciclopedia Italiana G. Treccani, Roma, pp 469–485
7. Ingegnoli V (2001) Landscape ecology. In: Baltimore D, Dulbecco R, Jacob F, Levi-Montalcini R (eds) Frontiers of life, vol IV. Academic, New York, pp 489–508
8. Ingegnoli V, Pignatti S (eds) (1996) L'ecologia del paesaggio in Italia. UTET-Città Studi, Milano
9. Pignatti S, Trezza B (2000) Assalto al pianeta. Bollati-Boringhieri, Torino
10. Ingegnoli V (2002) Landscape ecology: a widening foundation. Springer, Berlin, p 356
11. Ingegnoli V, Giglio E (2005) Ecologia del Paesaggio: manuale per conservare, gestire e pianificare l'ambiente. Sistemi editoriali SE, Napoli
12. Odum EP (1971) Fundamentals of ecology, 3rd edn. WB, Saunders, Philadelphia, PA
13. Odum EP (1983) Basic ecology. CBS College, Philadelphia, PA
14. Lovelock JE (1979) Gaia: a new look at life on Earth. Oxford University Press, Oxford
15. Giacomini V (1965) Significato e funzione dei Parchi Nazionali. In: I Parchi Nazionali in Italia. Ist. Tec. e Prop. Agraria, Roma, pp 7–37
16. Buchwald K, Engelhart W (eds) (1968) Handbuch fur landschaftpflegeund naturshutz. Bd.1 Grundlagen. BLV Verlaggesellschaft, Munich
17. Ingegnoli V (1971) Ecologia territoriale e progettazione: significati e metodologia. In: L'ingegnere di fronte ai problemi della sopravvivenza umana. European Congress of FEANI. Collegio Ingegneri, Milano, pp 398–400
18. Ingegnoli V (1991) Human influences in landscape change: thresholds of metastability. In: Ravera O (ed) Terrestrial and aquatic ecosystems: perturbation and recovery. Ellis Horwood, Chichester, pp 303–309
19. Walter H (1973) Vegetation of the earth in relation to climate and the eco-physiological conditions. Springer, New York

20. Lorenz K (1978) Vergleichende Verhaltensforschung: Grundlagen der Ethologie. Springer, Berlin
21. Forman RT, Godron M (1986) Landscape ecology. Wiley, New York
22. Odum EP (1989) Ecology and our endangered life-support systems. Sinauer, Sunderland, MA
23. Pignatti S (1994) Ecologia del paesaggio. UTET, Torino
24. Leser H (1997) Landschaftsoekologie. Ulmer, Stuttgard
25. Meffe GK, Carroll CR (1997) Principles of conservation biology. Sinauer, Sunderland, MA
26. Teilhard De Chardin P (1956) La place de l'homme dans la nature. Le groupe zoologique humain. Edition Albin Michel, Paris, trad. Il Saggiatore, Milano
27. Zonneveld IS (1995) Land ecology. SPB Academic, Amsterdam
28. Arcà A (2000) Agricultural landscapes in neolithic and copper Age engravings Valcamonica and Mt. Bégo rock Art. In: Nash G (ed) Signifying place and space. World perspectives of rock art and landscape, BAR International Series 902. Archaeopres, Oxford, pp 29–40
29. Mariano C (1958) Nuovo Dizionario Italiano-Latino. Soc. Editrice Dante Alighieri, Milano
30. Columella LJM (I Century) De Re Rustica I-IV. Translated by Ash HB, Harvard University Press, Cambridge, MA
31. Clark K (1976) Landscape into art. John Murray, London
32. Von Humboldt A (1845) Kosmos. Entwurf einer physischen Weltbeschreibung. Cotta, Stuttgard und Tubingen
33. Thoreaux HD (1854) Walden: life in the woods. Thicknor and Fields, Boston
34. Olmsted FL (1865) The great American Park of Yo-Semite. Yosemite Nature Notes (1954, reprinted)
35. Olmsted FL (1870) Public parks and the enlargement of towns. Cambridge University Press, Cambridge
36. Troll C (1950) Die geographische Landshaft und ihre Erforshung. Studium generale, n. 3, Heidelberg, pp 163–181
37. Troll C (1963) Ueber Landschaft-Sukzession, Vorwort des Herausgebers. In: Bauer HJ, Landschaftökologische Untersuchungen im ausgekohten Rheinischen Braunkohlenrevier auf der Ville. Arbeiten zur Rheinischen landeskunde 19
38. Zonneveld Is (1963) Landschapsecologie. In: Symposium uber Pflanzensoziologie und Landschaftoekologie, (Germany). Stichting voor Bodenkartering, Wageningen
39. Finke L (1971) Die Verwertbarkeit Bodenschatzungsergnergebnissefur die Landschaftsoekologie, Bochumer Geogr.Arb.10, 84 S
40. Leser H (1978) Landschaftsökologie. Uni-Taschenbucher, Stuttgart, 521
41. Ingegnoli V (1980) Ecologia e progettazione. CUSL, Milano
42. Ingegnoli V (1993) Fondamenti di Ecologia del paesaggio. CittàStudi, Milano
43. Farina A (1993) L'ecologia dei sistemi ambientali. Cluep, Padova
44. Manzelli P (1999) Il cervello, la percezione, il colore. Didattica Delle Scienze, n. 202, Firenze
45. Popper KR (1994) Alles Leben ist Problemlösen. Über Erkenntnis, Geschichte und Politik. R Piper & Co., München
46. Ruelle D (1991) Hasard et chaos. Éditions Odile Jacob, Paris
47. Gödel K (1931) Über formal unentscheidbare Satze der Principia Mathematica und verwandter Systeme. In: Mh Math Phys 38:173–198
48. Israel G (2010) Per una medicina umanistica. Apologia di una medicina che curi i malati come persone. Lindau, Torino
49. Monod J (1970) Le hazard et la nécessité. Ed. du Seul. Paris
50. Solzenicyn A (1963) Una giornata di Ivan Denisovic. It. Einaudi, Torino: One day in the life of Ivan Denisovich. Signet Classic, New York 1962.
51. Einstein A (1949) Autobiografisches, tr.it. Autobiografia scientifica, in. A. Einstein Opere scelte, Bollati-Boringhieri, Torino
52. Einstein A (1944) Einstein archives, pp 61–574
53. Bronowski J (1969) Nature and knowledge: the philosophy of contemporary science. Condon lectures, London
54. Ingegnoli V (2010) Ecologia del paesaggio: l'ecologia del paesaggio biologico-integrata. In: T. Gregory (Ed.) XXI Secolo, vol. IV. Istituto della Enciclopedia Italiana Treccani, Roma, pp 23–33
55. Ingegnoli V (2011) Bionomia del paesaggio. Springer, Milano
56. Prigogine I, Stengers J (1981) Order Out of Chaos. Bantam, New York. Trad. It. La Nuova Alleanza, Einaudi, Torino
57. Godron M (1984) Ecologie de la végétation terrestre. Masson, Paris
58. Lorenz K (1973) Die Rückseite des Spiegels. Versuch einer Naturgeshichte menschlichen Erkennens. R Piper & Co, München
59. Teilhard De Chardin P (1955) Le phenomòne humain. Edition de Seuils, Paris
60. Chaitin GJ (1990) Algorithmic information theory (Third printing). Cambridge University Press, Cambridge
61. Dawkins R (2009) The greatest show on Earth: the evidence for evolution. Free Press, New York
62. Meyer, SC (2013) Darwin's doubt: the explosive origin of animal life and the case for intelligent design. HarperOne, an Imprint of HarperCollins: New York

2 Landscape Structure

2.1 The Ecocoenotopes and Their Systems

2.1.1 The Landscape Subsystems

A landscape exhibits the same three fundamental characters of all living systems, like an organism or an agricultural system, and it represents a challenging research frontier. These three characters are structure, functions and transformation.

1. *Structure*: the organisation and the pattern of the distinctive ecosystems (better: ecocoenotopes) or "elements" forming the landscape.
2. *Function*: the interactions and processes among the structural elements, that is the flow of energy, materials, species and information among the component ecocoenotopes and the intrinsic behaviour of their complex mosaic.
3. *Transformation:* the evolution and alteration of the complex mosaic over time, both in structure and functions.

If we define the landscape as a complex system of ecocoenotopes, each type of landscape may be related to a particular configuration of interacting ecocoenotopes, that is, to a specific structural pattern. But these patterns share a similar structural model, as we will see later. Carl Stainitz, landscape planning professor at Harvard University [1], underlined the theoretical importance of the observation that all landscapes, from the wilderness to the central city and from the natural to the developed one, share a similar general structural model. We know that the reason for this behaviour lies in the fact that the landscape is a specific biological level of the ecological hierarchy, having its proper structure and functions.

In general, given a proper range of scales, the structural model of a landscape is concerned with the selection of the elements representing the minimum part which the landscape may be divided into. Then, it is necessary to analyse the configurations of these elements in a hierarchical sense, enhancing the ecological mosaics and other structures formed by them. Other larger units, or subsystems, can be detected; they arrive to form a principal landscape, as shown in Table 2.1.

Remembering the concept of ecological mosaic, the scientists use the Latin term "*tessera*" to express the basic element (ecocoenotope) forming a landscape unit. The integration of some *tesserae* (TS) in a peculiar geomorphic place and with a specific context role forms an ecotope, usually contained in a small area. A set of these ecotopes in a larger geomorphological scale and with a distinctive landscape character forms a simple landscape unit (LU) or, two or more LU strictly interconnected form a complex LU (c-LU). In summary, an elementary unit that is still a holistic unit at the proper range of scales of a landscape may be defined as a landscape element.

V. Ingegnoli, *Landscape Bionomics Biological-Integrated Landscape Ecology*,
DOI 10.1007/978-88-470-5226-0_2, © Springer-Verlag Italia 2015

Table 2.1 Proper landscape subsystems: ecotope, landscape unit, complex landscape unit

Scale	Biotic reference	Living entity	Peculiar systems	Usual surface (km^2)
Regional	Biome	Ecoregion	L. system (LS)	5,000–100,000
Territorial	System of community	*Landscape*	*Landscape (L)*	500–10,000
			Complex LU (c-LU)	50–1,000
			Landscape Unit (LU)	5–100
			Ecotope (ECT)	0.5–10.0
Local	Community	Ecocoenotope	Tessera (TS)	0.005–1.0

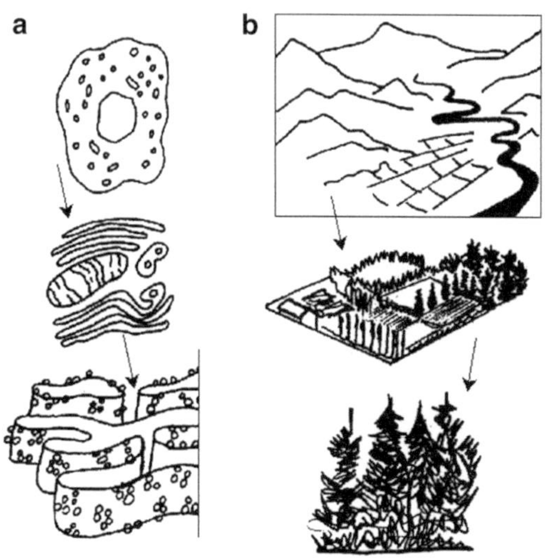

Fig. 2.1 Scale dependence of structure in physiological and ecological fields. Examples of two scale sequences. (**a**) From a cell to a mitochondrion and ribosomes. (**b**) From a landscape unit to a tessera. The spatial resolution in studying a landscape has to be enhanced, as in other biological disciplines

A hierarchy of the structural components of landscape can be proposed observing that the definition of landscape leads to a field of existence varying from a system of two-three tesserae to the ecoregional level. The spatial resolution in studying this large field has to be enhanced, as in other biological disciplines (Fig. 2.1).

For instance, if we have to study a cell we know that at a particular resolution scale it is possible to see only the cell as a whole (e.g. 200/1), but at a more detailed scale we may see the nucleus (e.g. 1,000/1) or even the Golgi apparatus, and finally, with an electronic microscope also the mitochondria (e.g. 20,000/1); or, by contrast, reducing the scale (e.g. 100/1) we may see cell types in their histologic tissue, and so on. A similar concept is pertinent to the landscape. A vegetated tessera of prairie can measure only 100 square metres and can be mapped on a 1/500 scale, a forested tessera can be observed on a scale 1/1,000 or 1/2,000, a mesoscale ecotope can be surveyed on a 1/5,000 map, a simple landscape unit from a scale of 1/10,000 to 1/25,000 and so on. An ecoregional unit can seldom be mapped under a scale of 1/250,000–1/500,000.

2.1.2 The Concept of Tessera

As we have seen, usually ecologists name the elements as *tesserae,* or patches, but many other expressions can be found, such as habitats, biotopes, geotopes, etc. Note that, generally, a patch is defined with spatial and ecosystemic characters, rarely with other biological attributes, and this is not sufficient. Remembering the proper and exportable characters of life organisation levels, we need to define a landscape element within this perspective.

The types of ecocoenotopes may be numerous as the so-called natural habitats, to which it is necessary adding also semi-natural (or agricultural) and human habitats. For instance:

Natural set of habitats [2]: (1) coastal and halophytic habitats, (2) coastal sand dunes and inland dunes, (3) freshwater habitats, (4) temperate heath and scrub, (5) sclerophyllous scrub (matorral), (6) natural and semi-natural grassland formations, (7) raised bogs and mires and fens, (8) rocky habitats and caves, (9) forests.

Agricultural set of habitats: (1) extensive cereal fields, (2) Mediterranean garden agriculture, (3) mountain mixed cultivations, (4) rural

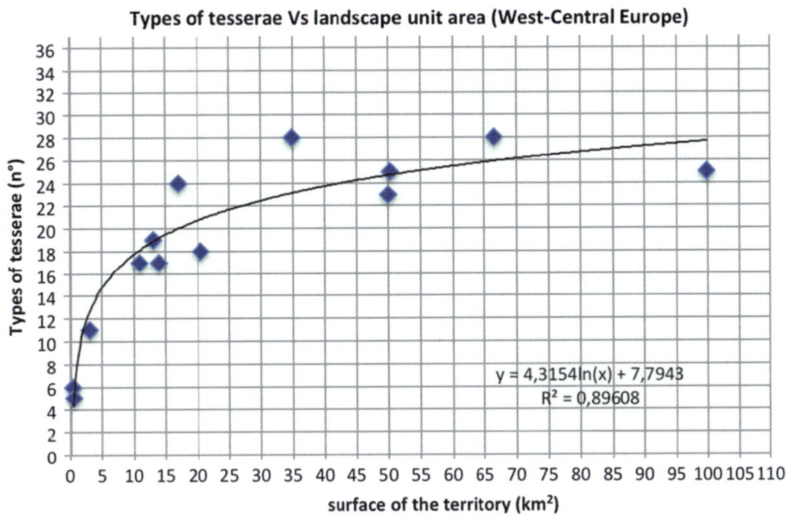

Fig. 2.2 The landscape units, from an ecotope to the entire landscape, are composed of a lot of elements which can be reduced to few types of tesserae. The increase of these types of tesserae is not linear, following a log correlation

suburban cultivations, (5) tropical fruit cultivations, (6) vineyard mixed cultivation, (7) terraced tea cultivations, (8) textiles (cotton, etc.) cultivations, (9) vegetables gardens.

Human set of habitats: (1) residential areas, (2) religious plots, (3) technological plots, (4) sport fields, (5) university campus, (6) industrial (commercial) areas, (7) parking areas, (8) urban parks, (9) private gardens.

Note that each set of habitats may contain at least from 10 to 30 more specific habitats; thus we have hundreds of types of tesserae and consequently thousands of types of ecotopes. If one type of tessera results to be dominant in a landscape unit, it characterises the type of the local landscape or (if repeated) even of the entire landscape. We know that, even before to acquire a scientific landscape classification, people is able to define many landscape typologies, just perceiving the dominance of the natural or human elements: forest landscape, rural vineyard landscape, industrial landscape, etc.

Anyway, the number of tesserae forming a landscape subsystem is generally very contained, as shown in Fig. 2.2.

This figure is based on the experience from the North Italian Regions of Lombardy and Trentino, within which about 40 types of tesserae form a lot of different landscape types. Even considering also mixed tesserae and sub-types, this number does not exceed 70–80, about 28–30 of which may be enough to form a landscape unit of 100 km^2. An ecotope may be composed only by 3–5 types of tesserae and increasing the scale of a landscape unit we did not find a linear correlation but a logarithmic one.

In this framework, to detect a tessera, a proposed series of analysis should be done as follows:

1. Plant association or sub-association
2. Other vegetational characters: growth phase and form, structure, biomass, etc.
3. Influence of the eco-mosaic
4. Geomorphologic characters
5. Natural (or human) boundaries
6. Land use and human artefact
7. Faunal habitat

The development phases of vegetation are particularly noteworthy, because to characterise a tessera other factors may be needed, even if the vegetational association is the same such as the animal distribution in an ecological mosaic, for instance in agriculture or forested ecotopes. For this purpose (Fig. 2.3) we can see the four typical phases (A, B, C, D) in a forested ecotope.

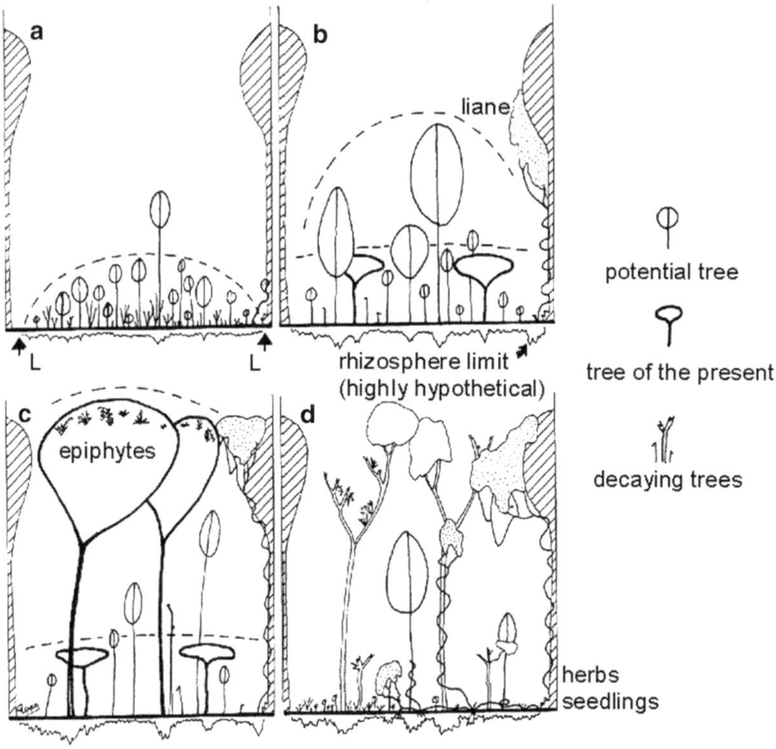

Fig. 2.3 Vegetation phases of growth in a forested tessera (from Oldeman [3], re-drawn). The innovation phase is **a**; the aggradation phase is **b**; the biostatic phase is **c**; the degradation phase is **d**

- The innovation phase is A, lens-shaped, in which the information of the propagule bank is operational. The highness is very limited in contrast to the density of organisms.
- The aggradation phase is B: the vertical structure is now differentiating, low trees survive and the rhyzosphere is deeper.
- The biostatic phase is C: trees have built structural ensembles that organise forest architecture and all other component behaviour. The canopy is again closed.
- The degradation phase is D: trees of the present decay, and forest architecture breaks down to liberate many other forest components. For instance, light and organic matter activate part of the propagule bank to prepare a following innovation phase.

2.1.3 Significance of Pattern Components of Landscape

At this point we have to note that in landscape ecology it is possible to find two different criteria in studying landscape structure.

The first criterion is independent of scale: e.g. the concept of patch/corridor [4, 5] or the concept of land unit [6, 7]. For instance, the definition of land unit is: "a tract of land that is ecologically relatively homogeneous at the scale level concerned". In the case where a hierarchical arrangement is needed, one should use by approximation the terms ecotope, micro-, meso-, macro- and megachore.

The second criterion is consistent with the above-mentioned observation on scale-dependent elements. We prefer this last criterion, because we are convinced that:

- We cannot pretend to have boundaries related to each levels of life organisation in a deterministic criterion; but there is no doubt that every biological system shows a structure formed by well-defined functional sets in a context of variable substrates, starting from cells up to ecosystems (ecocoenotopes), landscapes, regions and the ecosphere (see Fig. 2.1);
- Many patches may be considered relatively independent, but many others not, having an explicit strong linkage among them;
- In any case, it is better to refer to vegetational *tesserae* and ecotopes than refer, using informatic tools, to manifold clusters of various types of patches, the most part of which have no biological significance;

That is why we propose, as pattern components of a landscape, tessera, ecotope, simple landscape unit, complex landscape unit, landscape. This is similar to the Australian (D.O. S.) system: site, land facet, simple land system, complex land system, main landscape.

Moreover, the second criterion does not reject the first: it is possible to use "by approximation" the first criterion with the second. For example, we may speak of a patch of tesserae of the same typology, or a simple landscape unit may be viewed as a patch at a very synthetic scale (e.g. 1/200,000).

2.2 Configurations of Landscape Elements

2.2.1 Patches, Corridors and Landscape Matrix

2.2.1.1 Patches

The patch is defined as a non-linear surface area differing in appearance and/or in substance from its surroundings. The corridor is a narrow strip of land that differs from the surrounding elements of the landscape. Many observations have been made on patches and corridors, e.g. Forman and Godron [5], and it is better to try and synthesise them more in detail. It is possible to classify many types of patches in relationship to their origin:

1. Disturbance (isolated, chronicle, cyclic)
2. Remnant (associated or not with a disturbance)
3. Environmental resource (biotic or abiotic or both)
4. Human activity (direct or indirect)
5. Colonisation (natural or human)
6. Ephemeral patches (natural or human)

Generally, patches are formed by some environmental change and each patch transforms itself during time. The dynamics of the species is involved in this process, particularly extinction and immigration in relationship to the different origins of the patches. The persistence of a patch depends on the presence of a range of perturbations, and it is proportional to its stability, essentially for homeorhetic reasons.

Since the concept of landscape element is linked to a certain range of scales, we note that it is hierarchic. Many patches may be considered relatively independent, but many others are not, having an explicitly strong linkage. For instance, some forested patches in a sub-rural landscape could be independent; by contrast, on the same landscape, gardens and villas or ponds and reeds are linked.

The extension of a patch does not influence the flux of energy and nutrients of its ecosystem: there is a proportionality between the surface and these processes. The extension of a patch may influence its biomass, because of the difference between the border edge and the interior part of the patch. The influence of the borders of a patch is proportionally the inverse of its extension. In these marginal belts of a patch, biomass is denser than in the interior, where its species have less diversity. Anyway, large patches seem to have more species (especially animal) than small ones. A patch may be viewed as an isle, for instance a forested patch surrounded by a cultivated mosaic; therefore the species/area diversity generally follows what community ecology has expressed as:

$$S = f\ [(+)\ H_{habitat}, (-)\ Ds, (+)A, (+)Is, (+)\ age]$$

where : (+) = positive relation, (−) = negative relation, H = diversity, Ds = disturbances, A = area, Is = isolation, S = number of species. In these terrestrial islands we have to observe that the species diversity S is positively correlated with the surrounding mosaic and its heterogeneity and negatively correlated with the margin discontinuity of the patch. In a case like this, disturbances may have positive relationships.

The shape of a patch normally has ecological significance not only as a relationship between edge-interior (or littoral-lake), but even when the diameters are different or the shape is annular or dendritic. The shape affects the interaction of the patch perimeter with the landscape mosaic, the movement of species through the mosaic, the functional exchange with a patch archipelago, the type and number of species, and the role as an ecological attractor in the landscape. In the case of a peninsular shape we note particular effects, like the diminishing of the species toward the top, the changing of environmental gradient and a so-called *interdigitation* process.

The ecological influence of the shape of a patch may be observed also in a three-dimensional sense, as in marginal belts. A strong difference in light, temperature, humidity and species changes the structure of patch margins both in shape (e.g. more branched trees) and regarding the presence of more shrubs and herbs. Even in the case of patches of direct urbanisation (built-up areas) we have a margin effect, if the environmental relationships are not degraded.

2.2.1.2 Corridors

According to Forman and Godron [5] corridors may also be classified in relationship to their origin, but the main characters of corridors are the following:

1. Stream corridors (which are formed by a river or a canal)
2. Line corridors (patch boundaries, roads, paths, hedgerows, etc.)
3. Strip corridors (wider bands with interior species)

To study the structure of corridors it is necessary to analyse: curvilinearity, nodes, breaks, connectivity and their transverse section. Corridors are generally composed of multiple elements.

Stream corridors are characterised by the presence of the river bed and its banks, but often by a strip of forest, which controls water and nutrients fluxes and allows the movement of species. That is why it is necessary to preserve river corridors with all their structure, at least on one side. In a stream corridor, species change as the stream order increases and in relationship to the heterogeneity of the landscape mosaic.

Linear corridors show many types: hedgerows, dikes, roads, railroads, power lines, narrow ridges. They may act as through ways or barriers, depending on the landscape mosaic. Their micro-environment is particular, because of the effect of the winds, the light and the interactions with other contrasting elements of the mosaic.

Strip corridors may be of the same type as linear corridors, but they contain an interior environment. Bird communities of a forest differ generally with the bird community of a corridor, because of the length gradient. Also the plant species of a corridor may differ from those of the nearby forests, especially if the landscape matrix is very transformed by the human population.

Hedgerow corridors are particularly important since they are able to characterise agricultural landscapes. They may be planted or remnant and may have a large heterogeneity of trees and shrubs. In a wide garden a hedgerow acquires a function of refuge for natural species: in a case study, Ingegnoli and Giglio found 46 plant species in a *Prunus laurocerasus* hedgerow of about 100 m near Menaggio[1] (Lake Como).

Complex corridors were found in many regions of Europe, with a structure of a lane and two parallel hedgerows (Fig. 2.4), often with a

[1] In the garden of Villa Mylius-Vigoni, Deutsch-Italienisches Zentrum für Europäische Exzellenz, 2007.

Fig. 2.4 Example of ecological corridor along a semi-natural canal between two fields, North to Milan. The distance from one side to the other of the two lines of trees and shrubs measures about 22–25 m

ditch and a bank. They present many inner species of plants, being larger than 12 m, a measure which Forman and Godron assert to be a threshold for inner species. In the last 4–5 decades, the ratio of hedgerow per unit area is decreasing, as is the number of trees/100 m, e.g. from 5 to 2 in central France [8].

2.2.1.3 Landscape Matrix

The definition of landscape matrix is linked to the concept of the ecological characterisation of a mosaic; therefore a matrix is the most extensive and most connected element type present in a landscape and plays the dominant role in its functioning. It is possible to distinguish among three types of matrices:

1. Continuous, with a single dominant element type
2. Discontinuous, with a few co-dominant element types
3. Web-shaped, with connected corridors of prevailing functions.

Forman and Godron [5] cited three criteria available to distinguish a landscape matrix: relative area, connectivity, control over dynamic.

The relative area criterion pertains to the existence of the first or the second type of matrix. One type of element is considerably more extensive than the others when it covers in a connected way more than 50 % of a landscape. But this element must have a non-degraded ecological state. The percentage may be less than 50 % if additional characters indicate the strong ecological importance of the type of element: in cases like these, the highest relative frequency of the element in the mosaic is sufficient. A discontinuous type of matrix may be found.

The connectivity criterion identifies a matrix by evaluating the degree of connectivity. The matrix is the more connected landscape element type present. In fact, when one landscape element is completely connected and encircles most of the others, it has to be considered as a matrix.

The control over dynamic criterion defines as the matrix the type of elements able to control the transformation of the entire mosaic. Nonetheless, it is necessary to utilise all the three methods of evaluation of the landscape matrix.

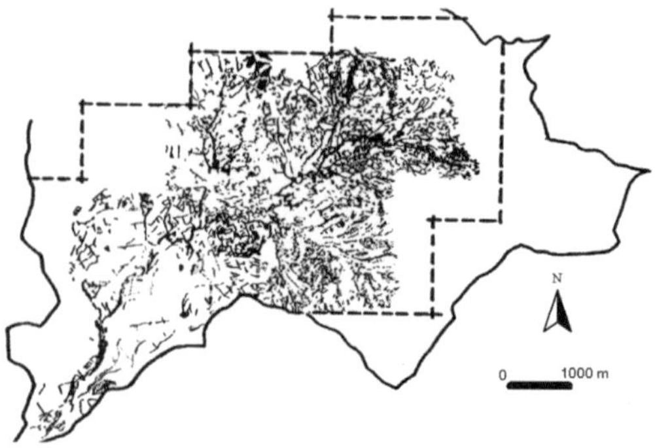

Fig. 2.5 Main characters of the landscape matrix: example of porosity of tree vegetation in a Mediterranean landscape unit (Magazzolo River, Sicily, Italy). Note the loss of structure in the central and south-western parts, due to human actions and clay instability of the hills

Even if the matrix is classified as continuous it contains some heterogeneity. Analysing a matrix becomes important to measure its *porosity*, that is, to measure the density of patches in a landscape. The type of patches to be measured depends on the case study, for instance forested patches in an agricultural landscape (Fig. 2.5) [9] or built-up patches in an urbanised one.

2.2.2 Ecotopes and Landscape Units

2.2.2.1 Ecotope

The minimum landscape unit, in the sense we have described in the first paragraph (Sect. 2.1.1), is the ecotope (Fig. 2.6) which is the minimum multidimensional system formed by two or more types of ecocoenotopes. As expressed by Leser [10] and Pignatti [11], an ecotope is formed by interdependent tesserae, for instance a small trait of a river with its bed, its banks and their vegetation complex.

In order to analyse an ecotope, Vos and Stortelder [13] suggested surveying the so-called *physiotopes* (geomorphologic and microclimatic characterised localities) and their *biotopes* (community-ecosystem characters). But this is not sufficient. Recurrence (topographical) and genetic (geomorphological) relations, in fact, are not always enough to locate an ecotope. For these reasons Ingegnoli [14] proposed a survey also of the *functional role* in the landscape unit: for example, along a river we may have groups of tesserae more applicable to contain a flood.

Moreover, remember that the main characters of an ecotope can be numerous and they have to be analysed on three scales: tessera scale, ecotope scale, simple landscape unit scale. They may be:

- *Basic mosaic characters*: ecological state of vegetation and its medium biomass; biodiversity degree; not-incorporated disturbances; land use functions; natural and human habitats; grain size; connectivity with the outside mosaic; potential transformability; landscape pathologies.
- Function characters, shared to upper scales: physiotope participation; role in hydrography; landscape apparatuses; role in regional landscape system; importance of ecotope for eurytopic animals; contribution in structural orientation of the landscape; presence of technological webs.
- Lower scale characters: see tessera

2.2.2.2 Landscape Unit

The landscape unit, or geo-bio-district, is intended as a sub-landscape, a part of a landscape which assumes particular characters or even functions in relationship to the entire landscape. In the old structural classification of Neef, following Troll [15], this concept is similar to the micro-mesochore and macrochore, which are horizontal arrangements of ecotopes. It is also similar to the D.O.S. Australian classification, as already discussed.

A (simple) landscape unit (LU) can be defined as an interacting disposition of recurrent and genetic (sensu geomorphology) ecotopes which assume a particular significance (function) in its own landscape. For example, the ecotopes around the lake of Doberdò, in the Karst landscape, or a suburban park in a Berlin urban landscape. In the first case, the formation of a karst lake produced a ring of geomorphic characterised slopes converging toward the lake (Fig. 2.7); in the second case, an abandoned field area among new expansion districts assumed a recreational function.

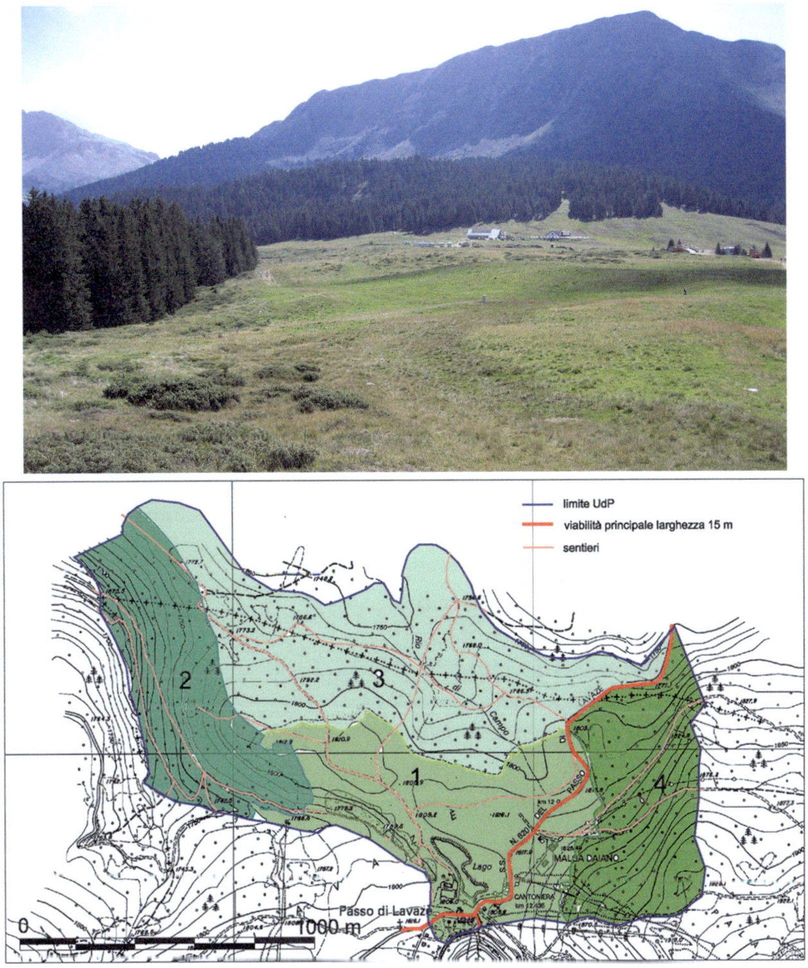

Fig. 2.6 Four ecotopes, forming a very small landscape unit at Lavazé Pass, 1,800 m, between Eggental and Fiemme, in the Alps (Dolomiti) [12]. Types of ecotopes: (1) prairie, with few buildings, some shrub patches and a small lake; (2) and (4) subalpine spruce forest on a slope soil; (3) subalpine spruce forest with *Pinus cembra* and some bog patches. The picture was taken from ecotope 1 towards 4; to the left the ecotope 3

Fig. 2.8 represents the municipality of Tesero, Fiemme Valley (Dolomites), a territory of 50 km^2. This area is clearly composed by three LU: the two lateral valleys (North and South) and the intersection with the main Fiemme Valley in the middle. Note that the administrative limits are in some traits out of the watershed lines. This set may represent a complex LU, because a *complex landscape unit* (c-LU) can be defined as an interacting disposition of a cluster of simple landscape units. It could represent a good proportion of an entire landscape, for example, a good trait of an alpine valley in a Dolomite landscape.

The structure of a landscape unit is not always immediately recognisable, because it is not always a simple arrangement of ecotopes, even if it forms a connected patch of ecotopes. An analysis of the range of functions which form the geo-bio-district is consequently necessary. A landscape unit (Fig. 2.8) is formed by the emergence of many and various functional components: (a)

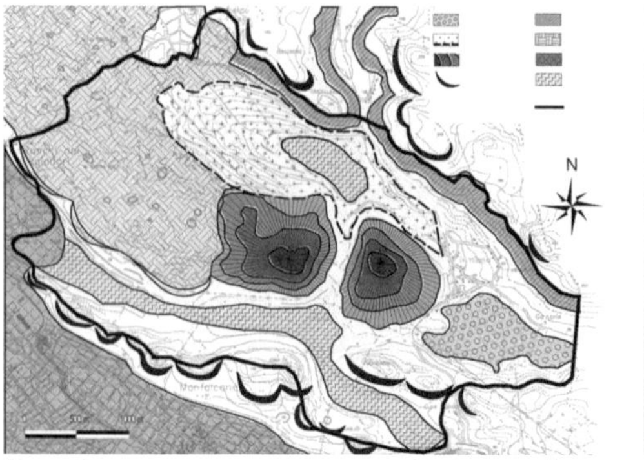

Fig. 2.7 The landscape unit (LU) of the Lake of Doberdò, near Trieste

geomorphologic characterisations, (b) ways and areas linked to the movement of permeant animals and/or men, (c) bio-functional effects produced by a community over other communities, (d) changing capacity of some ecotopes, (e) mesoclimate effects, etc.

Note that we may define other types of landscape units, not always directly linked to geomorphic substrate and ecologically related functions, but dependent on particular research needs. In this case, the sum of the surfaces of the operational landscape units may be greater than the area of the entire landscape. This fact is verifiable especially surveying animal habitats in a natural landscape, or human districts in urbanised landscapes. It is not ever a question of necessary approximation.

Thus, in summary, it is possible to distinguish many types of landscape units, most of which can be integrated into significant sub-landscapes. We may rank for instance the following:

- Subdivisions of a landscape in local sub-patterns
- Districts with a diverse range of naturalness
- Zoning of different criteria of human management
- Zoning of preservation value areas
- Pattern of specific agricultural and urban functions
- Districts of historical and geographic characterisation

Even the degree of alteration or some negative dynamics (e.g. degradation) may be utilised in order to distinguish landscape units. Anyway, note that in case of choosing an operative (merely working based) landscape unit, it is always necessary to aim for a better ecologically delimited one.

2.3 The Concept of Ecotissue

2.3.1 General Models to Study the Landscape Structure

Prior to 1960, general ecology usually emphasised spatial homogeneity. The reference was a sort of traditional *ecosystemic* model, in which—given a scale of interest—only a homogeneous spatial unit was considered, for example a wooded area in an agricultural environment (Fig. 2.9a). This unit could have been studied as a community (e.g. vegetational association) or as an ecosystem (e.g. a watershed). Outside this spatial unit, there was an indistinct (heterogeneous) environment. A more complete model considered the boundaries of that ecological unit: the edge belts, or the ecotonal belts if an environmental gradient separated the ecosystem from one's nearby.

Because theories assuming homogeneity developed earlier in the history of ecology, they had a powerful and persistent effect on how ecological systems were viewed. That is why E. P. Odum, in his famous book *Fundamentals of Ecology* [16], mentioned the landscape, but he was not able to properly represent it, even when writing on land use and the necessity of ecological planning, "undoubtedly the most important application of environmental science". Nevertheless Odum mentioned the book of Ian McHarg, *Design with Nature* [17], which showed how the natural landscape can provide guidelines for quality urban development. Mc Harg, a landscape architect, was able to represent the natural landscape as a patchy environment, probably on the basis of the vegetational maps made by botanists of the school of Braun-Blanquet.

Consequently, a new model of representing the landscape was elaborated by many authors,

Fig. 2.8 The municipality of Tesero (val di Fiemme) has a complex landscape units (LU), composed by 3 simple LU (*left*). (1) LU of Stava Valley, 20 km^2, with small built patches, some meadows and 51 % of forest; (2) LU of the central Fiemme Valley, 14 km^2, with the town of Tesero, some agriculture and 49 % of forest; (3) LU of Lagorai Valley, 16 km^2, 12 % pastures, 34 % of rocks and 46 % of forest and a lake. Geological substrates are very different in each LU. Each photograph represents a LU (1,2,3)

e.g. Forman and Godron [5], Haber [18] at the beginning of the 1980s. This second, mosaic model, has two aspects (Fig. 2.9b): the "patch-matrix" and the "*ecological mosaic*".

The first highlights the patches formed by the different types of ecological communities, at a given scale of interest, and the corridors composed of natural or human elements, within a sort of landscape matrix. The second emphasises the entire landscape mosaic, the elements of which we could name *tesserae* and "ecotopes": these are the delimitations of the principal types of ecosystems (i.e. biogeocenosis or, better, ecocoenotopes) and they constitute a

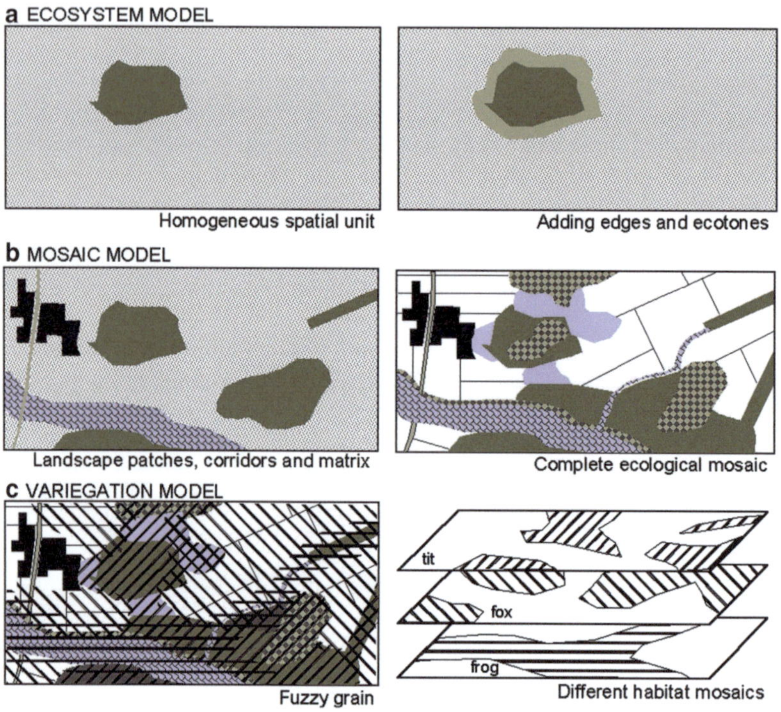

Fig. 2.9 Main structural models of the landscape. (**a**) ecosystemic model. (**b**) mosaic model. (**c**) variegation model. Each model may be represented in two versions, the second more detailed

sort of geographic map, apparently similar to the land use maps of the human territory, but with an ecological sense. These maps are more similar to the vegetational maps.

The concept of human fragmentation of the landscape implies that vegetation remnants are surrounded by unfavourable environments (e.g. a boreal forest patch in an arable fields mosaic). Consequently, as pointed out by Ingham and Samways [19], organisms may not be able to move across or inhabit the fragments. Much ecological research has been carried out on the basis of this model. However, the form and extent of the structure in landscape varies and the mosaic model may not always be applicable.

Many zoologists have observed that relatively few organisms perceive the landscape in a similar biological way, because of the existence of stenotopic and euritopic taxa. Moreover, the biology of an adult insect, for instance, may differ from that of the larvae. Therefore, a species may be dependent on two or more landscape elements for its life cycle or may live in one area but forage in another one. The landscape—thus—may act as a filter, separating out the species [20], or even becomes a species-specific environment [21]. So the landscape seems to disappear, evanescing into a sort of fuzzy-edged mosaic. Thus, a new landscape model emerges, called (Fig. 2.9c) "*variegation* model", in which the ecological mosaic is composed of tesserae having a variable conformation (e.g. fuzzy-edged boundaries); in other words, it is an overlapping series of different patch-matrix mosaics (proportionate to the main groups of steno-euri-topic habitats).

In summary, we have passed from the ecosystemic model—undoubtedly too simplified and unrealistic—to two divergent new models, the *mosaic* one and the *variegation* one, each one able to register the real situation, but in a typical complementary contrast (sensu Bhor and Heisenberg) resulting in a typical uncertainty; on the other hand, many researchers know that reality needs both the mosaic and the variegation model, e.g. when studying a mosaic of forested landscape and its bird communities.

The main limitation of the mosaic model is that differently scaled entities may be forced together in a rigid mosaic; other serious problems concern the variegation model too. As pointed out by Allen and Hoekstra [22], if organisms of different species on the same spot of ground respond to different scaled patches of landscape, then the concept of "organism" becomes inconsistent as a scaled unit. Alternatively, if we use the organism as the normalising framework, then the landscape becomes a curved space, because different types of organisms view it at their own scale of reproduction, dispersal, movement or home range.

This seems to be absurd. The theory of complementarity may lead to a misunderstanding of the problem of observation. As sustained by Popper [23] and Prigogine [24], the reality per se remains measurable. Therefore, we need to change our criteria if we want to add another step towards a true descriptive model of the landscape.

In fact, some observations regarding the variegation model are not completely true. No doubt that an euritopic species like one of *Orthoptera* or *Odonata* perceives the landscape differently from a stenotopic species like one of *Hemiptera* or *Carabidae (Coleoptera)*. An euritopic species forages indifferently in a patch of meadow or in the closest patch of trees, but its physiologic and behavioural responses will not be the same for each patch, because the microclimate (e.g. light, humidity) changes, the quality of food (e.g. amino acids, lipids, carbohydrates of leaves) changes, many predators change, etc.

Therefore, we may find diverse hierarchies of ecological factors, some of them depending on the mosaic model, others on the variegation one. Moreover, many animal species are definable as "permeants" (sensu Shelford in Odum 1971 [16]), simply to indicate their use of many types of ecosystems. These euritopic animals, which move, feed, protect, drink, nurse, with periodic rhythms, have a different perception than stenotopic ones, and they are even able to recognise the landscape as a complex mosaic in detail, from macro- to micro- scales (e.g. apes, wolfs, gooses, bees, etc.). Some of them are able to change directly some parts of the landscape, for example, beavers, but also ibexes, elephants, termites, etc.

2.3.2 The Ecotissue Model

As just discussed, the descriptive models of the landscape are mainly the "mosaic" model and the "variegation" model (see Fig. 2.9). These models represent two completely different outlooks, one static and one dynamic. But it is not completely correct to compare a mosaic of juxtaposed tesserae of landscape elements (*mosaic model*) with a mosaic of species-specific tesserae and variable geometry (*variegation model*). In reality, as we know, each biological system shows a structure made by well-defined functional groups in a context of substrates changeable in space and time (or variegated configurations over a fixed weft, and whose elements are not only juxtaposed, but also overlapped and intersected). Both Forman and Godron [5] and Naveh and Lieberman [25] talked about the interweaving of ecosystems in a landscape.

As a consequence, it is better to introduce the concept of ecological tissue (from the Latin *textus* or *textilis*; in English: textile) or *ecotissue*, which is a complex multidimensional structure represented by a basic mosaic and a hierarchic succession of correlated mosaics and attributes:

$$Ect = Mm \cup (\cap)\, M' \cup (\cap)\, M'' \cup (\cap)\, Mk$$

where: Ect = ecotissue, Mm = basic mosaic, M′ ... Mk = correlated mosaics and the set-theoretical ∪(∩) symbols of union and of intersection may be interchangeable or may even coexist, since they are often heterogeneous groups.

The basic mosaic is generally formed by the vegetational coenosis because the control of the flux of energy and matter and the capacity to create the proper environment pertain to it. This fact is in accordance with non-equilibrium thermodynamics. Whereas an energy concentration (i.e. photosynthetic plants) produces structure

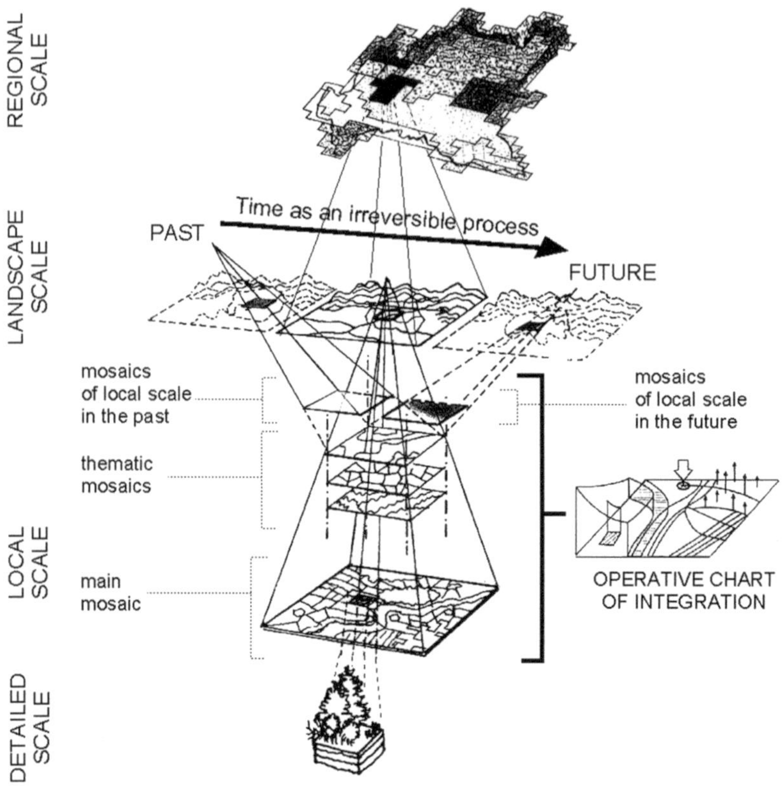

Fig. 2.10 The ecotissue model. The basic mosaic is generally the vegetational one. The complex structure of a landscape has to integrate diverse components: temporal, spatial, thematic. An operative chart of integration is therefore necessary to elaborate plans. Note that the integrations are intrinsic, that means they have to follow integration functions derived from the intrinsic characters of that level of life organisation

and organisation in a landscape matrix with increasing entropy, the order through fluctuation process (Cfr. 1.3.8) creates a patch, which acquires a specific landscape role. This may be the principal way by which ecological systems become heterogeneous (Ingegnoli [9, 26]; Forman and Moore [27]).

Anyway, all the other mosaics are correlated to the basic mosaic and compatible with its main scale. Trying to detect and gather information on organisms and communities outside this main scale of interest is generally a non-sense, because of the hierarchic organisation theory. We can study a self-organising system only through projections and sections of a hyper-space, so that it is possible to gather information to be integrated, step by step, in a hierarchic correlation with the basic mosaic.

It is necessary to consider constraints at different scales and intersections with specific thematic mosaics, so that they are compatible with the characters pertaining to the intrinsic level of life organisation of a landscape (Fig. 2.10).

The ecotissue model gives the right importance to the landscape and integrates the fundamental dimensions of the landscape:

1. A range of spatial scales, from regional to local configuration of elements
2. A set of thematic mosaics on species (biomass) and resources (energy) components
3. A range of temporal scales on developing processes, which permits the evolutionary dynamic of the landscape to be forecasted and reconstructed
4. A set of information contents, which permits to evaluate the level of the systemic organisation

This result may be configured in an operative chart of integration.

As we will see in the section on methodology, the importance of the concept of ecotissue can be found especially in the integration of elements and processes in a landscape. Actually the study of a multifunctional landscape is usually led by a certain number of different mosaics (geomorphologic, vegetational, zoological, agronomic, of land use, of human needs, etc.), which have to be integrated for a certain purpose. But this kind of integration is a sort of a posteriori process, obtained with a traditional multidisciplinary criterion, because the landscape is ultimately viewed as a support for biological and human systems: so, the landscape exists only as the result of an integration like this one.

By contrast, if a landscape is defined as a living system, we must refer to a more complex structural model in which the integration is made *intrinsically*. Remember that an ecologist has to act like a physician, what we called an "ecoiatra". This means that the mentioned thematic mosaics have to be related to the structure and behaviour which characterise that landscape. A clinical and pathological methodology is applicable only if we analyse the state of an organism knowing something about its anatomy and physiology. Similarly, we have to analyse the state of a landscape following the knowledge we have about its structure and functions.

2.4 Main Anatomical Components of the Landscape

2.4.1 Context Role Subsystems and Human Habitat

The ecotissue model, need to put in evidence the distributive modalities of specific systems of elements, choosing criteria of landscape functionality. For instance, the knowledge of the dynamic phases of each ecotope does not depend on the ecological state of patches and corridors, but especially on the spatial pattern of the landscape.

The definition of landscape apparatus concerns functional systems of tesserae and ecotopes forming specific configurations in the complex mosaic (i.e. ecotissue) of a landscape. As indicated by Ingegnoli and Pignatti [28] these structures are multi-functional (sensu Brandt et al. [29]; Naveh [30]) and can be named also "landscape context role subsystems".

They differ substantially from the landscape units, because they are not districts or sub-landscapes, but complex configurations of patches that may even be not-connected. These apparatuses are distinguished by a specific landscape function (and/or its range of sub-functions), not only by many local characters.

A first well known, general but important, landscape function is demonstrated by the survey of human habitat (HH) versus natural and semi-natural habitat (NH). Ecologically speaking, the HH cannot be the entire territorial (geographical) surface: it is limited to the human ecotopes and landscape units (e.g. urban, industrial and rural areas) and to the semi-human ones (e.g. semi-agricultural, plantations, ponds, managed woods). The HH can be defined as areas where human populations live or manage permanently, limiting or strongly influencing the self-regulation capability of natural systems.

The NH are the natural ecotopes and landscape units, with dominance of natural components and biological processes, without direct human influence and capable of normal self-regulation. Note that even near-natural ecosystems (i.e. little changed after human abandonment) are NH. Remember that, in landscape ecology, the management role of human populations, if not directed against nature, may be considered as semi-natural.

Following the ecotissue model, the presence of HH and NH is generally mixed: for instance, it is difficult to find a completely HH tessera or ecotope. In many margins of human tesserae and ecotopes we may find natural species, as confirmed in Sect. 4.1.2 (interface elements).

Therefore, even in urbanised landscapes it is generally possible to find NH patches and corridors. On the other hand, in natural landscapes many semi-managed patches may be found. Hence, the mosaic of NH and HH in a landscape is different from the land use mosaic.

2.4.2 Natural and Human Landscape Apparatuses

Rarely the following other functions have been linked to some basic concept of landscape ecology, such as to its structure and dynamic. In reality, it is possible to distinguish many types of landscape apparatuses:

- GEO = geologic (emerging geotopes or elements dominated by geomorphic processes)
- STB = stabilising (elements able to limit perturbations)
- RNT = resistant (elements with high metastability, e.g. forests)
- ETN = ecotonal (system of gradient belts)
- EXR = excretory (the fluvial web as landscape catabolite processing)
- SOU = source (ecological sources as centres of community expansion)
- DIS = disturbance (elements with a range of non-incorporating disturbances)
- CHG = changing (elements with high capacity of transformation)
- NH/hh = natural habitat functions within the human habitat
- RPD = reproductive set (reproductive and colonising elements)
- RSL = resilient (elements with high recover Capacity, e.g. prairies or shrub lands)
- CON = connective (elements with important connective functions in the mosaic)
- PRT = protective (elements which protect other elements or parts of the mosaic)
- PRD = productive (elements with high production of biomass)
- SBS = subsidiary (systems of human energetic and work resources)
- RSD = residential (systems of human residence and dependent functions)
- HH/nh = human habitat functions within the natural habitat

Note that many landscape functions are typical of human habitats (e.g. productive, residential, subsidiary), others are typical of natural habitats (e.g. source, resistant, stabilising, geologic) and others may be in common (e.g. protective, resilient, connective). Note also that even human habitat apparatuses are intended in an ecological sense, not in a geographical or urban planning sense (Fig. 2.11).

The *residential apparatus* is characterised by settlement and service functions of human populations, transmission of traditions, religion and culture and administrative centres. Small orchards, small gardens and urban street vegetation are components. Dense and dispersed urbanisation patches generally form its typical parts. The historical heritage of old towns and cities (and roads) has to be studied in order to know the process of formation of this apparatus within a landscape.

The *productive apparatus* is mainly part of the human landscape, even if there are often more productive patches (e.g. density of food resources) in a natural mosaic. The difference between the economic and urban definition of productivity is marked, because in ecology only communities with a high dominance of autotrophic species are truly productive. The degree of ecosystem stage in which respiration is not high plays an important role. Agricultural systems form the most typical productive patches, together with farms.

The *protective apparatus* is found both in natural and anthropic landscapes. Many ecosystems acquire a role of protection in their own mosaic, for instance resisting to slope disturbances at the base of a mountain and permitting the formation of other types of more sensitive ecosystems. In human landscapes, the hedgerow network in a field mosaic or the gardens around villas are typical protective elements.

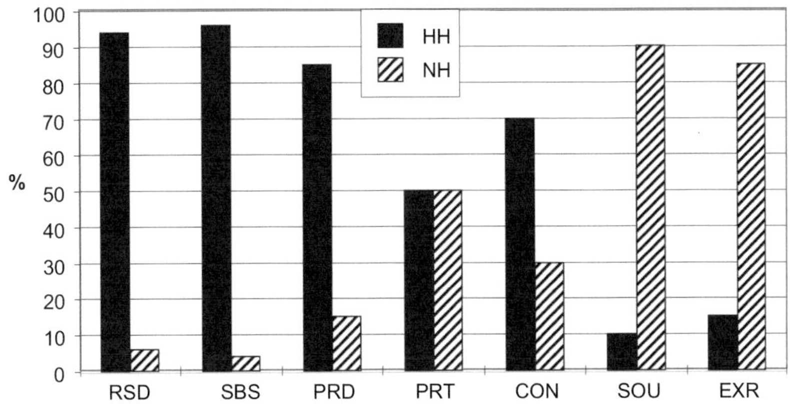

Fig. 2.11 Proportion of human habitat (*HH*) vs. natural habitats (*NH*) in tesserae and ecotopes with the following functions: residential (*RSD*), subsidiary (*SBS*), productive (*PRD*), protective (*PRT*), connective (*CON*), source (*SOU*), excretory (*EXR*). Values typical for Central Europe

The *excretory apparatus* is composed by the network of stream corridors (much more than the hydrologic network), very important because of the carrying potentiality, the buffer capacity, the cleaning capacity, the structural fitness with the mosaic, etc. Linking this web to landscape principles is not a matter of technical necessity, but is rather inherent to the landscape functioning. All the operations regarding the control and monitoring of water resources can thereby reach a higher level of preservation.

The *resistant apparatus* is composed of elements of high metastability, for instance mature forested patches, and plays a regulatory role in the mosaic. The resistance stability predominates in those elements. Generally mature fitted vegetation is particularly important, but even other ecosystemic stages may be crucial in difficult environmental conditions.

The *source apparatus* is intended as in source-sink theory, but is related to the potential expansion of centres of community: populations are rarely concerned. The connections with sink patches are very important and have to be differentiated as barriers, porous, open. Sink patches may also be differentiated as receptive or not. It is necessary in evidence supply, disposal, resistance and retention per patches.

2.4.3 The Peculiar Anatomic Components of a Landscape

The emergence of context role subsystems in the landscape configuration of elements brings to consider the presence of anatomical sets. As in any biological system, it is possible to recognise the following sets and their main structures (Table 2.2).

Let us observe that the landscape apparatus functions are important, but they are not sufficient to illustrate the 12 anatomical sets ranked in Table 2.2, especially in set n° 6, 9, 11 and 12. The study of landscape apparatuses has to be completed with considerations related to their distribution in a landscape unit and to their functional efficiency.

For instance, the configuration of spatial patches useful to human population is shaped in Euclidean geometry, while landscape structures are shaped in Fractal geometry. Therefore, this fact produces many interface fragments that only apparently have no importance (see Sect. 4.1.2). On the contrary, following the processes of ecotope reproduction (sensu Ingegnoli [9] and Ingegnoli [31]) we can see that these remnant fragments are frequently useful as propagule banks sites and ecological memory (see Sect. 3.6).

Table 2.2 The main anatomical sets in a landscape and their structures

	Anatomical components of a landscape unit	Main elements and structures
1	Geomorphic skeleton	Tesserae (TS) of GEO apparatus and their (hydro)-geomorphological pattern
2	High autotrophic biomass formations	TS and ecotopes of the RNT apparatus
3	Formations of autotrophic production	TS and ecotopes of the PRD apparatus
4	Heterotrophic formations	TS and ecotopes of the RSD and SBS apparatuses
5	Cybernetic and managing control centres	TS and spots of the RSD-management apparatus and zoological control niches
6	Connection and movement networks	River network, ecological corridors network, road & rail network
7	Energy and material producing areas	TS of the SBS apparatus and technical network, enlarged to natural resources
8	Excretion and sink areas	Sink and dump areas (natural and human)
9	Protective and recovery elements	Interface elements, undisturbed strategic areas, TS of the PRT apparatus
10	Reproductive and colonising elements	TS and ecotopes of the RPD apparatus
11	Morphological characters of the system	Grain, ecotonal network, contrast element, configuration design, fragmentation, etc.
12	Perturbation and alteration elements	Barrier, out of scale contrast or fragmentation, strong transformation areas, geological risk areas, etc.

For this purpose, the use of bionomic parameters and indexes is needed, as we will show in methodological chapters. Anyway, without the knowledge of the presence and type of these anatomical sets, the diagnostic check-up of the bionomic state of a LU remains not exhaustive. The presence and the evaluation of these anatomical components in a landscape unit are simply essential.

2.5 Landscape Classification

2.5.1 The Main Attempts to Classify a Landscape

As we discussed, it is impossible to study an ecocoenotope below the scale of the community and without knowing its component populations. When, enlarging the scale, other elements are added outside the structure and functions of an ecocoenotope, it is necessary to define a landscape. For instance, when a land mosaic is reached, new structures and new processes appear: ecotonal webs, connectivity, porosity, landscape matrices, landscape apparatuses, landscape dynamics, new strategies of metastability, etc. Other characters appear at a regional scale, impossible to be studied properly in a landscape, such as soil order, forest formation, fauna changes, river bio-gradients, climatic role in ecological differentiation, etc.

In fact, not only many characters but also many dominant controlling factors in an ecological system change with the enlargement of the scale, as suggested by the hierarchy theory. That is why it is necessary to define a landscape as limited to a specific range of space-time scales. It is not a question of a human scale perception, because even in a natural hierarchy of ecological systems every researcher can find different behaviours by changing some range of scales.

Note that at geospheric dimension, before the foundation of ecology, domains and regions were traditionally studied in biogeography and today they include global and regional ecology, together with the biosphere. Sometimes regions (ecoregions) are studied also in landscape ecology [32], but this is a normal overlapping margin between ecological fields, necessary when a framework is needed to analyse a landscape.

The main structures pertaining to a landscape, from the ecotissue concept to the anatomical components, have already been expressed.

Anyway, it must be clear that each type of landscape presents a peculiar behaviour, due to its peculiar characters. Therefore, the classifications of the landscape become indispensable to analyse landscape units.

Now it is possible to ask how to classify landscapes. Today there is not a complete taxonomy of the landscapes, but only some ordination methods. We can distinguish four principal methods: (A) dominance of man-made artefacts, (B) phytosociological criteria, (C) hierarchic factors criteria, (D) integrated landscape apparatuses.

(A) *Man-made artefacts* One of the most classical ordinations of landscape types has been proposed by Naveh and Lieberman [25], according to the definition of Total Human Ecosystem (ecosphere = biosphere + techno sphere).

Energy, matter, information inputs from bio- and techno-ecosystems are involved. On the x axis are reported the modification, conversion, replacement of natural bio-ecosystems. On the y axis (negative verse) we find the degree of dominance of man-made artefacts. There are seven major resulting types of landscape, disposed on the diagonal of the figure and divided into open and built landscapes (last three):

1. Natural landscape (natural bio-ecosystems)
2. Semi-natural landscape (natural bio- and rural techno-ecosystems)
3. Semi-agricultural landscape (natural bio-, agricultural bio- and rural techno-ecosystems)
4. Agricultural landscape (agricultural bio-, rural techno- and urban techno-ecosystems)
5. Rural landscape (rural techno- and urban techno- ecosystems)
6. Suburban landscape (rural techno- and urban techno- ecosystems)
7. Urban-industrial landscape (urban techno-ecosystems)

Ranked according to decreasing naturalness or increasing artificiality is the ordination of Wolfgang Haber [33]. There are five landscape types, divided into bio-ecosystems and techno-ecosystems (the last):

1. Natural (without direct human influence and self-regulated)
2. Near-natural (influenced by humans but little changed after human abandonment and capable of self-regulation)
3. Semi-natural (resulting from the use of natural and near-natural landscapes, changing significantly after human abandonment, requiring some management)
4. Anthropogenic-biotic (intentionally created by humans, fully dependent on management)
5. Techno (technical systems with dominance of artefacts for industrial, economic and cultural activities, dependent on human management and on surrounding ecosystems)

(B) *Phytosociology* Landscapes are classified on the basis of a survey of their vegetation complexes. A vegetation complex is a group of vegetational associations viewed as tesserae of a wider specific ecotope. This survey is divided into two phases : (I) registration of all the plant associations pertaining to the ecotope in percentage, (II) registration of all dominant elements of the surrounding landscape, but only as their presence. Note that the landscape elements are the most important in classifying the type of landscape [11].

Another method derives from the so-called synphytosociological and geo-synphytosociological analysis (Tuexen [34]; Géhu [35]; Rivas-Martinez [36]): it is possible to identify the relationships between the plant communities of serial and chain types, which appear inside the same dynamic succession, evolutionary or regressive, or between series in the same territory. It is possible also that vegetation integrates various aspects of the environmental system. This method is applicable even to identify sub-regions (landscape systems) at a wider scale. Vegetational associations interact dynamically and manifest steps of a regressive or evolutionary process (the so-called vegetation series or *sygmeta*).

In a typical South-European example, an abandoned pasture association (e.g. *Centaureo bracteate-Brometum erecti*) will change into a shrubby one (e.g. *Juniperus communis-Pyracanthetum coccineae*) which, in the future, will evolve into a forest (e.g. *Aceri obtusati-Quercetum cerridis*).

In landscape phytosociology a vegetation series (*sygmetum*) assumes the same role of an association in classical Braun-Blanquet [37] phytosociology. The sygmetum can be distinguished as climatic or edaphic series, depending on the water support: rain only or rain and soil origin. Among neighbouring associations a dynamic relation is not always possible, for instance because of the diverse soil potentiality. Their relationship will be only topographic or by soil chain. Therefore, the last level of classification emerges from the integration of diverse vegetation series, in what is called geosynphytosociology or landscape chain phytosociology. This analysis allows recognition of a geo-series in a homogeneous landscape unit, such as a valley or a mountain or a trait of a coast.

We should note that this so-called sigmasyntaxonomy method is not generally accepted (Naveh and Lieberman [25]; Zonneveld [7]; Pignatti [38]).

(C) *Hierarchic factors* As sustained by Forman and Godron [5] an important criterion of ordination is based on the assumption that it is necessary to create valid attributes in a proper hierarchy. No natural typology gives to all attributes equal priority, because of its latent hierarchy; in fact it is usually followed a descending hierarchy in five levels :

1. Zonal climates, principal climates of the biosphere, e.g. humid temperate
2. Climatic regions, specific sub-climates, e.g. Mediterranean ecoregion
3. Vegetational belts, bioclimatic units, e.g. sclerophyll vegetation (macchia)
4. Geomorphic units, soil structure in local areas, e.g. red ferallitic soils in terraced slopes
5. Human influences, agricultural fields, roads and settlements, e.g. scattered orchards, vineyards and villages

After this first ordering, it is possible to identify five distinct configurations of landscape on the basis of their structural pattern. Six types of landscape spatial pattern have been described by Forman [32]: (a) large patch, (b) small patch, (c) dendritic, (d) rectilinear, (e) checkerboard, (f) interdigitated. Moreover, a landscape may be locally:

- *Regular*, with uniform distance among the elements, even if the ecotopes are diversified, e.g. regular fields, large and small patches, webs of small corridors, etc.
- *Aggregated*, with specific clustered elements predominating, e.g. large patches of fields, series of different subparallel corridors, etc.
- *Linear*, oriented along a particular direction, following a valley, a big river, ridges, or, in the case of human landscape, a main road or a railway, etc.
- *Spatially linked*, with characteristic association of elements, e.g. fields and hedgerows, moss and small lakes or villas and gardens, etc.

Remember that a landscape is not the sum of the above attributes, but, on the contrary, it is an integrated complex system.

(D) *Integrated Landscape Apparatuses Criteria* Another possibility in ordering landscapes derives from their principal functional configurations of elements, before mentioned in 2.4.1, that is from their landscape apparatuses.

If we consider the nine most characterising landscape apparatuses and combine them with different proportions, it will be possible to note the emergence of at least 16 types of landscapes (Table 2.3). Using this criterion it is possible to classify most of the landscape types of the world, because they are compatible with the previous hierarchic ordination. In fact for each ecoregion (Table 2.6) of the ecosphere we may check the combination of the landscape apparatuses,

Table 2.3 Main composition of landscape apparatuses forming landscape types

Geo	Exr	Rnt	Rsl	Con	Prt	Prd	Rsd	Sus	Landscape types
+ +	−	−	+ −	−	−	−	−	−	Desert
+ +	−	−	+ +	−	+ −	−	−	−	Semi-desert
−	+ −	−	+ +	+ +	+ −	+ −	−	−	Prairie
+ −	+ −	+ −	+ +	+ −	+ +	+ −	−	−	Shrub-prairie
+ +	+ −	+ −	+ −	+ −	+ −	+ −	−	−	Shrubby
+ −	+ +	+ +	+ +	+ −	+ +	+ +	−	−	Open forested
−	+ −	+ +	+ −	+ +	+ +	+ −	−	−	Closed forested
+ −	+ −	+ +	+ +	+ −	+ −	+ +	−	+ −	Semi-natural, > bmass
+ −	+ −	+ −	+ −	+ −	+ −	+ +	−	+ −	Semi-natural, < bmass
+ −	+ +	+ −	+ +	+ +	+ +	+ +	+ −	−	Cultivated, protective
+ −	+ −	−	+ +	−	+ −	+ +	+ −	−	Cultivated, productive
+ −	I	−	+ −	−	+ −	+ +	+ −	+ −	Rural
+ −	+ −	−	+ −	−	+ −	+ −	+ −	+ −	Sub-urban, rural
+ −	+ −	I	−	−	+ −	−	+ −	+ +	Sub-urban, industrial
+ −	+ +	−	−	+ −	+ +	−	+ +	+ −	Urban, open
−	+ −	−	−	−	+ −	−	+ +	+ +	Urban, closed

Geo geologic, *Exr* excretory, *Rnt resistant*, *RSL* resilient, *CON* connective, *PRT* protective, *PRD* productive, *RDS* residential, *SUS* subsidiary, ++ full presence, + − partial presence, −− absence, *bmass* biomass

and finally, if the case, add some elements of the landscape spatial pattern.

If we put in evidence in Table 2.3 the presence (full or partial) or the absence of each landscape apparatus it is possible to recognise, in this example, 16 types of landscape: (1) desert, (2) semi-desert, (3) prairie, (4) shrub-prairie, (5) wild shrubbery, (6) open forest, (7) closed forest, (8) semi-natural L. with a high biomass, (9) semi-natural L. with a low biomass, (10) cultivated protective L., (11) cultivated productive L., (12) rural-agricultural L., (13) suburban-rural L., (14) sub urban-industrial, (15) urban L. open, (16) urban L. closed.

Table 2.3 is theoretically correct, but could be useful to complete the landscape classification, observing the existence of three order of landscape types: (I) natural, (II) semi-natural and (III) human, with the setting of the habitat types (Cfr. 2.1.2), dividing these three set of natural, semi-natural and human landscapes into "families" and then in types, as shown in Table 2.4.

In the first order of landscapes (natural) it is possible to rank at least 4 families: water Landscapes, desert L, prairie, forest L. In the second order (semi-natural) we can find semi-natural agricultural and touristic landscapes. In the third order (human) the agricultural, suburban and urban L. The landscape types linked to these families are at least 20, one half of which natural: riparian, coastal, marine, swamp, desert, semi-desert, prairie, steppe, bushed-prairie, open forest, closed forest, the others pertaining to semi natural (agro-forest, agro-prairie, marine-touristic, mountain-touristic) and to human L. (agro-protective, agro-productive, rural, sub-urban tech., open urban, closed urban).

Note that the local geographic specifications are essential in order to classify the landscapes; for instance a semi-natural mountain-touristic L. could be in the Alps, in the Pre-Alps or in the Carpathian mountains: the same type of landscape but with different regional characters. Let us present an example, from a case study of the landscapes of the Province of Monza (Lombardy), as shown in Table 2.5 and Fig. 2.12.

Table 2.5 ranks 16 landscapes by local geographic specification: 3 for the agricultural-productive type (14 municipalities), 4 for the

Table 2.4 Landscape classification: orders, families, types and local specifications

L. order	L. family	L. type	Local geography
Natural landscape	Water L.	Riparian	Local geographic specifications have to be added to the previous characters: e.g. Alpine (or Pre-Alpine) Valley; Padania Plain (or Rheinland Plain); Liguria Coast (or Kent's Coast); etc.
		Coast, marine	
		Swamp	
	Desert L.	Desert	
		Semi-desert	
	Prairie L.	Prairy	
		Steppe	
		Bushed-prairy	
	Forest L.	Open forest	
		Closed forest	
Semi-natural landscape	S-nat agr.	AGR-forest	
		AGR-prairie	
	S-nat touristic	Marine-tour	
		Mountain-tour	
Human landscape	Agricultural	AGR-PRT	
		AGR-PRD	
		RURAL	
	Suburban Urban	SB-urban tech.	
		Open urban	
		Closed urban	

Table 2.5 Example of classification of the landscapes of 55 municipalities in the Province of Monza (Lombardy)

Order	Family	Type	Local geographic specification	Municip.	Km^2	%
Human L.	Agricultural	*AGR-PRD*	Terraced Groane	4	33.58	8.29
			Morenic hills	6	38.44	9.49
			Terraced Molgora-Adda	4	31.72	7.83
				14	*103.74*	*25.61*
	Suburban		West edge Groane	1	5.15	1.27
		Suburban	Morenic hills edge	3	24.62	6.08
		Rural	Plain of Lambro-Molgora	3	29.12	7.19
			Terraced MOLGORA-Adda	7	27.60	6.81
				14	*86.49*	*21.35*
			Morenic hills	1	2.89	*0.71*
		Suburban	West edge Groane	3	22.47	5.55
		Technolog.	Terraced plain of meda	1	8.32	2.05
		Open-urban	Terraced plain of lambro	5	23.60	5.83
			Plain of lambro-molgora	3	24.09	5.95
				13	*81.37*	*20.09*
	URBAN	*Closed-urban*	Plain of Seveso-Lambro	10	73.78	*18.22*
			West edge Groane	1	11.44	*2.82*
			Plain of Lambro	1	33.08	*8.17*
			Plain of Lambro-Molgora	2	15.11	*3.73*
				14	*133.42*	*32.94*

In italic the totals per L. Type

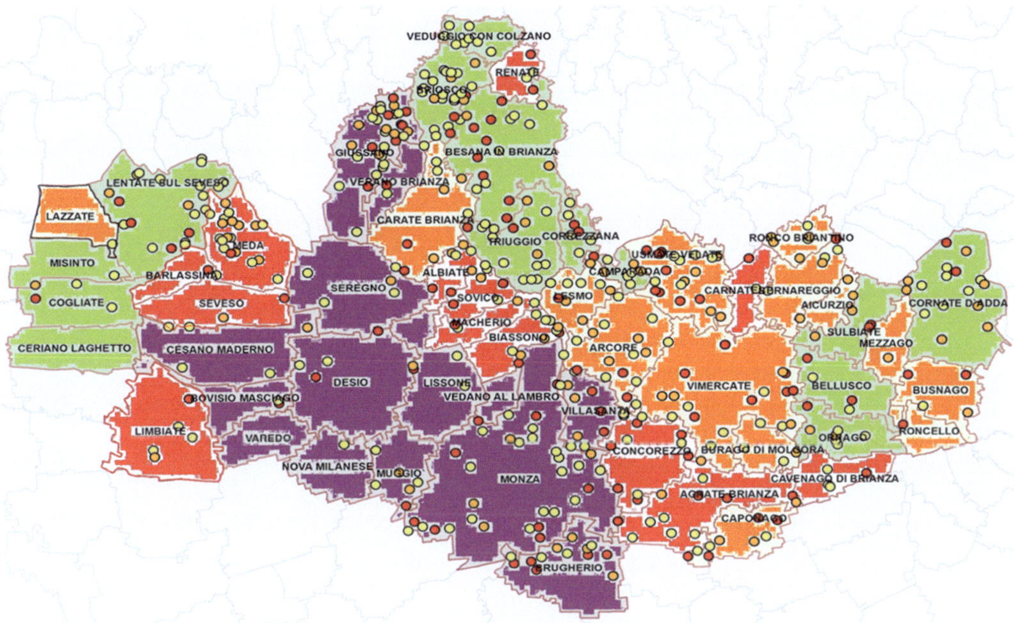

Fig. 2.12 A map of the province of Monza, near Milan. The four types of landscapes are distributed mainly depending on the attraction due to the city of Milan: dense urban landscape (*violet*), open urban landscape (*red*), suburban-rural landscape (*orange*), agricultural landscape (*yellow-green*). Monza is strictly part of the Metropolitan area of Milan. Coloured dots indicate the presence of the main old farms

suburban-rural type (14 municipalities), 5 for the open-urban type (13 municipalities) and 4 for closed-urban type (again 14 municipalities).

This classification allowed the drawing up of a map, indicating the distribution of the landscape types in the examined territory of Monza (Fig. 2.12). Limits of the Metropolitan area of Milan (sensu R.T.T. Forman [39]) are evident, composed by the 27 municipalities of closed and open urban landscapes, which measures 53 % of the entire territory, while the agricultural area is limited to 25 %.

2.5.2 Ecoregions and Landscape Systems

The hierarchy theory and the emergent property principle enhance the importance of the context in ecological analysis: that is why we need to mention regions. In fact, a region is defined as an ecological system composed of connected landscapes. Bailey [40] prefers to name those systems ecoregions. Regions occur in a wide range of scales, like landscapes, and they stand frequently in contrast with one another, while long-distance linkages connect them. The eco-climatic zones of the Earth are the basis of the criteria used in delineating ecoregion levels. Koppen used the composition and distribution of vegetation in his search for significant regional climatic boundaries, as did many other authors (Tricart and Cailleux [41]; Bailey and Cushwa [42]).

Where disturbances and not clear successional stages make regional boundary placement difficult, Bailey considered the patterns displayed on soil maps of broad regions, such as the FAO-UNESCO World Soil Map, because soils tend to be more stable than vegetation.

The ecoregions delineated by Bailey [39] are grouped in four domains, as shown in Table 2.6: (a) polar, where the frost action primarily

Table 2.6 Ecoregions of the continents (from R.G. Bailey 1996)

Rank n°	Ecoregions	Km^2	%
100	Polar domain	38,038,000	26.00
110	Ice cap division	12,823,000	8.77
110M	Ice cap regime mts	1,346,000	0.92
120	Tundra division	4,123,000	2.82
120M	Tundra regime mts	1,675,000	1.14
130	Subartic division	12,259,000	8.38
130M	Subartic regime mts	5,812,000	3.97
200	Humid temperate domain	22,455,000	15.35
210	Warm continental division	2,187,000	1.49
210M	Warm continental regime mts	1,135,000	0.78
220	Hot continental division	1,670,000	1.14
220M	Hot continental regime mts	485,000	0.33
230	Subtropical division	3,568,000	2.44
230M	Subtropical regime mts	1,543,000	1.05
240	Marine division	1,347,000	0.92
240M	Marine regime mts	2,194,000	1.50
250	Prairie division	4,419,000	3.02
250M	Prairie regime mts	1,256,000	0.88
260	Mediterranean division	1,090,000	0.75
260M	Mediterranean regime mts	1,561,000	1.07
300	Dry domain	46,806,000	32.00
310	Tropical/subtropical steppe division	9,838,000	6.73
310M	Tropical/subtropical steppe regime mts	4,555,000	3.11
320	Tropical/subtropical desert division	17,267,000	11.80
320M	Tropical/subtropical desert regime mts	3,199,000	2.19
330	Temperate steppe division	1,790,000	1.22
330M	Temperate steppe regime mts	1,066,000	0.73
340	Temperate desert division	5,488,000	3.75
340M	Temperate desert regime mts	613,000	0.42
400	Humid tropical domain	38,973,000	26.64
410	Savanna division	20,641,000	14.11
410M	Savanna regime mts	4,488,000	3.07
420	Rainforest division	10,413,000	7.11
420M	Rainforest regime mts	3,440,000	2.35

determines plant development and soil formation, (b) humid temperate, with a variable importance of winter frost, (c) dry, which comprises arid and semi-arid areas of middle and adjacent latitudes, (d) humid tropical, with persistent high moisture and high temperature.

The 30 macroscale regions listed in Table 2.6 are well defined due to climatic zones and macroscale vegetation. At mesoscale, regions are more difficult to define, because they are mostly dependent on landform differentiation and mesoscale vegetation. Anyway, at this mesoscale we cannot speak properly of landscape: rather, we can define landscape systems. To have an idea of landscape systems, it is necessary to refer to a subcontinental example, like the Italian peninsula or the British isles in Europe. Figure 2.13 is added to show the ecoregions of Europe.

A landscape system is therefore defined as an arrangement of landscapes which have significant geographical and vegetational features in a

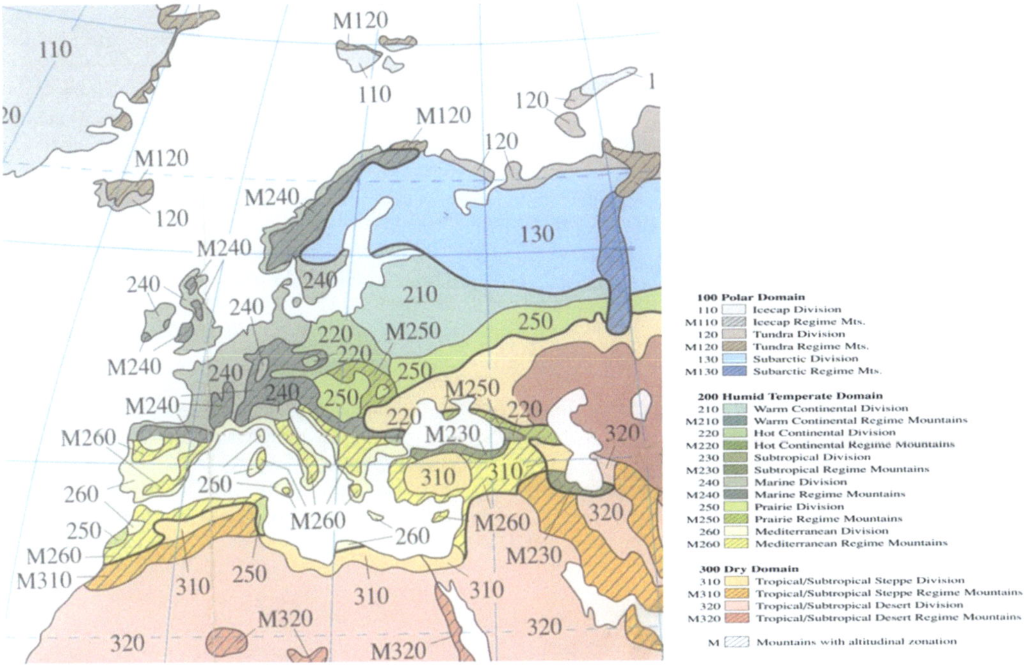

Fig. 2.13 The map of the ecoregions of Europe, from Bailey, 1996. Note the large variety of ecoregions, which help the historical development of the old continent

certain type of ecoregion (sensu Bailey). The Italian peninsula has been known for centuries as one of the most rich in landscape types in the entire world. It can be divided into 4 ecoregions:

I- Marine regime mountains, II Marine division, III- Mediterranean regime mountains, IV- Mediterranean division.

References

1. Steinitz C (2005) Padova and the landscape: alternative futures for the roncajette park and the industrial zone. Harvard graduate school of design. Massachusetts, Cambridge
2. EUR Commission DG Environment (2007) Interpretation manual of European Union Habitats. EUR27, Bruxelles
3. Oldeman RAA (1990) Forests: elements of sylvology. Springer, New York
4. Forman RTT, GODRON M (1981) Patches and structural components for a landscape ecology. Bioscience 31:733–740
5. Forman RTT, GODRON M (1986) Landscape ecology. Wiley, New York
6. Zonneveld IS (1989) Scope and concepts of landscape ecology as an emerging science. In: Zonneveld IS, Formann RTT (eds) Changing landscapes: an ecological perspective. Springer, New York, pp 3–20
7. Zonneveld IS (1995) Land ecology. SPB Academic Publishing, Amsterdam
8. Burel F, Baudry J (1999) Écologie du paysage: concepts, méthodes et applications. Technique & Documentation Ed, Paris
9. Ingegnoli V (2002) Landscape ecology: a widening foundation. Springer, Berlin
10. Leser H (1978) Landschaftsökologie. Uni-Taschenbucher 521, Stuttgart
11. Pignatti S (1994) Ecologia del Paesaggio. UTET, Torino
12. Ingegnoli V, Giglio E (2008) Landscape biodiversity changes in forest vegetation and the case study of the Lavazé Pass (Trentino, Italy). Annali di Botanica NS 8(2008):21–29
13. Vos W, Stortelder A (1992) Vanishing Tuscan Landscapes: landscape ecology of a Submediterranean-Montane area (Solano Basin, Tuscany, Italy). Pudoc Scientific, Wageningen
14. Ingegnoli V (1993) Fondamenti di Ecologia del paesaggio. CittàStudi, Milano
15. Troll C (1963) Ueber Landschaft-Sukzession, Vorwort des Herausgebers. In: Bauer HJ,

Landschaftökologische Untersuchungen im ausgekohten Rheinischen Braunkohlenrevier auf der Ville. Arbeiten zur Rheinischen landeskunde 19
16. Odum EP (1971) Fundamentals of ecology, 3rd edn. WB Saunders, Philadelphia, PA
17. Mac Harg I (1969) Design with nature. Natural History, New York
18. Haber W (1989) Using landscape ecology in planning and management. In: Zonneveld IS, Forman RTT (eds) Changing landscapes: an ecological perspective. Springer, New York, pp 217–232
19. Ingham DS, Samways MJ (1996) Application of fragmentation and variegation models to epigaeic invertebrates in South Africa. Conserv Biol 10:13–53
20. Wood PA, Samways MJ (1991) Landscape element pattern and continuity of butterfly flight paths in an ecologically landscaped botanic garden, Natal South Africa. Biol Conserv 58:149–166
21. Farina A (1993) L'ecologia dei sistemi ambientali. Cluep, Padova
22. Allen TFH, Hoekstra TW (1992) Toward a unified ecology. Columbia University Press, New York
23. Popper KR (1990) A world of propensities. Thoemmes, Bristol
24. Prigogine I (1996) La fin dès certitudes: temps, chaos et les lois de la nature. Editions Odile Jacob, Paris
25. Naveh Z, Lieberman A (1984) Landscape ecology: theory and application. Springer, New York
26. Ingegnoli V (1980) Ecologia e progettazione. Cusl, Milano
27. Forman RTT, Moore PN (1991) Theoretical foundations for understanding boundaries in landscape mosaics. In: Hansen AJ, Di Castri F (eds) Landscape boundaries, consequence for biotic diversity and ecological flows. Springer, Berlin, pp 236–258
28. Ingegnoli V, Pignatti S (2007) The impact of the widened landscape ecology on vegetation Science: towards the new paradigm. Rendiconti Lincei 18 (2):89–122
29. Brandt J, Tress B, Tress G (2000) Multifunctional landscapes: interdisciplinary approaches to landscape research and management. Centre for Landscape Research, Roskilde
30. Naveh Z (2000) Introduction to the theoretical foundations of multifunctional landscapes and their application in transdisciplinary landscape ecology. In: Brandt J, Tress B, Tress G (eds) Multifunctional landscapes: interdisciplinary approaches to landscape research and management. Centre for Landscape Research, Roskilde
31. Ingegnoli V (2011) Bionomia del paesaggio. L'ecologia del paesaggio biologico-integrata per la formazione di un medico dei sistemi ecologici. Springer, Milano
32. Forman RTT (1995) Land Mosaics: the ecology of landscapes and regions. Cambridge University Press, Cambridge
33. Haber W (1990) Basic concept of landscape ecology and their application in land management. In: Ecology for tomorrow. Physiology and Ecology Japan:131–146
34. Tüxen R (ed) (1978) Assoziationkomplexe. Ber. Intern. Symp. Veg., Rinteln. Cramer Verlag, Vaduz
35. Gehu JM (1988) L'analyse symphytosociologique et geosymphytosociologique de l'éspace. Tthéorie et métodologie. Coll Phytosoc 17:11–46
36. Rivas-Martinez S (1987) Nociones sobre Fitosociología, biogeografía y bioclimatología. In: La vegetacion de España. Universidad de Alcalá de Henares, Madrid, pp 19–45
37. Braun Blanquet J (1928) Pflanzensoziologie: Grundzüge der Vegetationskunde. Berlin
38. Pignatti S, Box EO, Fujiwara K, (2002) A new paradigm for the XXIth Century. Annali di Botanica vol II:31–58
39. Forman RTT (2008) Urban regions, ecology and planning beyond the city. Cambridge University Press, Cambridge
40. Bailey RG (1996) Ecosystem geography. Springer, New York
41. Tricart J, Cailleux A (1972) Introduction to climatic geomorphology. St. Martin's Press, New York
42. Bailey RG, Cushwa CT (1981) Ecoregions of North America. FWS/OBS-81/29. Washington, DC

3 Landscape Functions (Physiology)

3.1 The Study of Landscape Processes

We know that a landscape shows all the principal characters of any level of life organisation: structure, movement, reproduction, metastability, etc. So it is necessary to study the processes related to each character. In doing that, a complex articulation of landscape dynamics emerges, never completely examined up to now. So we prefer to suggest a well-articulated sequence, as exposed in Table 3.1.

Rather than recall what is described within the table, it could be interesting to divide the study of landscape dynamic into two chapters, because the questions concerning landscape control, evolution, transformation, alteration and pathology need to be separate as in medicine (see Chap. 4). It is extremely important to underline that, generally, the landscape components have a complex functional behaviour; therefore the processes ranked in the table have only a didactic and disciplinary significance. For instance, a corridor (e.g. a hedgerow) may have a structural function, but also an ecotonal one, or a movement one, etc. As pointed out by Brandt and Naveh [1, 2], we can properly speak of multifunctional landscapes. Note that the arguments ranked in Table 3.1 are congruent with the structural ones in Table 2.2, related to the main anatomical sets in a landscape; the first six arguments will be treated in this chapter while the other six in the next Chap. 4.

3.2 Geomorphic (Skeletal) System

3.2.1 General Processes

The most important interactions among the geophysical, climatic and biologic components of landscapes can be visualised in Fig. 3.1 in a very synthetic scheme. The main sequence of general processes may be described in six phases:

1. The *climate* influences directly the morphology of a landscape (erosion, transport, sedimentation, etc.), the soil formation and the vegetation.
2. The *vegetation* is influenced by climate, but plays a basic role in soil development and protection, as in the formation of ecotope mosaics and in the maintenance of local microclimates.
3. The *soil* interacts in return on the vegetation.
4. Climate, vegetation and soil constitute the *environmental conditions* for the establishment of animals.
5. The *animals*, in return, through their behaviour modify both soil and vegetation (pollination, grazing, seed dispersion, etc.) and complete the landscape formation.
6. With the evolution of *man*, inventive interactions integrate the precedents creating new landscapes. Feedback may influence almost all processes, but the possibility of a managerial control derives from these interactions.

V. Ingegnoli, *Landscape Bionomics Biological-Integrated Landscape Ecology*,
DOI 10.1007/978-88-470-5226-0_3, © Springer-Verlag Italia 2015

Table 3.1 Main articulation of Landscape physiology processes

Principal sections	Following subdivisions
1. Geomorphic (skeletal) system	1.1 General processes
	1.2 Geomorphic processes and soil formation
	1.3 Signs of hydrological processes
2. Autotrophic system	2.1 Bio-energetic of vegetation: the BTC function
	2.2 Efficiency of vegetation: the CBSt function
	2.3 New theoretical principle for vegetation dynamics
	2.4 Some vegetation processes in the landscape
3. Heterotrophic system	3.1 Populations and landscape: the vital space per capita
	3.2 The bionomic study of an urban region
	3.3 Signs of animal processes
4. Connection and movement system	4.1 Geo-climatic movements
	4.2 Biological movements
	4.3 Human movements
5. Reproductive and colonising system	5.1 Ecotope reproduction
	5.2 Colonisation processes
6. Main configuration processes	6.1 Ecotone network and delimitations
	6.2 Fragmentation and permeability
	6.3 Other landscape processes
	6.4 Landscape biodiversity
	6.5 General landscape metastability
7. Protective and recovery system	7.1 The protective system
	7.2 Interface elements
8. Cybernetic and control system	8.1 The ecological control of human culture
	8.2 The buffer effect of a cultural change on the bionomic state of a LU
9. Evolution	9.1 Limits of Neo-Darwinism
	9.2 Landscape and evolution
10. Transformation	10.1 Transformation modalities
	10.2 Control of the process
11. Landscape syndromes	11.1 Landscape alteration
	11.2 Landscape pathology
12. Landscape pathology and human health	12.1 New concepts of health damages
	12.2 Environmental stress
	12.3 Ecological damages and human health
	12.4 Health and territorial planning

These phases enhance the related processes belonging to three groups: geophysics-climatic (1); adaptive-evolutionary (2, 3, 4, 5); inventive-cultural (6). Major interactions among these three groups of processes take place not only on a macro-scale, but they are transmitted to various levels on lower scales too, even if the influences of each sub-process do not reach every level of the organisation, as stated by the hierarchic theory.

An in-depth treatment of general processes can be found in every text on general ecology. Here it is sufficient to remember the importance of the open thermodynamic dissipative process, starting from solar irradiation, which is absorbed in partial and differential ways into the atmosphere, generating wind circulation and all the climatic phenomena. Again the irradiation permits plant photosynthesis, heating and transpiration, and water extraction from the soil. Vegetal and animal communities form the soil, stabilise it and participate actively in its deposition. Contributions to soil formation derive even from alluvial sedimentation and from endogenous phenomena.

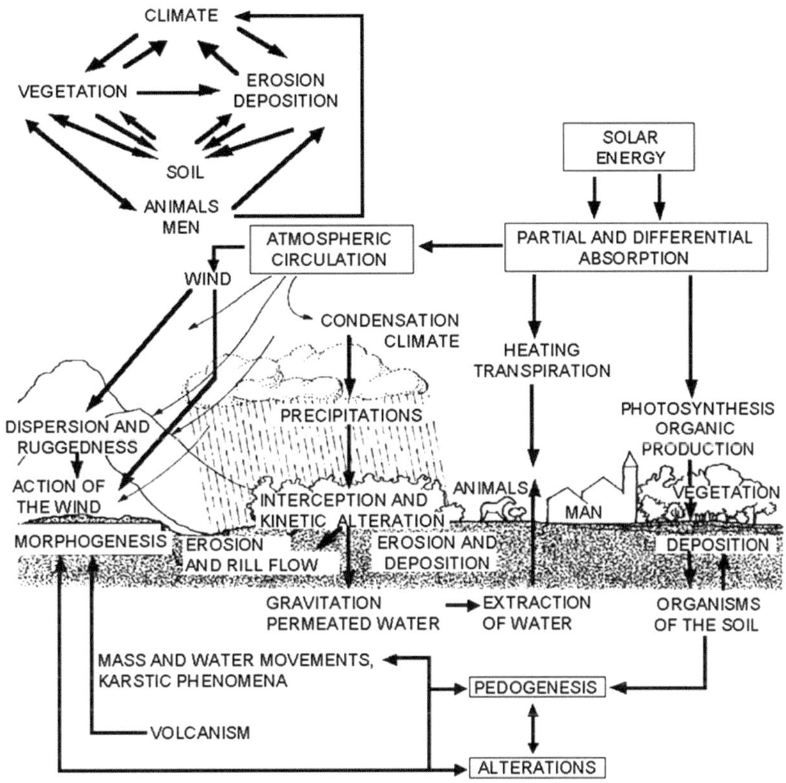

Fig. 3.1 Synthetic scheme of main general processes describing the interactions among the geophysical, climatic and biologic components of landscapes

3.2.2 Geomorphic Processes and Soil Formation

Geomorphic phenomena have two important functions in a landscape: (1) the generation of fluxes of matter and the consequent modification of the regional surface, which signifies energetic ties for a landscape; (2) the derivation of characters from the transport of eroded matter. Geomorphology is seen as a factor regulating many other landscape functions, so assuming a strong ecological significance. On the other hand, soil formation acts through four groups of processes:

1. *Humification processes*: formation of organic-mineral complexes, generally in well-drained soils with short cycles which have been altered bio-chemically in temperate and cold regions (e.g. brown soils).
2. *Seasonal conditioning*: formation of vertisoils by strong seasonal contrasts (wetting-drying) which alters the humification and the formation of distinct horizons in cold and continental regions.
3. *Geochemical alteration*: total hydrolysis in soils without organic soluble elements, which eliminates silicon and bases and accumulates aluminium and iron compounds through long cycles in hot climates, warm regions.
4. *Station conditioning*: production of physical and chemical environments mainly by water excess, which form oxidation–reduction processes and salt, e.g. pseudogley soils.

Tricart and Kilian [3] noted that *pedogenesis* (*soil formation*) and *morphogenesis* are not cyclic, as proposed by the rhexistatic theory of Erhart [4], but they are contemporaneous and interfere with each other. Therefore, it is necessary to distinguish between localised and diffuse processes, both on a spatial and a temporal scale, and especially to locate one of the three types of geo-dynamic environment: (1) stable, (2) intermediate and (3) unstable.

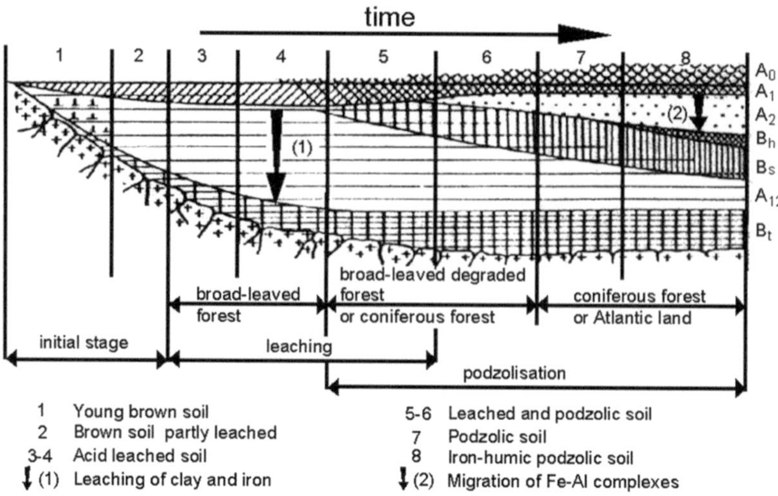

Fig. 3.2 Evolution and degradation of a typical brown soil in Europe. Soil evolution phases (1–8) show two main chemical processes (leaching and podzolisation) and three main types of vegetation which are adapted to this soil chain. When cultivated in the last phases, this acid soil is re-saturated by manure, eliminating its absorbent complex (degradation) (From Duchaufour [5])

(1) *Stable geodynamic environment*: slow evolution, tendency to a steady state, possible in regions with a weak internal geodynamic activity and a weak intensity of external processes. The balance is favourable to soil formation. Two forms exist:
   (a) Long-term stability, with surface evolution quite absent.
   (b) Almost recent stability, remnant forms due to Quaternary oscillations, where sometimes windy actions may be present.

   Here it is important to maintain a vegetational cover near to maturity, because these formations may be the basis of new soil transformation.

(2) *Intermediate environment*: contemporary interference between pedogenesis and morphogenesis, with a weak balance in favour of one of them. Morphogenesis assumes a great importance and permits two cases to be distinguished:
   (a) Diffuse pluvial erosion, surficial creep, with a possible development of soil base.
   (b) Diffuse superficial mass movements, soil-flux type, able to notch all the soil layers.

   Sometimes a partial deforestation may arrest the glides, because it diminishes the load and the water absorption, especially on clay soils.

(3) *Unstable environments*: predominance of morphogenesis. Two main causes:
   (a) Aggressive bioclimatic conditions, with irregular intense variations, transmitting a great amount of energy.
   (b) Impervious reliefs, high inclinations, intense internal geodynamic, no remnants possible.

Soil formation cannot develop; thus heterodynamic mosaics appear, with litho- and rhegosoils. Sometimes reforestation is not able to stop mass movements: lands are not cultivable and become marginal. A typical chain of soil evolution is shown in Fig. 3.2, regarding brown soils on a slime-sandy substrate (following Duchaufour [5]).

The soil is a natural environment, in continuous formation and evolution, which is extended to the superficial layer of the lithosphere. Soil evolution, in its history, follows three processes [5]: (1) rock decomposition and alteration, leading to the formation of an altered complex; (2) increase of the amount of organic matter due to colonising vegetation, with an equilibrium phase between fresh inputs and mineralised outputs; (3) transport of soluble or colloid elements through the profile, due to water flux, forming impoverished horizons A and enriched ones B.

Some bases of soil dynamics, i.e. the zonality law, have been enunciated along general lines beginning with Dokuchaev [6]. In every climatic zone, soils develop by different rocky substrates and are colonised by different ecological communities. When the ecosystems develop toward a steady state, soil and vegetation originating from different substrates converge toward a uniform ecological system. Even if the zonality law states that the climate is the first of all ecological variables, the local characters of the geosphere maintain an important role, particularly in human landscapes.

Therefore, in a region or in a landscape, some units dependent on geomorphology appear, called ecological sectors [7]. These units locate diverse soil mosaics within a uniform environment, and this may be very important in landscape analysis (e.g. for landscape unit delimitation). Concerning this, we have to underline that the integration of pedology (soil science) in the field of landscape ecology is often more difficult than that of geomorphology. In fact, a typical pedogenetic environment does not always produce the same soils everywhere, and this is in apparent contrast with the zonality law; in any case it remains much more interesting than the analytical knowledge of any type of soil located in a given landscape.

3.2.3 Signs of Hydrological Processes

A specific field of study includes the geomorphologic processes pertaining to water, which are too complex and detailed to be reported here: but we cannot avoid referring at least to fluvial characters, because rivers are generally the most important geobiologic elements in a region. Water courses due to water runoff are organised in corridors with a diverse drainage density and spatial distribution. These corridors result from the basic process of erosion and are regulated by diverse hydrologic regimes: glacial, snow-pluvial, pluvial.

Many scientists [8] have proposed a river zoning based on ecological and hydrological characters. The better known are: Hypocrenon (spring brooks), Epirhytron (mountains streams), Hyporhytron (valley torrents), Epipotamon (hilly rivers), Metapotamon (plan rivers), Hypopotamon (final rivers).

Anyway, we have to observe that in landscape ecology rivers assume diverse, more complex functions. They are:

1. Ecological systems with an accelerated landscape dynamic and a structure in continuous transformation, which many adapted ecosystems are connected to.
2. Excretory apparatuses of the catabolism of surrounding ecotopes, with a prevalent function of buffer and filtration and consequent cleaning.
3. Complex corridors, dependent on the scale both in function and structure, available even for a protective role and for landscape movements.
4. Polarised ecological systems regarding the ecological mosaics, with a function of biological conservation, thus particularly important in human landscapes.

In karst zones and some deserts with very high percolation rates, river corridors may be not present or result only temporarily. Nonetheless, water circulation remains an important landscape factor, as we can see in talwegs or huadys.

Furthermore, human colonisation has a strong influence on water courses and its consequences on landscapes. The example of the river Piave (near Venice) is well known. The “Serenissima Repubblica” of Venice modified the course of this river on sixteenth century, giving it two mouths, after the excavation of a new estuary to protect the Lagoon. The new mouth was diverted by the action of sea currents on the coast, then—after a strong flood—it was recently transformed into a small lagoon (Fig. 3.3), today a SCI area (Sites of Community Importance), see also 13.3.

Fig. 3.3 The recent lagoon formed by a diverted than abandoned new estuary of the river Piave, North to Venice (Italy)

3.3 Autotrophic System

Vegetation is the key-system also in the landscape, playing at least the triple role of "managing" energy, organising the landscape and structuring the chronotopes (three spatial dimensions + the temporal one) of the landscape. Thus, it should need a proper book to be exhaustively presented! Here, only some fundamental key concepts are proposed.

The energy managing role has to be related, first, to the metastability and the efficiency concepts.

3.3.1 Bio-Energetic of Vegetation: The BTC Function

The use of metastability concept enables

- To study vegetation through new perspectives.
- To evaluate landscape transformation in a proper way.

The physiology of vegetation leads to the concept of latent capacity of homeostasis of an ecocoenotope (i.e. the vegetation occurring on a tessera). It can be studied on the basis of:

(1) the concept of resistance stability; (2) its type of vegetation community; (3) its metabolic data (biomass, gross primary production, respiration, B, R/GP, R/B). Two coefficients can be elaborated:

$$a_i = (R/GP)_i/(R/GP)_{\max}$$
$$b_i = (\mathrm{d}S/S)_{\min}/(\mathrm{d}S/S)_i,$$

where R is the respiration, GP is the gross production, B is the biomass, ds/S is equal to R/B and is the maintenance/structure ratio or a thermodynamic order function [15, 16] and i are the principal ecosystems of the ecosphere.

The factor a_i measures the degree of the relative metabolic capacity of principal vegetation communities; b_i measures the degree of the relative antithermic (i.e. order) maintenance of the same main vegetation communities. The degree of homeostatic capacity of an ecocoenotope is proportional to its respiration [15, 16] and can be expressed as the flux of energy that the ecocoenotope must dissipate to maintain its condition of order and metastability. So the a_i and b_i coefficients, even related in the simplest way, give a measure which is a function of this capacity:

Table 3.2 Plant biomass (at the ecosphere scale: PB_b, at the ecocoenotope scale: PB_e), net primary production (NP), respiration vs. gross primary production rate (R/GP), respiration vs. biomass rate (R/B, where B = PB_e), and calculated BTC values (Mcal/m²/year), from Ingegnoli [13]

Main types of vegetation coenosis	PB_b	PB_e	NP	R/GP	R/B	BTC
Tropical rain forest	6–80	105	3.5	0.75	0.078	20.5
Tropical seasonal forest	6–60	80	2.5	0.70	0.072	12.8
Dry forest savannah	5–30	38	1.7	0.42	0.032	3.3
Shrub savannah	2–15	18	2.8	0.42	0.111	3.0
Graminoids savannah	1–4	4.5	2.6	0.40	0.377	1.9
Mangrove forest	5–45	55	1.2	0.65	0.040	6.2
Tropical marsh and swamp	2–30	36	4.5	0.39	0.083	4.7
Shrub semi-desert	0.1–4	4.2	0.25	0.45	0.059	0.5
Sand warm desert	0.01–0.2	0.25	0.02	0.45	0.140	0.05
Tropical yearly culture	0.5–2	2.8	1.8	0.41	0.464	1.4
Tropical perennial culture	2–12	15	1.9	0.55	0.190	4.4
Chaparral bush-land	3–15	20	1.4	0.65	0.130	5.0
Mediterranean forest	6–60	85	2.2	0.71	0.063	12.6
Dry temperate grass	0.5–2.2	2.8	0.8	0.40	0.179	0.6
Moist temperate grass	1–3	3.5	1.4	0.40	0.266	1.0
Temperate deciduous forest	6–60	80	2.2	0.69	0.061	11.7
Temperate woodland	6–40	50	1.9	0.69	0.054	8.7
Temperate marsh and swamp	2–15	18	3.3	0.39	0.117	3.0
Temperate yearly culture	0.2–2	3	1.4	0.41	0.333	1.1
Temperate perennial culture	2–10	15	2.3	0.60	0.227	5.5
Suburban green	0.5–5	6	0.8	0.55	0.167	1.6
Urban green	0.1–2	2.5	0.4	0.50	0.160	0.6
Boreal and alpine forest	6–55	73	1.8	0.71	0.060	10.7
Boreal open forest	5–35	40	1.0	0.71	0.062	6.2
Tundra and alpine vegetation	0.5–2.8	3	1.0	0.35	0.167	0.6
Cold desert	0.01–1	1.3	0.1	0.38	0.061	0.15
Peat bog	1–10	12	1.2	0.40	0.067	0.9
Lake and stream	0.01–0.4	0.5	0.1	0.40	0.160	0.1
Estuaries	1–15	18	2.0	0.39	0.072	2.2
Continental shelf	0.1–1.5	2	0.4	0.42	0.140	0.4

$$BTC_i = (a_i + b_i)R_i\, w,$$

where w is a variable necessary to consider the emergent property principle and to compensate the environmental constraints; putting $\Omega = (a_i + b_i)\, R_i$, the value of w results: $w = 0.89–0.0054\, \Omega$. Consequently:

$$\mathrm{BTC_i} = 0.89\,\Omega - 0.0054\,\Omega^2\,(\mathrm{Mcal/m^2/year}).$$

BTC or, better, the *biological territorial capacity* of vegetation [9–14] is one of the fundamental synthetic function of Bionomics, useful in evaluating the metastability of a vegetated tessera, of a LU and of an entire landscape. Reference values of BTC have been calculated on the 30 main types of zonal vegetation of the ecosphere, as shown in Table 3.2, where both natural and anthropogenic vegetation have been considered. The BTC evaluation allows the diagnosis of vegetated elements, like forest patches, estimating normal ecological conditions. Moreover, the BTC function becomes an index allowing the recognition of regional thresholds of landscape replacement (i.e. metastability thresholds) during time and, especially, the transformation modalities controlling landscape changes, even under human influence. It is useful

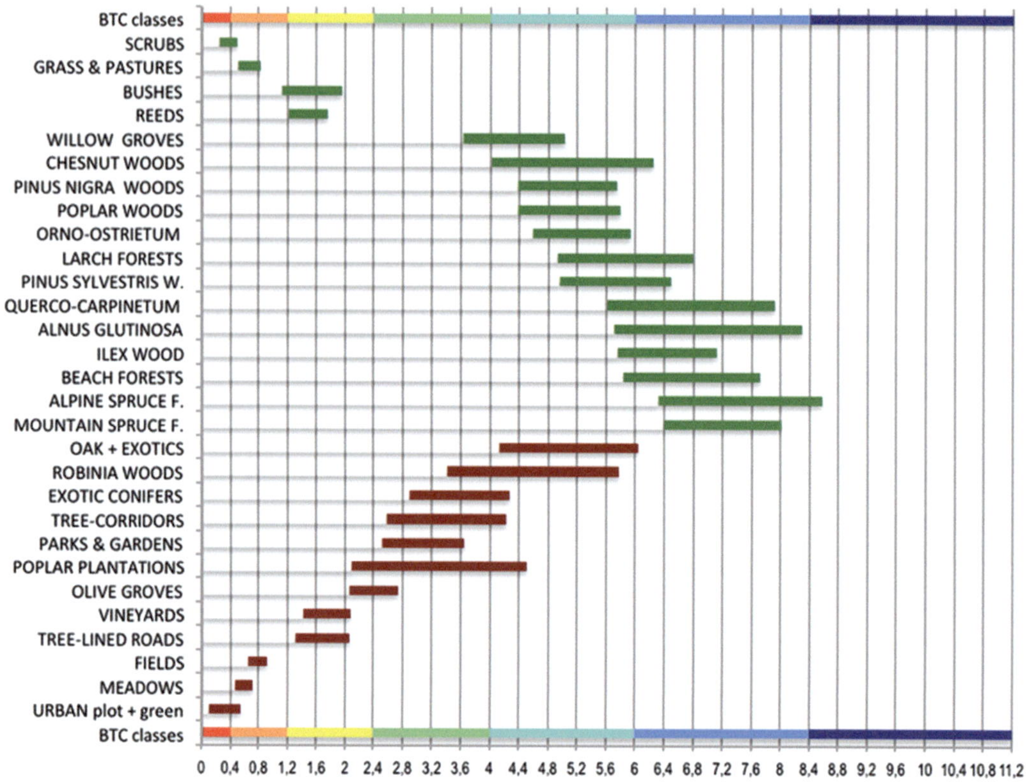

Fig. 3.4 Range of BTC values of 17 Natural (*green*) and 12 Synanthropic (*brown*) formations from a survey of 430 vegetation tesserae in the Regions of Lombardy and Trentino-South Tirol (North Italy). BTC measures (Mcal/m^2/year) are plotted in seven different classes (from *red* to *blue*)

to list the BTC values in nine classes, the first 7 of which are referred to boreal and temperate vegetation, while 8 and 9 are needed for tropical one. The landscape unit analysis through BTC classes lead to a better knowledge of their bionomic state.

In Fig. 3.4 we show the significant ecological situation of the main vegetation components of the Central European Regions of Lombardy and Trentino-South Tirol (North Italy). A research was made on 170 tesserae of human and 260 of natural vegetation, then grouped in 29 formations. Their statistical average present the Alpine Spruce formations (*Picea excelsa*) as the best (BTC = 7.44 Mcal/m^2/year ± 1.13), while oak forests of *Querco-Carpinetum* result similar to beach forests (BTC = 6.77 Mcal/m^2/year ± 1.15). To have an idea of the significance of the values of these forest formations, we may compare [13] the BTC values of the oak forest of Lombardy with the analogous formations of the Bialowieza Park (Poland), which reach a BTC = 8.65 Mcal/m^2/year.

Another interesting functional observation on vegetation is shown in Fig. 3.5. It is possible to assert that plant species biodiversity is not synonymous of order and efficiency of a forest tessera. Fifteen surveys (orange) present a BTC value >8.5 Mcal/m^2/year, but only three of them (orange dots) have a high biodiversity of species (>45). In fact, we can see the small value of R^2 (about 0.11) and the wide dispersion of the set of data. It signifies that the results of structural and functional organisation are more important that the species biodiversity.

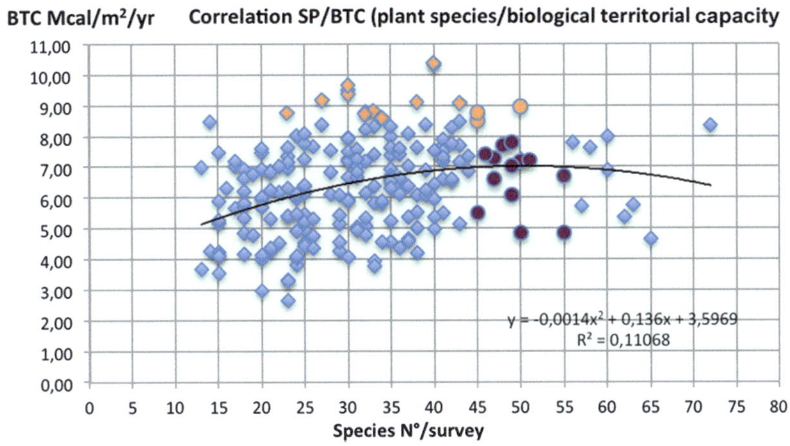

Fig. 3.5 Correlation SP/BTC from the mentioned research on the central European regions of North Italy in the decade 2000–2010. We observe the small value of R^2 and the wide dispersion of the set of data: therefore, structural and functional organisation results are more important than the species biodiversity. *Violet dots* correspond to the maximum of the curve

3.3.2 Efficiency of Vegetation: the CBSt Function and Its Correlation with BTC

The need to understand the main processes between vegetation and landscape imposes to consider both the metastability and the efficiency of vegetation. The evaluation of the biological territorial capacity of vegetation (BTC) represents a first good step, because of the correspondence with the concept of metastability. As already expressed, the BTC measures the degrees of the relative metabolic capacity and of the relative antithermic maintenance (i.e. order) of vegetation communities related to their respiration, being the degree of homeostatic capacity of an ecocoenotope proportional to its respiration.

Efficiency is another bionomic function, more precisely linked with the state of a vegetation patch, relating the maturity level of a vegetation coenosis (MtL) and its bionomic quality (bQ). As we will see, this function, named "concise bionomic state of vegetation" (CBSt) [17] is available to be applied in many theoretical and practical studies. Therefore, this function has to be designed as

$$\mathrm{CBSt} = (\mathrm{MtL} \times \mathrm{bQ})/100.$$

These data will be detailed in next Sect. 9.1 and 5.3.1 (bionomic evaluation of vegetation). Here we need to understand this function and its correlation with BTC, in order to put in evidence some processes of vegetation in relation with the landscape.

In Fig. 3.6, based on the same mentioned research on Lombardy and Trentino, the CBSt values estimated for natural formations are plotted in green, while the ones for human formations in red. A wide variability of CBSt appears at low BTC values (<3 Mcal/m^2/year), less in medium and high values. We can observe a clear complementarity between natural and human vegetation (see Table 9.1 for detailed data). The gap high-low BTC of natural vegetation types is filled by human formations, demonstrating the increase of landscape biodiversity and heterogeneity in a landscape unit, when in presence of some human vegetation. This fact has been previewed from theory, e.g. the curve of landscape general metastability [13], but still not fully proved.

Another observation is the proof that the vegetation communities with lower resistance strategy (hence lower BTC) reach a lower bionomic state (see 9.4.1). Only forest coenosis can arise the normal state, arriving to measure a CBSt from

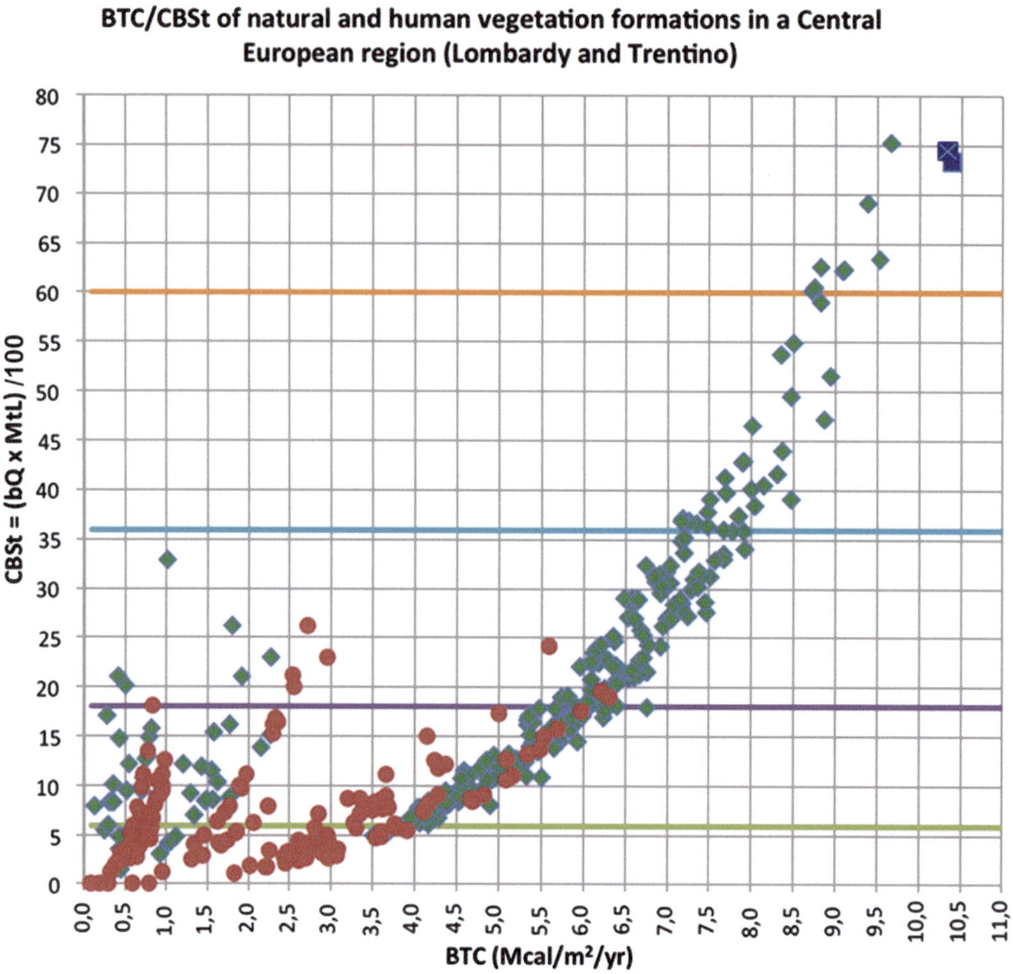

Fig. 3.6 Correlation between the biological territorial capacity of vegetation (BTC) and the evaluation of its concise bionomic state (CBSt), from 430 samples of natural (*green squares*) and human (*red dots*) vegetation, surveyed in the Region of Lombardy and Trentino (North Italy). The two *blue squares* represent a comparison with a case study in the forest of Bialowieza (Poland), one of the oldest natural forests in Europe. The four coloured horizontal lines define five range of CBSt, from the worst (bottom) to the better one (top)

40 to 65 % (good state). This seems to be impossible under BTC <7.25–7.50 Mcal/m^2/year.

3.3.3 New Theoretical Principles for Vegetation Dynamics

In landscape bionomics the concept of vegetation has to be extended to the *whole of the plants* of a landscape element, considered in their aggregation capacities and in their relations with environmental factors. Therefore, a cultivated tessera is to be considered as vegetation not only for its weeds (e.g. *Secalinetea, Chenopodietea*), but even for the cultivation itself (e.g. *Triticum aestivum, Hordeum vulgare*), without which the weeds does not succeed and the tessera does not become the habitat for many natural species (e.g.

Coturnix coturnix, Alauda arvensis), besides to be a crucial ecological component for human population.

The use of the concept of "potential natural vegetation" is not yet satisfactory for landscape bionomics, because the word "potential" is intended to represent undisturbed conditions in a not defined time. The proposal of Ellenberg [18], to distinguish among *zonal* vegetation, which expresses the responses of potential vegetation to climatic conditions, *extrazonal* vegetation, responding to local topoclimatic conditions and *azonal* vegetation, responding to soil moisture conditions, was a good step, but it is again not sufficient for landscape ecological theory, therefore even for vegetation science.

Ellenberg [19] already perceived the ecosystem and man's dual part in the structure of a landscape, and Naveh—citing Walter—proposed [20] to determine plant formations and types not only in their floristic aspect but also in stability, structure, human influence, diversity, productivity, etc. The reasons for this criticism derive from the self-organisation processes especially when the role of disturbances is seen as structuring and when transgressions in a linear succession are based on the interaction among landscape elements even in the same *zonal* area.

Deepening, the idea to measure a 'distance' between 'present' and 'potential' vegetation is not correct, because it implies the possibility that potential landscapes really exist, reduced to very few types of vegetation, sometimes only one or two. This is a projection of our mental scheme on the natural reality, in contrast with all the main processes and dynamics of the landscape and it is a sort of "virtual ecology"! For instance, as pointed out by Pignatti (personal communication) even the large zonal ecosystems (e.g. tropical forests, taiga, savannah, Australian deserts, etc.), or the best virgin forest in Europe (i.e. the Perucica), nearly undisturbed, *never* consist in a single association.

In fact, the concept of "potential natural vegetation" clashes with the non-equilibrium thermodynamics and the relative bifurcations of the state functions of a system in an instability field (see Fig. 1.15). Therefore, the concept of potential vegetation has to be strongly revised. It has to be defined not only for natural conditions, but in relation to the main range of landscape disturbances (including man) too, and with defined temporal limits. It must never be considered as the optimum for a certain landscape (or part of it), but only as a general indication (never to be widely reached) in relation to climate, soil and anthropization during a certain limited period of time. It could be better named *the fittest vegetation for...* [13, 21].

This new concept refutes the general notion of "potentiality" as the possibility of the coming into existence, in the absence of man and for large territories, of a deterministic, a priori fixed vegetation type, interpreted as the best condition for a place, independent of all other environmental and human factors in space and time. Moreover, no potential homogeneity can be a model for the development of a landscape. On the contrary, the concept of *the fittest vegetation for...* indicates the most suitable or suited vegetation for: the specific climate and geomorphic conditions, in a limited period of time and in a certain defined place; i.e. the main range of incorporable disturbances (including man's) under natural or not natural conditions. This could be a great change of perspective.

In addition, it signifies also to limit the validity of the concept of primary succession and a general revision of vegetation dynamics.

3.3.4 Some Vegetation Processes in the Landscape

Studying spruce forests in the Alps, we can find diverse typologies (see Sect. 12.3). According with Pignatti [22], these formations may be reduced to 11 groups (Table 12.13) the main of which are: (a) *Veronico urticifoliae-Piceetum* (930–1,700 m) or "mountain spruce forest" (*Picea abies* 96 %, *Larix europaea* 42 %), (b) *Homogino-Piceetum* (1,200–1,900 m) or "subalpine spruce forest" (*Picea abies* 100 %). After a

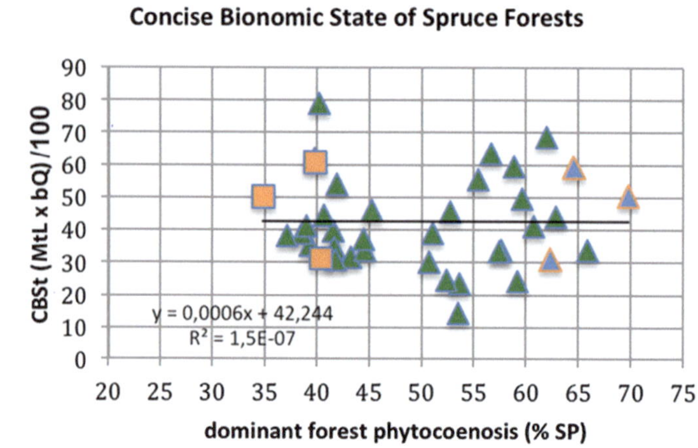

Fig. 3.7 The CBSt index, applied to the types of investigated forests, shows no correlation in relationship with the "purity" of the type of species formation. For instance, an evident dominance of Subalpine Spruce Forest association (*blue triangles*) does not present a better bionomic state in comparison with a mixed Spruce formation (*orange squares*)

preliminary vegetation survey (38 tesserae), the structure and composition of these typologies never result completely coinciding with the above syntaxonomies, but matching from 30–50 to 50–65 %. In fact, we find many other intermediate formations, due to environmental, biological and/or human disturbances (see Fig. 12.18). Thus, it is possible to list these different spruce forests in proportion of the dominance of the formation types.

These observations, resulting in a distance from "potential vegetation", should bring to underline a diffuse ecological alteration of alpine forests. Anyway, remembering the previous paragraph on the "fittest vegetation for", landscape bionomic principles do not accept this assertion. The same research on spruce forests in Central-Eastern Alps is confirming the fragile basis of "potential vegetation". Measuring the CBSt levels per dominance of the formation types (Fig. 3.7), we can see that to a low "purity" of spruce formations does not correspond a low CBSt.

The study of rural and agricultural landscapes near large towns presents known alterations in their vegetation, in function of the high human pressure. The degree of urbanisation (URB) is proportional to exotic species and to the rate of warm/cold species. A research on the vegetation of 10 rural-agricultural landscapes near Milan confirmed this observation.

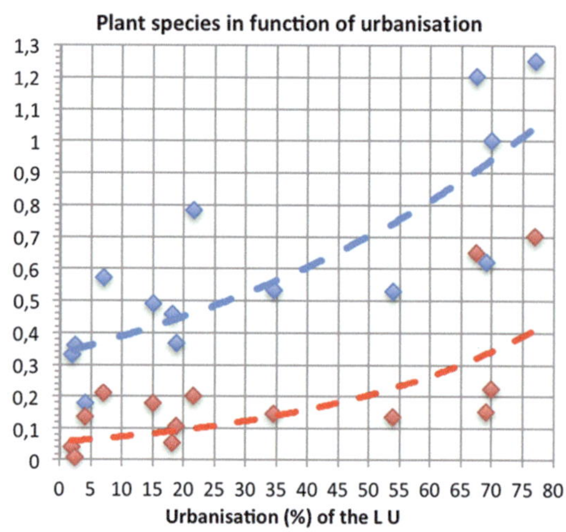

Fig. 3.8 The degree of urbanisation (built areas) is crossed towards (*blue*) the rate of warm/cold species of plant (using biotic indexes) and (*red*) the presence of exotic species. Note that generally when urbanisation grows over 50 % of a Landscape Unit, compare the UHI raise of temperature

In Fig. 3.8 we see two exponential curves, related (a) on the rate warm/cold plant species (steno and euri-Mediterranean vs. circum-Boreal

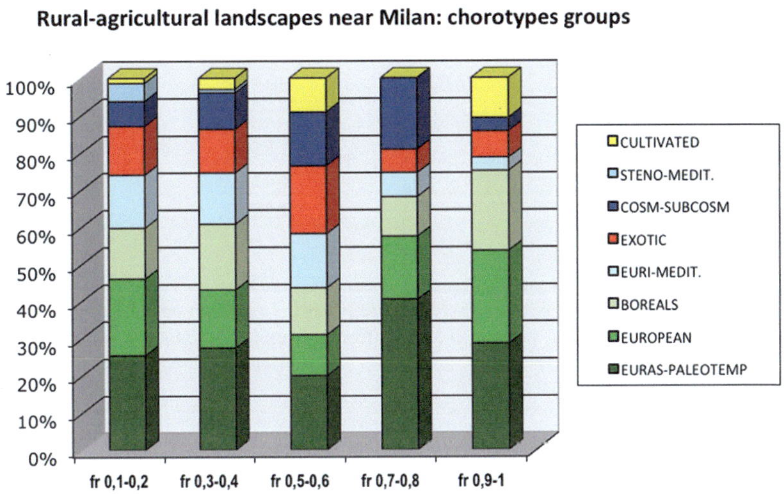

Fig. 3.9 A research on 10 rural-agricultural landscapes near Milan and their vegetation. The most common species (frequency of 0.9–1.0) did not present many exotic and cosmopolite chorotypes, while among less common species (frequency 0.7–0.8) cosmopolite become important; with frequency 0.5–0.6 exotic chorotypes are more numerous

and EU-Sibirian) and (b) on the exotic (and adventitious) plant species. We observe that the warm species, normally 1/3 of the cold one near Milan, can increase three to four times going towards a heavy urbanisation (URB >60 %) confirming the UHI effect (urban heat island), clearly enhanced after URB >50 %. We find also the presence of a constant background noise in the case of exotic species: even in absence of urbanisation we register about 4 % of exotic/adventitious species. In a metropolitan area it is possible to reach 60–70 % of non-native species. Falinski [23] reported plant 50–70 % of alien species in Poland cities vs. 20–30 % in forest settlements, and Kowarik [24] in Berlin indicated 41.5 % of alien plant species with an URB of about 55–60 %. In the urban parks of Wien, Berlin and Milan, the non-native species have been estimated by Ingegnoli [25] respectively: 20.9, 24.4 and 30.3 %.

As plotted in Fig. 3.9, comparing chorotypes vs. species frequency, among species with the highest frequency (=0.9–1.0), we don't find a sensible presence of exotic and cosmopolite chorotypes, while cosmopolite quantity becomes important for frequency of 0.7–0.8 and exotic one among that of lower frequency (0.5–0.6). Steno-Mediterranean species, normally lacking in the Po valley, compare only in the lowest class of frequency (0.1–0.2).

3.4 Heterotrophic System

3.4.1 Populations and Landscape: The Vital Space per Capita

The main linkage between heterotrophic components (animal and human populations) and their landscapes is in function of their vital space per capita. To express the processes behind this function we need to define the concept of "standard habitat per capita" (SH). SH is the inverse of the ecological (i.e. non-geographic) density of population, measurable in m^2/organism and intended as the set of portions of the landscape apparatuses (see Sect. 2.4.2) within the examined landscape unit (LU) indispensable for an organism to survive. Note that, even for the same species, SH may change in function of the bioclimatic belt and the landscape type.

For instance, in the case of animal populations, we will have a SH_N, that is a SH referred to the natural habitat (NH):

$$SH_N = (GEO + EXR + RSN + RSL + CON + PRT + PRD)/(n^\circ \text{ of animals} \times \text{species}) \left[\text{m}^2/\text{organism}\right].$$

Table 3.3 Theoretical minimum standard habitat per capita (SH*= vital space) related to the main climatic belts of the biosphere for human population

Climatic belts	Kcal/inhab[a]	SH* (m^2)	Min. (t°C[b])	PRD (m^2)
Arctic	3,500	2,500	−45.6 to −34.4	1,650
Boreal	3,100	1,850	−34.4 to −23.3	1,230
Cold-temperate	2,850	1,475	−23.3 to −12.2	1,050
Warm-temperate	2,750	1,360	−12.2 to −1.1	1,000
Sub-tropical	2,550	1,250	−1.1 to +10	900
Tropical	2,350	1,020	+10 to +21.2	730

PRD minimum field available to satisfy the edible energy per capita needed in 1 year
[a]Minimum edible Kcal/day per capita (very different from the 3,500 Kcal/day of USA population)
[b]Following USDA Plant Hardiness Zone Map for Cultivations (Average Annual Extreme Minimum Temperature 1976–2005)

In the case of human populations, we will have a SH_H, that is a SH referred to the human habitat (HH):

$$SH_H = (GEO + EXR + PRT + PRD + RES + SBS)/(n^\circ \text{ of peoples}) \ [\text{m}^2/\text{inhabitant}].$$

The ungulates usually have different home ranges in different landscapes, but applying the SH function such measures are reduced to a unique thresholds value in relation with the SH quantity needed per landscape type [26]. Note that this SH is referred to the landscape apparatuses which give a good bionomic state of the LU, going beyond the same trophic or climatic resources [27]. Elaborating these studies, we can expose some values of SH referred to European ungulates, as in Table 6.1. Note that ungulate SH results about 20–100 times wider than human SH. All this will be deepened in 6.1.1.

Studying landscape functions, to recognise the processes related with the vital space per capita, it is indispensable to calculate a "minimum theoretical standard habitat per capita" (SH*), both for human population and animal ones.

In Table 3.3 the SH* in relationship to the main climatic belts of the biosphere are exposed. Such magnitudes have been estimated in function of: (a) the minimum edible Kcal/day per capita [1/2 (male + female diet)], (b) the productive capacity (PRD) of the minimum field available to satisfy this energy for 1 year, taking into account the production of major agricultural crops, (c) an appropriate safety factor for current disturbances, (d) the need for semi-natural protective vegetation for the cultivated patches. The linkages with an average annual extreme minimum temperature (Min. t°C) are deduced from USDA "hardiness zone map" for cultivations [28].

The main reason for the concentration of human population in warm temperate, subtropical and tropical regions of the biosphere emerges immediately from these evaluations. In fact, applying the SH* of Table 3.3, a field of 1 ha in Russia, may support about 4.0–5.0 inhabitants, while in India 8.5–9.5.

The estimation of SH* in wide mountain systems needs a peculiar methodology. For instance, on the Alps (Fig. 3.10), to the variables in function of which SH* was to be calculated, we have to add the time of the snow cover and the decrease of the field productivity (q/ha) with the altitude. At an altitude of only 800 m a.s.l., SH = 1,700 m^2 and it reaches 2,600 m^2 at 1,400 m a.s.l., to arrive –theoretically- to 4,000 m^2 at 1,800–1,900 (the maximum high for an historical settlement).[1]

The ratio σ = SH/SH* (real vital space/theoretical vital space) allows the evaluation of the "*carrying capacity*" of a landscape unit, therefore the self-sufficiency of the human habitat, a basilar question for sustainability and ecological

[1] In the Alps, especially in Aosta Valley, Valais and Engadine some villages may reach 2,100 m a.s.l. (e.g. Cuneaz, Ayas Valley). But after the Small Ice Age (beginning on 1,600 to half 1,800) they were transformed in pastoral hamlets and today they are mainly touristic.

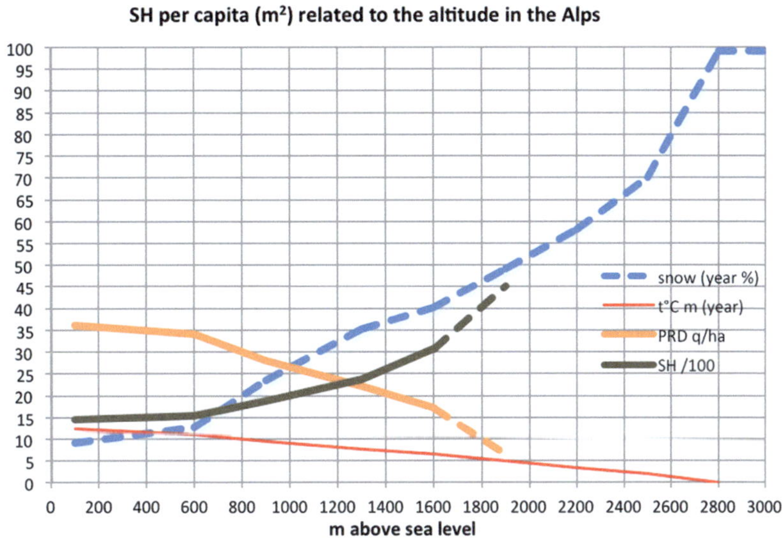

Fig. 3.10 The estimation of SH* in the Alpine valleys. Here are plotted the main variables in function of which resulted SH = 2,000 m^2 at 1,000 m a.s.l., to SH = 4,000 m^2 at 1,900 m a.s.l., the maximum high of an historical settlement

territorial planning. If such a rate is $\sigma < 1.05$ (i.e. 1 + 0.05 as safety factor) the LU results heterotrophic. Below to $\sigma < 0.20$, SH indicates a dense urbanised LU; on the contrary, if $\sigma > 1.60$, SH indicates an agricultural LU.[2]

The function relating $\sigma = SH/SH^*$ to HH follows an exponential curve, as plotted in Fig. 3.11. This correlation was based on experimental data, from about 40 LU of North Italy. A model may be built translating a bit of this curve towards an optimal condition, once known the disturbances of the case studies. The curve of the model is the dotted blue. Moreover, Fig. 3.11 shows the peculiar condition of agricultural (and rural) landscapes, which are thickened near the threshold of HH = 85 %. This fact brings to built a new branch to the model, because it is evident that the process of transformation of a LU dominated by agriculture is different from an analogous process dominated by technology and industry, at least after HH = 50 %.

[2] The threshold for an agricultural landscape in temperate regions depends from the local consumption of resources ($\sigma = 1$), the safety factor on agricultural productivity ($\sigma = 0.1$) and the exportable resources ($\sigma \geq 0.5$), therefore no less than $\sigma > 1.60$.

3.4.2 The Physiology of Urban Region

The heterotrophic systems of the landscape are today mainly characterised by "Urban Regions", sensu Forman [29]. Taking a cue setting of Forman, but considering the principles of landscape bionomics of Ingegnoli, we think it is necessary to express, albeit in summary, the following definitions:

1. An "*Urban Region*" is defined as the area capable of sustaining the ecological development of some conurbations in it through its own resources. Obviously the support must be understood not in a total way, but at least as a minimal level and, moreover, the metastability of the system must be guaranteed. Its perimeter must follow the historical reasons of transformation of urban areas and extend up to values of habitat carrying capacity and standard production (agriculture) $SH_{PRD}/SH_{PRD}^* \geq 0.55$ (0.75). In an Urban Region we may distinguish: (a) a hegemonic City, (b) a metropolitan area, (c) one or more satellites City (internal and external), (d) an extensive network of mobility, (e) any area of Nature Reserves or biotopes, (f) a matrix of agricultural support.
2. An *Hegemon City/Conurbation*: a single important city which is the motor development of a vast urban landscape, even if sometimes it can depend on multiple cities, as for

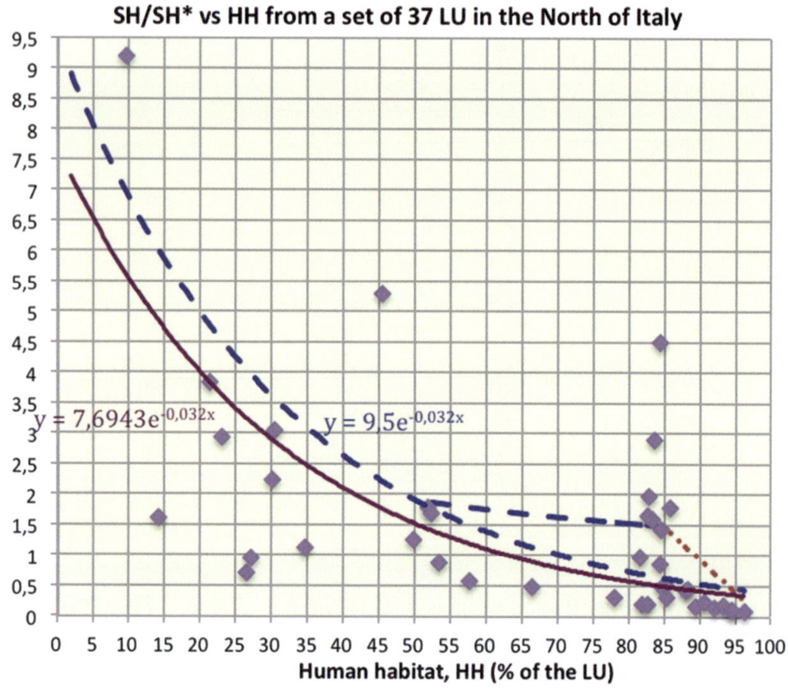

Fig. 3.11 The correlation between σ = SH/SH* and the human habitat in a LU is shown in this plot. The *dotted line* may be used as a model to compare a surveyed territory. Note that near to the threshold of HH = 85 % many case study are thickened, all regarding agricultural landscapes, which can no more exist if HH>85%. After this value they abruptly become suburban, then urban ones (brown dotted line)

the Rhur. The "old town" can be dominant over vast expanses of suburbia (e.g. Roma, Paris, London) or can be constituted by close welded suburban expansion of multiple nuclei, but the concept remains equally valid and necessary.

3. A *Metropolitan Area*: a zone of continuous urbanisation, though it may contain different types of tesserae (or to stains tags), such as: water bodies, gardens and fields residual urban green spaces (parks, gardens and public or private) and even some small biotopes. This area must also have infrastructure (e.g. highways, airports, power plants, green areas, etc.) at common. Furthermore, the human habitat (HH) should not be markedly lower than that of the City hegemon, but only slightly.
4. One or more *Satellites City* (internal and external): these cities, significantly smaller than the hegemonic, usually have complementary roles, at least in some respects. They can be internal or external to the Urban Region, and, in the second case, be themselves hegemonic districts of suburbs, often still smaller than the main one. If seriously growing in importance, we may have the nucleus of a megalopolis.
5. A vast *Network of Mobility*: the network of streets, squares, interchange centres, stations, airports, shipping ports if any, must have a configuration consistent with what has been said for an urban region and what you should add, both networks of energy transport and the hydrographical network.
6. Any area of *Nature Reserves* or biotopes: what matters, anyway, are not the areas of protection as such, because, unfortunately, they are often constraints "on paper" or linked to recreational and sporting purposes. It must, however, highlight the key natural areas, which are often part of park areas, but cannot be under guardianship. It must be protected above all the tesserae with high biological territorial capacity of vegetation (BTC).
7. A *Matrix for Agricultural Support*: being the urban development complementary to each landscape unit (LU) production (i.e. agriculture) adapted to it, it is essential to evaluate this presence in the complex system that forms an urban region. For these LU, the limit carrying capacity cannot be less than

Fig. 3.12 The "urban region" of Milan (*yellow boundary*), bounded on the basis 1:500,000 (Atlas Treccani, 2002) is formed by a terraced "upper part" to the North and a "low part" south of the line of springs (*risorgive o fontanili*)(*blue line*) and measures 425,000 ha. It contains the city of Milan (*orange*), the metropolitan area of Milan (*light pink boundary*) of 95,750 ha and has 4,800,000 inhabitants and four small satellites towns. Other eight small external Satellites City can be seen. The green surface indicates the "low wetted Po River Plain"

$\sigma = SH/SH^* \geq 1.55 \div 2.0$ (reduced to 1.45 in the case of neighbouring urban areas with high agricultural export capacity) for reasons of exportable agricultural productivity and thus for balancing, at least in part, the heterotrophic urbanised LU.

Urban regions often suffer from severe syndromes of altered environmental health, both concerning bionomic systems themselves and the human population. They are usually characterised by an increase in fragmentation, due to the intensification of the great territorial infrastructure (power lines, highways, railways, landfills, construction sites, etc.) (see 13.2). Cancellation or significant loss of ecological and geomorphological features of the area, due to abuse of technology and urban development, add problems to the previous ones. The climate is often distorted, due to the "heat island" (UHI, urban heat islands) with average annual temperatures of 2–3 °C higher than that in the surrounding area. The residual forest patches fail to achieve an adequate level of BTC, due to continuous disturbance and the invasion of alien species. Pollution of air and water are often chronic and the noise level is unbearable in many areas. Also, there are recurring problems of congested traffic.

Many naturalists (Lorenz [30]; Forman and Godron [31]; Ingegnoli [9, 11]) have compared these dysfunction ecological effects of the territory to the spread of cancerous tissue in histology. The adjustment of the territory to the phenomena to be managed is also necessary to monitor and seek to improve such multiple dysfunctions.

As an example, we can show the urban region of Milan, Italy. In this territory, which is the heart of the ancient Duchy of Milan and measures about 4,250 km^2, accounting for 17.8 % of the Lombardy Region (15.9 % of the Lombardy geographic region), live today approximately 4,800,000 people, ¾ of which in the metropolitan area of Milan, slightly higher than in Berlin, then the third of Western European Cities after Paris and London.

From Fig. 3.12 we see that the urban region of Milan comprises an Hegemon City, which

Table 3.4 The urban region and the metropolitan area of Milan and a comparison with the metropolitan area of Berlin (years 2007–2009)

Main ecological parameters	Milan, urban region	Milan, city	Metropolitan area, Milan	Metropolitan area, Berlin
Surface (ha)	425,000	18,200	96,000	88,900
Population	4,800,000	1,300,000	3,600,000	3,500,000
Urbanised areas (%)	28.5	68.3	54.5	54.5
Urban green (%)	5.0	5.5	4.5	12.0
Agriculture (%)	21.0	23.0	35.5	7.5
Woods (%)	5.6	1.5	4.2	17.5
Human habitat	85.5	96.3	93.4	82.5
BTC (Mcal/m^2/year)	0.94	0.43	0.70	1.24
SH (m^2/inhabitant)	757	132	249	210
Carrying capacity (SH/SH*)	0.54	0.094	0.17	0.14

prompted the formation of a metropolis of about 96,000 ha and with 3.6 million inhabitants - reaching out to the terraced plains north, to safeguard the area much more fertile of the "low Po River Plain" - in dialogue with: (a) the three (external) satellite towns of Como, Varese and Lecco, today hegemonic of the ancient urban region of Como, which measures about 330,000 ha with 1,700,000 inhabitants and (b) the Canton of Ticino, Switzerland, but traditionally and geographically Lombard and in fact linked to Milan. Other satellite cities are internal: Pavia and Vigevano, about 70,000 inhabitants, Lodi and Abbiategrasso of about 40,000 inhabitants, all in rural areas.

Considering the data presented in Table 3.4, we can verify the condition for the urban region through the ratio $SH_{PRD}/SH_{PRD}^* = 0.55$. It is a rather low value, but still acceptable given the ability to export foodstuffs from most of the land surrounding urban regions.[3] In the urban region of Milan the agricultural matrix of support, which has an area of 329,000 ha, was inhabited by 1.2 million people and reach a SH/SH* = 1.55.

At this point a comparison with the metropolis of Berlin (Berlin Raum) could be useful, having a similar size of Milan territory (89,000 ha, about 7.3 % less than Milan), a population of 3.5 million (−2.8 %) and the same urbanisation 54.50 %. The differences, however, are the parts in green: 7.5 % agriculture (−78.8 %), insufficient, and 17.5 % of forests (+426 %). It follows that the urban region of Berlin must necessarily be greater than that of Milan to have a matrix of agricultural support with SH/SH* = 1.5. In compensation, however, the BTC of the metropolitan area is better, being able to exceed 1.2 Mcal/m^2/year.

3.4.3 Signs of Animal Processes on the Landscape

We underlined that landscapes have their own characters: for instance they can support species that require forage in a grassland ecocoenotope and recover in a forest ecocoenotope, or can direct ecocoenotope transformation to maintain a steady state of the landscape itself. Thus, one of the most important criteria to evaluate faunal aspects in a landscape is to study the so-called *permeant species* [15] or *engineer species* sensu Sanderson and Harris [32]. These species use different ecocoenotope and change the landscape pattern: they cannot live in a single ecocoenotope; therefore they are typical *species of landscapes*. Note that man is one of these species, even if his ability and power is indeed incomparably much greater.

At the landscape scale these engineer species modify and create mosaics of habitat and enhance species richness and biodiversity. For instance, the transformation created by elephants in

[3] Consider that the only urban region of Piacenza, almost corresponding with its province, with only 300,000 inhabitants, has an order of magnitude $SH_{PRD}/SH_{PRD}^* = 4.50$.

African landscapes is sometimes impressive (see Fig. 6.1). Large organisms such as elephants, corals, beavers are known to be engineer, but there are many others. For example, wide-ranging predators, such as wolves, and migratory herbivores, such as wildebeest, might better be referred to as landscape engineer, because their activities strongly impact the environment.

Evaluating the presence of engineer species is certainly necessary in each study of landscapes, even in Europe. Some of these species are quite common (e.g. wild boars, red ants: *Formica rufa*); others are quite rare (e.g. ibexes). Many forested and/or wild landscapes present a serious lack of engineer species, sometimes even an absence, made worse by removing domestic herds (to preserve nature!).

An analysis of the relationships between animal populations and the landscape structure and functions is exposed in Sect. 6.1, to which we refer.

3.5 Connection and Movement Systems

3.5.1 Geo-Climatic Movements

Geo-climatic and biological movements pertaining to a landscape present many processes: it is not possible to discuss all of them. These movements depend on a small number of vectors: wind, water, flying animals, large animals, men and their vehicles. As underlined by Forman et al. [31, 33], all the ecosystems in a landscape are interrelated, with movement or flow rate of objects dropping sharply with distance, but more gradually for species interactions between ecosystems of the same type.

Vertical energy flow, interconnecting mosaic elements, is given by the different *reflections* of solar energy (i.e. albedo) by different surfaces, ecosystems and landscapes. Albedo is higher for smooth, light-coloured surfaces, but lower for rough dark surfaces (Table 3.5). This process results in a diverse energy budget among local ecocoenotopes even if the incoming energy in that landscape is constant.

The energy absorbed by the tesserae of an ecological mosaic heats them up and provides heat for metabolism, growth, decomposition and especially evapotranspiration too. The vaporisation of water mainly from soil and plant surfaces is called evapotranspiration (ET). Potential monthly ET (=EP) may be calculated by the well-known Thornthwaite expression:

$$EP = 0.1645\,(P/T + 12.2)^{10/9} \quad \text{(cm)},$$

where P is precipitation, T is temperature. EP is proportional to the medium temperature of a landscape and to its cloudiness: for instance we may find an annual EP of 60 cm in the Alpine border and an EP of 95 cm in Sicily and North Tunisia. ET is highest in the windbreaks and lowest on bare soil (a twofold difference). Among the fields, meadow has the highest ET and wheat the lowest.

The horizontal movement of energy or material carried by air is called *advection*. The contrasts in albedo and ET create energy gradients in the landscape. A 3–4 °C night temperature difference between forest and adjacent open patches means that, on a windless night, heat and air are moving horizontally to the cool clearing, while the cool patches are radiating heat, moving air masses vertically. These processes carry aerosols, gases, seeds, spores, and other particles (on a calm night).

It could be interesting [31] to compare wind patterns near dense vs. porous windbreaks (Fig. 3.13). Wind reduction is typically effective only in the 2 m to the ground or so next, and at a distance of about six times the height of the windbreak: in the case of a porous vegetated

Table 3.5 Albedo (reflectivity) values

Landscape elements	Albedo values (%)	Landscape elements	Albedo values (%)
Fresh snow	80–100	Deciduous forests	15–20
Sandy soils	20–40	Coniferous forests	10–15
Peat soils	0–10	Cities	10–25
Field crops	15–25	Lakes	0–10
Agricultural landscapes	20–27		

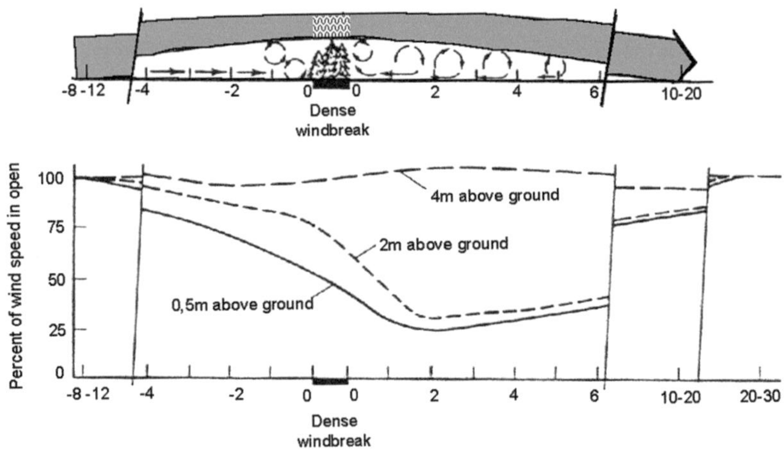

Fig. 3.13 The movement of the wind through a dense hedgerow. Distance from windbreak (*h*) is measured proportionally to the mean height of trees (the number on the horizontal axis times the height of the windbreak) (From Forman and Godron [31])

corridor we have less reduction (40 % of wind speed vs. 25 %) but less turbulence.

Water is transported by the planetary boundary layer (i.e. airflow affected below 1–1.5 km by friction) in the form of vapour, raindrops and snow. Shortly downwind of a source, such as a lake or a forest, is a plume of moist air. That is why in irrigated land surrounding an oasis, *ET* in the upwind portion causes moister air in the downwind part. In northern Japan (Hokkaido) a 100 m wide "fog-removing forest" collects enough droplets to clear the bottom 300 m of the planetary boundary layer, permitting solar radiation to reach rice paddies downwind. Briefly, water in all its forms, and the hydrologic cycle as a whole, is quite dependent on horizontal and vertical flows in the giant conveyor. Water is collected over a broad area and deposited in a concentrated form in a small area. Wide floods often extend greatly over a landscape.

Water velocity is generally low, but because the land has rarely experienced these floods, the removal of biomass and damage to landscape and human structures may be great. In addition, nutrient-rich sediment is spread in patchy deposits, and if the process is not too strong, may be in some cases positive.

Another moving process in the landscape is due to glaciers, both in arctic and mountain regions. This movement seems to be very slow, but anyway it is a significant motion, especially during warm period, when glaciers can reduce even tens or hundreds of meter/year. Figure 3.14 shows the example of a very small glacier in the Dolomites ("Pale of S. Martin" group, 3,192 m), which regulates the water flux of the natural park of Paneveggio and down to Predazzo. The movement is in act even under screes and moraines, as in this picture, where the crevasse is in the grey tongues under the "U"-shaped door of rocks. These moraines in few years have been colonised by vegetation (*Salix* and *Larix* sp.) even if the glacier tongues remain hidden (and protect) by gravel.

Fig. 3.14 Crevasse, due to the movement of ice, in the small glacier of Travignolo (Dolomites), about 2,500 m a.s.l. To the *right*, a sketch (*watercolour*) of the glacier and the position of the crevasse (*red dot*)

3.5.2 Biological Movements

The movement of species in a landscape has many reasons. Plant movement is linked essentially to dispersal processes (seeds), to the colonisation capacity (pioneer species) and to local expansion due to changes of the surrounding patches. Animal movements are much more complex, being linked to dispersal, home range, foraging, migration, territorial defence, invasion, etc. In many cases movements depend principally on the behaviour of the animals and evolved species may move even for exploration of a territory.

In a system of ecocoenotopes, some tesserae rich in resources and some populations living in favourable ecotopes (i.e. sources) export toward other areas and populations, in some case negative ecotopes (i.e. sinks). This is a corollary of the definition of a biological system as an open and dissipative one. Consequently we may talk about *metaclimax* and *metapopulation*, that is, interconnected sets of sub-units in different phases.

Each subpopulation is connected by the dispersal of individuals in a metapopulation, which changes in size over time and may persist for long periods. Metapopulation dynamics do not concern only movement: here we underline the local extinction and recolonisation process that is facilitated if the landscape resistance is low. The landscape resistance is determined mainly by the arrangement of the spatial elements, especially barriers, conduits and highly heterogeneous areas. Output and input fluxes of energy and information follow in a landscape the main law of ecology [34]: not too much, not too little, just enough. The flux is regulated by the following situations:

- Too much input (e.g. overfeeding resulting in poisoning).
- Not enough input (e.g. underfeeding resulting in deficiency).
- Not enough output (e.g. clogging resulting in constipation).
- Too much output (e.g. leakage resulting in depletion).
- Normal flux (e.g. proportional with distance).
- Indifferent flux (e.g. diffuse transitory species passage).
- Oriented exalting flux (e.g. species over-attracted by particular tesserae).

The flux, going from a source toward a sink, may find in the landscape configuration remedies for regulations, such as: supply (underfeeding), resistance (overfeeding), drainage (constipation), retention (depletion), new sources (too great distance), new orientations (few attractors), etc.

It is particularly important to remember the double interaction between the input–output fluxes among the landscape components and their mosaic context characters and behaviour. All ecosystems in a landscape are interrelated, with movement or flow rate of objects dropping sharply with distance, but more gradually for

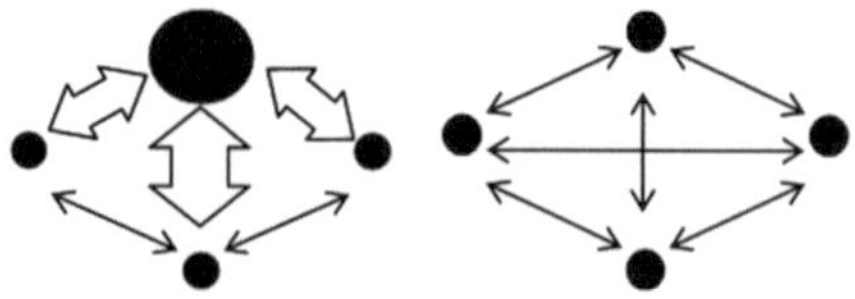

Subpopulation: Winking rate Local extinction rate Recolonisation rate	 Low Low High	 High High Low
Genetic: Inbreeding depression Strains disappear due to local extinction Strains disappear due to "swamping" (outbreeding)	 Low Low High	 High High Low
Differences among subpopulations Total variation of metapopulation	Low Low	High High
Adaptability to environmental changes Evolutionary rate	Low? Low?	High? High?

Fig. 3.15 Expected patterns from gene flow in two metapopulations composed of four subpopulations, living in different cluster of patches. *Thickness of arrow* represents relative amount of gene flow (From Forman [35])

species interactions among ecosystems of the same types. For subpopulations on separate patches, the local extinction rate decreases with greater habitat quality or patch size, and recolonisation increases with corridors, stepping stones, a suitable matrix habitat, or short inter-patch distance as written by Forman [35].

The transfer of new genes (alleles) and genetic combinations among subpopulations is called gene flow. Associated with the movements in a metapopulation, we can note and underline an important gene flow that is influenced by the spatial disposition of the patches and corridors of a landscape, and by the size of the sources. For instance, large populations on source patches provide individuals that recolonise empty patches: this is the process that maintains the metapopulation as a whole (Fig. 3.15).

Local extinction is frequent and normal for small subpopulations and a typical process in metapopulation dynamics. If the patches are of the same small size the inbreeding depression is high, the outbreeding low. Furthermore, with a high winking rate (i.e. the rate of local extinction followed by recolonisation), the subpopulations on different patches will be genetically quite different from each other [35]. Therefore, despite being composed of small subpopulations on separated patches, each threatened by inbreeding depression, the metapopulation as a whole maintains a high level of genetic variation.

Permeant species of animal (i.e. multi-habitat vertebrates) implies a daily use of different habitats, and alternatively different habitats may be used at different stages in the life cycle. Movement is the obliged consequence and in many cases, when these movements are repeated, the pattern of a landscape is structured with paths and places of standstill. Even if vegetation structure is a primary determinant of movement route, chemical signals applied by animals to plants, soil and air play also an important role for many species.

Human activities usually increase the rates of invasion by exotic and alien species, as well as population fluctuation and extinction. This process occurred several thousand years ago in the Mediterranean region, so the present biota in large areas represents those species that could survive there. Human societies need also to introduce their livestock into pastures. Livestock learn movement patterns and are creatures of habit. New herds introduced to a pasture first learn the boundaries, next locate water and then suitable forage. In many cases, livestock needs to be moved far away, e.g. in the mountains in summer and onto a plain in winter. Even if today the transhumance is less frequent, this process has had ecological consequences during the history of many countries and today may continue by road transportation.

The most spectacular movement of animals among different landscapes is probably the migratory flux. In its proper dynamic this process pertains to the ecosphere and regional ecology, but may influence also landscape ecology. One of the main problems is the presence of a series of landscape units (or at least ecotopes) favourable to the migratory routes. When human artefacts alter a landscape, the migratory movements of many bird species may change or disappear. Nonetheless, the adaptation of some animals is incredible in this sense and in Europe is not infrequent to see a stork nest on iron pylons of electrical plants (e.g. *Ciconia ciconia*, near Milan).

No doubt, the most impressive and important movement of living populations into landscapes is due to human transport systems: from airlines to navigation, and especially road webs and their infrastructures. The multitude of problems linking human transportation and landscape ecology cannot be discussed here, and they will be mentioned when opportune. It is meaningful to underline that the entire section of landscape ecology named road ecology is becoming a distinctive discipline [33].

3.5.3 Main Human Movements

In this recent period we may understand very well the meaning of population movements through the diverse territories, especially in Europe. In only 1 year (2011) there were an estimated 1.7 million immigrants to the EU-27. The United Kingdom reported the highest number of people from outside nations (566×10^3), followed by Germany (489×10^3), Spain (457×10^3) and Italy (386×10^3). These four countries accounted for 60.3 % of all Europe. Moreover, to the outside immigrants we have to add inner immigrants from low economy European regions and the distribution of this huge number of people is towards specific regions.

Immigrant movements bring many pressures on the landscape, for instance an increase in finding residences, in disturbing natural biotopes, but also in land abandonments. The Spanish regions show the highest net migration, while East Germany and some regions of Romania and Finland have a negative process.

The oldest process relating to human movement remains the building and managing of the road system, which has been interesting European landscapes since at least two millennia. About 100,000 km of paved roads allowed the movement of people and resources in the Roman world.

Today, especially after the formation of the rail-road network (since nineteenth century) and the raise of urbanism (twentieth century), this movement system increased the landscape fragmentation (see Sect. 3.7.4). The effective mesh size serves to measure landscape connectivity, i.e. the degree to which movement between different sites of the landscape is possible. It expresses the probability that any two points chosen (randomly) in a district are connected and barriers, such as transport routes or built-up areas, do not separate that.

The measure of mesh density is not conceptually difficult. If we divide a square in two parts ($A_1 + A_2$) and one of these again in two ($A_{2A} + A_{2B}$) we will have the probability of two points to be connected in A_1 in this way: $(A_1/A_{tot})^2 = (0.5)^2 = 0.25$ and in A_{2A}or A_{2B} the probability will be $(0.25)^2 = 0.0625$. Thus, the probability of these two points in the entire A_{tot} will be $0.25 + 0.0625 = 0.375$. If our square has a late of 2 km, its effective mesh density will be $m_{eff} = 0.375 \times 4\ km^2 = 1.5\ km^2$. The effective mesh density referred to 1,000 km^2 is $s_{eff} = 1/1.5 \times 1{,}000 = 666.7$ mesh per 1,000 km^2.

EEA/FOEN [36] elaborated a fragmentation map of Europe. The barriers are shown in black (built-up areas, roads, railways), and the colours indicate the sizes (km^2) of the remaining patches

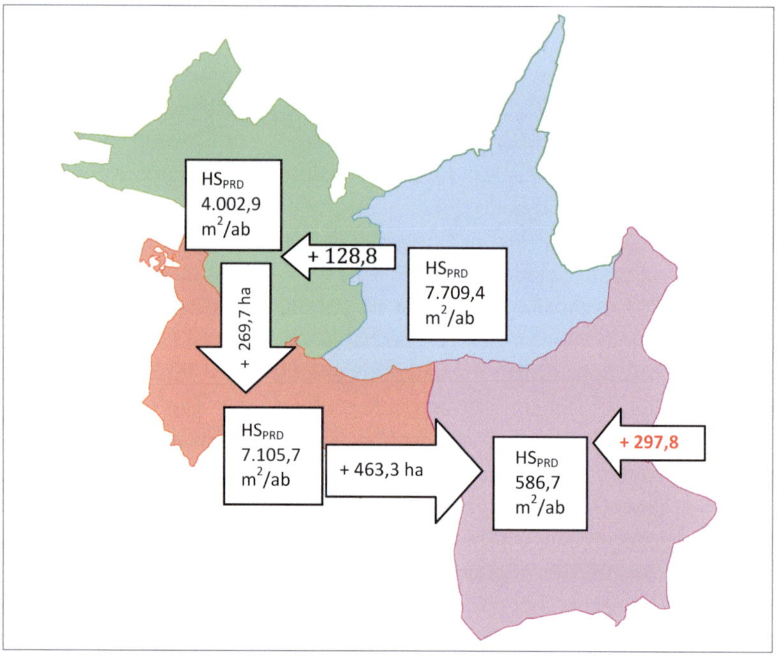

Fig. 3.16 Changes in the theoretical requirement of agricultural commodities (expressed in equivalent ectars) calculated in proportion to the areas of SH_{PRD} and existing PRD for each landscape unit (LU) of the municipality of Mori. LU1 Mori valley floor (*purple*), LU2 Loppio valley (*pink*), LU3 low val Gresta (*green*), LU4 Biaena South side (*blue*). Note that the system requires the contribution of the product of at least 300 ha of agricultural areas from the outside in order to balance the needs of the local minimum. It turns out that the LU 1 of Mori valley is clearly heterotrophic

in the landscape (red 0–10; orange 11–40; yellow 41–60; pale green 61–100; green >100 km^2).

These studies confirm more traditional observations on the interference of roads against big mammals, but are more real. If we measure the road density in km/km^2 we may note that these values change from 0.34 of Romania to 4.9 of Belgium. But over a limit of 0.45–0.60 km/km^2 the animal movement is very disturbed [33]. On the other hand, we note that Italy and UK have a similar value of road density: 1.6 km/km^2 even if their effective network density is different: about 3.8/1,000 km^2 in UK and 8.9/1,000 km^2 in Italy. This difference is still wider for France, which have 1.9 km/km^2 of road density but 29.5 of mesh density. These differences are mainly due to the interference between the road and the urban networks.

In the study of a landscape unit (LU) it can be useful to quantify the major movements in the two sub-systems of the human habitat (HH) and natural habitat (NH). These can be summarised as follows:

- HH= (a) the commuting population and tourism flow, (b) traffic on road and rail transport of passengers and goods, (c) volume of trade (import/export) of energy and raw materials, (d) input–output industrial products, (e) trade in agricultural products.
- NH= (a) surface and underground water flows, (b) movement (and direction) of animals, (c) forestry production, (d) transportation of mineral resources.

In particular, it is useful to have data on any surpluses (or shortages) farming, and because you can check the data on the move, or because you cannot control what should be the limits to the transformation of cultivated areas. Above all it is necessary to analyse the agricultural movements in the municipal territories formed by different LU (Fig. 3.16).

The example is based on the territory of Mori (Trento): using the indicators of standard productive habitat per capita SH_{PRD} [m^2/inh], we note that, among four landscape units, three are able to export agricultural commodities but, even if these were targeted to the consumption of Mori, it would not suffice, since in LU1 SH_{PRD} is equal to 586.7 [m^2/inh] instead of the minimum theoretical $SH_{PRD} = 1{,}050.0$ m^2/inh. Taking into account the local consumption per LU, with a safety factor of 1.4, we would have a budget

Table 3.6 Comparison of the main processes of reproduction of an animal population and of a landscape minimum unit

Population renewal	Reproduction process	Ecotope renewal
Predisposed gonads	Reservoir of information	Ecological memory (e.g. propagule bank)
Chromosome crossing over, coping and coding	Copying and coding	Local disturbances
Nest or parental cares	Young structure protection	Nursery niches
Competition and predation, biosemantic	Selection and correspondence	Competition and predation, biosemantic
Death, often deferred	Old structure death	Crucial disturbance

(year 2007) requiring for LU1, for additional production area (PRD), at least 761.1 ha. Potentially, the nearby LU should contribute to 463.3 ha, so the landscape need a movement of incoming agricultural products derived from approximately 297.8 ha from external territories.

In some case study it could be useful calculating the movement of the new urbanisation, which can be measured in ha/year. To continue the example of Mori, from 1860 to 2007 the urbanised area increased from 63.68 to 204.57 ha, at an average speed of 0.958 ha/year, that is 26.26 m^2/day: two rooms per day during 147 years! Note that in the same period the decrease of agricultural fields reached 172.4 ha vs. 140.9 ha of the urban increase. The difference of 31.5 ha is due to technologic infrastructures.

3.6 Reproductive and Colonising System

3.6.1 Ecotope Reproduction

Reproduction processes realise in different ways at each biological level. Let us focus on them comparing population renewal vs. ecotope renewal, for example. They still meet the reproductive functions available to all biological systems (Table 3.6). Given that the reproductive capacity of an ecotope is essentially dependent on the presence and the characteristics of the plant component, it is necessary to introduce the concepts of "zero event" [37] and "ecological memory" [38].

The first has the meaning of lethal disturbance (e.g. fire in small and medium scale, crashes or flush cut), the second is formed by two parts: (a) internal rules to the landscape element (e.g. propagule-bank) and (b) context relations (e.g. filters of dispersion). Among the major factors capable of producing zero events we recognised fire, water, wind, air humidity in some situations, earthquakes and earth movements, many human actions, the actions of the animals, the actions of microorganisms or the combined effects of more.

3.6.2 Colonisation

Although colonisation by the plant species (and sometimes animals) plays an important part in the reproduction of the landscape, as well as in managing human activities, unless this proves too much of robbery, the component-guide available to reproduce elements of a LU is the vegetation, which in fact, as we know, plays the pivotal role in the characterisation of a landscape. The colonisation depends on the size of the tesserae and their number, as well as on the intensity and amplitude of the triggering event, compared to the total size of the ecotope. So you will have:

(a) The *renewal* of the previous ecotope.
(b) The *creation* of a totally new ecotope (Fig. 3.17).

In the first case (a) the reproduction process falls within the regenerative phase of a vegetation that is in a situation of *fluctuation* sensu Falinski [39]; in the second case (b) the

Fig. 3.17 Fiemme Valley (Trentino-Alto Adige region). After a zero event due to a tornado during a strong storm, the reproduction of this forest tessera is going on dominated by *Larix decidua* species, even if the previous forest has been dominated by *Picea abies*

Fig. 3.18 Fiemme Valley. After a zero event near each stump of uprooted Spruce trees is growing a Swiss Pine (*Pinus cembra*)

reproduction process must be defined as "*ecological succession*" but in the new sense. As you will see in fact in the chapter on transformations, ecological succession can no longer be regarded as mechanistic, through predetermined stages, arriving at a climax stage. This process needs to be updated again, according to the non-equilibrium thermodynamic.

After a zero event an important role is acquired by the remnant strains (or dead plant biomass), because of their capacity to protect newborn elements and, in some cases, to contribute to their growth (e.g. newborn Spruce). In the same area of the Fiemme Valley shown in Fig. 3.17, near each stump of uprooted trees we find a young *Pinus cembra* (Fig. 3.18). While the larch trees grow everywhere and the old dominant *Picea abies* became very rare, the more frequent Swiss Pine is near each strain.

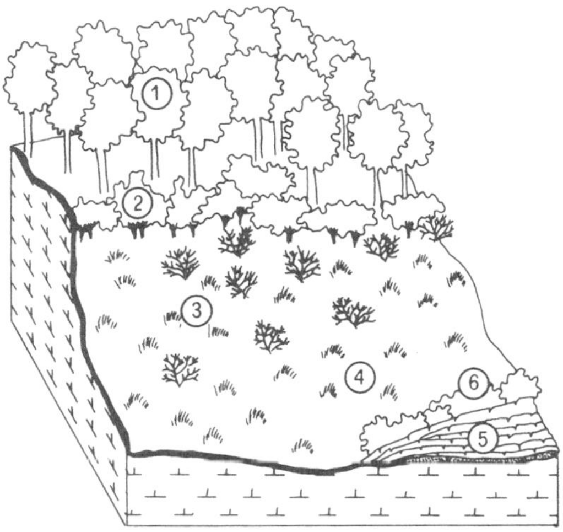

Fig. 3.19 Land abandonments underline the importance of the edge (the "mantle") of a forest patch, as illustrated by Biondi et al. [40] for the Central Apennine. The mantle is mainly responsible for plant colonisation. *To the left*: (1) *Quercus pubescens*, (2, 3) *Spartium junceum* and *Cytisus sessilifolius*, (4) *Brachipodium rupestre* and *Bromus erectum*, (5) fields, (6) *Ulmus minor*

Consequently, after a zero event very few wood can be removed from fallen logs.

While in the Fiemme example few remnant isolated Larches were the first responsible of the colonisation, in other cases due to land abandonment the main promoter of the colonisation is generally the forest mantle, that is the border edge. An example of mantle is given in Fig. 3.19, in the Central Apennines, in a formation of Downy oak (*Roso semperviventis-Quercetum pubescentis*).

You will notice that even human populations are colonising, through a method that is often repeated with minor variations and which is similar to the Roman *limitatio*, not surprisingly reused in the USA and in Brazil.

3.7 Main Configuration Processes

3.7.1 Delimitations and Ecotone Network

All landscape components may be distinct from their surrounding through their boundaries. This seems to be true even if each range of scale shows a particular type of boundary and if there is a distinction between a sharp margin and an ecotonal belt. Generally speaking, processes are held within surfaces. In fact, the limits of a process define where the surface shall be. As pointed out by Allen and Hoekstra [43], the multiplicity of devices that can be used to detect the surface of an entity that is resistant to transformation reflects the multiplicity of the processes responsible for that surface.

A surface can be tangible (e.g. tree bark) or intangible (e.g. thermocline of a lake) or mixed, such as the margin of a patch. A surface, or boundary, disconnects the internal functioning of entities from the outside world, but sometimes this fact may be not sufficient to warrant the correct design of the boundary and the surface remains arbitrary. In some cases not all the boundaries of a single element are definable. Therefore, it is necessary to evaluate the boundaries through observations derived from a higher level of scale.

Remembering the principles of hierarchical systems, we note that frequency and constraint are the most important criteria for ordering levels, because upper levels constrain lower levels by behaving at a lower frequency. This conditioning is a *passive* process, in which upper levels constrain lower ones by doing nothing, or refusing to act. That is why if we look at an ecological system at a too low level of scale, then the scope of observations will probably not extend to the principal constraints.

Thermodynamic theory suggests that an open system with energy input becomes spatially heterogeneous in two ways: (1) through gradual concentration gradients of the existing elements that make the system heterogeneous, but not patchy; (2) through the formation of a mosaic with boundaries. Due to inherent environmental patchiness and non-equilibrium thermodynamics, a landscape follows basically the second way, leaving the first only to partial or transitional phenomena (e.g. population invading an area).

The layers composed of vegetational biomass form the current boundary of an ecotope, but topographic and human barriers also play the

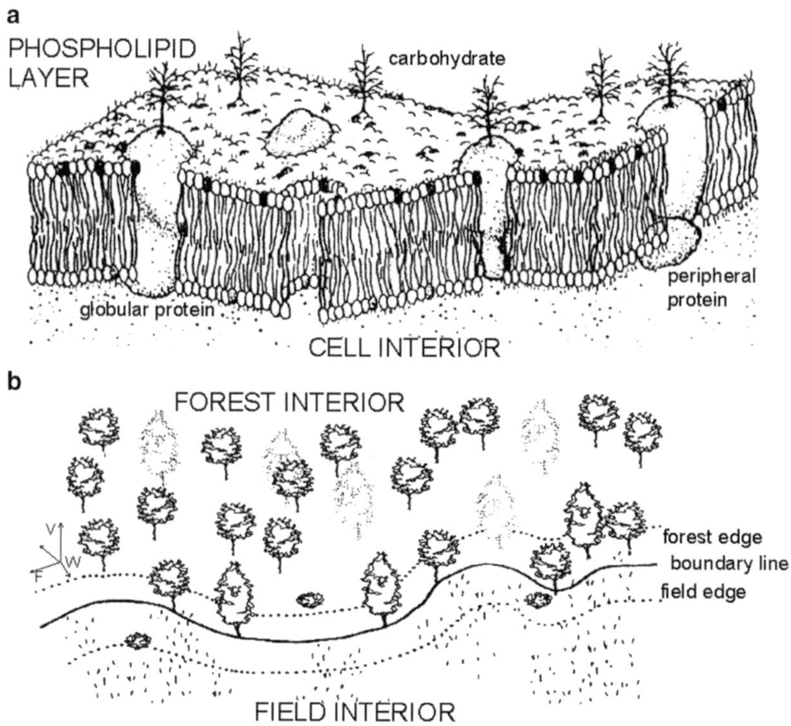

Fig. 3.20 Characters of a boundary of a landscape element: comparison with a cellular membrane (from Forman and Moore [44] re-drawn). (**a**) Molecular structure of a cellular membrane. (**b**) Structure of a forest-field boundary. The boundary zone includes the edges of both sides of a boundary line and its long three-dimensional structure, characterised by width (*W*), verticality (*V*), and form (*F*). Boundary surfaces, such as convex, lobe, and straight, may be in reference to either the forest or the field

same role. In this sense, human actions have had an evolutionary function in defining many boundaries once undetermined in natural ecotopes, therefore giving a higher functionality to landscape mosaics.

The delimitation of a landscape element plays an important and multiple functional role, first of all as filter. The internal structural characters of the boundaries are responsible for its permeability. With a correct theoretical analysis, Forman and Moore [44] claimed that landscape boundaries might be considered as functional analogues of cellular membranes (Fig. 3.20).

A series of statements of the membrane theory may be identified as pertinent to landscape boundaries: (1) boundaries appear and multiply spontaneously in open systems; (2) intense disturbance destroys boundaries, reducing the amount of variety within a system; (3) a boundary contains the record of interactions across itself; (4) strong interactions produce a rich textured (heterogeneous) boundary; (5) when boundaries move, they extend the space of the system with higher variety and decrease the space of the system with low variety.

Like a cellular membrane, energy and matter are concentrated at the edge of a patch, such as the plant biomass in the mantle of a forest edge. Herbivores and predators are often also concentrated at the edges.

Microclimatic effects are relevant in a boundary of different patches, such as from forest to open fields. The well-known edge effect has been defined in relation to the increase or the decrease of species in the margin belt of an ecotope: it is necessary to remember that this effect may be positive, void or negative. The width, verticality and form dimensions may determine functional roles of landscape boundaries. The principal functions are: (1) conduit, (2) filter or barrier,

Table 3.7 Amount of edge influence in rectangular forested patches

Patch size (ha) (m × m)	DEI (m)	Interior species available area (%)	Edge species available area (%)
5 ha	30	53.3	46.8
(250 × 200)	60	20.8	79.2
10 ha	30	64.6	35.4
(400 × 250)	60	36.4	63.6
20 ha	30	79.9	20.1
(500 × 400)	60	53.2	46.8
45 ha	30	82.1	17.9
(900 × 500)	60	65.9	34.1

DEI Depth-of-edge influence: we can assume about twice the height of the trees

(3) source, (4) sink, (5) habitat. Note that there is an essential affinity between boundaries and corridors, even if spatially the concepts differ and connectivity differs too.

Chen et al. [45] have shown that the fragmentation of the North American forests, which multiplied the margins of forest patches, influenced these patches up to a distance of 1–3 times the medium height of the trees. Depth-of-edge influence (DEI) is an important issue in the theory and practice, and we note (Table 3.7) that there is no "interior forest" for a patch <10 ha if the tree height is about 36 m (DEI = 72), or (same tree height) that we need a patch of 45 ha to have >50 % of interior species.

Ecotones are structural and functional discontinuities within landscapes, having their primary significance in presenting a gradient, thus the separation between two different ecotopes of landscape units is not sharp, but gradual. Being a transitional zone between ecological units, the ecotones represent places where interactions among patches are particularly characterised and they may modify flows between patches. Moreover, ecotonal network functions are linked with the transmission of information across a landscape. The main effects pertaining to the ecotones are:

- Gradient: progressive reduction of elements.
- Zoning: irregular distribution of flux and concentration of elements.
- Accumulation: retention of a great quantity of an element.
- Barrier: stopping the fluxes.
- Margin: maximum local biodiversity (edge effect).
- Neighbouring: influence of the ecotone on adjacent ecosystem.
- Transmitter: selection of information directed to nearby ecosystems.
- Ecotissue: landscape elements not simply juxtaposed.

One of the most important ecotonal networks is given by the set of river corridors which characterise a landscape. The main effects in this case are:

(a) Chemical filter, with a retention rate of about 80 (forest) or 8 % (meadow), for a 30 m large strip of riparian vegetation. Denitrification is particularly important in forested soils.
(b) Biological filter, e.g. when hydrocore plants, such as alders, etc. (*Alnus* spp., *Polygonum* spp., etc.), serve as a barrier in relation to the water–land interface, in contrast to the oaks and hazels (*Quercus* spp., *Corylus* spp.).
(c) Margin, e.g. when the heterogeneity of the ecotone may favour some animal species, such as the otter (*Lutra lutra*) if there are willows (*Salix* spp.) and oaks or ashes (*Fraxinus* spp.).

An evident ecotonal effect is produced in the peripheral belt between an urbanised landscape and its surrounding agricultural or natural landscapes. In a case like this, we have a large-scale delimitation which produces typical gradients. A chain of landscape units become structured by the ecotonal effect, with a series of passages from cultural and technical components to rural and natural ones. If not polluted or too much altered by human infrastructures, these belts

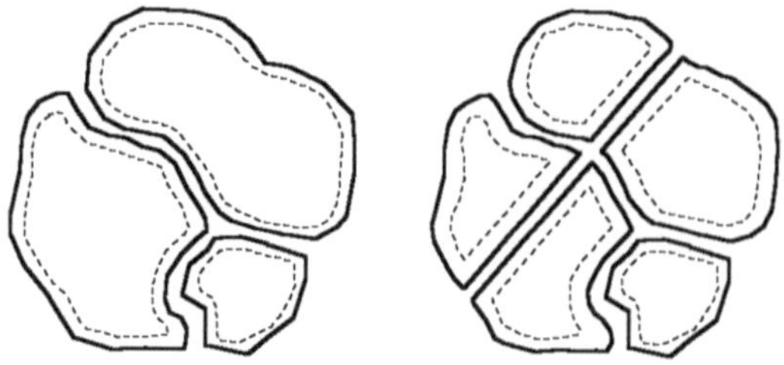

Fig. 3.21 Example of the effect of fragmentation on forested patches. Within an already fragmented former patch (*left*), the introduction of a new corridor (*right*) leads to an increasing of the number of the *small patches* and to a fall of the number of the interior species

Table 3.8 Effects of fragmentation on a medium size forested patch

Patch divisions	Patch No.	Forested area (%)	Corridor area (%)	Greatest interior area (%)	Interior species available area (%)	Edge species available area (%)
Entire patch	1	100	0	82.5	82.5	17.5
1st corridor	2	93.3	6.7	47.7	72.5	20.8
2nd corridor	3	90.6	9.4	43.0	65.1	25.5
3rd corridor	5	79.2	20.8	22.8	50.4	28.8
4th corridor	6	76.6	23.4	22.8	44.8	31.8

The *greatest interior area* identifies the part of the initial patch which remains untouched after the cuttings and the dimensions of which are so large as to be able to preserve interior species

can be rich in biodiversity and can be useful for many animal populations (e.g. starlings).

3.7.2 Fragmentation and Permeability

The breaking up of a habitat, patch or land type into smaller parcels and/or the dissection of them by roads or similar is usually called fragmentation. Normally fragmentation is considered as a negative process in landscape ecology, because: (1) it generates the loss of species (especially animals) more dependent on the size of patches, (2) it greatly disturbs species with a high standard habitat (low density species), (3) it produces an increase of alien species, especially in human landscapes, (4) it leads more isolated populations toward an inbreeding depression.

According to Forman [31, 35] the demonstration of the species loss due to fragmentation was done by observing the species number in a large forested patch, which is higher than in the same area composed of a set of smaller patches. Let us present a typical example. Given a forested patch of about 15 ha, we consider a depth-of-edge influence (DEI) limited to the tree height (20 m) and measure its interior-margin area in parallel with its fragmentation due to corridors of a width of 20 m (see Fig. 3.21 and Table 3.8).

The effects are very strong. After the opening of the first corridor, the loss of only 6.7 % of its area leads to an increase of the edge species of 3.3 %, to a decrease of the interior species of 10 % and of the greatest patch interior of 34.8 %. After a third corridor and the loss of only another 14.1 % of the forested area, the former forested patch changes its character. In fact, the greatest interior area loses another 25 %, the presence of interior species decreases from 72.5 to 50.4 % and one fifth of the total area becomes open to the entry of exotic species.

Nevertheless, sometimes fragmentation can be also a positive process, for instance when forest corridors fragment a rural landscape

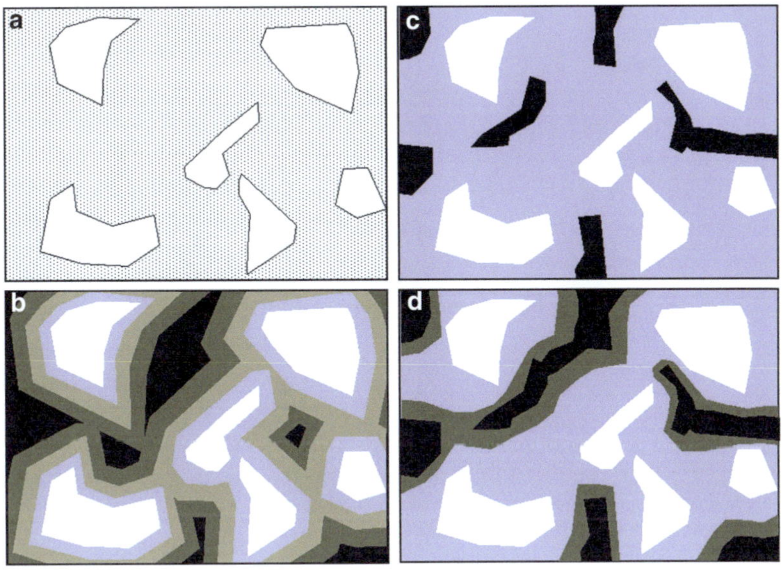

Fig. 3.22 Example of a connective landscape apparatus and its main dynamic (From Burel and Boudrie [46], re-drawn). Prairies are shown, as are the distances from them; to the *right* the roughness and the permeability of the landscape unit. (**a**) Prairies (*white*) and matrix. (**b**) Distance from prairies (*high-black; low-light blue*). (**c**) Roughness (*high-black; low- light blue*). (**d**) Permeability (*high-light blue; low-black*)

matrix, or when a forest landscape is partially opened by pastures. For example, the enrichment of habitat produced by opening small grazing pastures in a wide forested landscape in Folgaria (Trient) creates a more favourable habitat for capercaillies (*Tetrao urogallus*).

In a given landscape it is possible to recognise a connective apparatus (Fig. 3.22) composed of the elements, which have the capacity to connect different patches as an expression of their prevailing landscape function. A clear correlation between connected elements of the same biological type has been demonstrated by many scientists as Burel and Boudrie [46]. Note that the concept of permeability is linked to the concept of connectivity, thus to the dynamic of movement. The permeability may obviously be different for each animal group and also for man.

Corridors are considered as the principal elements of connectivity in a landscape. Fluvial corridors, canopy roads, windbreaks, greenbelts and hedgerows all play a role in the rural landscape, just as equestrian trails, wooded avenues and bicycle routes can be of value in urban areas. All represent linear connectors that permeate the landscape and play a role in an interconnected habitat system.

Anyway, we have to consider at least two other elements: (1) the ecotones between different ecosystems or ecotopes, with their peculiar structures and multiple functions, as buffer and filtering areas, (2) the connective tesserae among an ecological mosaic (ecotissue), with quite simple structures and the function of allowing the dynamics of other patches.

The evaluation of connectivity needs the definition of an optimum set of parameters, because the higher thresholds limit the change of the type of landscape, while the lower ones avoid a negative circuitry (see 7.1.4).

3.7.3 Other Processes Interacting with the Landscape

The landscape is a hyper-complex system; therefore it has a lot of interacting processes. We are perfectly conscious that the processes interacting with landscapes are very numerous, but in a book like this we have to give prominence to intrinsic basilar functions. Many of the other processes depend on facts by other disciplines (e.g. chemistry for geobiologic cycles) and contribute to the

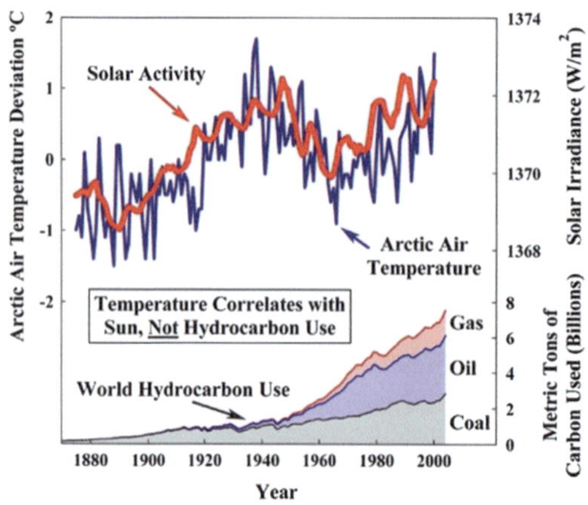

Fig. 3.23 NIPCC correlation between World temperature and solar activity shows an increase from 1880 while the World Hydrocarbon Use starts around 1955. Man does not seem to be the only responsible for Global Warming

environment in a wider sense. Moreover, many of these processes are not yet well comprised by scientists.

For instance, the climate change is no doubt evident, but nobody may seriously affirm what really is the contribution of man to global warming. The raise of CO_2 is worrying, the Kyoto Protocol has to be followed, but can we truly affirm that is mainly due to human actions? The IPCC false sequence of data (2001) is known [47]: they forgot the Medieval Warming Period, the Small Ice Age (1600–1840) and the beginning of temperature increase [48]. This last datum is very important, because temperature began to rise in the second half of nineteenth century, about 80–100 years before the rise of World Hydrocarbon use and consequent CO_2 increase (Fig. 3.23).

3.7.4 Landscape Biodiversity

What the concept of biodiversity is? The concept of biodiversity, defined by US Office of Technological Assessment [41], depends on two aspects:

(a) The diversity of the *components* of ecological systems.
(b) The diversity of their relations in the *organisation* of these systems, with other two aspects: (b1) *context subsystems* and (b2) *context LU*.

Therefore, biodiversity is also an attribute of an entire ecological system, i.e. of a landscape. To reach a better understanding of the ecological functioning of, we have to check:

(a) *Species* diversity, e.g. α, γ and β [42] and *landscape elements* diversity ψ, τ [26].
(b) *Landscape apparatus* diversity and *landscape complex* diversity (e.g. landscape unit), measuring the pattern of their ecological organisation.

We may begin this study considering *species* diversity in a landscape, for instance relative to the sequence: "*Quercetum ilicis*-Maquis-Garigue-Agricultural areas" in Mediterranean biomes, because it is a sequence that presents a growing number of species. As observed by Pignatti [22], the more evolved plant communities are constituted by a species biodiversity clearly less numerous that the other.[4] The necessity to reduce redundancy in well-ordered complex systems is evident. Anyway, a landscape cannot be formed only with low diversity ecocoenotopes.

Considering the *landscape elements* diversity, we have to note that it does not coincide with the diversity of land use, which does not see the difference in vegetation types as ecology needs. We must examine the *number of types of ecological tesserae* in a landscape unit. This biodiversity can be expressed as ψ, structural landscape diversity:

$$\psi = H \times (3 + D),$$

where H = Shannon heterogeneity, D = dominance.

But the diversity of landscape elements has to be checked also as a functional one, which can be expressed as τ, in the same way of the structural, but measured on the distribution of the BTC classes in the LU (see 3.3.1):

[4] In a vegetation survey of a patch of *Maquis* it is possible to find about 100–120 species, while in a *Quercus ilex* forest this number is mainly about 25–40 species.

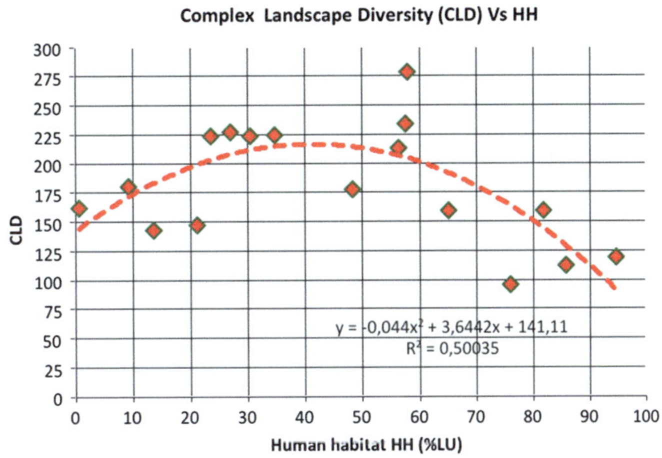

Fig. 3.24 Plotting the complex landscape diversity (CLD) vs. HH, on the basis of a set of landscape units of North Italy, the correlation shows that the maximum CLD is about 216.46 at HH = 41 %

$$\tau = H \times (3 + D).$$

The *landscape apparatus* diversity, or ω diversity, can be estimated through the same formula, measuring the extension of landscape apparatus types in a LU:

$$\omega = H \times (3 + D).$$

Finally, we arrive to the *complex* diversity of the entire landscape unit, or CLD (complex Landscape Diversity), which is composed as

$$CLD = \psi \times \tau \times \omega.$$

The study of landscape dynamics imposes to check this magnitude (CLD) on real case studies, correlating CLD to human habitat (HH) as independent variable.

In Fig. 3.24 we may see that the most natural landscapes (HH < 20 %) present a CLD from about 150 to 200 clearly higher than the urbanised ones (CLD = 90–150), but the maximum values of CLD = 216.46 is done at about HH = 41 %, corresponding to a agro-forestry landscape type.

3.7.5 General Landscape Metastability

In Sect. 3.3.1 the BTC function was indicated as related to the concept of metastability. This enables to study vegetation through new perspectives and to evaluate landscape transformation in a proper way. Following an opportune method (see Chap. 5), it is possible to estimate the BTC of an entire landscape unit (LU), calling it BTC_{LU}. This value is very important, as already underlined, but it is not available to express the true metastability of the LU under examination, because it does not consider the functional diversity of distribution of the BTC_{LU} and the amount of information linked to that composition. Therefore, if we want to find a bionomic index able to measure the metastability of a LU with an acceptable adequacy, we have to relate the two magnitudes of BTC_{LU} and τ. In fact, we know that τ = H × (3 + D), measuring H and D on the base of the BTC classes in which the landscape elements are represented in the LU.

Remember that the concept of *negentropy* allowed Shannon and Weaver to use the

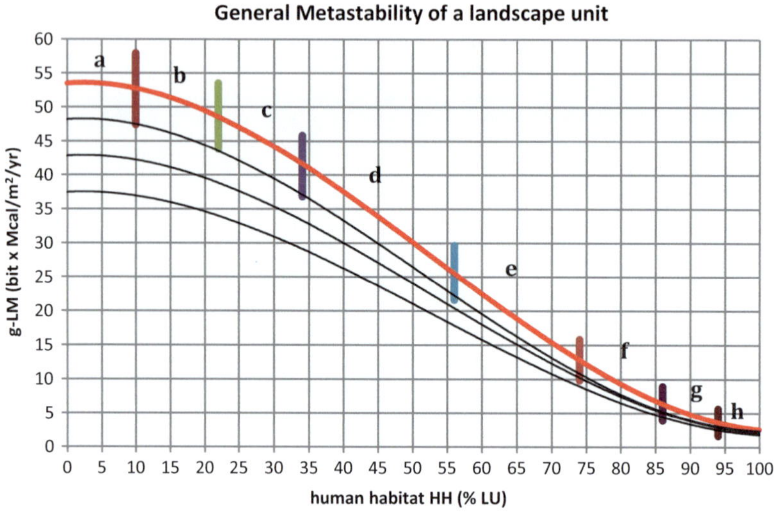

Fig. 3.25 Correlation of the general landscape metastability (g-LM) vs. HH, based on a set of LU samples and a set of landscape typological parameters. The *lower curves* represent case studies having part of the territory unusable. The *bars* divide the following types of landscape (from *left*): (a) natural-forested, (b) semi-natural-forested, (c) agro-forested, (d) agricultural-protective, (e) agricultural-productive, (f) suburban, (g) urban, (h) dense urban

Boltzmann entropy function as a measure of *information*. So the information content of a LU, for example, can be expressed by the diversity index H' (average mutual information) in a set of habitats:

$$H' = -\Sigma_{i=1}^{n} \log_2 P_i \text{ [bit]},$$

where P is a probability. For example, if we have to choose the right habitat from a set of 16, the probability P_0 of choosing the right one before receiving information is: $1{:}16 = 0.063$; but after receiving exact information P becomes 1, as the difference (certainty). Using a binary decision method, we measure I:

$$I = \log_2 P/P_0 \text{ thus } I = \log_2 16 = 4 \text{ bits.}$$

We have to note that Information is more than a physical dimension expressed by the previous formula, because a message with a high information content is a message selected from a large class of admitted ones or a message which contains a lot of chance. One of the reasons why, in the definition of τ, H was related with D is just to avoid to consider the information per se: a good landscape function must balance entropy and dominance.

Figure 3.25 shows the correlation of the general landscape metastability (g-LM) vs. the human habitat (HH). The curve is a polynomial:

$$g - LM = 0.0001x^3 - 0.0193x^2 + 0.1583x + 53.5.$$

It is based on a set of LU samples and a set of landscape typological parameters. The lower curves represent case studies, having part of the territory unusable (impracticable): in the figure the red curve is available when the use of the territory is 100 %, in the black ones, respectively, 90, 80, 70 %.

Note that the maximum g-LM corresponds to HH = 3.0 %, so the natural-forested landscape may be considered up to HH = 10 %, where the difference between the two g-LM$_{max}$ is only −0.025. The bars divide the main types of landscape in temperate countries: from left we find (a) natural-forested, (b) semi-natural-forested, (c) agro-forested, (d) agricultural-protective, (e) agricultural-productive, (f) suburban, (g) urban, (h) dense urban.

Remember that g-LM represents an information quantity per the flux of energy able to maintain the landscape level of organisation. Through the g-LM function it is possible to add a further deeper control on the bionomic state of a LU, as we will see in the chapters on application.

References

1. Brandt J, Tress B, Tress G (2000) Multifunctional landscapes: interdisciplinary approaches to landscape research and management. Centre for Landscape Research, Roskilde DM
2. Naveh Z (2000) Introduction to the theoretical foundations of multifunctional landscapes and their application in transdisciplinary landscape ecology. In: Brandt J, Tress B, Tress G (eds) Multifunctional landscapes: interdisciplinary approaches to landscape research and management. Centre for Landscape Research, Roskilde DM
3. Tricart J, Kilian J (1979) L'éco-géographie et l'aménagement du milieu naturel. Librairie François Maspero, Paris
4. Ehrard H (1956) La genèse du sol en tant que phénomène géologique. Masson, Paris
5. Duchaufour PH (1983) Pédologie et classification. (I) Constituants et propriétés du sol. (II). Paris, Masson
6. Dokuchaev VV (1898) Writings (in Russian) reprinted 1951 vol 6, Akad Nauk, Moscow
7. Godron M (1984) Ecologie de la végétation terrestre. Masson, Paris
8. Ghirardelli E (1981) La vita nelle acque. UTET, Torino
9. Ingegnoli V (1980) Ecologia e progettazione. CUSL, Milano
10. Ingegnoli V (1991) Human influences in landscape change: thresholds of metastability. In: Ravera O (ed) Terrestrial and aquatic ecosystems: perturbation and recovery. Ellis Horwood. Chichester, UK, pp 303–309
11. Ingegnoli V (1993) Fondamenti di ecologia del paesaggio. CittàStudi, Milano
12. Ingegnoli V (1999) Definition and evaluation of the BTC (biological territorial capacity) as an indicator for landscape ecological studies on vegetation. In: Windhorst W, Enckell PH (eds) Sustainable landuse management: the challenge of ecosystem protection. EcoSys: Beitrage zur Oekosystemforschung, Suppl Bd 28:109–118
13. Ingegnoli V (2002) Landscape ecology: a widening foundation. Springer, Berlin, p 356
14. Ingegnoli V, Giglio E (1999) Proposal of a synthetic indicator to control ecological dynamics at an ecological mosaic scale. Annal Bot LVII:181–190
15. Odum EP (1971) Fundamentals of ecology, 3rd edn. WB Saunders, Philadelphia
16. Odum EP (1983) Basic ecology. CBS College, New York
17. Ingegnoli V (2013) Concise evaluation of the bionomic state of natural and human vegetation elements in a landscape. Rend Fis Acc Lincei. doi:10.1007/s12210-013-0252-2
18. Ellenberg H (1974) Zeigerwerte der Gefässpflanzen Mitteleuropas. Scripta Geobotanica 9, Göttingen
19. Ellenberg H (1978) Vegetation Mitteleuropas mit den Alpien. Oekologischer Sicht Ulmer
20. Naveh Z, Lieberman A (1984) Landscape ecology: theory and application. Springer, New York
21. Ingegnoli V, Pignatti S (2007) The impact of the widened landscape ecology on vegetation science: towards the new paradigm. Rend Fis Lincei s IX XVIII:89–122
22. Pignatti S (1998) I boschi d'Italia: sinecologia e biodiversità. UTET, Torino
23. Falinski JB (1986) Vegetation succession on abandoned farmland as a dynamic manifestation of ecosystem liberal, of long continuance anthropopression. Wiad Bot 30.1:21–50 (part 1); 30.2:115–126 (part 2)
24. Kowarick I (1990) Some responses of flora and vegetation to urbanization in Central Europe. In: Sukopp H, Hejny (eds) Kowarick I (co-ed) Urban ecology. SPB Academic, The Hague, pp 45–74
25. Ingegnoli V (2011) Bionomia del paesaggio. L'ecologia del paesaggio biologico-integrata per la formazione di un "medico" dei sistemi ecologici. Springer, Milano, pp XX–340
26. Ingegnoli V, Giglio E (2005) Ecologia del Paesaggio: manuale per conservare, gestire e pianificare l'ambiente. Sistemi editoriali SE, Napoli
27. Reimoser F, Gossow H (1996) Impact of ungulates on forest vegetation and its dependence on the silvicultural system. For Ecol Manag 88
28. USDA Agricultural Research Service (2012) Plant hardiness zone map. PRISM Climate Data, Oregon State University
29. Forman RTT (2008) Urban regions, ecology and planning beyond the city. Cambridge University Press, Cambridge, UK
30. Lorenz K (1981) L'etologia. Bollati Boringhieri, Torino
31. Forman RTT, Godron M (1986) Landscape ecology. Wiley, New York
32. Sanderson J, Harris LD (eds) (2000) Landscape ecology: a top-down approach. Lewis, Boca Raton, FL
33. Forman RTT, Spierling D, Bissonette JA, Clevenger AP, Cutshall CD, Dale VH, Fahrig L, France R, Goldman CR, Heanne K, Jones JA, Swanson FJ, Turrentine T, Winter TC (2002) Road ecology: science and solutions. Island Press, Washington, DC
34. Zonneveld IS (1995) Land ecology. SPB Academic, Amsterdam
35. Forman RTT (1995) Land Mosaics: the ecology of landscapes and regions. Cambridge University Press, Cambridge, UK
36. EEA, FOEN (2011) Landscape fragmentation in Europe. European Environment Agency, Joint EEA-FOEN report

37. Oldeman RAA (1990) Forests: elements of sylvology. Springer, Berlin
38. Bengtsson J, Angelstam P, Elmqvist T, Emanuelsson U, Folke C, Ihse M, Moberg F, Nistrom M (2003) Reserves, resilience and dynamic landscapes. Ambio 32(6):389–396
39. Falinski JB (1998) Dioecious woody pioneer species in the secondary succession and regeneration. Supplementum Cartographiae Geobotanicae 8. Phytocoenosis 10 (N.S.)
40. Biondi E, Baldoni M (1994) The climate and vegetation of peninsular Italy. Coll Phytosoc XXIII:675–721, Bailleul
41. US Congress, Office of Technology Assessment (1986) Assessing biological diversity in the United States: data considerations—background paper #2, OTA-BP-F-39 U.S. Government Printing Office, Washington, DC
42. Whittaker RH (1975) Communities and ecosystems, 2nd edn. Macmillan, New York
43. Allen TFH, Hoekstra TW (1992) Toward a unified ecology. Columbia University Press, New York
44. Forman RTT, Moore PN (1991) Theoretical foundations for understanding boundaries in landscape mosaics. In: di Hansen AJ, Di Castri F (eds) Landscape boundaries: consequence for biotic diversity and ecological flows. Springer, New York, pp 236–258
45. Chen J, Franklin JF (1992) Vegetation responses to edge environments in old growth douglas-fir forests. Ecol Appl 2(4):387–396
46. Burel F, Baudry J (1999) Écologie du paysage: concepts, méthodes et applications. Technique & Documentation Ed, Paris
47. IPCC (2001) Climate change 2001: the scientific basis. Cambridge University Press, Cambridge
48. Idso CD, Carter R, Singer SF (2011) climate change reconsidered: 2011 interim report of the nongovermental panel on climate change. The Heartland Institute, US

4 Transformation and Pathology of Landscapes

4.1 Protective and Recovery System

4.1.1 The Protective System

In Sect. 2.4.2 we defined the protective landscape apparatus (PRT) as composed by elements available to protect other elements or parts of the mosaic. Such *apparatus* is found both in natural and anthropic landscapes. In fact, many ecosystems acquire a role of protection in their own mosaic (a) directly or (b) potentially: for example,

(a) An ecotope resisting slope disturbances at the base of a mountain and permitting the formation of other types of more sensitive ecosystems. In human landscapes, the hedgerow network in a field mosaic or the garden around house are typical protective elements.
(b) An area, with peculiar natural characters, protecting vital resources (e.g. water springs). In human landscapes, the system of interface areas (see later) becoming a refuge for natural species.

Even if these two aspects of the protective apparatus may be overlapped, as in many conserved areas, it is basilar to underline the concept of *reserve*. Reserve is a supply of something that is not actually being used at present but that is available for use when needed, sensu Collins Dictionary [1]. A flooding preserve belt around a river may represent this concept.

In these last 50 years, the strong improving (in number and square miles) of the so-called "protected areas", passed from about 2 million km^2 in 1960 to about 12 millions in 2000, is known. No doubt that the main reason for this incredible increase had been the conservation of Nature, largely erased by human consumes after the huge population growth. Therefore, it was the need for biological conservation, and more strictly the need of survival, that moved to limit the destruction of Nature.

Actually, we should change the term "protected areas" into "*protecting* areas", because these areas—properly speaking—are just *reserves*, but with a wider sense derived from the landscape bionomics principles, i.e. the concept of protective apparatus. These areas have to protect their inner natural resources, but even the landscape in which they play a role. They are needed by the system of ecocoenotopes forming a landscape because it cannot survive without a protective apparatus.

It is a question of homeostasis: if parts of a landscape must be transformed (indispensable for humans), suitable other parts have to be preserved (and/or fulfilled with high BTC formations) to compensate these changes. Pay attention that it isn't a question of surface extension replacement: following bionomic principles, "compensation" concerns ecological attributes and/or parameters, functions and/or roles played within the system, before and after the transformation, through a surface that could be larger but less extensive too. It is also a question of disturbance incorporation and recovery: without reserves of spatial resources, with adequate

V. Ingegnoli, *Landscape Bionomics Biological-Integrated Landscape Ecology*,
DOI 10.1007/978-88-470-5226-0_4, © Springer-Verlag Italia 2015

bionomic capability, it is simply impossible to resist or recover perturbations.

What the complex system has to conserve are essential functions as the so-called General Landscape Metastability (g-LM) as we have seen, or the correlation between HH and BTC for a typical landscape, sensu Ingegnoli [2]. We must underline that the existence of the protective apparatus is mainly in function of the structure and the transformations of the landscape and so it is possible to evaluate the need of minimum theoretical protective areas as a proper landscape process. It could be interesting to estimate this value starting from the World scale, the entire eco-bio-geo-sphere.

The landscape bionomic parameters, necessary to evaluate the theoretical protective (PRT) area, are exposed in Table 4.1. As we will better see in Sect. 4.4, the elaboration is based on five periods of time: 1882, 1900, 1962, 1990, 2010, principally related with FAO and Eurostat statistics [3–6]. In these 128 years, the human population grew from 1.5 billion to 6.97 (+465 %). The average BTC of the emerged land had a strong decrease (−24 %) in the first 80 years, passing from BTC = 2.95 to 2.24 Mcal/m^2/year, and then remained near constant, with a small decrease in the last 20 years. In the entire period, the human habitat (HH) changed from 21.9 to 26.8 % (+22.3 %). Note that the increase of croplands, from 11.6 to 14.8 million km^2, is contained in + 27.6 % due to the "green revolution" which augmented agrarian production of 600 % in 130 years.

To calculate the theoretical PRT, we have to evaluate the compensation of BTC due to the main human transformations of natural habitat (cropland and urbanisations) in relationship to the average world biological territorial capacity (BTC_W) of all the landscape systems of emerged lands. Considering to maintain a theoretical BTC_W minimum value[1] at 2.20 Mcal/m^2/year and to balance the total SH with a BTC proportional to the forest cover and types, we arrive to find the protective SH_{PRT} values. For examples, in 2010 we measure a cropland SH_{prd} equal to 2,123 (m^2/inhab) and an urbanised SH of 631 (m^2/inhab); their deficit BTC results[2] (2,123 × 1.45) + (631 × 2.05) = 4,373 (Mcal/inhab/year); being the balancing BTC = 5.13 Mcal/m^2/year, it results SH_{PRT} equal to 853 (m^2/inhab).

Multiplying these SH_{PRT} per the number of inhabitants and using a security coefficient to count even the rise of disturbances, we obtain the minimum theoretical PRT, in million km^2. This value should have passed from 3.23 to 7.72 million km^2 in 128 years (+239 %). In fact, after having calculated the productive standard habitat per capita SH_{PRD} and the urbanised SH, we measured the amount of transformation made by man in each period. This quantity should be balanced preserving the presence of natural and semi-natural vegetation available to cover the BTC deficit due to the needed changes.

A surprise derives from these data: after an unquestionable lack until 1970, the total World protected areas [7] are today about 1.8 times broader than the theoretical one, as plotted in Fig. 4.1. This abundant increase is due, at least since 1980, to new recreational and especially touristic use of the protected areas and "Nature Sanctuaries", becoming an important economic source in many countries of the developing nations. Moreover, we must realise that many preserved areas do not entirely reach a protective function, because a conspicuous part of them does not contain any bionomic resource.

In Fig. 4.1 the impressive increase of protected areas in the World (emerged lands) is compared with the minimum theoretical protective PRT system (landscape apparatus). The population growth and the average BTC are also plotted. The increase of the protected areas aroused after the strong BTC descent (1905–1960), mainly due to deforestation, and the change in the growing rate of human population in the years 1960 (starting of the "baby boom"). In only 20 years (1960–1980) the amount of preserved areas

[1] The most recent homeostatic plateau, since 1960, may be estimated around BTC = 2.10–2.25 Mcal/m^2/year.

[2] 1.45 Mcal/m^2/year is the cropland BTC deficit value, while 2.05 Mcal/m^2/year is the urbanized BTC deficit value.

Table 4.1 Landscape bionomics parameters available to evaluate the protective standard habitat per capita (SH_{PRT}) and the minimum theoretical of protected areas (PRT) at global (World) scale

Landscape parameters	1882	1900	1962	1990	2010
Population (billion)	1.50	1.65	3.05	5.26	6.97
Forests (million km^2)	54.60	53.60	41.32	41.68	40.33
Other natural formations (million km^2)	32.40	31.95	42.70	40.45	40.07
Grassland (million km^2)	25.00	24.90	24.50	25.50	26.00
Cropland (million km^2)	11.60	12.30	14.62	14.42	14.80
Urbanised (million km^2)	0.90	0.95	1.66	3.10	4.40
Deserts (million km^2)	24.50	24.35	24.20	23.95	23.70
BTC (Mcal/m^2/year)	2.95	2.89	2.24	2.20	2.11
HH (% land)	21.9	22.4	25.0	25.8	26.8
SH (m^2/inhab)	21,853	20,228	12,213	7,308	5,729
SHprd (m^2/inhab) cropland	7,733	7,576	4,786	2,741	2,123
SHrsd + SH sbs (m^2/inhab) urbanised	600	576	544	589	631
BTC deficit (Mcal/inhab/year)[a]	12,443	12,166	8,055	5,183	4,373
Balancing BTC[b] (Mcal/m^2/year)	6.07	6.04	5.66	5.24	5.13
SH_{PRT} (BTC deficit/balancing BTC) (m^2/inhab)	2,050	2,016	1,415	989	853
Min. PRT (million km^2)	3.08	3.33	4.31	5.20	5.94
Security coefficient[c]	*1.05*	*1.06*	*1.10*	*1.20*	*1.30*
Theoretical PRT (million km^2)	3.23	3.53	4.74	6.24	7.72

[a]BTC deficit is due to the difference between PRD, RSD, SBS average values of BTC (BTC_{PRD} = 0.75; $BTC_{RSD+SBS}$ = 0.15) and the level of BTC considered as the theoretical minimum (in this case BTC = 2.20 Mcal/m^2/year)
[b]World forest BTC average value: these values derived from the amount and type of forest cover
[c]Depending on population increase, traffic volume and technologic disturbances

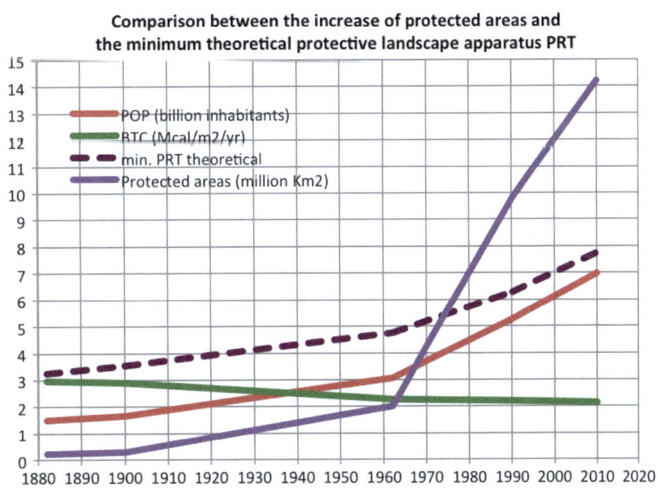

Fig. 4.1 The spectacular increase of protected areas in the World (emerged lands) is compared with the minimum theoretical PRT system (landscape apparatus). The population growth (*red*) and the average BTC (*green*) are plotted. Note that the increase aroused after the BTC descent and the new surge of population

passed from 2.0 to 7.0 million km^2, reaching the minimum theoretical level of PRT, and then continuing up today to about 14.1 million km^2, as already underlined.

To better explain the abundance of protected areas at large scale, as already underlined, we can try to understand what happens at local scale. We report two examples near Milan: the municipalities of Triuggio (a) and Cusago (b), both interested by regional parks preserve areas.

The municipality of Triuggio, a 839 ha rural landscape in low moraine hills (25 km N–E of Milan), with 7,800 inhabitants and a HH of 66.5 %, presents an SH equal to 715.3 m^2/inhab and a $SH_{PRT} = 162.3$. So it needs a protective area of at least:

$$\begin{aligned} PRT &= 162.3 \times 7,800 \\ &= 126.59\,\text{ha}\ [15.1\ \%\ \text{of its territory}]. \end{aligned}$$

Triuggio is part of the Lambro Valley Regional Park and is covered by 28.8 % of forest patches, thus a protective apparatus exists.

On the contrary, the municipality of Cusago, a 1,100 ha suburban-rural landscape in the plain 5 km W of Milan, (3,600 inhab; HH = 85.9 %), presents an SH equal to 2,624.7 m^2/inhab and a $SH_{PRT} = 511.3$. Therefore, it needs a protective area of at least:

$$\begin{aligned} PRT &= 511.3 \times 3,600 \\ &= 184.07\ \text{ha}\ [16.7\ \%\ \text{of its territory}]. \end{aligned}$$

Cusago is part of the Regional Agricultural Park South Milan, but only 2.8 % of its territory is covered by forest. Thus, Cusago is located within a protected area, but it has not sufficient resources (e.g. natural habitat, both in extension and level of BTC) to ensure an adequate landscape protective apparatus (PRT).

That is why we insist on the importance of realising that many preserved areas do not entirely reach a protective function, because a conspicuous part of them does not contain any bionomic resource or they are in bad ecological conditions.

4.1.2 The Importance of Interface Elements

Since the first chapter we have been underlying that the evolution is going towards an ever more increasing complexity. The recent (Holocene) large mutualism between natural systems and human population brought many types of human and semi-natural landscapes and enhanced the amount and the exchanges of information [8–10]. The actions directed to landscape transformation implied the emergence of rules of correspondence between the complex structure of natural landscapes and the formation of new anthropic structures. The high complexity of the landscape expresses relationships allowing the maintenance or the increase of a good level of metastability through processes that produce subsystems of "*interface* elements".

As underlined treating the anatomic components of the landscape (Sect. 2.4.3), the spatial patches, useful to human population, are shaped following an Euclidean geometry, while landscape structures are shaped in Fractal geometry. Therefore, this fact produces many interface fragments that only apparently have no importance. On the contrary, following the processes of ecotope reproduction, sensu Ingegnoli [2, 10] we can see that these remnant fragments are frequently useful as propagule banks sites and ecological memory. An example is given in Fig. 4.2, where a remnant patch of semi-natural vegetation resulted among a road, a school and a canal in Milan. Another good example of the significance of the emergence of systems of interface elements regards the formation of a garden near a house or of a park in a city. This process is mainly due to *protection* from disturbances and metastability *compensation* and sometimes it is not a completely conscious choice.

Landscapes present a modality of transformation led by ecological laws, which may cause in some way even a change in the culture and the ethology of man in order to maintain a metastable equilibrium when landscapes suffer a heavy changing pressure.

Actions of protection from disturbances and edge formation in boundaries of landscape elements result in buffer belts or in vegetated corridors. Fragmentation and patches configurations enhance this process of interface both in human and natural elements. The layers composed of vegetated biomass form the current boundary of an ecotope, but topographic and human barriers also play the same role. In this sense, human actions have had an evolutionary function in defining many boundaries once undetermined in natural ecotopes, therefore giving a higher functionality to landscape mosaics.

Through similar processes, the emergence of rules of correspondence between the complex structure of natural landscapes and the formation

Fig. 4.2 An example of interface patch in Milan, near the Olona canal

of new anthropic habitats leads to a complex land use which goes beyond the classical four land use categories:

(1) Residential (urban and services).
(2) Subsidiary (industry and transportation).
(3) Rural (agriculture and forestry).
(4) Conservation (natural systems).

Consequently, to these four systems must be added the (5) system of interface elements that presents crucial bio-semantic values, as emergent structures in a complex adaptive system like a landscape.

So, this more complex land use system finds its significance in the concept of landscape as a biological level—that is as a complex, self-organising, dynamic and dissipative system—and consequently in its structural model, which cannot be anymore an ecomosaic but must be an *ecotissue*—that is a multidimensional structure created by an intrinsic hierarchical integration of mosaics, attributes and information, spatially and temporally related [2, 10]. Thus, we can investigate about the *ecological role of each interface* within its proper landscape unit (LU), not only about its functions.

Gulinck et al. [11] gave some definitions of landscape interfaces, as land uses:

- *Hybrids* (H): any type of land use that can be seen as a deliberate "offspring" of two or more contrasting parent land use types.
- *Gardens* (G): any type of mainly unsealed tract of land, intimately linked to a built unit, or to an urban neighbourhood, and which has a range of functions of direct benefit to the occupants of these built or urban units.
- *Reaches* (R): in this context, zones of influence because of neighbour effects between two contrasting land uses. Reaches can be natural, such as coastal beaches and cliffs, or river banks, but they can also have a human origin or imprint.
- *Buffers* (B): strips or zones intentionally allocated and designed to mitigate the negative effects from one land use on a neighbouring land use or landscape element.
- *Commons* (C): any type of open space characterised by relatively unconstrained right of use by members of a community.
- *Guest* (U): any type of land in which an activity occupies the place of another land use type which is allocated to this place by rule or by convenience.
- *Fallows* (F): any tract of land that shows some sign of under-use, lack of management, temporary abandonment or severe degradation. In EU agricultural policy, it has been recuperated as a tool to restrict overproduction.
- *Residuals* (M): the margins of actively used land. Take the example of arable land.

Gulink notes that a complementary way of synthesis is to see the different types of interfaces not as independent, but rather as overlapping categories, in the sense of the "interface of interfaces".

4.2 Cybernetic and Control System

4.2.1 The Ecological Control of Human Culture

Again we insist to recall you that the actions directed to landscape change imply the emergence of rules of correspondence between the complex structure of landscapes and the formation of new structures. During these transformations, the high complexity of the landscape expresses relationships allowing the maintenance or the increase of a good level of metastability through cybernetic processes. We have seen that the configuration of human parcels is shaped in Euclidean geometry, while landscape structures are shaped in Fractal geometry. Therefore, this fact produces many interface fragments. These remnant fragments are useful, e.g. as propagule banks sites, ecological memory and refuge of natural species. Moreover, actions of disturbances protection and edge formation in boundaries of landscape elements result in buffer belts or in vegetated corridors.

Similarly, chains of interacting organisms and communities behave as information networks in an ecotissue, so that through them it is possible to maintain a certain level of metastability. The landscape is an information system essential for the processes of co-evolution and group selection, because the genetic characterisation is linked to three scale levels: cell, population and landscape.

Many animal populations in a new environment tend to lose genetic variability. For example, the rate of heterozygosis is only 7 % for vertebrates, 13 % for invertebrates, 17 % for plants [9]. Moreover, we note that the balance of evolution through anagenesis and cladogenesis is a process more related to environmental and genetic causes. In fact, the selection is conservative: each species, living in a certain environment, also explores marginal environments and thus the exploration of different biotopes favours polymorphism, mainly because of the competition.

That is why even human landscapes present a modality of transformation led by ecological laws, which may cause in some way even a change in the culture and the ethology of man in order to maintain a metastable equilibrium when landscapes suffer a heavy changing pressure. Keep in mind, in fact, that many phenomena of reorganisation, both in the design of the territory and in population movement, apparently decided by man, actually are controlled at the highest hierarchical level and man remains a mere executor of an unconscious ecological necessity (e.g. land abandonment, natural-shaped gardens, natural remnants among the fields, interface plots, etc.).

In summary, the beginning of the new era of the Holocene signified principally a change of the rate of information among biological systems, due to the expansion of human populations, both geographically and culturally. The rise of human ecosystems, especially agricultural ones, meant first of all a large-scale mutualism among species and a restructuring of entire landscapes and regions. A second phase was the passage from villages to cities. The presence of a membrane around the most important villages (walls) coincided with the regulation of the exchanges with their landscapes. It was the origin of a differentiated web of circulation, system of transmission and memorisation of information, control of a vast territory, long-distance commerce and specialised districts [9].

Consequently, the rise of landscape heterogeneity after the Holocene is incredible, and it is dependent just on information. Moreover, the higher degree of landscape organisation and the lowest instability of ecological systems are proportional [12] with the evolutionary indexes of human population, as D.T.I. (demographic transition index), which considers the growth rate and the longevity of populations.

We have to remember that an urban LU may form and develop only in dependence on other landscape units of its territory. When human perturbations do not reach a level of danger (e.g. rupture of cultural-natural control), we could affirm that, in general, the evolutionary strategy of biological systems is to develop human components to reinforce their cybernetic webs through land management.

Fig. 4.3 An example of the change in garden design during the transformation of the landscapes in Lombardy due to industry and monoculture. Villa Gallarati Scotti, Oreno village, near Milan (Italy). *Above*, the formal garden in the eighteenth century; *below*, a photo of the English park in nineteenth century. Note the island configuration of the park in an agricultural landscape without any more hedgerows and woods

Countries with higher proportions of urban civilisation show higher ecological metastability. The nature of the present man–environment crisis confirms this observation, being a development syndrome linked with cultural degradation. On the other hand, the ecological evaluation of the negative changing parameters of the environment is another man–nature information flux tending to ecological control through the condemnation of excess deforestation, CO_2 release, human population growth and industrial N fertiliser use.

4.2.2 The Buffer Effect of a Cultural Change on the Bionomic State of a LU

Let us examine, for example, the change in garden planning criteria during the industrial revolution. We may observe that, when rural landscapes were structured as "gardens" with remnant forest patches and a web of hedgerows in a very heterogeneous field matrix, the theory of garden design was strictly "formal", that is geometric, terraced and with leisure buildings (French or Italian gardening); but when rural landscapes, especially around towns, were changed by industries and agriculture began to increase monoculture, a new theory–that of English gardening- became dominant, following natural landscape criteria (Fig. 4.3). On the other hand, the development of human settlement systems in natural regions and landscapes may be seen as the creation of cybernetic webs applicable to territorial management.

Figure 4.3 shows the sharp difference between the wide formal garden of the Villa Gallarati-Scotti at Oreno (near Vimercate, 25 km N–E

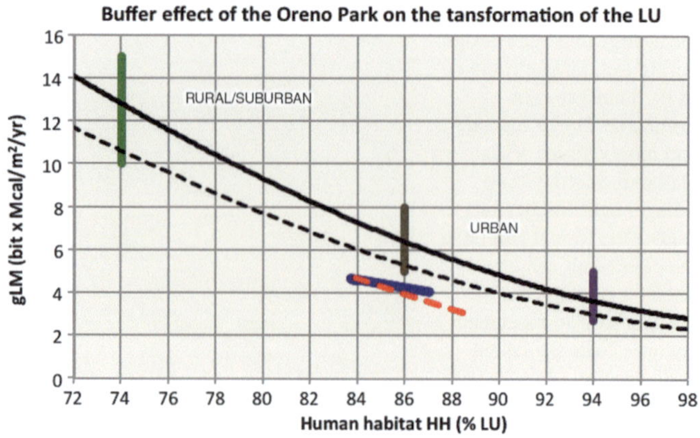

Fig. 4.4 The demonstration of the buffer effect due to the Oreno park (Fig. 4.3) on its LU is given by this plot, based on the g-LM. The *blue segment* represents the real transformation (1805–2005), the *red* one the change if the formal garden should not have been developed in a natural-shaped park

from Milan) in eighteenth century and the transformation of that garden in an English Park, since the nineteenth century. As already underlined, landscapes present a modality of transformation led by bionomic laws, which may cause a change in the culture and the ethology of man in order to maintain a metastable equilibrium, inducing a buffer effect, when landscapes suffer a heavy changing pressure. The cultural change of the shape of gardens during the Industrial Revolution followed this unconscious modality.

The demonstration of this buffer effect in dependence by a cultural change, apparently based on aesthetics, is today possible applying the principles of landscape bionomics. In this case study, we considered the LU of the village of Oreno in the period 1805–2005. It is a small LU, 140 ha, in which the productive (PRD) elements covered 87.02 % and the urbanised (RSD + SBS) 6.65 % in 1805, while the same components become respectively 53.9 and 39.7 % in 2005. The wide garden covers reaches 6.3 %. Thus, the change was very strong because the urbanisation has grown 600 %!

As plotted in Fig. 4.4, we can use the g-LM function (see Fig. 3.25) to value the difference between the real transformation of the LU (blue segment) and the potential change in absence of the cultural passage from formal to natural gardening (red segment). It is evident the small descent from 4.5 to 4.0 bit × Mcal/m^2/year in contrast with the potential descent from 4.5 to 3.0. This signifies a distance from the g-LM curve of −0.11 vs. −0.33. The LU is today only at the edge of the urban type of landscape, maintaining many aspects of rural-suburban, in the other case impossible.

4.3 Signs on Evolution

4.3.1 Limits of Neo-Darwinism

As all living entities, the landscape can participate to evolution process. This fact brought to the necessity to put in evidence the limits of Neo-Darwinism.

In the first chapter we enhanced that Konrad Lorenz [13] defined life as "a process of knowledge" and Karl Popper [14] wrote: "From the beginning, life must have been equipped with a general knowledge, which we usually name knowledge of the natural lows." This expression results very far from the current Darwinian definition of adaptation, which asserts that adaptation is the capacity of an organism to become more suited to an environment only through the process of natural selection. But without the mentioned a priori knowledge it is impossible for any eventually spontaneous form of life to survive.

Moreover, since the possible decrease of entropy generates order, applying a quantitative and reductionist conception of science, Neo-Darwinians claim that through this negentropic process, albeit gradually and very long, very complex levels of order can be achieved, as is the case of living systems. But this is another illusion.

The "principle of emergent properties" (see Sect. 1.3.3) demonstrates that each hierarchical level in natural systems acquires greater complexity and order than the previous one, and we must not forget that, in fact, there are different types and levels of "order", incomparable with each other. As again underlined in the first chapter (see Sect. 1.4), we must distinguish between the concept of *energetic order* and the concept of *information order*. For example, it is evident that information (incompressible), such as DNA, cannot be read through the output of energy alone. Since a processor may interpret messages only if it shares encoding, a functional protein must get along with its interpreter, hence it cannot be generated randomly.

Coming back to natural selection, we see that it is due to chance variations in the hereditary characters and is based therefore on the process of molecular copying. When a process of copying is repeated indefinitely, copying mistakes become inevitable and in a world of limited resources not all changes can be implemented, which means that a process of selection is bound to take place. That should be how natural selection came into existence.

But, as observed by Barbieri [15], this means that natural selection would be the sole mechanism of evolution if molecular copying were the sole basic mechanism of life. The discovery of the genetic code, however, has proved that there are at least two distinct molecular mechanisms at the basis of life, *copying* and *coding*. The discovery of other organic codes, furthermore, allows us to generalise this conclusion because it proves that coding is not limited to protein synthesis. Even the actions directed to landscape change imply the emergence of rules of correspondence between the complex structure of landscapes and the formation of new structures.

Copying and coding, in other words, are distinct processes even at molecular scale and this suggests that they give origin to two distinct processes of evolution, granted that an evolutionary process is the long-term result of a molecular one. More precisely, *copying* leads, in the long run, to natural selection and *coding* to natural conventions. Again, Barbieri notes that there are three major differences between copying and coding: (1) copying can only produce relative novelties whereas coding brings absolute novelties into existence; (2) copying acts on individual objects whereas coding acts on collective sets of objects and (3) copying is about biological *information* whereas coding is about biological *meaning*. Copying and coding are profoundly different processes of molecular change. Evolution, in short, took place—at least—by natural selection and by natural conventions.

All these limits are very tangibles, as indicated by the history of biology. For instance, the hypothesis "mutation and selection" has not been able to explain the arising of mitochondrion, found by Lynn Margulis as a process of symbiosis. Kurt Gödel, the most important mathematician of the past century, demonstrated that the evolution in geological times of a human body from the simplest forms of life, starting from a random distribution of elementary particles and of an energy field, is so improbable as the random separation of the atmosphere into its chemical components. Moreover, the Neo-Darwinian hypothesis is not able to explain the "Cambrian Explosion" of life. In less than 5 Ma (530–525 Myr B.P.), that is about 0.1 % since the beginning of life, not only the metazoan animals but all the animal phyla emerged, whose "body plans" never changed till now!

Another concept contrasting with Neo-Darwinism is the existence of the *irreducible complexity*: the impossibility to remove a part without eliminating the functions of a system. Complex behavioural innate programs of, e.g., some insects[3] are based on irreducible complexity, which cannot be explained through the Darwinian concept of adaptation. That is why Piattelli-Palmarini and Fodor [16] wrote that "if we insist that natural selection should be the only way to follow, will never arrive to a naturalistic explanation".

[3] For instance a wasp able to attract and then immobilise a coleopteran in order to nourish their future larvae....

Till more advanced was G. Sermonti [17], studying some proteins (e.g. prions) changing from a configuration to the other and the electro-magnetic forces able to modify protein regime conditions. This dynamic set of vector forces can be named "morphogenetic field" and, through it, the form of an organism is built. We may observe that those are not the genes that choose the nascent form, but is the nascent form which chooses the genes recruiting them into his program. Information and knowledge are the driving forces.

4.3.2 Environment, Landscape and Evolution

The criticisms of two major epistemologists and naturalists, Konrad Lorenz and Karl Popper [14, 18], are of great interest. They observed that concepts such as "struggle for existence" or "natural selection" are *metaphors*, not theories (talking about things that simply do not exist); then Darwin's hypothesis should be expressed in the form: "Individuals better adapted are more likely to have offspring." You may well notice more clearly the limits of Darwinism. It must assume, in fact, that there are individuals adapted, at least "some extent" adapted: and this is a problem that is linked to the origin of life, in the sense that the mere appearance of life still does not solve any problems.

The improbability of a meeting of life with an adapted environment has the same high improbability of the birth of life. In fact, the adaptation of life to its environment is based on reciprocity and is a kind of knowledge. But a certain stability of the environment is necessary in order to have adaptation and, thus, a certain knowledge: from the beginning life has to "know" the environment. Adaptation is a form of knowledge a priori, because life is essentially a cognitive process, as we remembered at the beginning of Sect. 4.3.1.

To understand this statement, let us remember what was suggested in the first chapter: life is manifested as self-organising hyper-complex dissipative system and is able to receive, store, process and transmit information, to reproduce, to belong to history, to pursue a project and be the main actor of evolution. Thus, life is not limited to an organism: it cannot exist without the environment, because it cannot do without the exchange of matter, energy and information between an organised entity and its specific environment. This exchange is so important that the emergence of life on Earth has radically altered the evolution of the entire planet, from the atmosphere to different types of rocks. We repeat: the adaptation of life to its environment is based on reciprocity and is a kind of knowledge. The concept of environment therefore has been misinterpreted in the study of evolution, because this reciprocity and this knowledge are not considered as appropriate. Returning to the organism, too, it's Fodor and Piattelli-Palmarini again to rightly note that all organisms that are neither extinct nor virtual shall ipso facto be adapted to the environment. The theory of evolution is not even necessary to explain the adaptation of the phenotype of an organism to its ecological niche. Since niches are identified in retrospect, in reference to phenotypes living there, if the organisms were not there, there would not be even niches.

The ecotopes are naturally "territorial niches" and this is very significant. A LU is, in fact, a complex system of biological organisation and is able to reproduce, through the renewal of its constituents ecotopes (see Chap. 3). This process is not identical to that of the level of organism, but is verifiable. Furthermore, chains of interacting organisms and communities behave as information networks in an ecotissue, so that through them it is possible to maintain a certain level of metastability.

Godron [9] points out that the growth of agro-ecosystems in the Holocene represents a new mutualism that produced the birth of landscapes with high circulation of information. It might even be said that the evolutionary strategy of biological systems has developed human components to strengthen its networks with cybernetic management and planning, providing, however, that man does not alter the relationship between nature and culture. Remind, in fact, that

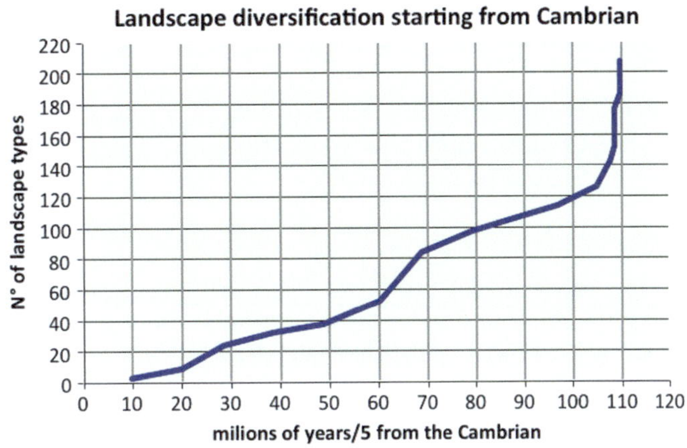

Fig. 4.5 The diversification of landscape types became during Cambrian Era and had a strong increase in the last 10 Ma

Fig. 4.6 Pictures of two primeval landscapes: Devonian (*left*) and Jurassic (*right*), from Wikipedia

many phenomena of reorganisation, both in the design of the territory and in population movement, apparently decided by man, are actually controlled at the highest hierarchical level, and man is at times a mere unconscious executor of an ecological necessity.

The landscapes are able to evolve, improving their level of metastability, are even able to control, within the limits of disturbances incorporation, the behaviour of populations and communities biologically evolved as human ones. The landscape is able to integrate the processes that may be defined as natural and cultural, because it represents the level of integration par excellence of natural and human components. The evolution of culture, that is the formation of a "noosphere", has increasing complexity: it is precisely the hyper-complex systems. All these phenomena cannot be hidden under a general evolutionary process seriously considered.

As you can see from the graph in Fig. 4.5, after referring to the 16 classes of landscapes that we mentioned previously (Sect. 2.5.2)[4] and after considering their variants in the three climatic zones (cold, temperate, warm) and taking account of 3–4 classes of BTC (bio-territorial capacity of the vegetation) and their average permanence (10^3 years B.P.), we were able to reasonably estimate the presence of landscape biodiversity in the exosphere with a logarithmic graph, starting from the Cambrian. It will be observed that there will be a sudden increase in landscape biodiversity starting by the appearance of man. A couple of types of landscapes of past ages (Devonian and Jurassic periods) are shown in Fig. 4.6.

[4] In order of time (i.e. of appearance on Earth): marine pelagic, marine coast, semi-aquatic, semi-desert, desert, paleo-forest, open neo-forest, closed neo-forest, grasslands, savannahs, shrubby savannahs, woodlands semi-anthropogenic, semi-agricultural, agricultural-rural, urban, suburban technological.

4.4 Landscape Transformation

4.4.1 Transformation Modalities of Landscape Systems, from Global to Local Scale

Before considering the transformation processes of the landscape, it is necessary to expose the dynamics of the main bionomic parameters from global scale (World and continental landscape systems, LS) to local scale (landscape unit, LU) in the last 130 years (about 1880–2010). We know that the ecological crisis manifested during the strong growth of crop production and technology, which allowed the population growth and the depletion of natural resources.

We'll take care of the transformation of the geo-biosphere landscape systems of the emerged lands (149,000,000 km^2), the continental LS of European Union—24 states[5]—(3,600,000 km^2), the regional LS of Lombardy (23,865 km^2) and the local LU of the suburban Low Brianza in the Seveso-Lambro plain near Milan (102.5 km^2).

The transformation of the emerged lands LS, from 1882 to 2010, is exposed in Table 4.2. The so called "primary forests"[6] registered the wider decrease, particularly from 1900 to 1962—the period of the World Wars: these formations lost 12.28 million km^2 in only two generations, a surface of about 40 times Italy or 54–55 ha/day, mainly in subtropical and tropical regions. Since 1962–2010 the destruction was less drastic, but it is of about 15–16 ha/day. Anyway, if we take into account all the world forested lands the transformation is more sustainable: during the first doubling of population (1.5–3.05 billion) from 1882 to 1962 the depletion of forested landscapes was of 13.28 million km^2 in 80 years (45.7 ha/day) while during the second doubling (3.05–6.97) the depletion of forests decreased to about 0.99 million km^2 in 48 years (5.6 ha/day).

In the same periods, arable crops increased 3.0 million km^2 in 80 years (1882–1962) and then only 0.2 million km^2 in 48 years. Remember, as already underlined, that since 1882 up today (130 years) the HH has been having small changes, from 21.9 to 26.8 %, while the population has been growing, passing from 1.5 to 6.97 billion: thus the increase of HH is only 122.4 % vs. the population increase of 464.7 %.

We must deduce that the landscape reorganisation represented the main ecological process of this period. The increase of urbanisation has been near the same of arable crops, about 3.5 million km^2 in 130 years, but it is the role of "attractor" of urban landscapes that is becoming important. The urbanised population, about 30 % in 1960, is today 51 % and it is continuing to grow. As underlined in the Sect. 4.2, this is another process guided by bionomic exigencies much more than economic.

All the transformations, from Global LS to local LU, are exposed in Fig. 4.7 to facilitate the comparison. The transformation of the continental landscape system LS of EU—24 states—is concerned with no more than 2.4 % of the previous LS. The dynamics result more constant even for forests, except the very recent increase as a reaction of land abandonment and consequent diminished cropland. But in Europe the HH reached a high value since the Roman Empire: up to nineteenth century the European HH was more than twice the World HH, only in the last 50 years the ratio passed from 2.16 to 1.87. Moreover, the area of grassland is still inferior of the area of cropland, the opposite of the global tendency.

If we compare the EU transformation with the one of Lombardy region, we can see more dynamics. The Lombard HH is still over the European one (+18.4 %), but it followed a similar low decreasing trend. The difference between grassland and cropland is more strong in Lombardy, with an enlarging tendency. Very impressive is the growth of urbanised landscapes in this region, which increased from 1 to 3 times the European average during 90 years.

[5] Statistical information of EU being found for the time do not consider the last four countries admitted. So, in this text, EU is always intended as EU 24.

[6] Remembering the concept of "the fittest vegetation for...," it should be better to avoid the term "primary forest," here used being the formal expression of FAO World statistics.

Table 4.2 Transformation of the main landscape systems, World scale, in the period 1880–2010

World land systems	1882 (million km^2)	%	1.900 (million km^2)	%	1962 (million km^2)	%	1990 (million km^2)	%	2010 (million km^2)	%
Primary forest	30.00	20.13	28.60	19.19	16.32	10.95	16.00	10.74	13.58	9.11
Other forests	24.60	16.51	25.00	16.78	25.00	16.78	25.68	17.23	26.75	17.95
Forest	54.60	36.64	53.60	35.97	41.32	27.73	41.68	27.97	40.33	27.07
Arable and crops	11.60	7.79	12.50	8.39	14.62	9.81	14.42	9.68	14.80	9.93
Meadows and pastures	25.00	16.78	24.90	16.71	24.50	16.44	25.50	17.11	26.00	17.45
Agricultural	36.60	24.56	37.40	25.10	39.12	26.26	39.92	26.79	40.80	27.38
Savannah	15.50	10.40	15.00	10.07	14.80	9.93	14.00	9.40	13.50	9.06
Other lands	16.00	10.74	16.95	11.38	26.24	17.61	26.45	17.75	26.57	17.83
Other lands	31.50	21.14	31.95	21.45	41.04	27.54	40.45	27.15	40.07	26.89
Urbanised land	0.90	0.60	0.95	0.64	1.66	1.11	3.10	2.08	4.40	2.95
Extreme desert	8.00	5.37	8.00	5.37	8.20	5.50	8.15	5.47	8.20	5.50
Ice lands	16.50	11.07	16.35	10.97	16.00	10.74	15.80	10.60	15.50	10.40
Deserts	24.50	16.44	24.35	16.34	24.20	16.24	23.95	16.07	23.70	15.91
Human habitat HH (%)	21.9		22.4		25.0		25.8		26.8	
BTC (Mcal/m^2/year)	2.95		2.89		2.24		2.20		2.11	
Population (billion inhab.)	1.5		1.65		3.05		5.26		6.97	
SH*(m^2/inhab.)	3,150		2,985		2,240		1,700		1,330	
SH (m^2/inhab.)	21,853		20,228		12,213		7,308		5,729	
Carrying capacity SH/SH*	6.87		6.81		5.45		4.30		4.31	

Data from: Goldsmith and Allen [34] for 1882, Buringh and Dudal [35] for 1900; FAO Reports for 1962, 1990, 2006 [3, 4]

This is due to the population growth in Lombardy, passed from 3.55 to 10.05 million inhabitants in the last 130 years, a population size more than that of Austria, the same of Hungary or Portugal. Anyway, the capacity of this regional landscape system to incorporate disturbances derived from population is—until today—indubitable, as attested by the near constant BTC. In fact, the forest landscapes increased 1.4 times starting from the end of the First World War, while population increased 4.6 times.

It could be interesting to compare the dynamics of the different landscape systems even evaluating their carrying capacity ($\sigma = SH/SH^*$), as shown in Fig. 4.8. At European Union scale the $\sigma = 2.5$ values are near constant and over the threshold of $\sigma = 1.6$, that is over the limit of an agricultural landscape. At regional scale—in the case of Lombardy, one of the more overcrowded territory in Europe—σ was inferior to 1.6 since 1880, passing from 1.43 to 1 in the first 80 years, then oscillating between 0.95 and 1.0, just a bit over the threshold of suburban (and heterotrophic) landscape ($\sigma = 0.8$). At local scale, the LU of Low Brianza presented a $\sigma < 0.8$ since 1880, which arrived near the threshold of a dense urban landscape after 1990. The dynamic at global scale is different, presenting a jump from the homeostatic plateau (about $\sigma = 7.0$) to the new plateau (about $\sigma = 4.2$) in only 100 years.

Anyway, even if we assume a security coefficient of 2 (considering the limits due to cropland), we will see that the carrying capacity $4.2 \times 0.5 = 2.1$: it means that the World population should (theoretically) reach the number of about 14–15 billions before to destroy the ecosphere. But pay attention: the UN forecast indicate 10 billion inhabitants in 2080, thus the time to prevent irreversible damages is very short!

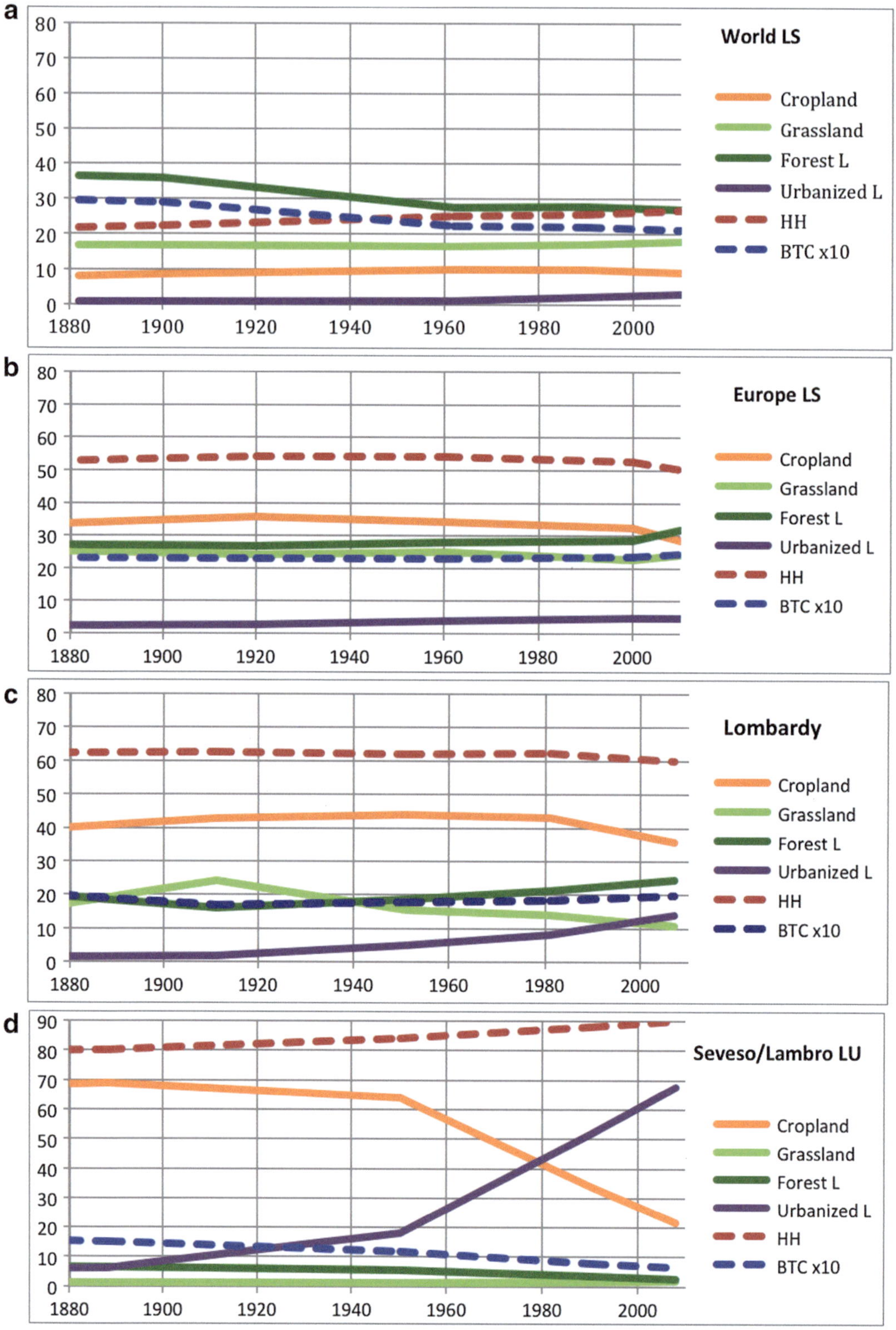

Fig. 4.7 Dynamics of the main landscape elements and parameters at four spatial scales: (*from above*) (**a**) world L systems (LS), (**b**) Continental LS (EU24), (**c**) Regional LS (Lombardy), (**d**) local LU (suburban Brianza).

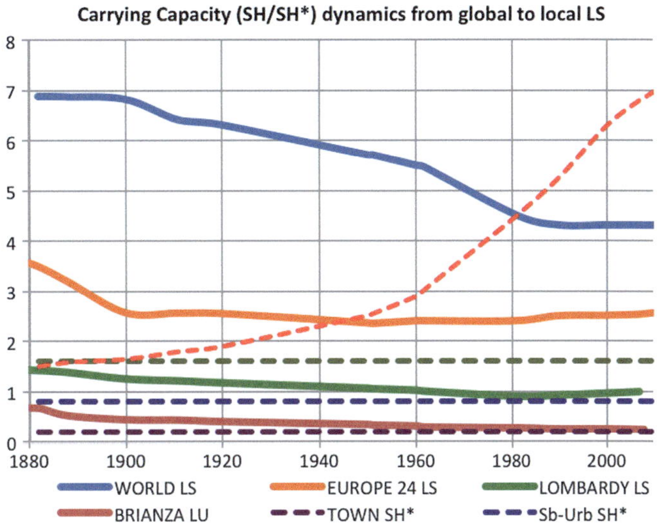

Fig. 4.8 The transformation of the carrying capacity (SH/SH*) of landscape systems, from global to local scale. Note that the population growth (*dotted red curve*) induced an evident change at global scale (emerged LS) after the increase of 1960. The green dotted line indicates the Agricultural SH*

4.4.2 The Transformation Processes

In a landscape or in its subsystems (i.e. Landscape Units, LU) the principal transformation processes depend on the hierarchical structuring of an ecological system and its non-equilibrium dynamics, metastability, coevolution, evolutionary changes and ecological reproduction. Let us review the main steps, essential even to revise some basic concepts of vegetation science:

- *Hierarchical structuring*. The behaviour of an ecological system is limited by: (a) the potential behaviour of its components on the lower level of scale, (b) the environmental constraints on the upper level of scale. This set of conditions represents the existence field within which the system of ecosystems must reside.
- *Non-equilibrium dynamic*. Thermodynamic bonds may determine an attractor, in its proper existence field, that represents a condition of minimum external energy dissipation. Possible macro-fluctuations produce instabilities, which move the system toward a new organisational state. These new states permit an increase of dissipation and move the system toward new thresholds to reach a new attractor. This could be represented as a cybernetic process of "order through fluctuation".
- *Metastability*. An ecological system can remain within a limited set of conditions, but it may show alterations if these conditions change. The system may cross a critical threshold, approaching even radical changes. For example, different types of landscapes or their parts may be correlated with diverse levels of metastability.
- *Coevolution*. The history of the interactions among the elements of a landscape in a given area shows a particular dominion that is characterised by the coherence of their reciprocal adaptation. This process leads to the stabilisation of different homeostatic and homeorhetic capacities of a landscape, which may be expressed with a particular degree of metastability of the entire system.
- *Evolutionary changes*. The structuring of every biological system may be pursued, that is the information may be transmitted, only if the final state of the considered system is less unstable (i.e. more metastable) than its initial state. The modalities by which these processes are realised may be different and not limited to a single scale.
- *Reproductive processes*. Each level of life organisation presents typical reproductive processes: (a) system available to maintain information, (b) mutation phase, (c) protection of new elements, (d) selection phase, (e) crucial disturbance eliminating the old structure. Following previous points and

ranked processes, each level of life have to renew: note that both assembly rules and dispersal filters need also a context.

Let us focus on ecological succession, which, in general ecology, is the most important process related to transformation. Classical theory assert that through serial stages, an ecosystem changes in a predictable way toward a final stage, called climax. After an outside perturbation (or partial substitution of inner components), succession returns the ecosystem to the climax. For instance, an abandoned field near a forested patch is recolonised from the forest edge (i.e. the mantle) and, in a given time, after the regrowth of shrubs and then of trees, the succession restores the climax. Succession is a concept of primary importance in ecological theory, it has become the basis for dynamical explanations of many ecological phenomena, such as in phytosociological sygmeta. But this kind of succession is incompatible with the scientific principles underlined in the first chapter.

It should be sufficient to remember the non-equilibrium thermodynamic with branching points after the instability threshold, or the concepts of landscape metastability. In the first case, the history becomes the leading criterion of transformation. In the second, it is evident that, even when a succession to a climax may be considered valid at a single ecocoenotope scale, certainly it is not valid at landscape scale.

Succession does not work as linear and mechanistic; according to Pignatti [19], in the vegetational phytocoenosis of *Cytisus villosus* which follows after a fire of a *Viburno-Quercetum ilicis* patch, for instance in central Italy, or the recolonisation of *Picea abies* on abandoned alpine pastures in Central Europe, two cases in which normally succession is present, if more than one key factor becomes dominant, the ecological system and its transformation become unpredictable.

Remember that self-organising processes have to be considered at least on three scales: the one of interest, the upper (constraint) one and the lower (significance). If some components of an autocatalytic set have been excluded, the system will appear as linear. It is what happens to the classical theory of succession, because, e.g. the landscape is never considered as a basic parameter. Therefore, in landscape ecology the importance of ecological succession as linear and divided into primary and secondary phases is drastically reduced.

It is possible to visualise the regional thresholds of landscape replacement in two coordinated fields, monitoring changes in anthropic and natural sets of landscape types, measured on the same scale of BTC (Fig. 4.9).

The possibility of being able to analyse large-scale changes allows us to study the past of a landscape, to control the present state and to guide future management. Note that the BTC can be measured also by estimation procedures. As we may observe in Fig. 4.9,

- In the HH the transformation modalities generally follow a cluster of parabolic functions crossing a series of thresholds, from semi-natural protective type of agricultural landscape to the most urbanised one. The mosaic sequence remains that expressed by Richard Forman. Even if the opposite is theoretically possible, in human landscapes these transformations tend to be unidirectional.
- In the Natural Habitat the transformation modalities are more complex. An ecological succession from a near desert type of landscape to a high BTC mature forest type is theoretically possible, but certainly it is not the main modality and does not follow a straight line. The role and the range of disturbances generally lead the way to change.

4.5 Landscape Pathology

4.5.1 Landscape Alteration

The first signs of ecological alteration in a landscape are expressed at the ecocoenotope level and are generally correlated with the most sensitive populations. If there is a sufficient redundancy, other ecosystems may occupy the landscape degraded niches, but when the alteration influences the entire ecotissue level, then a degradation process is full in action.

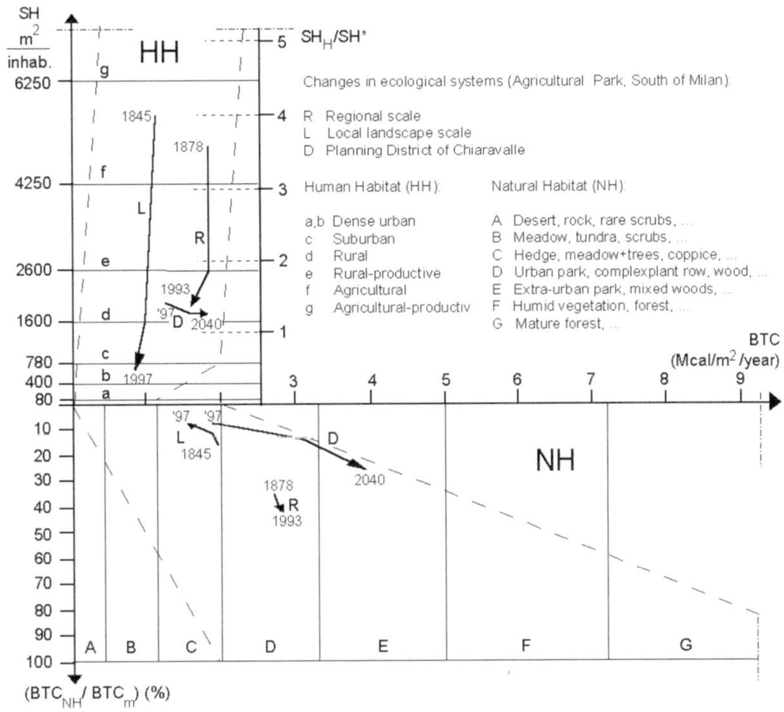

Fig. 4.9 Transformation modalities of a landscape: model of control based on the BTC function. We can see three spatial levels of change: R regional scale (Lombardy), L local landscape unit scale (suburban landscape of South Milan), D district of Chiaravalle Abbey; *Dotted lines* define the more frequent movement field of the system. Note the different directions of District *arrows* vs. L and R ones within the HH field and of all the *three arrows* within NH

Note that the worst degradation occurs in a landscape when the greater part of its dynamic is frozen, that is, when its evolutionary processes have been stopped. This happens when (1) the metastability level of the upper scale systems is no more able to incorporate the disturbances of their subsystems and/or (2) when the biological potentiality of the lower scale components is destroyed, and/or (3) when permanent alterations are caused to the main structures and functions of the landscape.

Many researches demonstrated that a landscape crosses critical thresholds [20, 21] in correspondence of which its ecological processes present dramatic qualitative changes: for example, rapid changes in the number and the length of ecotonal edges near a critical threshold and the influence of these modifications on the behaviour of many species, or a drastic change of the BTC_W (i.e. weighted average BTC) indicating a change in the landscape from a natural, semi-natural or agricultural to a suburban or urban one.

Let us consider four levels of scale and two series of hierarchically correlated processes for each scale (Table 4.3):

If the geomorphic equilibrium (erosion-sedimentation) is altered, for instance G4 is exalted and linked with B2, we will note a lack of incorporation in the third level, thus the production of an "out of scale" disturbance, which may grow as a dangerous perturbation. In cases like these, even more complex and involving many structures in a landscape, an out of scale disturbance may disrupt the organisation of the system of ecocoenotopes; processes of slow frequency, linked with a high degree of landscape organisation, will tend to disappear. Therefore, the components of the system will begin to operate in the fastest way. When the upper scale constraints are lost, the hierarchic organisation of the landscape falls to pieces.

If landscape disturbances are incorporated at a regional scale, the BTC_W of the region remains almost constant during a very long period of time, even under strong landscape changes. A good example is the landscape transformations verified in Lombardy from 1878 to 2008, where urban landscapes increased from 1.5 to 14.0 %, the agricultural landscapes increased from 39.8 to 44.0 % (1951) then decreased to 35.9 %, but the main BTC

Table 4.3 Example of hierarchically correlated processes in a landscape

Correlated processes (biological and geomorphic)		Scales
B1	Transformation of forested ecotopes	Landscape unit
G1	Soil formation within the ecotopes	(1st level)
B2	Growth of arboreal vegetation complex	Ecotope
G2	Soil pedon processes	(2nd level)
B3	Single vegetation association processes	Tessera
G3	Humus formation	(3rd level)
B4	Growth of the micro-biota within the litter	Dot
G4	Soil micro-grain transportation	(4th level)

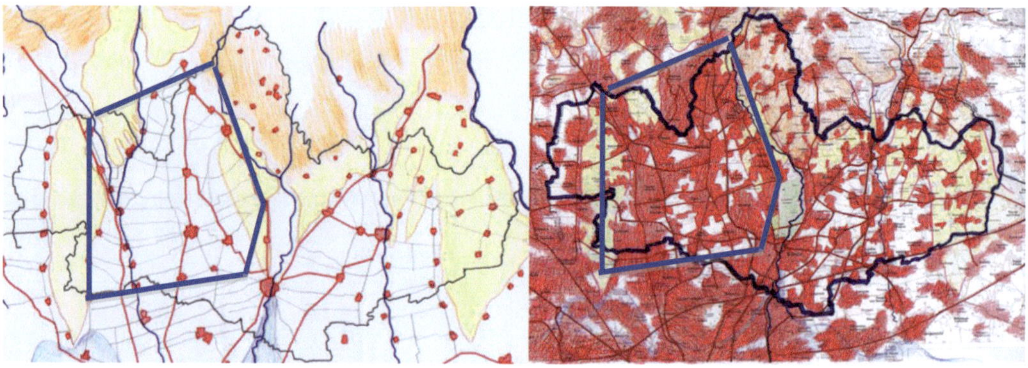

Fig. 4.10 The landscape unit (*blue*) of the Seveso-Lambro Plain, about 10–20 km North of Milan, in 1782 (*left*) and 2010. The *red colour* indicates the urban elements; the *yellow tongues* are glacial remnants from the pre-Alpine Mountains. The urbanisation was very aggressive

remained almost constant, about 1.95–2.05 Mcal/m^2/year (Fig. 4.7c). Anyway, as we have seen in Fig. 4.7d, at LU scale the transformation can be very disruptive as shown in Fig. 4.10.

In this case study of the Seveso-Lambro LU we passed from a very exploited rural-agricultural landscape to an urban one in about two centuries. Figure 4.7d and the data in Table 4.4 indicate that the process has been having an acceleration since 1950, but the ratio σ = SH/SH* affirmed that this LU was heterotrophic since 1850. This is the main reason of the dynamic of changes, which express a g-LM continuously low as plotted in Fig. 4.11 (blue dotted).

Anyway, if we consider the g-LM values in 1841 and in 2008, the first one, even if lower than the normality curve, remains within the tolerance range, thus being liable to rehabilitation; the second one results to be out of this range of tolerance, even if apparently not too much: is this altered situations recovering or not?

To check this fact it is necessary to introduce the concept of "potential of bionomic rehabilitation" (PoBR). This ecological index evaluates the changing potential of each landscape element, supplying the theoretical capacity of re-naturalisation of a LU. This potential depends on the structure and functions of the landscape[7] and it was synthetised in this case study as: PoBR = 0.15 (cropland), 0.50 (grassland), 0.20 (fruit cultivations), 0.05 (urban areas), 1.00 (forest). Therefore, PoBR decreased from 21.2 % in 1841 to 11.0 % in 2008. The g-LM* (i.e. the bionomic rebalance potentiality) is thus proportional to these values:

[7] Few L. elements may return in natural condition: e.g. only 10 to 20% of the cropland or 2 to 8% of the urban areas, while 95 to 100% of forests can. So the potential of bionomic rehabilitation must be evaluated after a good analysis of the possibilities of each type of L. element in a given LU.

Table 4.4 Main bionomic parameters characterising the transformation of the landscape unit (LU) of the Plain Seveso-Lambro (1841–2008). Note the decrease of the Potential of Bionomic Rehabilitation (PoBR)

Seveso/Lambro LU (102.5 km^2)	1841	PoBR-41	1888	PoBR-88	1950	PoBR-50	2008	PoBR-08
Population (10^3 inhab.)	34.0		66.7		138.6		277.3	
Cropland (% of LU)	66.70	10.05	69.10	10.37	64.30	9.65	21.79	3.27
Grassland (% of LU)	1.50	0.75	1.48	0.74	1.28	0.64	2.19	1.1
Other cultivation (% LU)	15.90	3.98	13.40	3.35	7.90	1.98	2.80	0.70
Forest L (% of LU)	6.95	6.95	6.57	6.57	5.72	5.72	2.70	2.70
Urbanised L (% of LU)	5.76	0.29	6.34	0.32	18.35	0.92	67.70	3.39
PoBR (% of LU)		21.17		20.67		22.11		11.00
1 + (PoBR/100)		1.21		1.20		1.19		1.11
g-LM (bit/Mcal/m^2/year)		8.67		7.82		6.09		3.08
g-LM* (bit/Mcal/m^2/year)		9.98		8.96		6.86		3.25
HH (% LU)	79.28		80.06		83.94		90.27	
BTC (Mcal/m^2/year)	1.67		1.55		1.20		0.65	
SH (m^2/inhabitant)	2,448.4		1,260.3		635.9		340.8	
SH* (m^2/inhabitant)	2,842		2,569		1,860		1,418	
SH/SH* (=σ)	0.86		0.49		0.34		0.24	
SHprt (m^2/inhabitant)	438.1		236.9		141.7		92.7	
Theor. PRT as % of LU)	14.9		15.8		19.6		25.6	

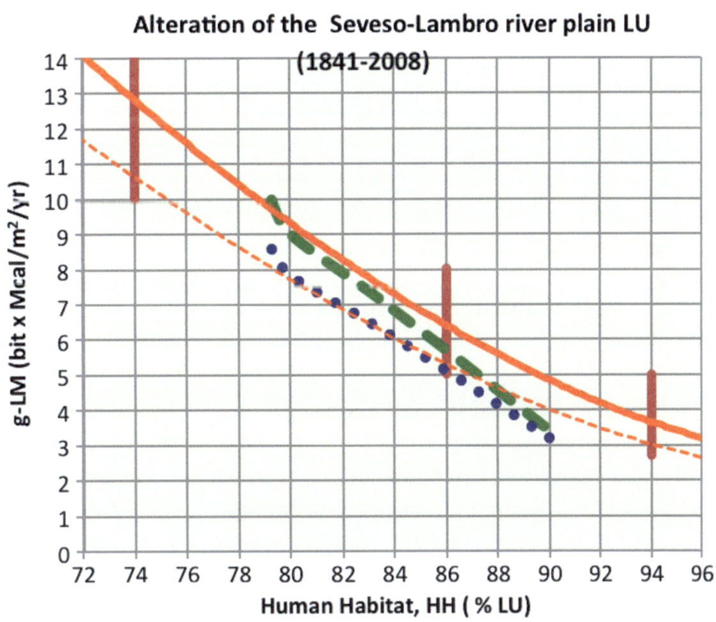

Fig. 4.11 If we measure g-LM, the transformation of this LU (*dotted blue*) seems to be proportional to the normal g-LM curve (*orange continuous*), even if lower (out of the tolerance range, *orange dotted*). What underlines the alteration is the loose of the potential of bionomic development (PoBD) expressed in g-LM* (*green broken line*)

$$g - LM^{*} = (1 + PoBD/100) \times 0.95\ g - LM.$$

As plotted in Fig. 4.11, the transformation of the LU follows the normal g-LM curve (orange), at a short distance from the tolerance range (orange dotted). So the difference between the present ecological state and the state in 1888 seems to be not dramatic. But the green line, expressing the g-LM*, shows the true situation: in 1888 g-LM = 7.82, and potentially 8.96 (just on the line of normality); in 2008 g-LM = 3.08 and potentially 3.25 (only 81% of the normality). A complete potential of rehabilitation results impossible. A true alteration is in act.

Table 4.5 Pathogenic scheme of the agrarian industrialisation syndrome of temperate agricultural landscape of the plains

Original state	**Traditional agricultural landscape "a Bocage"**	
Permanence: 3–18 centuries	(BTC = 1.2–2.1 Mcal/m^2/year; HS/HS* = 2.5–6.0; HH = 50–75 %) heterogeneity, connectivity and circuitry = good	
Main human cause of alteration	Socio-economic pressure due to growing agrarian production	Increase of help for agrarian technologies
	Specialisation of cultivations	Canalisation of small rivers
	Increase of arable land and of chemical fertilisers	Cutting of tree lines and hedgerows, mechanical irrigation
Positive feedback processes	Destruction of forest patches,	
	Increase of crop pests,	
	Increase of chemical pesticides,	
	Decrease of fauna,	
	Soil depletion	
	Enlarging field area, increase road network	Rupture of geomorphologic constraints,
Altered State	**Open mono-cultural landscape**	
Permanence: 20–80 years	(BTC = 0.9–1.3 Mcal/m^2/year; HS/HS* = 4–9; HH = 70–85 %) heterogeneity, connectivity and circuitry = weak, partial	
	Structural weakness,	
	Increase system fragility,	
	Attraction for highways,	
	Attraction for industrial areas,	
	Increase of fragmentation	
Disordered State	**Suburban-rural landscape**	
permanence?	(BTC = 0.7–1.1 Mcal/m^2/year; HS/HS* = 0.8–2.7; HH = 80–90 %) heterogeneity = increasing; connectivity and circuitry = disrupted	
	Loose of functionality,	
	Decrease of agrarian production	

4.5.2 Pathological Symptoms

Some considerations on landscape pathology. The definition of landscape as a specific level of life organisation becomes a challenge for environmental evaluation, first of all because man has to pass from a discipline related to technology, economy, sociology, urban design, visual perception *and* ecology, to another related to biology, natural sciences, medicine *and* traditional disciplines. Consequently, as we underlined in Chap. 1, analysis, evaluation (and intervention or planning) of the environment require changed methodologies: from engineering, economical and aesthetical rearrangement, to biological diagnosis and therapy.

Let us remember that the study of the pathology of any biological system, independently from the levels of scale and of the organisation, needs a basic diagnostic methodology, which cannot be avoided. This is true also for landscape dysfunction, and it may be articulated in six phases:

1. Survey of the symptoms.
2. Identification of the principal causes.
3. Analysis of the reactions to pathogen stimuli.
4. Risks of ulterior worsening.
5. Choice of therapeutic directions.
6. Control of the interventions.

Like in medicine, environmental evaluation needs comparisons with "normal" patterns of behaviour of a system of ecosystems. Therefore, the main problem becomes how to know the normal state of an ecological system, and/or, at the same time, how to know the levels of alteration of that system.

In medicine it is the physiology/pathology ratio which permits a clinical diagnosis of an individual, here is the same ratio which permits a clinical diagnosis of an ecocoenotope or a landscape.

For example, dysfunctional landscapes will have less patchiness than normal ones, and any remaining patches will have lower concentrations of soil nutrients, lower water infiltration rates, lower levels of biological activity and lower production cycles as sustained by Tongway and Ludwig [22]. We may add [10]: higher transformation deficit, decreasing BTC, natural habitat loss, incorrect ratio between HH and NH, decreasing correlation between heterogeneity and information, higher fragmentation, loss of connectivity, incongruent landscape apparatuses, higher landscape resistance, etc. Nevertheless, it is sometimes very difficult to perform a correct diagnosis, because some pathologies have "low" symptoms, or apparently not alarming ones.

A forested landscape, for instance, may present a high biomass volume and quite high BTC_W, but it may have a too low reproductive rate, too few patches far from the dominant state and a too homogeneous structure of the landscape main mosaic thus presenting a hazardous senescence [23]. On the other hand, we have just seen the huge capacity of disturbance incorporation on a regional scale, which seems to be reassuring.

The identification of the main causes producing landscape syndromes needs a good knowledge of anatomy and physiology of the complex system of ecocoenotopes, of their pathologic disturbances and also of a good anamnesis. Semeiotic must be added, in which even perceptive studies may help. Moreover, remember that—as physicians—the etiopathogenesis of a syndrome is mainly due to interpretation. An example may be done by the case of agrarian industrialisation (Table 4.5) in Central European Regions.

4.5.3 Landscape Syndromes

A more complete framework of clinical diagnosis of the landscapes is needed. Even if a true classification of the main syndromes has not been elaborated yet, we think that the principal types of landscape dysfunctions can be articulated in six categories, each one of these divided into subcategories and all related to a particular type of landscape.

The landscape types can be summarised at least in six main classes: high BTC natural (hN); natural (N); semi-natural (sN); agricultural (Ag); sub-urban (sU); urban. A schematic framework may be given, remembering structural and dynamic components (Table 4.6).

A sort of synthetic check-up list for each main landscape syndrome category should be useful in diagnostic evaluation. The present list does not pretend to be an exhaustive work, but it could be

Table 4.6 Main landscape syndromes categories and sub-categories

Main landscape syndrome categories	Sub-categories of syndromes
A—Structural alterations	A1—Landscape element anomalies
	A2—Spatial configuration problems
	A3—Functional configuration problems
	A4—Multiple structural degradation
B—Functional alterations	B1—Geobiological alterations
	B2—Structurally dependent dysfunctions
	B3—Delimitation problems
	B4—Movement and flux dysfunctions
	B5—Information anomalies
	B6—Reproduction problems
	B7—Multiple dysfunctions
C—Transformation syndromes	C1—Stability problems
	C2—Changing process dysfunctions
	C3—Anomalies in transformation modalities
	C4—Complex transformation syndrome
D—Catastrophic perturbations	D1—Natural disasters
	D2—Human-made destruction
E—Pollution degradations	E1—Direct pollution
	E2—Indirect pollution
F—Complex multiple syndromes	F1—Acute
	F2—Chronic

good to have an idea of the variety of problems in studying landscape pathologies and to underline possible critical phenomena.

A. Structural Alterations

A1—Landscape Element Anomalies.

- Lack of ecocoenotope structure formation within a tessera, e.g. vertical strata of the vegetation.
- Too geometrical form of landscape elements, e.g. rectangular shape of agricultural fields.
- Abnormal grain of the tesserae or the ecotopes, e.g. too large, in comparison with the same of nearby landscape units.
- Lack of correspondence of form-function in the structure of the elements, e.g. fields contrasting with the water flow direction.

A2—Spatial Configuration Problems.

- Low heterogeneity of tesserae and ecotopes, e.g. type and form.
- Excess of road density, e.g. >600–800 m/km^2 (see Sect. 3.5.3).
- Lack of patchiness among landscape elements, e.g. when all the delimitations of L. elements are fuzzy.
- Lack or excess of grain contrast in the main ecological mosaic, e.g. industrial zoning in a urban or forest landscape.
- Sharp difference in matrix porosity of vegetation patches, e.g. disposition of wood patches or hedgerows.

A3—Functional Configuration Problems.

- Excess or insufficiency of network density, e.g. fast decrease of vegetation corridors network.
- Presence of some anomalous type of landscape apparatus, e.g. introduction of RSD L. apparatus in forest natural landscape.
- Lack of natural habitat, e.g. NH <10–15 % in PRD landscapes.
- Lack or excess of ecotissue subsystems, e.g. some PRT apparatus in urban landscape.

A4—Multiple Structural Degradation.

- Loss of congruence between geomorphologic traits and ecotissue structure (even with different levels of scale), e.g. destruction of halophiles vegetation and canal rectification in a lagoon landscape.
- Chaotic structure, e.g. difficulty in detecting landscape units or patches of ecotopes.
- Structural subsystems not compatible with the functional role of the land unit within its landscape, e.g. presence of technical networks and industrial tesserae in agricultural-protective landscapes.

B. Functional Alterations

B1—Geobiological Alterations.

- Predominance of morphogenesis, particularly excessive, abnormal instability, e.g. mountain slopes subject to landslides.
- Excessive, abnormal of soil degradation processes, e.g. lose of humus layer.
- Insufficient or excessive drainage network, e.g. patches of aridity or turned into swamp.
- Erosions and alterations by extractive areas, e.g. digs opening in conservation areas.
- Over-bridling of rivers, e.g. with barriers to the salmon swimming up.

B2—Structurally Dependent Dysfunctions.

- Excessive, abnormal fragmentation of landscape components, e.g. remnant patches without core areas.
- Excessive, abnormal landscape resistance to key species, e.g. barriers to badgers and foxes movement in agricultural landscapes.
- Irregular presence/absence of corridors and connections, e.g. network gaps.
- Negative variation of influence fields among ecotopes or landscape units, e.g. limitation of influence fields into a protected area.
- Problems in strategic points of a landscape matrix, e.g. alteration in a node area among different ecotopes.
- Critical thresholds of landscape habitat per capita (in HH or NH), e.g. insufficient SH in an agriculture landscape.

B3—Delimitation Problems.

- Alteration in the boundary formation, e.g. rectification of the margins of a forest patch.
- Alteration of the ecotonal web and/or ecotonal margins, e.g. gradient simplification.
- Loss of functional characters of landscape boundaries, e.g. decrease of the porosity of a boundary.

- Too high boundary crossing frequency, e.g. density of fence in agricultural landscape.

B4—Movement and Flux Dysfunctions.

- Irregular operations in the main inter-ecosystemic fluxes, e.g. barriers to the movement of catabolites in water networks.
- Presence of perturbation in the advection process, e.g. disturbances to seed movement due to industrial emissions.
- Flood anomalies and obstructions, e.g. buildings near the fluvial bed.
- Cleaning capacity loss in the excretory apparatus, e.g. destruction of riparian reeds.
- Not enough input or output of species and matter in ecotopes or landscape units, e.g. destruction of patches available for metapopulations.
- Invasion of exotic species, e.g. introduction of allocthonous species near natural elements or protective areas.
- Limitations to the inter-regional seasonal movement of species, e.g. barrier in main corridors.
- Deviation of migratory flux of species, e.g. change in attractor habitats near towns.
- Too high interference by human transport systems, e.g. highway opening in a residential area.

B5—Information Anomalies.

- Loss of genetic variability, e.g. due to decrease of landscape heterogeneity.
- Loss of natural signs for orientation of moving populations, e.g. due to repeated similar elements in suburban landscapes.
- Divergence between human culture and ecological laws, e.g. managing criteria not following ecological needs of peculiar landscapes.
- Lack of any network for landscape monitoring, e.g. pollution sensors in urban landscapes.
- Degradation of landscape capacity of order accumulation, e.g. too simplified structure of urban parks.

B6—Reproduction Problems.

- Anomalies in reproduction processes of the landscape components, e.g. propagule banks reduction.
- Obstacles to zero event cyclical occurrence, e.g. too drastic burning reduction in fire adapted ecosystems.
- Absence of any senescence phases in the ecotopes and tesserae, e.g. forest management without maturity patches.
- Interferences and obstacles to recolonisation processes, e.g. biotopes enclosed in urban landscapes.

B7—Multiple Dysfunctions.

- Spreading out of biotic pathologies into the landscape, e.g. forest tesserae altered by insects.
  Lack of congruence among inter-scale processes, Cfr. Table 4.3.
- Presence of not-incorporable range of disturbances (out-of-scale), e.g. important road and rail networks in forest landscape.
- Insufficient potential of bionomic rehabilitation (PoBR), Cfr. Fig. 4.11.

C. Transformation Syndromes

C1—Stability Problems.

- Break in homeorhetic cycles of landscape maintenance, e.g. in a semi-natural landscape, if seasonal grazing is stopped.
- Lack of potentially changeable ecotopes, e.g. in a suburban landscape, when no element is available to be converted for bionomic rehabilitation.
- Too homogeneous strategy of balance maintenance of each landscape element, e.g. if only human cutting and harvesting is the maintenance force for an entire landscape.
- Lack of high metastability components, e.g. when no natural patch and corridor can reach the maturity level (especially in forest landscape).

C2—Changing Process Dysfunctions.

- Sudden loss of heterogeneity in the main ecological mosaic, e.g. when the most part of interface elements are eliminated.
- Break of connectivity networks in the landscape, e.g. if removed corridors are not substituted by other one.

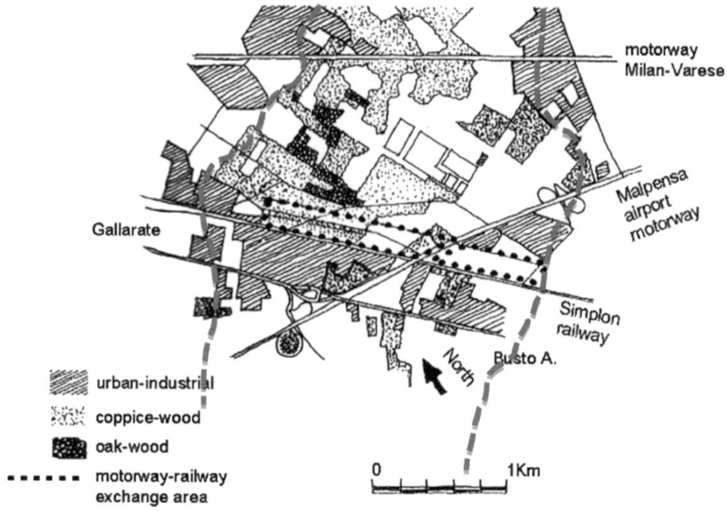

Fig. 4.12 A wide corridor (about 2 km) limited by the two green dotted lines between Gallarate and Busto Arsizio, officially protected by the Ticino River Regional Park, is altered with a barrier of urban-industrial buildings which follow the railway Milan-Simplon-Paris

- Too accelerated perforation, fragmentation, shrinkage processes, e.g. human activities in building a suburban landscape.
- Impossibility to follow the optimum ecological sequence of transformation, e.g. actions of landscape change without mitigation and compensation.

C3—Anomalies in Transformation Modalities.

- Remnants of transformation deficit in the ecotopes, e.g. for insufficient actions of mitigation and compensation.
- Break in coevolution processes of landscape elements, e.g. when the contrast of land use patches is too elevated.
- No gradual change through landscape replacement thresholds, e.g. a direct change from forest to industrial areas.
- Excess of transformation frequency of the main elements, e.g. in an agricultural landscape, if only the cropland elements are built up.
- Sudden excess of human technology plants or buildings, (e.g. Fig. 4.12).
- Anomalies in correlation between BTC and plant biomass change, e.g. when wooded patches are planted with allocthonous species.
- Land abandonment with unbalanced distribution of human population, e.g. cropland and grassland strong reduction decrease the SH.

C4—Complex Transformation Syndrome.

- Drastic change of typical levels of metastability, e.g. g-LM sudden decrease.
- Improper role of the ecotissue vs. the upper scale landscape/region, e.g. a LU under protection (regional park) showing a BTC below the regional average.
- Drastic or unbalanced changes in the landscape apparatuses, e.g. when landscape functions are eliminated without any compensation.
- Loss of historical references and trends (natural or human), e.g. if a landscape is renewed without any remnant of the previous structure.
- Drastic change in the surrounding landscape, e.g. a LU acquiring new landscape functions due to the drastic change of its surrounding landscapes.

4.5.4 Perception of Landscape Pathologies

According to environmental history, the first observations on landscape destruction, in the nineteenth century, aroused only an aesthetically based reaction. It was probably a consequence derived from the German *Naturphilosophie*, well expressed by von Humboldt in his famous books, such as *Kosmos* [24], in which aesthetical

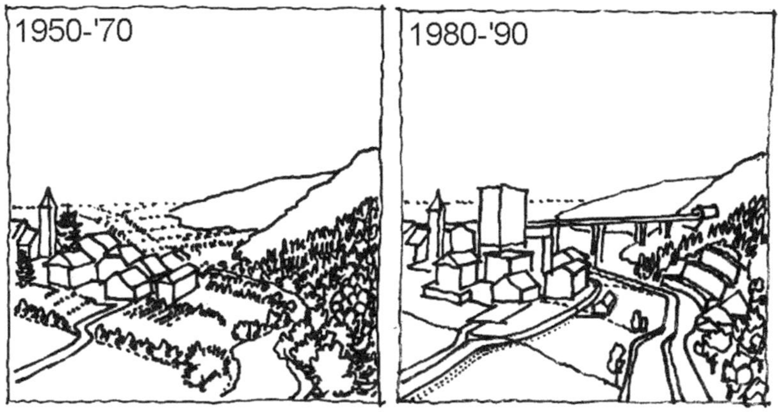

Fig. 4.13 Perception of the landscape alterations in a typical pre-Alpine village and its LU. The sketches help to underline the semiotic information

considerations were mixed with scientific knowledge. The vision of the aspect of wild natural systems and the knowledge of the laws of nature were said to produce joy. Even if the American ambassador in Italy, George Perkins Marsh, noted, in 1864, that man can be considered in many cases as a disturbing agent regarding natural equilibrium, especially the landscape, a scientific notion of landscape degradation is quite recent.

Note the environmental impact statement (EIS) that emerged in 1969, in the U.S.A. It tried to value and prevent the destruction of the environment in a more scientific way, but the evaluation of the landscape was again mainly based on visual analysis or it was limited to just few components. For example, deforestation is considered negative, causing landslides or the loss of a rare bird species: but we know that a deforested ecotope is a basic metastable landscape element. On the other hand, there is no doubt that visual perception remains very important, as we can see in Fig. 4.13.

It represents a typical European village within its territory (for instance, a pre-Alpine locality in northern Italy, France, Switzerland, southern Germany, Austria): the first view is from the years 1950–1970, the second 30 years after (1980–2000). The first drawing represents a particular landscape condition, the stability of which, excluding small technological changes, has increased over many centuries, until it was destroyed by war. The coevolution with nature had created a well-managed landscape, whose equilibrium is visually immediately perceivable. In the second drawing the same view gives a completely different perception. In a single generation or less, the same territory dramatically changed. The landscape was altered in many components, from the forest and the river, to the fields and the village. The presence of a highway gives the sensation of an out of scale disturbance, as does the new tower built in the centre of the old village.

The visual disfigurement of the landscape moves, in every civilised country, the Association for Monument and Heritage Protection (like the famous British National Trust), as well as the Botanical or Zoological Conservation Associations if rare species are threatened to protest. But nowadays this may be not sufficient. If we remember the landscape ecological principles, we need something more!

Visual perception is very important, and it has to do with aesthetics, as architects know. The etymology of aesthetic, following Aristotle, is "what is endowed with sense". Therefore, the visual perception should be very useful also to any ecologist. But this affirmation needs some precise statements.

An ecologist cannot be an aesthete: even if beautiful per se exists in nature [13], the ecologist has always to find the sense connected with the aesthetic information. In medicine this fact is well known and the formation of a "clinical eye" has to pass through the medical semiotic [25], that is the capacity to find the sense of what observed, analysing the information derived from the signs expressed by a biological system.

This system is the human body, but it can be also the landscape. Semiotics are necessary to find out all the symptoms and to address the research: analysis and diagnosis.

It is useful to remember that the character of a complex system is given by a specific set of dominant and rare signs, both important in finding the sense of the structure. Moreover, an altered state of a system is not always perceived through its clear disfigurement. This kind of degradation may not be visible, but the syndrome can be serious. The alteration may be perceived by noticing a peculiar lack of information in comparison with what was expected, or noticing the beginning of disorder where not expected.

The alteration of an ecological system could even be discovered by the observation of its uncommonly beautiful aspect, not compatible with the structure and the functions of the examined ecotissue: for instance, the emergence of luxuriant patches of herbs in a prairie landscape, in reality due to the degradation by overgrazing. The transformation of a forest LU into a golf course may be another case (difficult to limit because people living in the neighbourhood look only to aesthete aesthetics).

4.6 Landscape Pathology and Human Health

4.6.1 Health Damages cannot be Limited to Landscape Pollution

The relations between landscape pollution and human health are well known. Landscape pollution is known to produce toxins, chemical substances that cause any of a wide number of adverse effects in living organisms. The effects may be acute and/or chronic, including changes in living tissues physiology, mutations and consequently cancer [26]. Figure 4.14 shows the distribution of air pollutants in Europe, from a recent study of ESA [26]. Ruhr, Benelux, North Italy (Po valley) and England present the most polluted landscapes. The impact of air pollution was estimated in 22 months detracted to local life expectancy and in €31.5 billion of sanitary and social costs!

Anyway, the possibility to reduce many air pollutants is concrete [27] and we reached a good success, e.g. for calcium and sulphate ion emissions in both EU and USA (Fig. 4.14) where in 10 years the decrease reached about 2/3 of the previous level.

But we know that landscape pathologies are not limited to pollution (Table 4.6): the most important landscape syndromes derive from structural and functional disorders. Therefore, it is necessary to check the eventuality that even these landscape pathologies should be dangerous for human health. It is a very important question, because any scientific demonstration of these threatening linkage may change, deeply, our responsibility and our actions to protect our health.

Moreover, the recreational factors of the environment changed in the last decades, being today generally far from residential areas and expending to reach. Many environmental components are today altered at wide scale (e.g. an entire LU) and an alarming stress condition is more diffuse, often in an unconscious way. For these reasons, the spontaneous rebalance of the stress has become more difficult and many illnesses are growing.

Oncological epidemiology indicates that in metropolitan suburbs LU the illness frequency is higher than in country LU. Suburbia present generally many landscape pathologies, but it is difficult to distinguish between the prominent role of pollution and the environmental stress due to landscape structural dysfunctions. Insufficient epidemiologic research seems to be done in this direction. So, psychic disorders may be more indicative: for instance, major depression disorder (MDD) in metropolitan LU is known to be higher than in country LU [28] and its frequency is near twice in suburban LU.

In the aetiology of coronary syndrome it has been demonstrated [29] that, when the sympathetic response exceeds the limits of control, it may generate a sequence of episodes like: medial neuritis post-atherosclerotic, spasms, regional asynergy, hypokynesia and increase of ventricular pressure, extra-vas compression, hematic block and necrosis infraction, hart-brain-endocardial reflex interaction, catecholamine myonecrosis of the infraction/sudden death,

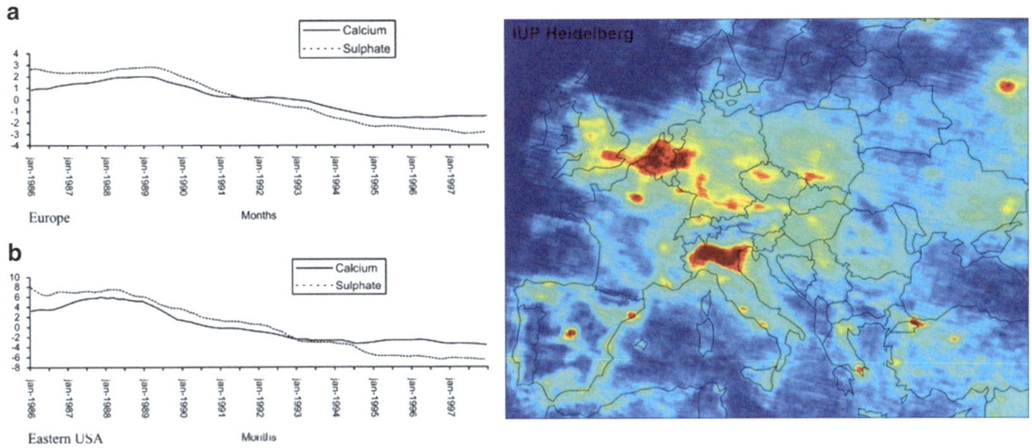

Fig. 4.14 The reduction of calcium and sulphate emissions in EU and USA (*left*, from Gimeno et al. [27]) and the distribution of air pollutants in Europe (form ESA-IUP Heidelberg [32])

arrythmogen hypersensitivity, ventricular fibrillation/cardiac arrest. This sequence is very important because it demonstrates another way to cardiac infraction besides the classic thrombogenetic one. The catecholamine myotoxicity has been found at histological level with high frequency (78 %).

From the above exposed aetiology, it results that a sympathetic response may exceed the capacity of control in dependence of an environmental stress. Even if (for the moment) it is impossible to quantify how much of this stress should be due to landscape pathologies, it is also impossible to exclude this component. No doubt that the main origin of the stress is due to the working environment, but the role of the natural environment remains crucial both in re-balancing or in the aggravation. Consequently, if the adrenergic stress may be partially caused by a landscape structural dysfunction, the bionomic rehabilitation of our habitat reaches a new level of primary importance.

4.6.2 Main Processes linking the Organism to the State of the Environment

On the other side, we know that any living organism needs to reach the best environmental fitness and is predisposed to do it. The genotype brings to the best phenotype possible in a given environment. That is why mammals at the moment of birth do not present a complete brain, but their neural structure requires a period during which the organism interacts with the environment. In humans this period is particularly important, characterising all the young phases till the adult age, about 20 years!

The help to medical research given by magnetic resonance imaging (MRI) is able to produce accurate pictures of brain anatomy and physiology. This fact has opened a new window into the biology of the brain, e.g. how its tissues function and how particular mental or physical activities change blood flow. Because the MRI does not use ionising radiation, it is well suited for paediatric studies (Fig. 4.15) launching a new era of neuroscience.

Neuroimaging research in adolescents indicates [30] at least three processes: (a) Brain cells, their connections and receptors for chemical messengers called neurotransmitters peak during childhood, then decline in adolescence; (b) Connectivity among brain regions increases; (c) The balance among frontal (executive-control) and limbic (emotional) systems changes. The neuronal connections grow during the first 5 years, as to register environmental inputs whose redundancy is reduced in more structured information and coding through a "pruning effect", probably guided by feasible interactions.

Being inevitable that landscape structure and functions represent the environment in which

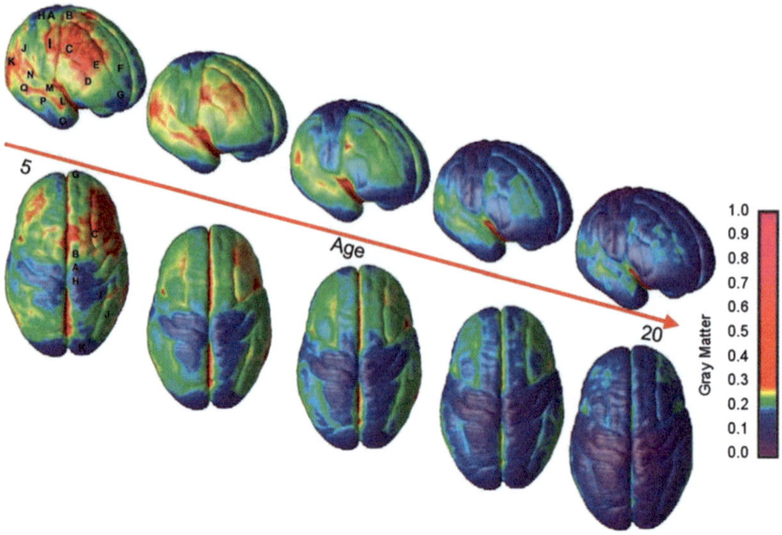

Fig. 4.15 During adolescence, different parts of the brain mature at different rates. Neural connections *increase* after the birth to about 5 years. This picture indicates an average *decrease* in "grey matter" volumes between ages 5 and 20, thanks to the "pruning" of neural connections. Areas that mediate "executive functioning" mature later than areas responsible for basic functions (from Gogtay et al. [30])

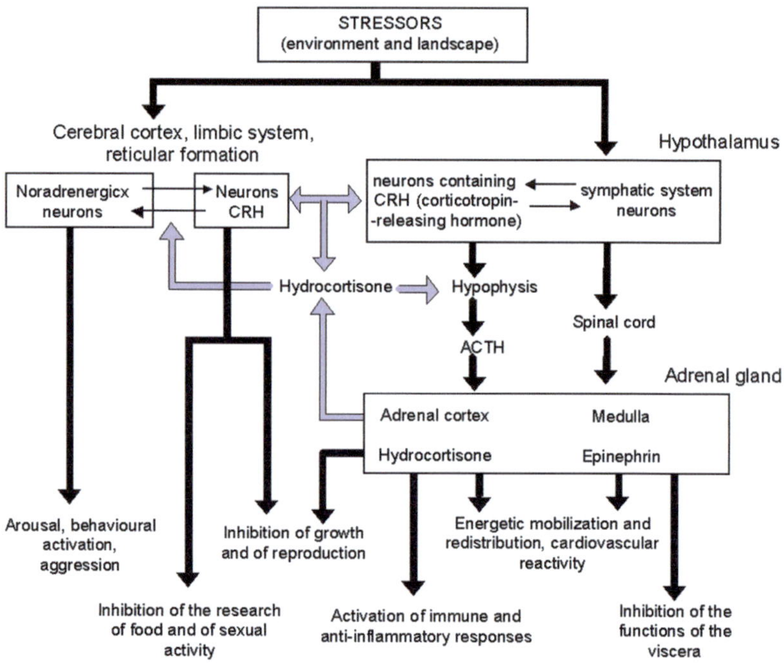

Fig. 4.16 The sympathetic nervous system and the hypothalamus-pituitary-adrenal axis mediate the integrate answers of the organism to stress. The negative feedback acted by cortisol (hydrocortisone) functions to limit too strong reactions, which can be dangerous for the organism (from Berne and Levy [31] modified). Many of these stressors are due to landscape structural dysfunctions, even in absence of pollution

man has to live, no doubt that a wide part of the input modelling the brain concerns landscape conditions. We have to ask what happens if these conditions should be altered. Anyway, besides the neuronal growth and pruning, our organism is related with the environment even through a system of stress alarm (Fig. 4.16), described in Physiology since 1992 by Berne and Levy [31].

As shown in Fig. 4.16, the human body presents very efficient ways to stress alarm, both via the sympathetic nervous system and the hypothalamus–pituitary–adrenal axis. The negative feedback acted by hydrocortisone functions to limit too strong reactions, which can be very dangerous for the organism. It is inevitable that many of these stressors are due to landscape structural dysfunctions, even in

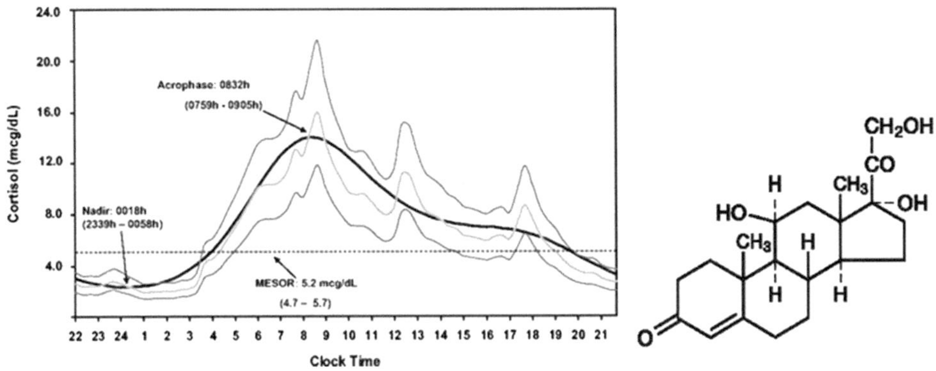

Fig. 4.17 Physiological cortisol circadian rhythm [33]. Cortisol has a distinct circadian rhythm with a peak of 15.5 μg/dl (95 % reference range 11.7–20.6) occurring at 08.30 h and a nadir less than 2.0 μg/dl (95 % reference range 1.5–2.5) at 00.15–00.30 h. To the right the chemical formula of hydrocortisone

Table 4.7 Main health damages and their linkages with environmental alterations

Syndromes	Reasons	Environmental origins
Adrenergic hyperactivity	Environmental alarm	Perception of the structural and functional disorders of the landscape;
Ischemic heart diseases	Non-compensated environmental stress	Chronicity of ecological alterations;
Immune system anomalies	Hypothalamus-adrenal gland disturbances due to environmental stress	Lack of natural recreational elements; Too high ecological density (i.e. small SH) Environmental alarms
Allergies	Environmental maladjustment	Disaccustom to natural vegetation; Pollution
Pulmonary syndromes	Meso- and micro-climatic alterations	Urban heat island; Lack of protective vegetation; Too high ecological density (i.e. small SH) Pollution
Oncological syndromes	Immune disorders and carcinogens due to environmental dysfunctions	Environmental and food pollutions; Stress due to landscape dysfunctions, reducing the immune system
Intestine dysfunctions	Environmental maladjustments and insufficient movement	Mechanical transports and too simplified pathways.
Irascibility and aggression	Environmental psychic and behavioural constraints	Landscape structural degradation; Lack of landscape heterogeneity;
Sexual and food disorders	Environmental dysfunctions	Too high ecological density (i.e. small SH) High chronic noise

absence of pollution. Remember that cortisol follows a circadian rhythm (Fig. 4.17).

4.6.3 Health Damage due to Landscape Pathology

Note that the environmental stress alters and enhances the circadian rhythm of cortisol bringing negative consequences on inhibitory effects exerted by this hormone on the excess of catecholamine. Among these dangerous consequences we have to underline the limit of the efficiency of immune and anti-inflammatory systems. Consequently, the adrenergic stress may be dangerous both to ischemic and oncologic syndromes. In Table 4.7 are ranked the main syndromes potentially linked to environmental

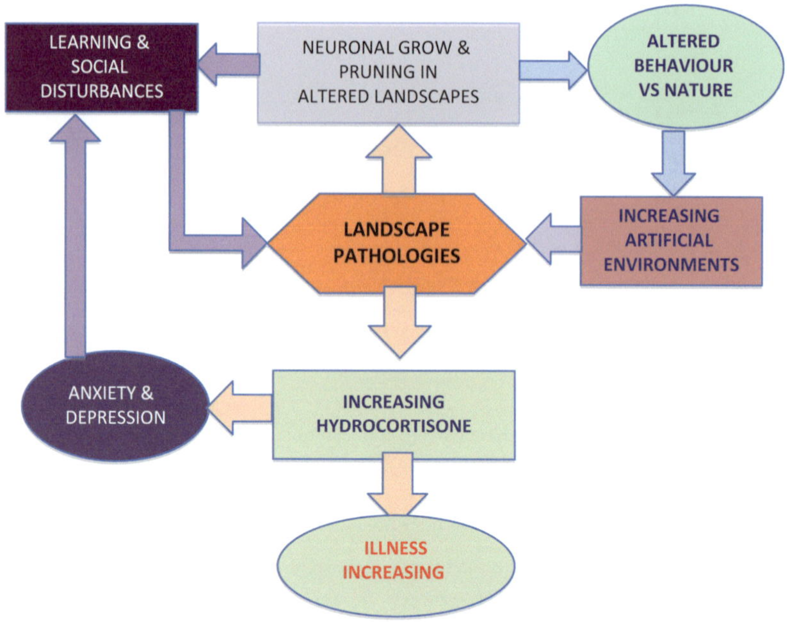

Fig. 4.18 Flow diagram on the main consequences of landscape pathologies. Note the formation of two feedback loops which can aggravate the landscape syndromes, therefore the consequences on illness

stress, derived by landscape structural dysfunctions and pollution.

Let us observe that the two processes we exposed before, the "growth and pruning neural synapsis"[8] and the "environmental stress alarm", have to be obviously integrated in relation to landscape pathologies and human health. In Fig. 4.18 a flow diagram tries to explain their main linkages.

If neuronal growth and pruning process operates in altered landscapes it can result in an altered behaviour vs. Nature, so that bringing to increase artificial environments and this intensifies landscape pathologies. A typical positive feedback loop. But another behavioural feedback loop appears when an increase in hydrocortisone brings to increase arousal and aggression, then learning and social disturbances lead to act against natural lows, thus degrading ecological components and producing a new intensification of landscape pathologies. Chronically elevated cortisol levels have been linked to problems including abdominal fat gain, cognitive decline, and compromised immune function. Note that learning and social disorders are reinforced by the altered neuronal growth and pruning.

The increase of illness depending from landscape pathologies is theoretically available to reach levels of danger, especially because the capacity of natural rebalancing of natural ecotopes is strongly reduced. A first preliminary demonstration that the increase of the mortality rate is linked to the landscape alterations can be suggested in Fig. 4.19, concerning a research of Ingegnoli on the Monza-Brianza province, near Milan. These 55 municipalities were divided in nine groups, from agricultural to dense urban landscape types and plotted in the HH/BTC model. Four of these landscape types does not follow the curve of normality presenting heavy structural dysfunctions. Relating the mortality rate (M-r %) with HH, we can note that the increase before the tolerance limit (HH from 65–81) results +5.4%, while the M-r increase in the second interval (HH = 81–97) is +16.7%. This exceeding increase in M-r is due to bionomic

[8] Pierre Vanderhaeghen and HJ Cheng [36] note that axon pruning and neuronal cell death constitute two major regressive events that enable the establishment of fully mature brain architecture and connectivity. Although the cellular mechanisms for these two events are thought to be distinct, recent evidence has indicated the direct involvement of axon guidance molecules, including semaphorins, netrins, and ephrins, in controlling both processes.

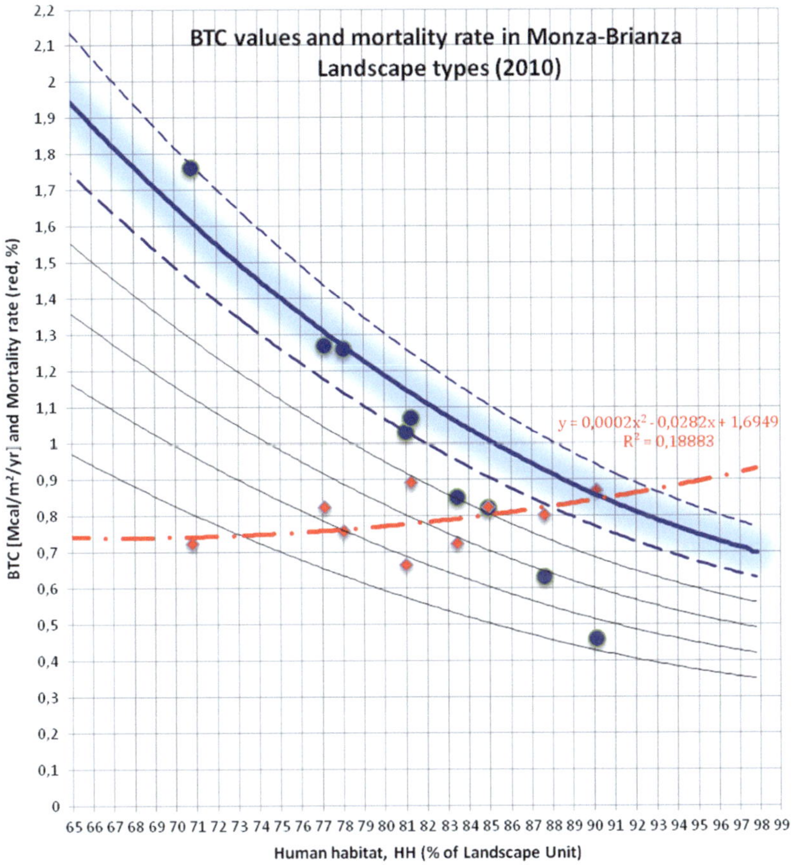

Fig. 4.19 The 55 municipalities of the Monza province were divided in 9 landscape types (from agricultural to dense-urban) and plotted in the HH/BTC model (*blue*)(see Fig. 7.9). Only four-five of them can be considered in a normal bionomic state (*dotted curves*), the other resulting altered, gradually from the *blue curve*. Their mortality rate (*red*) increases 0.74–0.78 before the tolerance limit, but from 0.78 to 0.91 after this threshold (same interval)

alterations, an evident correlation between human health and landscape structural degradation. Deeper investigation is following.

The strategic environmental assessment must be obliged to enclose health evaluation even in absence of pollution, in the direction of bionomic rehabilitation. The figure of the "*ecoiatra*" [2] must be seriously considered.

References

1. Sinclair J, Hanks P, Fox G, Moon R, Stock P (1988) Collins cobuild English language dictionary. Collins, London
2. Ingegnoli V (2011) Bionomia del paesaggio. Springer, Milano
3. Marklund LG, Batello C (2008) FAO datasets on land use change, agriculture and forestry and their applicability for national greenhouse gas reporting. FAO, Helsinki, Finland
4. FAO (2010) Global forest resource assessment: Main report. FAO
5. Eurostat (2013) The EU in the world, EU Commission. doi 10.2785/35119
6. Eurostat (2013) Eurostat regional yearbook, EU Commission, DOI 10.2785/44451
7. BIP (Biodiversity Indicator partnerships) (2010) Coverage of protected areas. NEP, WCMC
8. Ingegnoli V (1980) Ecologia e progettazione. CUSL, Milano
9. Godron M (1984) Ecologie de la végétation terrestre. Masson, Paris
10. Ingegnoli V (2002) Landscape ecology: a widening foundation. Springer, Berlin, p 356
11. Gulinck H, Marcheggiani E, Lerouge F, Dewaelheyns V (2013) The landscape of interfaces: painting outside

the lines. In: UNISCAPE conference Landscape and Imagination, Paris, 2–4 May 2013
12. Ingegnoli V (1986) Considerazioni sul rapporto fra transizione demografica e crisi ecologica. In: Ecologia e Longevità, Atti Conv. Naz. di Ecologia Umana, Firenze, Boll. SIEU, pp 53–63
13. Lorenz K (1978) Vergleichende Verhaltensforschung: Grundlagen der Ethologie. Springer, Berlin
14. Popper KR (1994) Alles Leben ist Problemlösen. Über Erkenntnis, Geschichte und Politik. R Piper & Co, München
15. Barbieri M (1985) The semantic theory of evolution. Harwood Academic, London, p 188
16. Piatteli-Palmarini M, Fodor J (2010) Gli errori di Darwin. Feltrinelli, Milano
17. Sermonti G (2005) Why is a fly not a horse? Seattle. Tr. It. Dimenticare Darwin: Perché la mosca non è un cavallo? Il Cerchio Iniziative editoriali, Rimini
18. Popper K (1989) Intervista alla RAI. In: Enciclopedia multimediale delle scienze filosofiche
19. Pignatti S (1996) Some notes on complexity in vegetation. J Veg Sci 7:7–12
20. Turner MG, Gardner RH (eds) (1991) Quantitative methods in landscape ecology: the analysis and interpretation of landscape heterogeneity. Springer, New York
21. Ingegnoli V (1991) Human influences in landscape change: thresholds of metastability. In: Ravera O (ed) Terrestrial and aquatic ecosystems: perturbation and recovery. Ellis Horwood, Chichester, pp 303–309
22. Tongway DJ, Ludwig JA (1997) Nature of landscape dysfunction in rangelands. In: Ludwig J, Tongway D, Freudenberger D, Noble J, Hodgkinson K (eds) Landscape Ecology: function and management. CSIRO, Australia, pp 35–48
23. Ingegnoli V, Aquila C, Padoa-Schioppa, E (1995) Rapporto preliminare sullo studio dell'ecomosaico forestale del Gariglone. Atti 5° Workshop "Progetto Strategico Clima Ambiente e Territorio nel Mezzogiorno" Amalfi 1993, Tomo I, Guerrini editore, C.N.R., Roma, pp 397–419
24. Von Humboldt A (1845) Kosmos. Entwurf einer physischen Weltbeschreibung. Cotta, Stuttgard und Tubingen
25. Lorenz K (1973) Die Rückseite des Spiegels. Versuch einer Naturgeshichte menschlichen Erkennens. R Piper & Co, München
26. Kampa M, Castanas E (2008) Human health effects of air pollution. Environ Pollut 151(2):362–367
27. Gimeno L, Marin E, del Teso T, Bourhim S (2001) How effective has been the reduction of SO2 emissions on the effect of acid rain on ecosystems? Sci Total Environ 275(1–3):63–70
28. Kessler RC, Chiu WT, Walters EE et al (2005) Prevalence, severity, and comorbidity of twelve-month DSM-IV disorders in the national comorbidity survey replication (NCS-R). Arch Gen Psychiatry 62 (6):617–627
29. Baroldi G (2003) Patologia cardiovascolare: ruolo nell'iter diagnostico-terapeutico. Edizioni Primula, Pisa
30. Gogtay N, Giedd JN, Lusk L, Hayashi KM, Greenstein D, Vaituzis AC, Nugent TF (2009) Dynamic mapping of human cortical development during childhood through early adulthood. Proc Natl Acad Sci USA 101(21):8174–8179
31. Berne RM, Levy MN (1990) Principles of physiology. The CV Mosby, St. Louis
32. ESA (2012) Reorganisation of the SCIAMACHY Level 2 data set at D-PAC, Earth Online
33. Grossman AB (2010) The diagnosis and management of central hypoadrenalism. J Clin Endocrinol Metab. doi:10.1210/jc.2010-0982
34. Goldsmith E, Allen R (1972) La morte ecologica. Laterza, Bari
35. Buringh P, Dudal R (1987) Agricultural land use in space and time. In: Wolman MG, Fourier FGA (eds) Land transformation in agriculture. SCOPE; Wiley, Chichester, pp 9–43
36. Vanderhaeghen P, Cheng HJ (2010) Guidance molecules in axon pruning and cell death. Cold Spring Harb Perspect Biol 2(6):1–18

Landscape Bionomic Analysis of Vegetation

5

5.1 Renewing Vegetation Science

5.1.1 The Reasons of the Renew

In the late nineteenth and early twentieth century, science has begun to present crucial changes that have been putting in crisis many scientists. Physics has lost his mechanistic confidence, due to the progression of the physics of sub-atomic particles (Quantum Mechanics), his faith in a time and space as "absolute", because of the Theory of Relativity and in the reversibility of processes, because of studies on the thermodynamics (e.g. Boltzmann). Moreover, the positivist belief that the data of experience can result in scientific knowledge per se was decidedly disproved. In fact, as Einstein proved [1, 2], this view is wrong and can lead to a catalogue, not a theory.

As other fields of ecology, vegetation studies carried out by Braun-Blanquet were undoubtedly useful and important, but they failed to keep pace with the upheavals of scientific paradigms. For these reasons, studies of phytosociology, sensu Braun-Blanquet [3], have obvious advantages with regard to the description (cataloguing) of the vegetation, but also safe limitations, being based on a positivist epistemology and on ecological concepts today surpassed.

Let's take a quick overview of the main limitations of phytosociology (Naveh and Lieberman [4]; Pignatti, Box and Fujiwara [5]; Ingegnoli [6, 7]): (a) reference to a concept of naturalness that excludes humans in any event; (b) dynamics based on the concept of ecological succession mainly understood as linear and deterministic; (c) reference to an "ecological space" that does not consider the principle of emergent properties (Fig. 5.1); (d) use of the concept of "potential vegetation" that does not consider the role of disturbances in ecological systems; (e) ignorance of complex system and scale-dependent functions, claiming to be able to study the landscape with the deterministic approach of the concept of "sygmetum" and "geo-sygmetum" by Tüxen [8] and Rivas-Martinez [9].

To all this we must add that: (f) the level of organisation to which we refer is different: the community/ecosystem in one case, the landscape units in the other [10]; (g) there are limits to the capacity of species to be "environmental indicator", due to genomic redundancy [11]. Moreover, (h) the current method of estimating the ecological distance between potential and actual vegetation assumes that the optimal landscape was homogeneous, implying that the set of all the components of a vegetation series has reached the climax stadium, which is contrary to all the principles of landscape bionomics and to the concept of biodiversity and metastability.

Following phytosociology, some basic questions in the study of the landscape remain, therefore, without answer. It is worthwhile to mention some of these issues: (1) how to consider the contribution of a vegetated tessera to the

V. Ingegnoli, *Landscape Bionomics Biological-Integrated Landscape Ecology*,
DOI 10.1007/978-88-470-5226-0_5, © Springer-Verlag Italia 2015

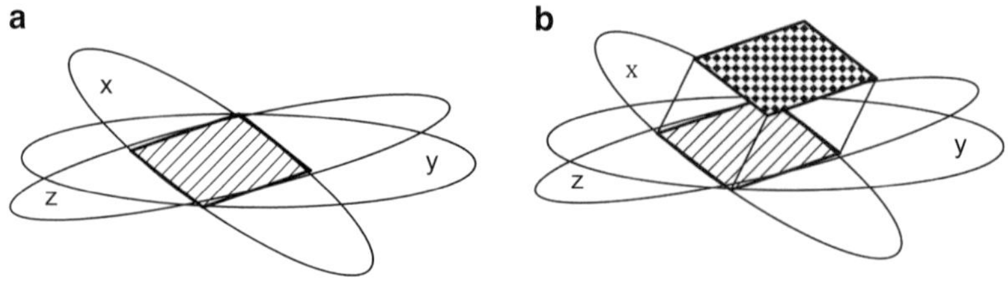

Fig. 5.1 Representation of an "ecological space" in the study of vegetation. (**a**) in the phytosociological model, (**b**) in the bionomics model of the landscape. Note that the principle of emergent properties acquires ecological space characters that go beyond the sum of individual species (from Ingegnoli, 2002)

general metastability (g-LM) of a landscape units; (2) how to compare ecological data of a forested patch with those of another vegetation type; (3) how to use the ecological characters of the different types of vegetation to reach a diagnostic assessment related to a certain landscape unit; (4) how to integrate other ecological parameters of a landscape unit (HH, SH, etc.) with those relating to the vegetation.

The answers, consistent with the theory of landscape bionomics, start with the proposal of the new concept of "*the fittest vegetation for...*", as exposed in the third chapter (see Sect. 3.3.3), to overcome to *potential vegetation*. This reinterpretation of the concept of potential vegetation indicates "the vegetation most fitting in specific climatic and geomorphic conditions, in a limited period of time, in a certain defined place, in function of the history of the same place and with a certain set of incorporable disorders (including those human) in natural and not natural conditions". In fact, in the absence of a range of incorporable disorders, a landscape fails to structure and to endure.

The implications are different: for example, the concepts of primary or secondary vegetation lose meaning; also, in heavily populated regions, given the environmental changes in act, to compare the actual vegetation with the potential one loses sense. The concept of "the fittest vegetation for...", combined with the principle that "... the behaviour of a system depends not only by its component elements, but also by the way in which they are assembled and arranged ..." (Emergent Properties Principle) shows a large change of perspective in applications too. Let us summarise the main implications:

1. The zonal vegetation, even in the absence of anthropogenic disturbances, is structured in a complex of associations and the configuration of this complex affects the functionality of the system
2. Interactions among different components of a same zonal vegetation can bring to a change along a line of succession
3. In a complex adaptive system, multifunctionality is not expressed in all its aspects if a system of constraints doesn't work in order to fully activate all its functions, given the complementary nature of many of them (e.g. without the constraints due to fragmentation, connectivity and circuitry functions cannot exert)
4. The ecological role of anthropogenic vegetation must be recovered, at least as long as it does not lead to "out of scale" disturbances. Being able to put into play such recovery is crucial if you want to manage the land in the sense of conservation biology, since many natural species have been coevolving for thousands of years of relationships with human populations
5. The dynamics of the vegetation is to be studied again, surpassing the reductionism still present in phytosociology. It is necessary to take into account the principles of landscape bionomics, too, not only the principles inherent in the species and the plant communities
6. The concept of primary or secondary vegetation loses much of its meaning. The theory of Non-equilibrium Thermodynamics and the Order through fluctuations in fact does not admit these reductionist aspects

Fig. 5.2 Forest of Gariglione, in the Park of Sila Piccola, Calabria (Italy): *Asyneumati-Fagetum Abietetosum* according to Gentile. Studies undertaken for the ecomosaic does not reveal a mixture of only two, but of 12 different types of tesserae [10]

7. It is no longer possible to talk about recovery and/or restoration of ecological systems of vegetation. In fact, the return to conditions quo ante is unthinkable even in relatively short periods of time and with the removal of anthropogenic disturbances (see Sect. 1.3.8 and 1.3.9)

These aspects have been appearing since the studies carried out by Ingegnoli [12] on the Forest of Gariglione in the Park of the Sila Piccola (*Asyneumati-Fagetum Abietetosum*) (Fig. 5.2).

First of all, it came out that while the phytosociological map showed only two types of tesserae, dominated by *Fagus sylvatica* or *Abies alba*, in reality 12 types of tesserae could be well identified, many of which derived from the ancient habit of charcoal (Fig. 5.3) which, after centuries of use, have changed the soil to such an extent as to create suitable conditions almost exclusively for patches of Poplars (*Populus tremula*).

Fig. 5.3 Not frequent example of a charcoal pile in a beech forest of Calabria. The working area is quite wide and hosts three of these buildings in oven

5.1.2 Landscape Bionomic Survey of Vegetation (LaBiSV)

These premises, albeit synthetic, enhance the urgent need to develop a method to study the vegetation following the landscape bionomics principles. That's why Ingegnoli [10, 7], Ingegnoli and Giglio [13], Ingegnoli and Pignatti [14] have proposed a new methodology, called LaBiSV (Landscape Bionomic Survey of Vegetation), whose theories are summarised as follows: (1) reference to the concepts of ecocoenotope and ecotissue as structural entities of the landscape; (2) use of biological territorial capacity of vegetation (BTC) as the main integrative function; (3) drawing up of development models of different types of vegetation (time-BTC) based on an exponential and logarithmic functions and (4) the possibility of comparison between the ecological status of natural and man-made vegetated tesserae, according to the principles of landscape bionomics; (5) ability to determine the state of normality of the ecological parameters of different types of vegetation; (6) ability to measure the concept of biodiversity at the landscape level (diversity of biological organisation of the context).

5.1.2.1 Steps of the Method

For the study of the vegetation of a landscape unit (LU), this methodology can be divided into at least five to six phases, which are reported as follow:

Phase I—identification of elements of the landscape. Once established the boundaries of the ecological landscape unit under consideration (LU), to highlight the ecotopes and vegetated tesserae (Ts). It should be noted that the types of vegetation do not always coincide with the phytosociological associations, e.g. in case of ecotonal belts.

Phase II—study of the geographical and historical character of LU. To collect the geographic data of LU (e.g. phytoclimate, soil substrate, morphology, etc.). To study the historical transformations, also anthropogenic, through old maps and land uses in the past. Particularly important are the historical presence of forested patches and/or historic farms or monasteries.

Phase III—survey of sample tesserae for vegetation parameters. To choose the tesserae/samples for each type of vegetation and, for each "tessera" (Ts), to detect four groups of parameters: (1) characteristics of the tessera = T, (2) characters of the phytomass, that is plant biomass or (above-ground) = F, (3) characters of the ecocoenotope = E, (4) relationships tessera/landscape unit, that is among the element and its landscape parameters = U. The parameters for each T, F, E, U sets range from 3 to 12, thereby reaching the number of 26-33 (28 when related to forests, as shown in Table 5.1 and 5.2).

Phase IV—ordering and evaluation of bionomic parameters. To order the above parameters into 4 classes, summarised in a standard form (e.g. Table 5.1, 5.2; see Sect. 5.2.5) that allows to evaluate them for columns with scores dependent on the different models of development of each vegetation types [7, 10], as we'll see ahead. It is thus possible to evaluate the bionomic quality (bQ) of the parameters and/or estimating the BTC of the Ts with model equations related to the type of vegetation present [13].

Phase V—eco-issues of landscape and choice of normal ranges. Referring to the principles of landscape bionomics, to process the diagnostic aspects needed to know the ecological status of LU in question. Indeed, the mentioned parametric standard form has been designed to check the organisation level and to estimate the metastability of a tesserae, considering both general ecological and landscape ecological characters.

Phase VI—criteria and guidelines for intervention. To identify a therapeutic rehabilitative function of the examined vegetation, taking into account the principles of landscape bionomics.

This method is very helpful in checking the difference from the so-called "*optimal state*[1]" for each specific type of tessera, that is the difference between the esteemed bionomic quality of the tessera and the maximum one for the relative vegetation type: it's the first diagnostic goal.

5.1.2.2 The models of development of forest vegetation

Remembering the well-known relationships among gross productivity, net productivity and respiration in vegetation ecosystems [16–18], the development of a vegetation community may be synthesised in (a) the growing phases from young-adult to maturity, expressed by an exponential process; (b) the growing phase from maturity toward old age, expressed by a logarithmic process (Fig. 5.4). The function able to value these processes has been recognised as BTC, the mentioned biological territorial capacity of vegetation (see Sect. 3.3.1). Therefore, it was possible to design a model (Fig. 5.4) and to calculate

[1] Note that the concept of 'optimal' state concerns the maximum degree of quality and health reachable by the type of vegetation of the examined tessera, regarded as a complex system related to its LU.

Table 5.1 LaBiSV parametric form for boreal and temperate forests: ecological evaluation of the tessera quality (bQ) and estimation of biologic territorial capacity of vegetation (BTC)

Boreal forest	1	5	14	25	Score
Temperate forest	1	5	12	22	Score
T. Tessera characters (Ts)					
T1—Vegetation height (m)	<9	9.1–18	18.1–29	>29.1	Canopy Hc = ...m
T2—Cover of the canopy (%)	<30	>90	31–60	61–90	Ts surface
T3—Structural differentiation	Low	Medium	Good	High	Age, space groups, etc.
T4—Interior/edge (%)	None	<30	31–89	>90	(% Ts)
T5—Management	Simple coppice	Coppice	Wood	Natural forest	Or similar
T6—Permanence (years)	<80	81–160	161–240	>240	Old trees
F. Vegetational biomass (aboveground)					
F1—Dead plant biomass	Near 0	>10	1–5	5–10	% of living biomass
F2—Litter depth	Near 0	<1.5	1.6–3.5	>3.5	cm
F3 b—Biomass volume (m^3/ha)	<200	201–500	501–950	>950	pB = ...m^3/ha
F3 t—Biomass volume (m^3/ha)	<150	150–350	350–600	>600	pB = ... m^3/ha
E. Ecocenotope parameters					
E1—Dominant species (n°)	>3	3	2	1	as pB volume
E2—Species richness	<15	16–30	31–40	>40	n° sp./Tessera
E3—Key species presence (%)	<5	6–40	41–75	>75	Phytosociological
E4—Allocthonous species (%)	>10	10–4	<4	0	From other ecoregions
E5—Infesting plants %	Near all	>25	<25	0	Covering area on Ts
E6—Diseased plants	Evident	Suspect	risk	0	Even acid rain damage
E7—Biological forms (n°)	<3	4–5	6–7	>7	Cfr. Box [15], mod.
E8—Vertical stratification	2	3	4	>4	Traditional
E9—Renewal capacity	None	Intense	Sporadic	Normal	Dominant species
E10—Dynamic state	Degradation	Recreation	Regeneration	Fluctuation	Cfr. Ingegnoli [10]
U. Landscape unit (LU) parameters					
U1—Similar veg. contiguity	0	<25	26–75	>76	% of perimeter
U2—Source or sink	Sink	Neutral	Partial	Actual source	Species & resources
U3—Functional role in LU	Reduced	Minor	Evident	Important	Context & typology
U4—Disturbances incorporation	Insufficient	Scarce	Normal	High	Local disturbances
U5—Geophysical instabilities	Evident	Partial	Risk	None	On the physiotope
U6—Permeant fauna interest	Low	Medium	Good	Attraction	Key species
U7—Transformation modalities of the Ts	Strong disturbances	Gradual changes	Temporal instabilities	Fluctuation	Today + tendency
U8—Landscape pathology interference	Serious	Near chronicle	Easy to incorporate	None	Coming from surroundings ecotopes
U9—Permanence of analogous vegetation (years)	<100	100–300	300–1200	>1200	Historical presence
Results of the survey					
Total score Y (= h+j+k+w)	h = ...	j = ...	k = ...	w = ...	Y = ...
Bionomic quality of the Ts (bQ)	bQ_b = Y/700 (b) bQ_t = Y/616 (t)				**bQ = ...** [%]
Estimation of the BTC Boreal (conifer) forest	**BTC** (b) = 0.01339 (y–28) + 0.12 (pB/70)				**BTC = ...** [Mcal/m^2/year]
Estimation of the BTC Temperate (deciduous) forest	BTC (t) = 0.01667 (y–28) + 0.13 (FM/65)				**BTC = ...** [Mcal/m^2/year]

an exponential-logarithmic curve of development sensu Ingegnoli [10] having an adequate temporal dimension. The related equations, respectively for temperate deciduous and boreal coniferous forests, are:

$$\text{Temperate forest}: \text{BTC} = \text{T}^{0.515} - 0.77, \quad (1)$$

$$\text{Temperate forest}: \text{BTC} = 2.363 \ln \text{T}, \quad (2)$$

$$\text{Boreal forest}: \text{BTC} = \text{T}^{0.50} - 0.70, \quad (3)$$

$$\text{Boreal forest}: \text{BTC} = 2.17 \ln \text{T}, \quad (4)$$

where T = time (years).

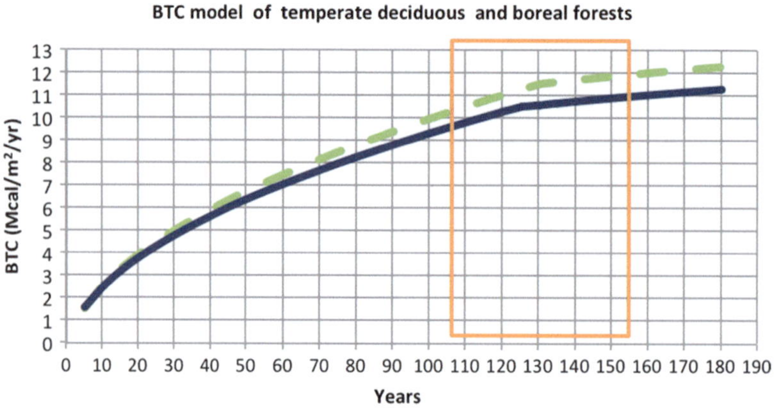

Fig. 5.4 Model of the development of temperate (*green broken line*) and boreal (*blue continuous*) forests. Theoretical and field studies on these types of vegetation show these two exponential-logarithmic curves of development. The threshold between adult-mature and mature-old phases is respectively 10.5 or 11.5 Mcal/m^2/year (in the middle of the *orange rectangle*). After about BTC equal to 12 other unpredictable decreasing curves may express the beginning of a senescent phase

The flex of each curve presents its own BTC value, defined after a control through the field study of critical points referred (for instance) to plant biomass relations, structural and ecological relations, etc.

Defining an analogous model[2] for each one of the main types of vegetation types of Boreal-Temperate zone, we obtain the equations and the flex values showed in Table 5.5.

In Table 5.6, we reported 13 main vegetation types available in Europe, their maximum of BTC values, the thresholds of maturity available to build their models and the BTC evaluation equations (Mcal/m^2/year). For each type of vegetation, the maximum value of BTC has been subdivided in four intervals of the same amplitude, corresponding to four evaluation classes [10]. Thus, the derived values are the weight (scores) to be coupled to the T, F, E, U survey ranks of parameters to catch the evaluation of the bionomics quality of a tesserae, as previously mentioned (see Tables 5.1 and 5.2).

[2] Note that, because of the high complexity of the ecological system of vegetation and its context, it is impossible to survey the vegetation following the main scientific paradigms concerning living systems and their self-organization and metastability, without building an appropriate model.

5.2 The LaBISV Main Parametric Standard Forms

5.2.1 The Analysis of the Characters of a "Tessera"

T1: Average Height of the Canopy (m). The height of the vegetation should be considered as a weighted average of the heights of all the members of all the plant species concurring to create the "canopy" of the concerned tessera: you do measure the height of at least 20 % of individuals for each plant species present in the dominant layer (individuals which are already representative of the average of the species in that specific tessera, taking care to avoid the highest and the lowest), then you must count how many individuals of each species are present in the tessera in question and calculate the weighted average.

T2: Canopy Cover (%). The covering of the canopy should be considered with respect to the whole extension of the tessera in question (usually from 1 to 12 ha) and not to a single plot (such as in phytosociological reliefs). It considers only the covering of the tree component.

T3: Structural Differentiation. The age difference among trees is measured by the

Table 5.2 LaBiSV parametric form for Mediterranean pine and schlerophyll forests: ecological evaluation of the tessera quality (bQ) and estimation of biologic territorial capacity of vegetation (BTC)

Mediterranean pine forest	*1*	*5*	*13*	*23*	*Score*
Medit. sclerophyll forest	*1*	*5*	*11*	*22*	*Score*
T. Tessera (*Ts*) characters					
T1p—Vegetation height (m)	<9	9.1–17	17.1–25	>25	Canopy Hc=....m
T1s—Vegetation height (m)	<8	8.1–15	15.1–21	>21	s. canopy trees
T2—Canopy cover (%)	<30	>90	31–60	61–90	*Ts* surface
T3—Structural differentiation	Low	Medium	Good	High	Age, groups
T4—Interior ratio (%)	Absent	<30	31–89	>90	vs ecotope
T5—Management	Simple coppice	Complex copp.	Wood	Natural forest	or similar
T6—Permanence (years)	<80	81–160	161–240	>240	*Ts* real age
F. Vegetational biomass (aboveground)					
F1—Dead plant biomass in *Ts*	Near 0	>10	1–5	5–10	% of pB
F2—Litter depth of the *Ts*	Near 0	<1.2	1.2–3.0	>3.0	cm
F3p—pB volume (m^3/ha)	<160	161–450	450–750	>750	pB = ...m^3/ha
F3s pB volume (m^3/ha)	<120	120–300	300–500	>500	pB = ...m^3/ha
E. Ecocoenotope parameters					
E1—Dominant species	>3	3	2	1	pB of trees
E2—Sp Diversity	>65	<13	14–35	>35	n° sp./*Ts*
E3—Key species presence %	<10	11–40	41–75	>75	Botanical
E4—Allocthonous sp. (%)	>10	10–4	<4	0	Not regional
E5—Infesting plants	Near all	>25	<25	None	% of covering
E6—Diseased plants	Evident	Suspect	Risk	0	Even acid r.
E7—Plant forms (n°)	<3	4–5	6–7	>7	Cfr. Box 1987
E8—Vertical stratification	2	3	4	>4	>25–35 % Ts
E9—Renewal capacity	None	Intense	Sporadic	Normal	Dominant sp.
E10—Dynamic state	Degrading	Recreation	Regeneration	Steady attractor	Transformation process
U. Landscape unit (*LU*) parameters					
U1—Boundary connections	0	<25	26–75	>76	% perimeter
U2—Source vs surroundings	Sink	Neutral	Partial	Actual source	Sp & resources
U3—Role in the *LU*	Reduced	Minor	Evident	Important	Contest etc.
U4—Disturb incorporation	None	Scarce	Normal	High	Local disturb.
U5—Geo-physical instability	Evident	Partial	Risk	None	Physiotope
U6—Permeant fauna interest	None	Medium	Near good	Attractive	*Ts*/key sp.
U7—Transformation reason of the *Ts* as landscape. element	Strong disturbance	Temporary instability	Gradual change	Steady attraction	*LU* trend
U8—Landscape pathology interference	Extremely serious	Near chronic	Easy to recover	None	Coming from surroundings
U9—Permanence (years)	<100	100–300	300–1200	>1200	Age of *LU*
R. Results of the survey					
R1—Total score Y (= a+b+c+d)	a = ...	b =.....	c =.....	d =.....	Y = ...
R2—Quality of the Ts (%)		bQ_s = Y/616	bQ_p = Y/644		**bQ** = ...
R3—BTC estimation (Mcal/m^2/year) Mediterranean pine forest	BTC(p) = 0.0151 (y–28) + 0.12 (pB/65)				**BTC = ...** [Mcal/m^2/year]
R3—BTC estimation (Mcal/m^2/year) Schlerophyll forest	BTC(s) = 0.01705 (y–28) + 0.13 (pB/60)				**BTC = ...** [Mcal/m^2/year]

distribution of the diameters on the examined tessera. As regards the spatial arrangement (or horizontal structure, or texture), we consider the main types of species. It is recalled that both entries should be verified by reference, firstly, to all the plant species of trees, whether belonging to the dominant tree layer or to the dominated one and, secondly, to the species components the high shrubs layers. The evaluation will correspond to a consideration of the status of the tessera. Therefore, it is considered

- Well-structured (*high*) a tessera in which individuals are uneven-aged stands and normally aggregated and this condition is checked both by tree species and for those shrubs; it is attributed

- The value of *good* when the uneven-aged stands are not very strong combined with a spatial arrangement with hints of aggregation, or with certain species able to aggregate and not others, or with regular arrangement of aggregates; it is attributed
- The *mean* value when we found one of the two conditions, i.e., alternately uneven-aged stands or the spatial arrangement in groups; it is attributed
- The *low* value when, within the tessera in question, almost the same age occurs and with random arrangement of the plants or, conversely, the tessera show a precise pattern of planting, then adjust, all for both components

T4: Internal/Margin (%). We consider the ratio between the surface of the tessera identified as "internal" (depending on the type of plant and animal species and on the behaviour of plant species) and the surface of the tessera identified as the "margin". In this regard, please note that the band edge is deep from one to two and a half times the average height of the canopy species. It is noted that there are tesserae all pertaining to the margin, recognisable by a cover of the canopy quite sparse, from a posture of tree species with the development of the foliage more similar to that of isolated trees (therefore with ramifications foliate also along the trunk),[3] by a number of higher species that also includes species of the types of their neighbours' tesserae: it is emphasised that, actually, the tesserae of margin are not always located in the marginal band of the forest, but they can also be located within forest ecotopes.

Similarly there are tesserae of all internal, in this case always located in the inner parts of forest ecotopes (or in those areas that were internal, before recent deforestation or selective cuts of surrounding land) and recognisable for the specificity of the flora and the fauna that are found there, as well as for the behaviour of the trees, characterised by bare trunks and foliage very reduced and concentrated in the upper part of the canopy. Remember that there are plant communities that are in themselves "closed formations" and others that are "open formations": thus these latest will never reach the maximum score.

T5: Type of Management. Remember that this refers to the actual type of human management and not to similarities of vertically or horizontally structuring with these utterances types. So,

- Are considered equivalent to a *simple or sapling coppice* all those tesserae of thicket or "planting trees" (e.g. poplar groves, or young fir) or similar, which will be cut after a certain number of years or cutting wheel (i.e. a part of the young at regular intervals) and in which the trees are all peers (simple coppice), or with the presence of plants of a generation older, fulfilling the function of saplings (sapling coppice); moreover, the ground between the rows of individual trees is often weeded; there are neither herbaceous layer, shrub layer nor other layers excluding the arboreal plant; propagation occurs primarily by vegetative reproduction
- Are considered equivalent to a *complex coppice* all those tesserae of arboreal vegetation which, following the rhythm of the cuts, assume a trend uneven-aged (base of coppice with saplings of a shift, two shifts, three rounds, four rounds, of five rounds): the cuts are subsequent (i.e., the trees will all be eliminated in the course of 2–5 successive cuts) or occasional for the high forest component, flush for the component coppice; in addition, we consider here vegetated tesserae in which collection of wood gathering or charcoal take place, or in which a selective cutting of some species is carried out
- Are considered *equivalent to a wood* those tesserae governed as high forest, in which the propagation occurs predominantly by the

[3] As Piussi reported [19]: "... the plants grown in an isolated state have a different behaviour from that of plants grown in the wood in contact between them. In the first the branches remain alive even in the lower part, often up to the foot of the plant, and increase greatly in length and diameter. In the second lower branches dry out more or less rapidly and the development side of the canopy is hindered ...".

sexual reproduction, the management is carried out with the purpose of timber production or for purposes of tourism-recreation or protection: therefore collecting dead wood that falls to the ground or felling of dead trees still standing, or of mature trees, senescent or diseased (renewal cuts), or making cuts interlayers or thinning, or eliminating shrubby understory species over a certain height are periodically carried out; in addition, one can recognise the presence, on the edge of the tessera, of forest roads or places of collection of the piles of wood or sheds collection of tools or for overnight of woodcutters, traces of the passage of heavy vehicles also tracked for the removal of timber, fire lines, cable cars etc.

- Are considered equivalent to a *natural forest* management those tesserae of woody vegetation in which the propagation occurs solely by sexual reproduction, dead woods on the ground are never collected, nor dead trees still standing; or mature ones or senescent sick are cut down, neither any kind of cut is possible, nor elimination of the species of the undergrowth or weed species (whether herbaceous, shrub or tree); in addition, those tessera where forest roads (or comparable) and places of stable human presence are absent or no longer actively used (except in exceptional cases). Therefore, the tesserae of simple or sapling coppice and those of the forest in which the management regime has been changing for at least 3–5 decades may fall into this category

T6: Residence Time (Years). It is considered under this item the average age of the tessera, evaluated according to the average age of older individuals within the same. This assessment can be done quite precisely with coring and counting of the rings (subject to the comments of consequential damage to the plants themselves) or the age of the individuals can be estimated according to the diameter, measured at the height of 130 cm from the base: you have to remember, however, that in this case the datum may be wrong by default, considered the possibility that the plants pass even 50 or 70 years of their lives in the form of young seedlings waiting for a fall to give them the opportunity to complete their development (Fig. 5.5). This datum should then be calibrated with an assessment of favourable or unfavourable climatic conditions that occurred during the time in question. In this connection there are tables accretion processed by the foresters, to which it may apply a tolerance value of 10 %. On the other hand, please note that the standard form does not need an exact value but the location within range of four fairly distinct age, and this allows you to bypass some objective difficulties.

Fig. 5.5 Section of the trunk of a spruce (*Picea abies*) in the Alps. The inner radius A (2.5 cm) counts 50 years exactly like the segment B (15 cm)

5.2.2 Analysis of the Aboveground "Plant Biomass"

F1: Amount of Dead Phytomass. We look at the *deadwood*, whether it consists of trunks on the ground or standing still, cable or solid, by branches or bark, including the bases of the logs cut, excluding the leaves that will be counted in the litter. The value as a percentage of the volume should be compared to the calculated aboveground phytomass alive.

F2: Depth of Litter. Recalling that the term *litter* indicates the set of more or less decomposed organic residues (in particular leaves and needles, twigs, etc..) in which the original structure is still recognisable and there are no obvious signs of decomposition, and

Fig. 5.6 Relascope "mirror" by Biitterlich, University of Vienna. In the *center* of the instrument, the optical sights to the *left*, while to the *right* an example of a station relief of tree phytomass from the point of the instrument station is shown

which is directly superimposed on the mineral soil (which it supplies), it must be measured in 5 to 10 different points of the tessera, always away from the trunks of trees, and the final value will be a weighted average of the measured values.

F3: pB Volume (m^3/ha). The item refers to the estimated aboveground plant biomass (pB) per hectare, calculated not by weight but by volume, a value that is more explanatory of the employment space for the purpose of Landscape Bionomics. For the estimation of phytomass see again Ingegnoli and Giglio [13]. In cases of multi-layered forests (that is forests in which the total basal area is composed of individuals of—at least—two very different height but comparable diameter length), it will be necessary to consider the height-weighted average between the two layers involved in the calculation of phytomass, instead of just the single canopy. In some tables representing the LaBiSV standard form, this item is twofold to provide the ranges of phytomass peculiar to two types of similar vegetation, as *temperate* and *boreal* forests in Table 5.1.[4]

Relascope The method most widely used today for the relief of the volume of phytomass is based on a tool called "spiegel-relascope" of Bitterlich (Fig. 5.6). It is an optical instrument [20] working with a system of optical sights combined with 4 possible BAF (basal area factor) through circular surveys.[5]

It follows that the equation of Relascope for the determination of the basal area in m^2/ha is:

$$G = K \times N^*,$$

where $K = BAF, N^* = [No. trees > (bK) + 1/2$ n° $trees = (bK)]$, *being* (bK) *broadband K in the instrument.*

Allometric Coefficients The study of the shape of a tree or a shrub, understood as the architectural model, is of considerable importance as the outward appearance is evidence of a specific adaptive significance of changing conditions and demands of the environment, i.e. the "... result of the project overall growth of the individual encoded by its genotype, common to all individuals of that species, but filtered through the action of the environment ..." [19].

In bionomics of the landscape, such knowledge is used for two specific purposes:

[4] In case of mixed forests of two different types quite equal in dominance, like a Boreal coniferous and Temperate broadleaves forest (Table 5.2), or a Mediterranean Pine and Sclerophyllous forest, when filling in the standard form, use a reasonable average between the two rows F3. In doubt, choose a worse value column.

[5] For all the detailed information on the scientific functioning and utilisation of the relaskop refers to [20] or to www.relaskop.at

Table 5.3 Allometric coefficients for plant biomass evaluation, referred to European vegetation (main species) (from Susmel [21] modified and integrated). These values can be used only as a reference, because each forest may present allometric variation. (Ø = average diameter of the trunk)

Plant species	Greenery/trunk (mass ratio)	Allometric coefficient (A)	(A) for an isolated tree	Specific weight
Abies alba Miller	Ø 20=0.167	0.47	0.51	0.44
	Ø 50=0.152	0.46		
	Ø 95=0.127	0.45		
Picea abies (L.) Karsten subsp. *abies*	Ø 20=0.500	0.55	0.6	0.44
	Ø50=0.467	0.54		
	Ø95=0.445	0.53		
Larix decidua Miller	Ø 50=0.450	0.57	0.62	0.6
Pinus sylvestris L.	Ø 50=0.300	0.46	0.5	0.53
Fagus sylvatica L.	Ø 20=0.267	0.71	0.95	0.74
	Ø50=0.528	0.84		
	Ø95=1.450	1.35		
Prunus avium L.	Ø 30=0.186	0.73	0.8	0.66
	Ø50=0.410	0.7		
Robinia pseudacacia L.	Ø 30=0.170	0.65	0.75	0.78
Populus alba L.	Ø 50=0.450	0.77	0.85	0.5
Alnus glutinosa (L.) Gaertner	Ø 30=0.180	0.6	0.7	0.56
Quercus robur L.	Ø 50=0.560	0.85	1.05	0.75
Quercus pubescens Willd.	Ø 30=0.190	0.7	0.8	0.96
Quercus ilex L.	Ø 30=0.250	0.75	0.88	0.9

1. The first is to aid the identification of the characteristics of a vegetated tessera respect to the factor “margin” of woodland or shrubland (with carriage therefore more similar to that of isolated tree) or “internal” woodland or shrubland.
2. The second is in the development of appropriate indices, called *allometric* indices (Al I), useful as a correction factor in the calculation of total phytomass of a vegetated tessera. Indeed, the volume of a tree, simply calculated as the product of basal area (base surface of the trunk) per height, would provide the volume of a cylinder, regardless of the shape and texture of the branches, instead of fundamental importance. The correction factor takes into account the relationship between its greenery and trunk (Table 5.3, Fig. 5.7). As expressed by Piussi [19] the biomass proportion and its NP (net production/year) of a tree of 990 kg (*Fagus sylvatica*) are the following:

- Trunk = 700 kg (70.7 %), with NP = 15.5 kg/year
- Treetop = 150 kg (15.1 %), with NP = 6.5 kg/year
- Branches = 130 kg (13.2 %), with NP = 8.0 kg/year
- Foliage = 10 kg (1 %), with NP = 10 kg/year

Calculation of the Volume of Phytomass We have now the possibility to establish the equation which allows the calculation of phytomass for each test species in m^3/ha:

$$pB = G \times Hc \times Al \left[m^3/ha\right],$$

where G = basal area, Hc = height of the canopy, Al = allometric coefficient.

The former director of the Geobotanical Station of Bialowieza, prof. Falinski (personal communication) suggested to reduce the above measure of a value between 5 and 10 %, given that in the reliefs with Relascope it tends to overestimate the trees of size limit (which are recorded as 1/2 n ° trees).

Fig. 5.7 Example of monumental Beech in the Pre-Alps. The two figures (age 3 and 7) with a 1.2 m calibrum may give a proportion of the shape of this tree, whose allometric coefficient is >1

5.2.3 Analysis of Ecocoenotopic Characters

E1: Dominant Species (n°). The dominant species are not established in relation to the value of coverage, as in phytosociological relevés, but depending on the volume of aboveground phytomass. Therefore, when more than one species in the canopy exists, the procedure is as follows: once calculated the total volume of phytomass, it must be divided by the number of species found in the canopy, in order to obtain the value of equitability, i.e. the quantity of phytomass that would compete to each species in the case in which all had the same importance. This value is then compared with the actual values of phytomass measured for each species: in the case where one or more species will end up with a phytomass clearly superior to the average value of equitability, these species are considered dominant. For this purpose, a threshold of significance has to be not lower the order of magnitude of 20 % (i.e. the dominant must have a percentage of phytomass at least 20 % greater than the others). Consider, for instance, a canopy with 5 species whose equitability is 20 % of phytomass: if a species has 40 % of phytomass and the other four levels ranging between 10 and 20 %, (for a total of 100 %), that species can be considered dominant because its value exceeds phytomass of at least 20 % the value of the other classes. If there were two species with about 30 % each (and with the remaining amounted to around 13 %), already the situation at the time of the survey would not be sufficiently clear and you should talk to more than 3 dominant species.

E2: Species Richness. Species richness refers to the total number of plant species found on the total surface of the tessera (and not only those of the phytosociological relevés) considering all the three layers of tree, shrub and herbaceous species with musk and lichens too: that is mosses and liver present on the soil (and not bark lichen), at least as an esteem of possible species and, in addition, if very numerous, the species of mushrooms (whereas those with visible carpophores). In practice, the possible more complete list of plants.

E3: The Presence of Characteristic Species (%). The answer to this voice requires the performance of at least one phytosociological relevé for each type of formation recognisable within the tessera. It should be noted, in fact, that the criteria for the delimitation of a tessera according to landscape bionomics does not coincide with those of "pure formation" of phytosociological criteria. Therefore, a tessera can be located in a position of the gradient between two or more different phytosociological associations, or present within small clearings with herbaceous vegetation, or even be a tessera of margin with interdigitation of two types. The relevé according to the method of Braun-Blanquet has the purpose to provide an indication of the Phytosociological Association (or phytocoenosis) to which the vegetation of the tessera can be referred (or among which the dynamic interactions are in progress or will activate). Done this identification, it is necessary to highlight the species that are described as characteristic of that/those specific association/s, plus any applicable differentials facies, and to count what percentage they constitute, as a whole, in the complete floristic list of that tessera.[6]

E4: Allocthonous Species (%). From an ecological point of view we have to consider as "allocthonous" those species pertaining to an ecoregion different from the local one. Therefore, the allocthonous species will be not only the exotic one, but also the steno-Mediterranean referred to the Central-European region (or vice-versa). The percentage of these species has to be evaluated on the complete floristic list considering the species frequencies.

E5: Infesting Plants (%). Even one key species may become infesting if, within its layer, its expansion may cover and limit other species usually pertaining to the layer itself or limit other depending layers. An example in degraded temperate forests could be the bramble (*Rubus sp.*).

E6: Diseased Plants. When it is recognised, inside the examined tessera, the presence of dead individuals, or sick, or signs of a state of imperfect health, considering the state of the bark, trunk, branches and leaves and the so-called "new type of damage". The item is important to consider risk damage from acid rain, even when not yet made themselves, or from other sources of pollution, but also for instability that could create precisely a risk to the vegetation of the tessera in question. Remember to pay attention to insect damages.

E7: Biological Forms (n°). Biological forms, from the ecological point of view, must be evaluated not on the basis of the division of Raunkiaer [22], but on that of Box [15] as modified in Ingegnoli [10]. This is to better match the shape of the plant actually occurring on the tessera and also its ecological role in different stages of life (e.g. a seedling beech does not have the same ecological role in the tessera that an individual of young beech, or the one that an individual adult beech can have). Please pay careful to do not confuse the recognition of biological forms (hence the shape of the plants) with that of the vertical structure of the tessera in question, as the next step. Remember that the term *krummholz*, for which there is no corresponding neither English nor Italian, indicates the bearing shrubs with their twisted and/or prostrate branches, typical of mountain pine.

The resulting forms are presented in Tables 5.4.[7]

E8: Vertical Stratification. From the point of view of Landscape Bionomics, a layer is considered to be present when it is represented by a number of individuals able to give a coverage of at least 20–30 % of the surface of the tessera. Otherwise, sign in the marginal notes the presence of a certain number of individuals referable to a certain height, but does not consider an effective layer. The vertical stratification adopted for these tesserae detection is summarised as follows (Fig. 5.8):

- Layer of the trees of the first magnitude = height greater than 30 m
- Layer of the trees of the second magnitude = height between 20 and 30 m
- Layer of the trees of the third size and high shrubs = height between 5 and 20 m
- Shrub layer and tree-saplings = height between 1 and 5 m
- Herbaceous layer or low shrubs = height between 0.45 and 1 m

[6] The complete floristic list must be caught through a floristic survey/phytosociological relevees for each relascope station (n) plus others ($n + 1$) in the paths linking them. The calculation of species frequency must be derived by these data.

[7] Tallophyte and bryophytes, epiphytes (including parasitic), lianas, ferns, graminoids (i.e. species with behaviour similar to that of *Poaceae*), other herbaceous (herbaceous species not graminoids, including hemicryptophytes and geophytes), reeds, shrubs washer (with the leaves at ground level, often succulent), or semi-shrubs (camephyitae, renewals, seedlings and seedlings of woody species), succulents (perennials with or without woody skeleton), krummholz (twisted shrubs, with one or more barrels less than one metre in height—except for the main stem—prostrate, branched, e.g. mountain pine in the pine woods), arborescent shrubs (with one or more drums height of more than one metre, but the form of indeterminate growth), 'trees tuft' (tree ferns and similar, but with a trunk without bark and with the leaves arranged in a tuft at the top), trees (or even young samplings).

Table 5.4 Dichotomic sequence of plant forms (from Box 1987, modified and integrated)

1.	Vascular plants (kormophyte)	2
	Non-vascular plant (cryptogam)	Thallophytes (incl. mosses, algae, liverworts, lichens)
2.	Plant rooted in ground	3
	Plant not rooted in ground, sitting on other plants	Epiphytes (incl. parasitic)
3.	Plant self-supporting (at least at maturity)	4
	Plant permanently sprawling, climbing, or otherwise not self-supporting	Vines/Lianas
4.	Plant with stems	5
	Stems essentially absent above ground (ab.gr.); leaves in terminal rosette at ground level (often succulent)	Rosette-shrubs
5.	Stems not permanently woody or succulent (ab. gr.)	6
	Stems permanently woody (ab. gr.) or succulent	10
6.	Plant totally herbaceous	7
	Plant not totally herbaceous	9
7.	Spermatophyte (especially angiosperm)	8
	Pteridophyte	Ferns
8.	Plant grass-like	Graminoids
	Plant not grass-like	Forbs (incl. geophytes)
9.	Plant perennial with progressively sclerotic culm, >0.8 m tall (or similar)	Reeds
	Plant perennial from woody *xylopodium* (or similar)	Semi-shrubs (incl. trees *plantulae* or seedlings)
10.	Stems succulent, with or without woody skeleton; plant perennial	Stem-succulents
	Stems permanently woody (ab. ground)	11
11.	Plant with multiple main stems, often with none dominant; plant generally <5 m tall	12
	Plant with a single main stem (trunk); plant generally >5 m tall	13
12.	Plant with 1 or more main stems, <1 m (except for emergent trunks), prostrate, usually highly branched	Krummholz (e.g. twisted shrubs, incl. cushion-shrubs, dwarf-shrubs)
	Plant with 1 or more main stems, >1 m; growth form indeterminate (e. g. overgrown bushes, scandents)	Arborescent shrubs (incl. cushion-shrubs, dwarf-shrubs)
13.	Plant growing from terminal bud, with wood produced secondarily, no bark; rarely branching; leaves in rosette(s), usually terminal	Tuft-trees (incl. tuft-treelets, tree-ferns)
	Plant with true wood growing outside; usually with bark, plant usually branching	Trees (incl. treelets, small trees)

- Small herbaceous layer and/or very small scattered shrubs = medium height between 0.15 and 0.45
- Musk layer and the low herbs = height between 0.1 and 0.15 m

Recall that, in this case, what is important is the actual height of the species at the time of the survey and in the considered tessera. A transect transect of vegetation may help in this situation (see Fig. 7.20) [10, 13].

E9: Renewal Capacities. This capacity is measured on the renewal of dominant species on the tessera at the time of the study in the form of sexual propagation, that is, from seed. If the dominant species do not present renewal, but there is abundant renewal of another species, also tree, you have to mark the left-most column. The renewal of this other abundant species will instead be indicative of the dynamic state of the tessera, in the next step.

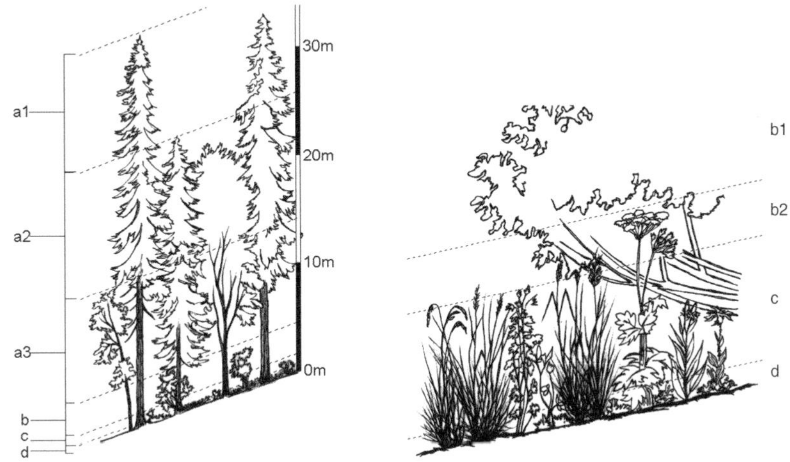

Fig. 5.8 Example of the vertical structure of a mixed forest (boreal and temperate). The six main layers are indicated with *dotted lines*. To the *right*, a detail for the high herbs. (from Ingegnoli, 2002)

E10: Dynamic State. The presence of a strong renewal of a species different from the dominant is indicative of the possibility of a change. If such renewal is accompanied by the presence of the same species in different vertical layers of the tessera, it indicates that the change is in place; otherwise, that there are some factor, or complex of factors, which are opposed, i.e. are able to incorporate this change, regardless of any evaluation of the positivity or otherwise of the "incorporation". The recognition of these phases on the field is not simple and is favoured by knowledge, or by some inspection also on tesserae surrounding the one directly concerned. In a very schematic way it can be stated that:

- *Degradation:* the characteristics detectable in the tessera are indicative of a *degradation* process, in place, in the conclusion or metastable, when non-incorporable disturbances are present: for example, when there is the advance or the penetration into the tessera of non-native or invasive plant species that take a consisting cover or biomass or are characterised by a renewal abundant, able to win the competition with native species; or if the tessera is reaching a state of widespread senescence, for lack of incorporable disturbances, or in the case of disruptive events; or when we are witnessing a gradual deconstruction (vertical or horizontal) for both natural and human events; or the strong reduction of biological forms present or even to a state of widespread disease of plants. With regard to the species richness, remember that even a high biodiversity may indicate degradation, while a low value of specific biodiversity does not necessarily indicate forest degradation (always refer to the average number of species per relief, then relate to the tessera, as indicated by different type of wood in Pignatti [23]).
- *Recreation*: the characteristics detectable in the tessera are indicative of a phase of *recreation* in progress or nearing completion when the tessera is changing in floristic composition or structure and organisation under the influence of forces from the surrounding tesserae or ecotopes. Generally this phase is the one that follows a "zero event" or a natural strong disturbance (e.g. floods, landslides, volcanic eruptions), or even a sudden, sharp, modification of multiyear balances and therefore the range of disorders to which the tessera appeared to have been adapted (for example, consider the case of the sudden cessation of a millennial silvopastoral activity managed by man, following the establishment of a protected area, or of a certain type and mode of cultivation, or even the sudden abandonment of certain areas of hill or mountain for urbanization). In all these cases also the "seed bank" contained in the soil or the forces of the

tessera may not be able alone to trigger a process of recovery. So the tessera must reorganize itself so completely. In practice it is recognisable by the presence of the quotas of characteristic species attributable to three or more different phytosociological associations or plant communities, usually from the surrounding tesserae so that, at the time of detection, it is not possible to predict where it will take the road taken, or from the behaviour weed taken by one or more of these, by strong characterisations of margin. In practice, the situation is nearest the concept of "succession" of traditional ecology.

- *Regeneration*: the characteristics detectable in the tessera are indicative of a phase of *regeneration*, in place or in phase of conclusion, when the tessera is recovering from an incorporable disturbance or by a phase of degeneration not particularly strong, or, again, it was in a phase of fluctuation having been followed by only a slight change of the range of disorders. The dynamics of the tessera is moved only by forces internal to the tessera itself, by its potential (think of the abandonment of a coppice of a species belonging to the floristic cortege of the vegetation of reference for that specific area or when it starts to high forest). The "plantulae" and seedlings belong in this case to the dominant species, the vertical structure denotes the reconstitution of a layer as missing or discontinuous, ... A distinctive feature is the fundamental absence, or low incidence, of species coming from different plant associations respect to the reference vegetation (*fittest vegetation*) of that area. In practice, only a change of order of the system occurs. Falls in this point also the case of a tessera in phase of normal growth.
- *Fluctuation*: the characteristics detectable in the tessera are indicative of a phase *fluctuation* at maturity in place when the tessera oscillates around a metastable position, without going into the field of instability. It is recognised by the change of only marginal characters, as in the case of a regular alternation of dominance between the two dominant species in the kind of vegetation (e.g. *Fagus sylvatica* and *Abies alba* in a beech forest) or for the input, or the output of some species, even gradual and recurrent, without changing the relations of dominance and alternation.

5.2.4 Analysis of the Vegetation Related to a Landscape Unit (LU)

U1: Contiguity with Similar Vegetation. It measures the percentage of the perimeter, on the total one, which is located in close or direct contact with vegetation of a similar type. To this end, it should first be determined which vegetation can be considered similar to those present in the tessera in question: all, but only those, pertaining to the phytosociological Alliance of reference.

- Other wooded tesserae referring to different phytosociological Alliances -a *Larici-Cembretum* (*Rhododendro-Vaccinion*) compared to an *Orno-Ostrietum* (*Orno-Ostryon*) or to a chestnut wood- are not considerable as similar to the tessera in question.
- Other similar shrubby vegetation or grassland or pasture or farming of any order, nor urban, suburban or industrial, are not considerable as similar vegetation (remember we are dealing with a forest tessera!).
- The roads that can be found in delimitation of the examined tessera constitute impediment to contiguity in case their width prevents contact between the crowns of the canopy of the two tesserae between which we examine the contiguity, or in case they are found widely trafficked, even if not asphalted.
- Fences are a hindrance to the contiguity in case they are continuous and and closed; they are not so if discontinuous or open (such as networks).

U2: Characteristics of Source or Sink. This item refers to the theory of source-sink [24] (see Sect. 3.5.2).

- *Sink*: The tessera is treated as *sink* if it is in an ecological state that only acts as a receptor of species or resources: it is generally (but each case must be assessed directly) proper to tesserae very young or very degraded; or coppice of land in recovery after a recent cut or tesserae traversed by fire; or survivors of an epidemic or an attack of insects, fungi or parasites, or even pieces of woody plants when held without undergrowth (e.g. poplar), or with a high number of non-native species, few biological forms and a few key species, etc.

- *Neutral*: The tessera is considered to be *neutral* if it is in an ecological state that will not be able to export neither species nor resources to the surrounding tesserae and it does not need them, or it is in conditions of substantial equality between import and exports: this pertains usually to tesserae partially degraded in recovery, or tesserae from extensive deforestation or coppice of age, or even pieces of woody plants managed in a more "environmentally friendly" way or even to tesserae in transition from youth to adult phases in good condition; or tesserae adults but degraded, or with a high number of cosmopolitan species or not characterizing ones in relation to the low number of characteristic species; or tesserae with individuals with suspected disease.
- *Partial source*: The tessera is treated as *partial source* if it is in an ecological state as to be able to export partially both species and resources to the surrounding tesserae: what happens to adult tesserae in good condition or to mature but partially degraded tesserae, or to tesserae with an average number of key species, biological forms and chorologically native species, not high ratio internal/margin (voice T4) or higher ratio but with internal signs of degradation.
- *Actual source*: The tessera is considered as *actual source* if it lies in ecological conditions such as to be a source of species and other resources for the surrounding tesserae: it pertains to mature or senescent tesserae or in part in good conditions, in the dynamic phase of fluctuation or regeneration, with a good vertical stratification, a large number of key species present, scarsity or absence of non-native species, large numbers of biological forms, high percentage of surface extension in good ecological conditions compared to the margin.

U3: Functional Role in L.U. The role of the tessera in question must be directly evaluated in function of the LU of belonging (i.e. the landscape unit, *not* of the ecotope of belonging), but should not be confused with the function of the source of point U2. It is rather to be evaluated considering the landscape functions that the tessera can play (or, for simplicity, the various landscape apparatuses to which could be ascribed). This role is a function primarily of the position of the tessera (or of the ecotope of belonging) within the ecological mosaic: the role of a forested tessera will be stronger as the matrix of LU will be of a different type (ascending: matrix forestry-pastoral, agricultural, rural, suburban, urban); in subordinate way, it is a function of the overall ecological status of the tessera as assessed by the previous entries: the best ecological status, the greater the role in context. It is necessary to emphasise that, since the role of the tessera is considered in the context of LU, more tesserae of the same forest ecotope with similar characteristics will have the same important, or even evident, role in respect of LU, even if they would have a minor role, or different between them, if considered rather to the same ecotope; in addition, also a tessera of low ecological quality may play an important role, or even remarkable, if inserted in a small wooded ecotope but belonging to a suburban or urban matrix.

U4: Incorporation of Disturbance. The capacity of disturbances incorporation at the level of tessera considers the local disturbances of the environment in which the tessera itself is inserted, in particular meteorological events (e.g. storms, tornadoes, hail, lightning, rain of different frequency and intensity) and geophysical events (e.g. soil creeping, small landslides, floods) recurring, albeit on a multi-year cycles, but the order of magnitude of which is comparable to that of the development of a forest tessera; or periodic or chronic anthropogenic activities (e.g. management interventions, wood collection, charcoal, steps or stationing of people for leisure or sporting activities or study); or, again, disorders related to the activities of wildlife, livestock and domestic (e.g. holes and removal of turf made from wild boar or the elimination of all young shrubs by goats or deer or fawns). If these disorders are located, not too frequent and affect only limited areas are incorporable. Otherwise, or in the case of events outside the temporal or spatial scale, the incorporation takes place at a scale greater than that of tessera.

U5: Geophysical Instability. We highlight the possible factors of instabilities linked to climatic and hydrogeological events and activities, in a geophysical sense, with cycles of recurrence also very long, to which the tessera is potentially subject, and that can be evaluated on the basis of information derived from reports and maps relating to geological, hydrological, hydrogeological "risks of instability" (geologists) or "risks of meteorological events" (agronomists).

U6: Interest of Permeant Fauna. The interest that the tessera has for the permeant fauna is generally a function proportional to the structural and functional characteristics of the tessera itself (when inserted in a forest ecotope), but also to the distribution of tesserae of the same type in the matrix (in particular, the interest may increase over time when the tessera constitutes almost an *unicuum* in the matrix, as in the case of one or a few tesserae of poplar wood cultivated in an agricultural matrix). For a proper assessment, it is necessary to understand the elements of attraction or interest for the characteristic fauna of that particular environment (e.g. different ecological niches, the place of supply of food, refuge from predators, nesting or shelter for night, nursery . . .)

U7: Reasons for Transformation of Ts. The dynamics of a tessera as part of the landscape is influenced by the context conditions, that is what happens within the ecotissue of a LU. They relate both to the reasons that led to the current state of the tessera, as well as those which will guide the transformations in the near future. They are identified as:

- *Loud noise* if concerning a net intervention performed directly on the tessera itself or on neighbouring tesserae (e.g. clear cutting, selective drastic cut, fire, attack of various diseases or parasites or other), or a strong transformation that took place on the edge not directly neighbouring but whose influence can invest the tessera in question (e.g. construction of a major road or of a railway or of an industrial or commercial infrastructure, sudden change of land use of adjacent estate to build new living quarters or services of various kinds or recreational area, or reservoirs for hydroelectric purposes).
- *Gradual change,* if there is a gradual and slow evolution of the ecotissue as a whole, within which also the transformation of the tessera is driven by the internal reorganisation of the system in search of a new threshold of metastability, which could be not higher than the previous. In practice, a gradual change takes place when the system modify step by step following the change in the range of disorders to which it was first submitted, slowly adapting its organisation to the new parameters. But this change can be addressed both to higher levels than toward lower levels of metastability and organisation (e.g. a same forest tessera can belong to both a forestry-pastoral landscape gradually evolving towards an opened forest landscape, then closed forest for abandonment of management activities, or to a forestry-pastoral landscape in gradual change towards a semi-agricultural one, then to industrialised agriculture).
- *Temporary instability* to mark the presence of an event that has created a disturbance in the LU as a whole, lowering the level of metastability, organisation and order, that is as a perturbation necessary, however, to a wider reorganisation of the total LU into a higher level of metastability (at the time not yet accomplished) and whose effects are also being felt on the tessera in question, although not being directly involved (because for this

tessera the perturbation could probably have assumed destructive effects). An example can be constituted by an accidental fire in a prairie, which does not substantially alter its composition nor flora nor fauna, but can change the dominance relationships among the species of both plants and animals that inhabit it.

- *Fluctuation* in the situation of an ecotissue that has reached a threshold of high metastability and that, therefore, oscillates around its attractor, allowing the tessera of ecotopes that compose it to evolve as a function of their intrinsic potential.

In case of discrepancy between the event that produced the current situation and the event that guides the transformations in the near future (as estimated on the basis of the knowledge possessed at the time of drafting of the board) sign the column corresponding to the event more negative between the two.

U8: Interference with Surrounding Landscape Pathologies. This item shows the real diseases of the surrounding LU that may have an influence on the tessera in question. They are differentiated according to their intrinsic gravity and in relation to the effects it may have on LU and then on the tesserae and the surrounding (where the radius of influence is different from case to case depending on the type of disease). A very important aspect to be verified in this item is the presence or absence of bark lichens on trees: according to the protocol parameters ANPA [25], the absence (desert lichens) or scarcity (under 15 species for *forophyta*) of the number of different species indicates a strong deterioration of air quality, maybe from the surrounding LU, even in forests in apparent good condition (see what species of trees are more favourable to the installation of lichens).

U9: Permanence of the Same Type of Vegetation (yr). We consider here the number of years, possibly reconstructed through historic maps, for which -on an on-going basis- that tessera has been being covered by tree vegetation, even if attributable to different phytosociological associations or plant communities. The ecological significance of this assessment is to be found in the importance that the availability of a time interval, appropriate to correct forest paedogenesis, plays for the quality and the health of a forest tessera. Therefore, this value can be totally different from that reported in Section T6: in fact, you can have a very young tessera (less than 80 years) on a ground which had been being covered by forest vegetation for thousands of years (more than 1200 years), or on the contrary a very mature tessera, even of more than 300 years of life, but in a position where previously there was an arable crop or maybe a suburban area, so the two values coincide.

Finally, remember that each Standard Form must be written out on the field but, then, it must be corrected and completed after deeper/wider studies.

5.2.5 The Main Vegetation Types and Their Ecological Parameters

The researches linked to the LaBiSV methodology allow its application in Boreal, Temperate and Mediterranean landscape systems, because the development of correct parametric standard forms in tropical regions had no financial support. We hope in next future.

After the presentation of the LaBiSV methodology and Tables 5.1 and 5.2, concerning Temperate deciduous plus Boreal-Alpine conifer forests and Mediterranean Pine plus Mediterranean Sclerophyllous forests, we expose in next pages the other Tables 5.7, 5.8, 5.9, 5.10, 5.11 and 5.12, representing the formations shown in Tables 5.5 and 5.6.

Each type of vegetation should need some notes to better explain the significance of the ecological parameters and their analysis on the field. But this book is not an application manual,[8] so we will write the principal observation useful to these tables in a general way, even because we are convinced that the most part of students and professionals interested in vegetation survey have a good knowledge of the procedures needed

[8] A previous explication, in Italian language, can be found in [13].

Table 5.5 Equations of the models of BTC development ranked for the main vegetation types available in Europe and maximum score values for parametric standard forms

Vegetation types	Flex value of BTC ($Mcal/m^2/year$)	Model maturity years (MmY)	Max Score (Ymax)	Equations of the model	
				Exponential	Logarithmic
Sclerophyllous forest	12.18	135	616	$BTC = T^{0.525} - 0.85$	BTC = 2.485 ln T
Temperate forest	11.50	130	616	$BTC = T^{0.515} - 0.77$	BTC = 2.363 ln T
Boreal alpine forest	10.48	125	700	$BTC = T^{0.50} - 0.70$	BTC = 2.170 ln T
Medit. pine forest	10.61	120	644	$BTC = T^{0.51} - 0.8$	BTC = 2.234 ln T
Corridor with trees	9.08	105	1,023	$BTC = T^{0.49} - 0.7$	BTC = 2.008 ln T
Urban Parks & gardens	7.59	80	1,170	$BTC = T^{0.48} - 0.6$	BTC = 1.793 ln T
Wooden agrarian	4.51	40	812	$BTC = T^{0.45} - 0.75$	BTC = 1.223 ln T
Tall shrubs	4.02	37	960	$BTC = T^{0.445} - 0.9$	BTC = 1.1112 ln T
Low shrubs	2.42	30	840	$BTC = T^{0.315} - 0.5$	BTC = 0.712 ln T
Agricultural fields	2.01	20	806	$BTC = T^{0.23} - 0.72$	BTC = 0.67 ln T
Prairie and pasture	1.30	25	928	$BTC = T^{0.20} - 0.8$	BTC = 0.339 ln T
Salt marshes prairie	1.05	15	336	$BTC = T^{0.20} - 0.9$	BTC = 0.303 ln T

Table 5.6 Evaluation equations of BTC ranked for the main vegetation types in Europe

Vegetation types	Max BTC value ($Mcal/m^2/year$)	Maturity thresholds (years)	BTC Evaluation equations ($Mcal/m^2/year$)
Sclerophyllous forest	12.5	120–150	0.01705 (y–28) + 0.13 (pB/60)
Temperate forest	12.0	120–140	0.01667 (y–28) + 0.13 (pB/65)
Boreal alpine forest	11.0	120–130	0.01339 (y–28) + 0.12 (pB/70)
Medit. pine forest	10.5	100–130	0.01510 (y–28) + 0.12 (pB/65)
Corridors with trees	9.8	90–120	0.0072 (y–33) + 0.10 (pB/75)
Urban parks & gardens	8.1	70–90	0.00526 (y–30) + 0.10 (pB/45)
Wooden agrarian	5.0	25–45	0.00575 (y–29) + 0.15 (pB/35)
Tall shrubs	4.2	30–40	0.00344 (y–30) + 0.10 (pB/17)
Low shrubs	2.6	25–35	0.00247 (y–30) + 0.03 (pB/0.2)
Reed thicket	2.2	36–48	0.0023 (y–29) + 0.04 (pB/0.3)
Agricultural fields	2.3	10–20	0.00192 (y–26) + 0.09 pB
Prairie and pasture	1.35	20–30	0.001335 (y–29) + 0.02 (pB/0.14)
Salt marshes prairie	1.1	15–20	0.0026 (y–28) + 0.10 (pB/1.4)

Data from: Ingegnoli 2002; Ingegnoli &Giglio 2005
Note that some descriptors (as Corridors with trees or Tall shrubs) are physiognomic, because the differences among vegetation types in these cases do not change the BTC evaluation in a too sensible way. In case of landscape unit in which prevail these physiognomic types, they can be expressed in specific vegetation types, e.g. Rhododendron shrubs, etc. after adjusting some ecological parameters

by these standard forms. It could be necessary to propose a formula available to suggest the minimum number of Relascope stations (R,St) in a forest survey, related to the dimension (ha) of the tessera (Ts):

$$n(\mathrm{R,St}) = 2\sqrt{\mathrm{ha}}(\mathrm{Ts}) + 1$$

A more accurate survey needs 1 more R,St:

$$n(\mathrm{R,St}) = 2\sqrt{\mathrm{ha}}(\mathrm{Ts}) + 2$$

Remember that at a statistical level of 95 % the percentage error cannot exceed 10–15 % in a speed survey and 5–10 % in a detailed one.

Table 5.7 LaBiSV parametric form for tall and low shrubs: ecological evaluation of the tessera quality (bQ) and estimation of biologic territorial capacity of vegetation (BTC)

Tall shrubs (a)	1	6	16	32	Score
Low shrubs (b)	1	5	13	28	Score
T. Tessera characters (Ts)					
T1a—Height (m)	<0.8	0.81–2.0	2.01–3.5	>3.5	Dominant shrubs
T1 b—Height (m)	<0.2	0.2–0.6	0.6–1.0	> 1	Dominant shrubs
T2—Cover of vegetation (%)	<30	>90	31–60	61–90	Related to Ts surface
T3—Grass patches (%)	50–65	36–50	21–35	>20	If <65 % of the Ts
T4—Tall/low shrubs (%)	35–50	25–35	15–25	>15	Ratio
T5—Structural differentiation	Low	Medium	Good	High	Age, space, Sp. groups, …
T6—Interior/edge (%)	None	<30	31–89	>90	(% Ts)
T7—Management	Planted	Managed	Near-natural	Natural	Or similar
T8—Permanence (years)	<20	21–80	81–120	>120	Oldest shrubs
F. Vegetational biomass (aboveground)					
F1—Dead plant biomass	Near 0	>10	2–6	6–10	% on living biomass
F2—Litter depth	Near 0	<1.5	1.2–3.5	>3.5	cm
F3a—pB volume (m^3/ha)	<5	5–25	25–75	>75	Pl. biomass (pB)
F3b—pB dried (kg/m^2)	<1.5	1.5–3.5	3.5–5.0	>5.0	Pl. biomass (pB)
E. Ecocoenotopes parameters					
E1—Dominant species (n°)	Not clear	1	2	>2	On pB value
E2—Species richness	>15	16–30	31–44	>45	n° sp./Tessera
E3—Key species presence (%)	<5	6–20	21–75	>75	Phytosociological
E4—Allocthonous species (%)	>10	10–4	<4	0	From other ecoregions
E5—Infesting plants (%)	Near all	>25	<25	0	Covering area on Ts
E6—Diseased plants	Evident	Suspect	Risk	0	All pathologies
E7—Biological forms (n°)	<3	4–5	6–7	>7	See Box 1987, modified
E8—Vertical stratification	1	2	3	4	From musk to small tree
E9—Renewal capacity	None	Sporadic	Normal	Intense	Dominant species
E10—Dynamic state	Degradation	Recreation	Regeneration	Fluctuation	Transform. process
U. Landscape unit (LU) parameters					
U1—Similar veget. contiguity	0	<25	26–75	>76	% of perimeter
U2—Source or sink	Sink	Neutral	Partial	Actual source	Species & resources
U3—Functional role in LU	Reduced	Minor	Evident	Important	Context & typology
U4—Disturbances incorporation	Insufficient	Scarce	Normal	High	Local disturbances
U5—Geophysical instabilities	Evident	Partial	Risk	0	On the physiotope
U6—Permeant fauna interest	Low	Medium	Good	Attraction	Key species
U7—Transformation modalities of the Ts	Strong disturbances	Gradual changes	Temporal instabilities	Fluctuation	Today + tendency
U8—Landscape pathology interference	Serious	Near chronicle	Easy to incorporate	0	From landscape
U9—Permanence of analogous vegetation (years)	<50	50–150	151–300	>300	Historical presence
Results of the survey					
Total score Y (= h+j+k+w)	h = …….	j =…….	k = …….	w = ……	Y= ………………
Bionomic quality of Ts		**bQ$_a$** = Y/960	**bQ$_b$** = Y/840		bQ = …………….[%]
Estimation of the BTC	**BTC** (a) = 0.00344 (Y–30) + 0.10 (Fm/17)				BTC = ……….[Mcal/m^2/a]
	BTC (b) = 0.00247(Y–30) + 0.03 (Fm/0.2)				

Note: *T3*—Patches of grass (%). When the shrubberies contain grass patches, these herbaceous formations must not exceed 35 %

T4—Presence of different types of shrubs (%). In mixed shrubberies (tall and low patches) we have to see the dominance as covered surface

Table 5.8 LaBiSV parametric form for grass/prairies and reed thickets: ecological evaluation of the tessera quality (bQ) and estimation of biologic territorial capacity of vegetation (BTC)

Grass & pastures (p)	1	4	13	32	Score
Reed thickets (c)	1	6	17	37	Score
T. Tessera characters (Ts)					
T1p—Vegetation height (m)	<0.2	0.2–0.6	0.6–1.0	>1.0	Dominant layer
T1 c—Vegetation height (m)	<0.7	0.7–1.5	1.5–4.0	> 4.0	Dominant layer
T2—Vegetation cover (%)	<30	>90	31–60	61–90	Ts surface
T3—Shrub presence (%)	<5	5–15	15–25	>25	<34 % of TS
T4—Structural differentiation	Low	Medium	Good	High	Age, space, Sp. groups, ...
T5—Edges	None	>30	30–10	<10	(% Ts)
T6—Management	Artificial	Near natural	Nat. grazed	Natural	Or similar
T7—Permanence (years)	<20	21–60	61–120	>120	See also old shrubs
F. Vegetational biomass (aboveground)					
F1—Dead plant biomass	>60	60–21	<20	Quasi 0	% on living biomass
F2p—Litter depth	Quasi 0	<1.5	1.2–3.5	>3.5	cm
F2c—Litter depth	Quasi 0	<5	5–10	>10	cm
F3p—Plant biomass (kg/m^2)	<0.6	0.6–1.2	1.2–2.0	>2.0	Dried pB
F3c—Plant biomass (kg/m^2)	<1	1–2.5	2.5–4	>4	Dried pB
E. Ecocoenotope parameters					
E1p—Dominant species (n°)	Not clear	1	2–3	>3	Cover
E1c—Dominant species (n°)	Not clear	>2	2	1	Cover
E2p—Species richness	>15	16–25	26–40	>40	n° sp./Tessera
E2c—Species richness	<10	10–20	20–30	>30	n° sp./Tessera
E3—Key species (%)	<5	6–20	21–75	>75	Phytosociological
E4—Allocthonous species (%)	>10	10–2	<2	0	From other ecoregions
E5—Infesting plants (% Ts)	All	>25	<25	0	Covering area on Ts
E6—Diseased plants	Evident	Suspect	Risk	0	All pathologies
E7—Biological forms (n°)	<2	3	4	>4	See Box 1987, mod.
E8—Vertical stratification	1	2	3	4	From musk to shrub
E9—Renewal capacity	None	Sporadic	Normal	Intense	Dominant species
E10—Dynamic state	Degradation	Recreation	Regeneration	Fluctuation	See Ingegnoli 2002
U. Landscape unit (LU) parameters					
U1—Similar vegetation contiguity	0	<20	21–80	>80	% of perimeter
U2—Source & sink	Sink	Neutral	Partial	Actual source	Species & resources
U3—Functional role in LU	Reduced	Minor	Evident	Important	Context & typology
U4—Disturbances incorporation	Insufficient	Scars	Normal	High	Local disturbances
U5p—Geophysical instabilities	Evident	Partial	Risk	0	On the physiotope
U5c—Hydrological instability	Floods	No water	Irregular water	Seasonal variation	Water flooding
U6—Permeant fauna interest	Low	Medium	Good	Attraction	Key species
U7—Transformation modalities of the Ts	Strong disturbances	Gradual changes	Temporal instabilities	Fluctuation	Today + tendency
U8—Landscape pathology interference	Serious	Near chronicle	Easy to incorporate	0	From landscape
U9—Permanence of analogous vegetation (years)	<70	70–150	150–300	>300	Historical presence
Results of the survey					
Total score Y (= h+j+k+w)	h = ...	j =.......	k = ...	w = ...	Y= ....
Bionomic quality of Ts	**bQ$_c$** = Y/1073		**bQ$_p$** = Y/928		bQ = ...[%]
Estimation of the BTC	**BTC** (p) = 0.001335 (Y–29) + 0.02 (Fm/0.14)				BTC = ...[Mcal/m^2/a]
	BTC (c) = 0.0023 (Y–29) + 0.04 (Fm/0.3)				

Note: *T5*—Edges. A marginal belt of about 5–10 times the mean height of the herbaceous formation is generally characterised by the presence of species from surrounding tesserae

F3—Phytomass. In herbaceous formations the plant biomass is measured in kg/square- meters. So, if we do not find opportune data, we need to mow some sample areas aboveground, dry hay and weigh it

Table 5.9 LaBiSV parametric form for corridor with trees: ecological evaluation of the tessera quality (bQ) and estimation of biologic territorial capacity of vegetation (BTC)

Corridor with trees	1	7	17	31	score
Cd. CORRIDOR CHARACTERS (Cd)					
Cd1—Wide of the corridor (Wc)	<2.5	2.6–10	10.1–20	>20	Canopy expansion (m)
Cd2—Canopy height (h)	<3	3–12	12–24	>24	Weight average (m)
Cd3—Tree cover	<30	30–60	60–90	>90	(%) of the tract
Cd4—Presence of stream (Str)	0	Canal	Near natural	Natural	Str wide <Wc
Cd5—Road presence (Rd)	Traffic	Paved	Rural	0	Rd wide <Wc
Cd6—Interruptions (>h)	>4	3–4	1–2	0	N° in the tract
Cd7—Linearity	Rectilinear	Near-rectilinear	Mixed	Irregular	Tract
Cd8—Management	Strong cut	Pruning	Light pruning	None	Human control
Cd9—Permanence (years)	<60	60–120	120–180	>180	Old trees
F. PLANT BIOMASS Aboveground					
F1—Dead plant biomass	Near 0	Too much	Medium	Normal	% on living biomass
F2—Litter depth	Quasi 0	<1.5	1.2–3.5	>3.5	cm
F3—Plant biomass in volume	<100	100–300	300–500	>500	pB = m^3/ha
E. ECOCOENOTOPE PARAMETERS					
E1—Dominant species (n°)	1	1–2	2–3	>3	On pB value
E2—Species richness (n°)	>15	16–29	30–44	>45	Tract
E3—Key species (%)	<5	6–20	21–75	>75	Phytosociological
E4—Allocthonous species (%)	>10	10–4	<4	0	From other ecoregions
E5—Infesting plants (% Ts)	All	>25	<25	0	Covering area on Ts
E6—Diseased plants	Evident	Suspect	Risk	0	All pathologies
E7—Biological forms (n°)	<3	4–5	6–7	>7	See Box 1987, modified
E8—Vertical stratification	1	2	3	4	Musk to tall Trees
E9—Renewal capacity	None	Sporadic	Normal	Intense	Dominant species
E10—Dynamic state	Degradation	Recreation	Regeneration	Fluctuation	Transform. process
U. LANDSCAPE UNIT (LU) PARAMETERS					
U1—Internal species presence	0	Sporadic	Few	Many	e.g. forest Species
U2—Lichen presence	0	1–15	16–35	>35	N° of bark species
U3—Source & sink	Sink	Neutral	Partial	Actual source	Species & resources
U4—Connections	0	1	2	>2	With other corridors
U5—Network participation	None	Potential	Partial	Effective	Today situation
U6—Bird fauna exchanges	None	Reduced	Normal	High	With the landscape
U7—Mammal fauna presence	None	Reduced	Normal	High	Around the corridor
U8—Apparatus functions	Not clear	1	2	>2	Landscape apparatuses
U9—Corridor presence in LU	<80	80–160	160–240	>240	years
U10—Disturbances incorporation	Insufficient	Scars	Normal	High	Local disturbances
U11—Landscape type of LU	Urban	Suburban	Rural	Near-natural	Landscape uunit (LU)
RESULTS OF THE SURVEY					
Total score Y (= h+j+k+w)	h = ...	j =.......	k = ...	w = ...	Y= ....
Bionomic quality of the Ts	bQ = Y/1023				bQ = ... [%]
Estimation of BTC	BTC (cd) = 0.0072 (Y-33) + 0.10 (Fm/75)				**BTC =** ....[Mcal/m^2/a]

Note: *Cd1*—Corridor width. We have to measure the distance between the expansions of the tree canopies (of the marginal trees, in case of multi-row corridors)
Cd6—Corridor interruptions. If we find breaks >*h* (mean height of the trees), they have to be counted in a stretch of corridor = 10 *h*

Table 5.10 LaBiSV parametric form for urban parks and gardens: ecological evaluation of the tessera quality (bQ) and estimation of biologic territorial capacity of vegetation (BTC)

Urban Parks and Gardens	1	7	20	39	Score
T. GARDEN CHARACTERS					
T1—Canopy height (m)	<6	6–12	12–24	>24	Weight average
T2—Canopy cover (%)	<25	25–50	50–75	>75	Ts
T3—Underground	Built	Terraced	Mixture	Natural soil	See also percolation
T4—Garden design	Technical	Formal	Mixture	Natural	general composition
T5—Peculiar species	Exotic	Rare exotic	Local	Monumental	characterizing garden
T6—Management	Strong pruning	Light pruning	Marginal	None	or similar
T7—Permanence (years)	<50	50–100	100–250	>250	Oldest trees
F. PLANT BIOMASS Aboveground					
F1—Litter depth	Quasi 0	<1.5	1.5–3.5	>3.5	cm
F2—Plant Biomass volume	<100	100–200	200–300	>300	pB = (m^3/ha)
E. ECOCOENOTOPE PARAMETERS					
E1—Dominant species (n°)	Not clear	3	2	1	On pB value
E2—Species richness	>10	11–25	26–40	>40	n° sp./Garden
E3—Key species (%)	<5	6–20	21–50	>50	Phytosociologic
E4—Allocthonous species (%)	>30	11–30	5–10	<4	Outside the ecoregion
E5—Diseased plants	Evident	Suspect	risk.	0	All pathologies
E6—Biological forms (n°)	<3	4–5	6–7	>7	See Box 1987, modified
E7—Vertical stratification	2	3	4	>4	from musk to high tree
E8—Structural differentiation	Low	Medium	Good	High	age, groups
E9—Renewal capacity	None	Sporadic	Normal	Intense	Dominant species
E10—Green proportion	<0.3	0.31–0.50	0.51–0.75	>0.75	
U. LANDSCAPE UNIT (LU) PARAMETERS					
U1—Similar vegetation contiguity	0	<10	11–50	>50	% of perimeter
U2—Internal species presence	0	sporadic	few	many	e.g. forest Species
U3—Lichen presence	0	1–15	16–35	>35	N° of bark species
U4—Ecological network	none	neutral	potential	effective	garden participation
U5—Apparatus functions	not clear	1	2	>2	Landscape apparatuses
U6—Functional role in LU	reduced	minor	evident	important	Context & typology
U7—Disturbances incorporation	insufficient	scars	normal	high	Local disturbances
U8—Geophysical instabilities	evident	partial	risk	0	On the physiotope
U9—Permeant fauna interest	low	medium	good	attraction	Key species
U10—Transformation modalities of the Ts	strong disturbances	gradual changes	temporal instabilities	fluctuation	Today + tendency
U11—Landscape pathology interference	serious	near chronicle	easy to incorporate	0	From landscape
RESULTS OF THE SURVEY					
Total score Y (= h+j+k+w)	h = ...	j =.......	k = ...	w = ...	Y= ....
Bionomic quality of the Ts	bQ = Y/1170				bQ = ... [%]
Estimation of the BTC	BTC (vu) = 0.00526 (Y−30) + 0.1 (Fm/45)				**BTC =** .... $[Mcal/m^2/a]$

Note: *T1*—Mean height of vegetation (m). We have to measure only the height of the trees present
T4—Garden design. Remember that the term “technical” is referred to sport green, in which sport facilities dominate the design. The term “formal” indicate a so-called Italian or French garden design. The term “near-natural” or naturaliforme is typical of the so-called English garden
U4—Green/urban ratio. This ratio measures the examined garden in relation with the green areas of the entire urban (or suburban) ecotope, comprise forest remnants

Table 5.11 LaBiSV parametric form for wooden agrarian: ecological evaluation of the tessera quality (bQ) and estimation of biologic territorial capacity of vegetation (BTC)

Wooden Agrarian	1	6	15	28	Score
T. TESSERA CHARACTERS (Ts)					
T1—Canopy height (m)	<1	1–2.5	2.5–4.5	>4.5	Canopy Hc=....m
T2—Canopy cover (%)	<15	15–30	30–50	>50	Ts
T3—Field shape	Geometrical	Polygonal	Irregular	Natural	Ts
T4—Grass presence	0	<20	20–50	>50	(% Ts)
T5—Management	Industrial	Near industrial	Traditional	Biological	Cultivation
T6—Technical structures	Plastic	Metal	Wood	None	Sustains & irrigation
T7—Permanence (years)	<20	21–80	81–120	>120	Oldest trees
F. PLANT BIOMASS Aboveground					
F1—Dead plant biomass	Quasi 0	>10	1–5	5–10	% on living biomass
F2—Litter depth	Quasi 0	<1.5	1.2–3.5	>3.5	cm
F3—Plant biomass in volume	<25	26–75	76–125	>125	pB = (m^3/ha)
E. ECOCOENOTOPE PARAMETERS					
E1—Species richness (n°)	<10	10–20	20–30	>30	n° sp./Tessera
E2—Allocthonous species (%)	>10	10–2	<2	0	Outside the ecoregion
E3—Presence of natural phytocoenosis	0	Sporadic	Marginal	Patches	Remnants
E4—Trees & shrubs	0	1	few	Many	Non cultivated plants
E5—Diseased plants	Evident	Few plants	risk	0	All pathologies
E6—Genetic characters	Transgenic	Allocthonous	Current	Traditional	Cultivars
E7—Biological forms (n°)	<3	4–5	6–7	>7	See Box 1987, modified
E8—Chemicals	>4	3	1–2	0	All types
E9—Soil fertility classes	IV	III	II	I	Cfr. Land capability
E10—Presence of margins	0	<50	50–80	80–100	% perimeter
U. LANDSCAPE UNIT (LU) PARAMETERS					
U1—Contagion with natural elements	0	<10	11–50	>50	% del perimeter
U2—Ecological network	0	Marginal	Partial	Effective	Presence
U3—Source & sink	Sink	Neutral	Partial	Actual source	Species & resources
U4—Functional role in LU	Reduced	Minor	Evident	important	Context & typology
U5—Soil preparation	Technical	Mixed	Marginal	0	e.g. tilling
U6—Geophysical instabilities	Evident	Partial	Risk	0	On the physiotope
U7—Permeant fauna interest	None	Medium	Near good	Attractive	*Ts*/key sp.
U8—Landscape pathology interference	Serious	Near chronicle	Easy to incorporate	0	From landscape
U9—Permanence of analogous vegetation (years)	<50	50–150	150–300	>300	Historical
RESULTS OF THE SURVEY					
Total score Y (= h+j+k+w)	h = ...	j =.......	k = ...	w = ...	Y= ....
Bionomic quality of the Ts	bQ = Y/812				**bQ** = ... [%]
Estimation of the BTC	BTC (p) = 0.00575 (Y–29) + 0.15 (Fm/35)				**BTC** = ...[Mcal/m^2/a]

Note: See Sect. 5.2

Table 5.12 LaBiSV parametric form for agricultural fields: ecological evaluation of the tessera quality (bQ) and the estimation of biologic territorial capacity of vegetation (BTC)

Agricultural Fields	1	5	15	31	score
T. TESSERA CHARACTERS (Ts)					
T1—Vegetation height (m)	<0.5	0.5–1.0	1–2	>2	Weight average
T2—Field shape	Geometrical	Polygonal	Irregular	Natural	Ts
T3—Tree presence	0	1	Few	Many	Even at the edge
T4—Management	Industrial	Near industrial	Traditional	Biological	Cultivation
T5—Irrigation	Technical	Near technical	Canal	Near natural	Main method
F. PLANT BIOMASS Aboveground					
F1—Dead plant biomass	0	Low	Medium	High	On soil
F2—Litter depth	Quasi 0	<1.5	1.5–3.5	>3.5	cm
F3p—Plant biomass in volume	<1	1–2	2–3	>3	Dried pB = Kg/m^2
E. ECOCOENOTOPE PARAMETERS					
E1—Species richness	>10	11–20	21–30	>30	n° sp./Tessera
E2—Allocthonous species (%)	>10	10–2	<2	0	Outside the ecoregion
E3—Natural phytocoenosis	0	Sporadic	Marginal	Patches	Species presence
E4—Diseased plants	Evident	Few plants	Risk	0	All pathologies
E5—Genetic characters	Transgenic	Allocthonous	Current	Traditional	Cultivars
E6—Chemicals	>4	3	1–2	0	All types
E7—Soil alterations	Patches	Spots	Marginal	0	Ts
E8—Soil fertility classes	IV	III	II	I	Cfr. Land Capability
E9—Soil preparation	Technical	Mixed	Marginal	0	e.g. tilling
E10—Presence of margins	0	<50	50–80	80–100	% perimeter
U. LANDSCAPE UNIT (LU) PARAMETERS					
U1—Contagion with natural elements	0	<10	11–50	>50	% del perimeter
U2—Ecological network	0	Marginal	Partial	Effective	Presence
U3—Functional role in LU	Reduced	Minor	Evident	Important	Context & typology
U4—Permeant fauna interest	None	Medium	Near good	Attractive	*Ts*/key sp.
U5—Geophysical instabilities	Evident	Partial	Risk	0	On the physiotope
U6—Landscape pathology interference	Serious	Near chronicle	Easy to incorporate	0	From landscape
U7—structural & functional congruence with LU	Contrast	Poor	Near adequate	Good	Main parameters
U8—Permanence of analogous field (years)	<25	26–100	101–200	>200	Even diverse cultivation
RESULTS OF THE SURVEY					
Total score Y (= h+j+k+w)	h = ...	j =......	k = ...	w = ...	Y= ....
Bionomic quality of the Ts	Q = Y/806				**bQ** = ... [%]
Estimation of the BTC	BTC (ca) = 0.00192 (Y–26) + 0.09 Fm				**BTC** = ...[$Mcal/m^2/a$]

Note: *E1*—Species richness. Three components: (a) the cultivated species, (b) the so-called pest species, (c) the species of the edges (e.g. a grass belt). Any isolated trees must be counted
E8—Soil capacity classes. We can follow—even in Europe—the classical USDA Land Capability Classification

5.3 Other Vegetation Analysis

5.3.1 The Bionomic State of Vegetation

We wrote that an integrated evaluation is essential to check landscape pathology. As already underlined, the BTC function is very useful, but in some analysis we need to consider both the metastability and the efficiency of vegetation. In this frame, it emerges the necessity to refer to the new bionomic index, the "concise bionomic state" (CBSt) of vegetation [26], already introduced in Sect 3.3.2, linked with the state of a vegetation patch, relating the maturity level of a vegetation phytocoenosis (MtL) and its bionomic quality (bQ).

The CBSt of a vegetation patch should be reached considering: (1) the significance of the surveyed BTC of the patch in relation with the "maturity level" (MtL) of its vegetation coenosis and (2) its bionomic quality (bQ) always resulted from the parametric survey. Therefore, as already written, this function has to be designed as:

$$\mathrm{CBSt} = (\mathrm{MtL} \times \mathrm{bQ})/100. \quad (5)$$

The variable bQ of the eq. (5) derives from the parametric standard form, being the ratio between each surveyed parametric scores (Y_{surv})[9] and its maximum (Y_{max}), ranked in Table 5.5:

$$\mathrm{bQ} = Y_{surv}/Y_{max} \quad (6)$$

In Table 5.13, the example of the tessera of spruce forest (Lavazé Pass)- containing the plot TRE-1 surveyed in 2013, 9 years after the CONECOFOR programme- results in bQ = 592/700 = 84.57 %.

The variable "maturity level" MtL is not so simple, being related to the exponential equation of the models of BTC development during time, changing in accord to the different types of vegetation, as shown in Table 5.5. Generalising Eqs. (1) or (3), we have:

$$\mathrm{BTC} = T^a b \quad (7)$$

where BTC [$\mathrm{Mcal/m^2/year}$] is the biological territorial capacity of an examined vegetation tessera (Ts), *T* is time [years] and *b* is an opportune coefficient, from which

$$T = (\mathrm{BTC} + b)^{1/a} \quad (8)$$

Note that T represents the theoretical age of maturity.

If we want to compare the surveyed BTC (sBTC) with the flex of the model to measure the real level of maturity (MtL), we have to solve the following equation:

$$\mathrm{MtL} = sT/(\mathrm{MmY}) \times 100, \quad (9)$$

where MmY is the model maturity (years) from Table 5.5 and sT is the surveyed age of the tessera.

In the case of a boreal-alpine conifer forest (Table 5.1), the Eqs. (8) and (9) become:

$$sT = (\mathrm{sBTC} + 0.70)^{1/0.50} \text{ and} \quad (10)$$
$$\mathrm{MtL} = (sT/125) \times 100;$$

Thus

$$\mathrm{MtL} = (\mathrm{sBTC} + 0.70)^{1/0.50}/(125/100). \quad (11)$$

Therefore, applying the equation to the Lavazé example (Table 5.13), it results:

$$\mathrm{MtL} = (9.36 + 0.70)^2/1.25 = 81.29\ \%.$$

The meaning of the evaluated maturity levels (MtL) may be understood observing the curve of the model (e.g. Fig. 5.4). It is possible to read 5 phases of development: (a) first juvenile (0–19 years); (b) juvenile (19–44); (c) pre-adult (44–88); (d) adult (88–125); (e) mature (>125). Every class corresponds to maturity ranges, as shown in Table 5.14. The examined tessera of Lavazé presents a MtL = 81.29 which corresponds to an age of about 105–110 years

[9] An example referred to table 5.13 give this result: Ysurv = (h × 1) + (j × 5) + (k × 14) + (w × 25) = 0+5+ (8 × 14) + (19 × 25) = 592. In case of mixed forests of two different types quite equal in dominance, like a coniferous and broadleaves forest, use a maths average between the two values of each score, that is for example (25 + 22)/2 = 23.5 and so (w × 23.5).

Table 5.13 LaBiSV parametric form for boreal forests: survey of the CONECOFOR TS of Lavazé Pass (Trentino-Sud Tirol) spruce forest, altitude 1,780 m, August 2013

Boreal forest	1	5	14	25	Score
T. Tessera characters (Ts)					
T1—Vegetation height (m)	<9	9.1–18	18.1–29	**>29.1**	Canopy Hc = 31.50 m
T2—Cover of the canopy (%)	<30	>90	31–60	**61–90**	Ts surface
T3—Structural differentiation	Low	Medium	Good	**High**	Age, space groups, etc.
T4—Interior/edge (%)	None	<30	31–89	**>90**	(% Ts)
T5—Management	Simple coppice	Coppice	Wood	**Natural forest**	Or similar
T6—Permanence (years)	<80	81–160	**161–240**	>240	Old trees
F. Vegetational biomass (aboveground)					
F1—Dead plant biomass	Near 0	>10	**1–5**	5–10	% of living biomass
F2—Litter depth	Near 0	<1.5	1.6–3.5	**>3.5**	cm
F3 b—Biomass volume (m^3/ha)	<200	201–500	501–950	**>950**	pB = 1052 m^3/ha
E. Ecocenotope parameters					
E1—Dominant species (n°)	>3	3	2	**1**	As pB volume
E2—Species richness	<15	16–30	**31–40**	>40	n° sp./Tessera
E3—Key species presence (%)	<5	6–40	**41–75**	>75	Phytosociological
E4—Allocthonous species (%)	>10	10–4	<4	**0**	From other ecoregions
E5—Infesting plants %	Near all	**>25**	<25	0	Coverage on Ts
E6—Diseased plants	Evident	Suspect	Risk	**0**	Even acid rain damage
E7—Biological forms (n°)	<3	4–5	6–7	**>7**	See Box [15], mod.
E8—Vertical stratification	2	3	4	**>4**	Traditional
E9—Renewal capacity	None	Intense	Sporadic	**Normal**	Dominant species
E10—Dynamic state	Degradation	Recreation	Regeneration	**Fluctuation**	see Ingegnoli [10]
U. Landscape unit (LU) parameters					
U1—Similar veg. contiguity	0	<25	**26–75**	>76	% of perimeter
U2—Source or sink	Sink	Neutral	**Partial**	Actual source	Species & resources
U3—Functional role in LU	Reduced	Minor	Evident	**Important**	Context & typology
U4—Disturbances incorporation	Insufficient	Scarce	**Normal**	High	Local disturbances
U5—Geophysical instabilities	Evident	Partial	Risk	**None**	On the physiotope
U6—Permeant fauna interest	Low	Medium	**Good**	Attraction	Key species
U7—Transformation modalities of the Ts	Strong disturbances	Gradual changes	Temporal instabilities	**Fluctuation**	Today + tendency
U8—Landscape pathology interference	Serious	Near chronicle	Easy to incorporate	**None**	From landscape
U9—Permanence of analogous vegetation (years)	<100	100–300	300–1200	**>1200**	Historical presence
Results of the survey					
Total score Y (= h+j+k+w)	h = **0**	j = **1**	k = **8**	w = **19**	Y = **592**
Bionomic quality of the Ts (bQ)	bQ = Y/700 (b) [616] (t)				**bQ = 84.57** [%]
Estimation of the BTC Boreal (conifer) Forest	BTC (b) = 0.01339 (y–28) + 0.12 (pB/70)				**BTC = 9.36** [Mcal/m^2/year]

(average), confirmed by the survey: the oldest trees found are about 200 years old.

Coming back to equation 5, the measure of CBSt is now available. In our case study of Lavazé, it results: CBSt = (84.57 × 81.29)/100 = 68.98 for the tessera. The meaning of the evaluated concise bionomic state (CBSt) may be understood comparing 5 classes of CBSt (Table 5.15): (a) precarious or juvenile (0–6), (b) weak (6–18), (c) normal (18–36), (d) good (36–60), (e) optimum (60–90). The example of the Lavazé tessera shows an optimum bionomic state.

Note that the two scales of evaluation are different, because bQ and MtL have not similar ranges of values. As exposed forward, the eq. (5) gives the possibility to evaluate the level of organisation of vegetation, both for natural and human formations.

Table 5.14 Evaluation classes of maturity levels (MtL) in temperate and boreal forests

Levels of maturity (MtL)	MtL classes	MtL % of maturity	Model time (years)	Oldest trees
First development (or degradation)	A	0–15	0–19	28
Juvenile phases (or alteration)	B	15–35	19–44	66
Adult phases	C	35–70	44–88	132
Adult–Maturity	D	70–100	88–125	198
Maturity–Old	E	>100	125–160	240

Table 5.15 Evaluation classes of concise bionomic state (CBSt) in vegetated elements

Concise bionomic state (CBSt)	CBSt classes	CBSt (%)
Precarious (or juvenile) state [− −]	A	0–6
Weak state [−]	B	6–18
Normal state [=]	C	18–36
Good state [+]	D	36–60
Optimum state [++]	E	60–90

We will see on vegetation evaluation (chapter 9), the comparison between the CBSt of Bollate (Brianza) and Asse (Brabant).

5.3.2 Chorological Analysis and Biological Indexes

Chorology. To the floristic list of each tessera, joins, for each species, the attribution of chorology. On the basis of the mere presence/absence of species (i.e. regardless of their coverage in the relief or class of presence at first), the presence percentage in the survey itself is calculated for each chorological type.

It must, however, reduce the number of chorologic types, for merging the affine area of origin. A typical sequence for the pre-alpine environment may be: (1) Euro-Siberian, (2) Circum-boreal, (3) Alpine endemic, (4) Orophyite, (5) Steppic, (6) Euro-Caucasian, and Eurasian paleo-temperate (7) Atlantic, (8) Euri-Mediterranean, (9) Steno-Mediterranean, (10) Sub-Cosmopolitan and Cosmopolitan, (11) Exotic.

Look at the comparison (Fig. 5.9) among some typical distribution of the association or alliance of reference. Very useful, in northern Italy, it may be the calculation of the value of the ratio Mediterranean/Circum-boreal. This ratio is much higher as the detected species are better adapted, among other aspects, to high annual temperatures.

As suggested by Biondi and Baldoni [27], the comparison of chorologic data of a LU with the chorologic spectra at syntaxonomic alliance level can be useful in verifying the differences of local plant communities with the European statistical averages by type of alliance.

Bioindicators of Ellenberg. Completion of the work done by Pignatti [28] in order to adapt the biological indexes of Ellenberg to the Italian flora, they can also be used for interesting comparisons between floristic lists of different LU. It should be noted that these indicators are: (L) light, (T) temperature, (C) continentality, (H) moisture, (R) reactions of the soil, (N) soil nutrients (Fig. 5.10).

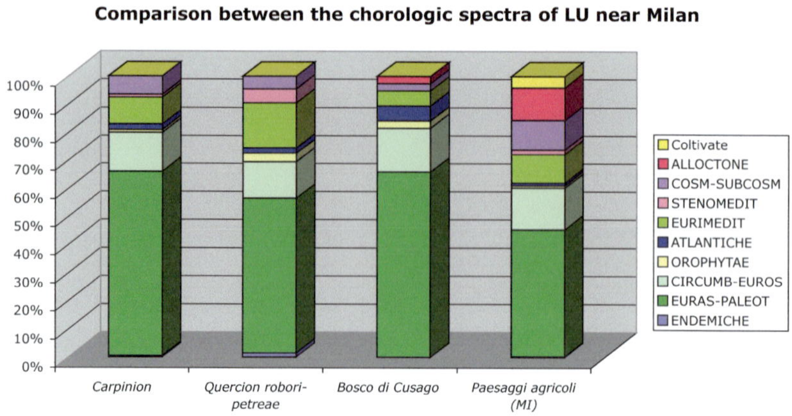

Fig. 5.9 Comparison between the chorologic spectra of a remnant forest in Cusago and the average of 10 agricultural LU near Milan (Lombardy). The spectra of the two phytosociological alliances refer to theoretical formations reported by Biondi and Baldoni [25]

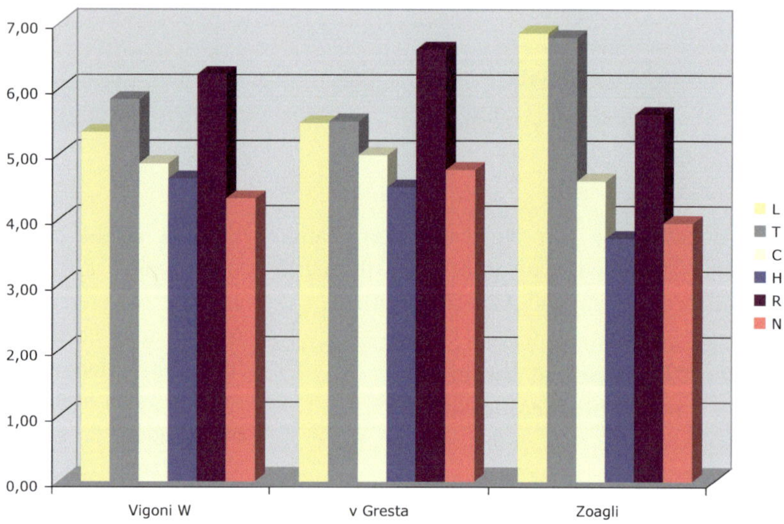

Fig. 5.10 Ellenberg biological indexes of forest vegetation on the lower Val di Gresta, compared with a similar foothills of the Alps (Woods of Villa Vigoni, Loveno-Menaggio) and with a LU Mediterranean (Zoagli). Despite the influence of Lake Garda and the trend of warming climate, the Gresta valley results very different from the sub-Mediterranean example

From Fig. 5.10 we see the comparisons between the bio-indicators of Ellenberg measured in the woods of Val di Gresta (Trento, not far from Garda Lake) and still in the woods of Loveno (Lake of Como) and Zoagli (Genova). You can notice the good correspondence of Val di Gresta to the subalpine forests of Loveno-Menaggio (similar quotas and latitudes) and the clear distance with the band Gallo-Mediterranean of Zoagli. With this, we can debunk the assertion that the surroundings of the great alpine lakes are sub-Mediterranean environment: they are thermophilic, but Central European.

References

1. Einstein A (1918) Motiv des Forschens, tr. It. La ricerca scientifica. In: Come io vedo il mondo, Newton Compton 1975, Roma
2. Einstein A (1923) Einstein and the Philosophies of Kant and Mach. Nature 112:253
3. Braun Blanquet J (1926) Etudes phytosociologiques en Auvergne. Mont-Louis G, Clermont Ferrand
4. Naveh Z, Lieberman A (1984) Landscape Ecology: theory and application. Springer, New York
5. Pignatti S, Box EO, Fujiwara K (2002) A new paradigm for the XXIth Century. Annali di Botanica II:31–58
6. Ingegnoli V (1999) Definition and evaluation of the BTC (biological territorial capacity) as an indicator for landscape ecological studies on vegetation. In: Windhorst W,

Enckell PH (eds) Sustainable Landuse management: The challenge of ecosystem protection. EcoSys: Beitrage zur Oekosystemforschung, Suppl Bd 28:109–118
7. Ingegnoli V (2005) An innovative contribution of landscape ecology to vegetation science. Isr J Plant Sci 53:155–166
8. Tüxen R (1956) Die heutige potentielle natürliche Vegetation als Gegenstand der Vegetationkartierung. Angew. Pflanzensoziologie Stolzenau/Weser 13:5–42
9. Rivas-Martinez S (1987) Nociones sobre Fitosociología, biogeografía y bio-climatología. In: La vegetacion de España. Universidad de Alcalá de Henares, Madrid, pp 19–45
10. Ingegnoli V (2002) Landscape Ecology: A Widening Foundation. Springer, Berlin
11. Campos-De Quiroz H (2002) Plant genomics: an overview. Biological Research 35: 3 4 doi:10.4067/S0716-97602002000300013
12. Ingegnoli V, Aquila C, Padoa-Schioppa E (1995) Rapporto preliminare sullo studio dell'ecomosaico forestale del Gariglione. Atti 5° Workshop "Progetto Strategico Clima Ambiente e Territorio nel Mezzogiorno" Amalfi 1993, Tomo I. Guerrini editore, C.N.R., Roma, pp 397–419
13. Ingegnoli V, Giglio E (2005) Ecologia del Paesaggio: manuale per conservare, gestire e pianificare l'ambiente. Sistemi editoriali SE, Napoli
14. Ingegnoli V, Pignatti S (2007) The impact of the widened landscape ecology on vegetation science: towards the new paradigm. Rend Lincei Sci Fis Nat, s.IX, XVIII:89–122
15. Box EO (1987) Plant life form and mediterranean environments. Annali di Botanica XLV:7–42
16. Odum EP (1971) Fundamentals of ecology, 3rd edn. WB Saunders, Philadelphia, PA
17. Odum EP (1983) Basic ecology. CBS College, New York
18. Duvigneaud P (1977) Ecologia. In: Enciclopedia del Novecento, vol II, Enciclopedia Italiana Treccani, Roma
19. Piussi P (1994) Selvicoltura generale. Utet, Torino
20. Hellrigl B (1990) Relaskop scala metrica CP, il relascopio a specchio. FOB, Salzburg
21. Susmel L (1980) Normalizzazione delle foreste alpine. Liviana Editrice, Padova
22. Raunkiaer CC (1934) The Life Forms of Plants and Statistical Plant Geography. Oxford University Press, Oxford
23. Pignatti S (1998) I boschi d'Italia: sinecologia e biodiversità. Utet, Torino
24. Pulliam R (1989) Sources and sink complicate ecology. Science 477–478
25. ANPA (2001) I.B.L. Indice di biodiversità lichenica. Centro Tematico Nazionale-Atmosfera, Clima, Emissioni. Roma
26. Ingegnoli V (2013) Concise evaluation of the bionomic state of natural and human vegetation elements in a landscape. Rend Fis Acc Lincei. doi:10.1007/s12210-013-0252-2
27. Biondi E, Baldoni M (1994) The climate and vegetation of peninsular Italy. Colloques Phytosociologiques XXIII, Bailleul, pp 675–721
28. Pignatti S, Menegoni P, Pietrosanti S (2005) Bioindicator values of vascular plants of the Flora of Italy, vol 39. Braun-Blanquetia, pp 3–9

Landscape Bionomics Analysis of Animal and Human Populations

6

6.1 Landscape Bionomics Analysis of Fauna Component

6.1.1 Animal Populations and Landscape Structure

A first introductory consideration regarding the landscape and its fauna component, following the principles of landscape bionomic, leads to novelty, both conceptual and methodological, although on this approach we have, at present, a more limited number of examples.

The structure of the landscape is due, as we know, at first to the relationship between geomorphological processes and vegetation distribution. Yet animals play a role of great importance in the structuring of the landscape. People who has had the opportunity to observe the behaviour of the most important populations of animals in the great African parks or in other European nature reserves realise that animals act on the structure of the landscape especially in two ways:

(a) As *users of resources* in complex systems of ecocoenotopes, as animals are able to take advantage of the greater resources of the plant components (e.g. grasslands), but also to contribute to their maintenance (preventing the widening of forests, spreading fertilisers manure, transporting seeds, etc.)

(b) As *engineer species*, because animals are able to change the environment where they live to their advantage, forming peculiar structures or retaining other, creating paths or tracks, links to bodies of water (e.g. tracks of elephants drinking, Fig. 6.1), building dwellings (ants, termites) and ponds or dams (beavers), or niches and trails on steep slopes of high mountains (e.g. ibex).

The effects on the structure of the landscape are usually enhanced by a series of correlations with the vegetation, so that even if the alterations of certain populations or communities of animals seem to be modest or marginal, they are capable of great influence on the structure. Elephants are very active in structuring their landscape, due to their strong body and their intellective capacities: they are able to change or destroy patches of Acacias but also big baobabs (*Adansonia digitata*), as affirmed by the zoologist Fiorenza De Bernardi (personal communication).

On this argument, the case of the beaver (e.g. *Castor fiber*, *Castor canadensis*) is the most obvious: a dam on a small stream is not spatially relevant in a valley, but it means regular water and continuous use of the timber, and soon, as a consequence, they widen the whole landscape units (LU), the constituent elements of which change within a few tens of years [1], up to form a basic ecomosaic entirely different from the previous, so as to give the name to entire valleys (e.g. "val Bévéra", Lombardy and Liguria).

Similarly, the case of fire ants (*Formica rufa*) on a subalpine landscape of spruce forest is important (Fig. 6.2). Please note that a nest like

V. Ingegnoli, *Landscape Bionomics Biological-Integrated Landscape Ecology*,
DOI 10.1007/978-88-470-5226-0_6, © Springer-Verlag Italia 2015

Fig. 6.1 Changes in the landscape due to engineer species such as African elephants. Study of the Author in the Serengeti (Tanzania), where there are two flights of stairs to get down on the banks of the river. From Ingegnoli [17]

Fig. 6.2 Two nests of *Formica rufa* sp. more than 1 m height and 2 m wide in the forest of Lavazé Pass (Trentino). The nest is also dug into the ground and may contain a few hundred thousand ants with more than 100 queens. A nest like this may destroy about five million insects in a season

this one exceeding 1.3 m in height, is also dug into the ground, so it is very vast and it may contain a few hundred thousands of individuals with more than 100 queens [2]. A nest, as that in the figure, however singularly rich in herbaceous cover (e.g. *Calamagrostis villosa*), can in a season destroy about five million insects.

The presence of populations of large herbivorous mammals is also closely linked to the structuring of the landscape, especially the effects of grazing and fertilisation. Even here, however, there are special relationships with the vegetation that increase their influence on the structure of the landscape: for example, the animals may induce felting of the pastures, which selects the reproduction of certain plant species of trees (including shrubs and herbs) acting on biodiversity in a positive way, but also in a negative one.

Herbivores, however, succeed in enriching in nitrogen content lawns that live on incomplete soil, with low humus content, as happens in the mountains. Sometimes it become also important

Fig. 6.3 Example of stone pine (*Pinus cembra*) in "val de Cornon" (*above* Panchià di Fiemme), 2,100 m. Formations are in harmony with the people of nutcrackers (*right*, from Figuier, 1926). Archive of the Author

the symbiosis between the trees and birds or big mammals, as seen in the Alps with the known links between the nutcrackers (*Nucifraga caryocatactes*) and stone pine (*Pinus cembra*) (Fig. 6.3), an example of which is shown in upper Val de Cornon (Fiemme), or the expansion of larch in conflict areas with pastures.

In the first case, the bird is able to extract pine seeds from the cone and to hide them in small holes (e.g. near rhododendrons), distributing the plants in the marginal open space (forest mantle). In the second case the expansion of larch depends on the breaking of the pasture sward, which allows the growth of larch plantulae. Only large mammals (e.g. ungulates), such as deer (*Cervus elaphus*) or Alpine cows (which can graze in semi-wild over the limit of forests, 2,200 m), can affect the thick sward.

With the domination of man over ecosphere, there are many alterations, including structural relationships between tesserae or vegetated ecotopes and animals, which are not always replaced by the presence of herds of herbivores allowed. This too often induces to forget the role of animals as a constructive disturbance in the landscape, disorder that we know to be of primary importance for the structuring of complex systems of ecocoenotope. A comparison with the African parks is a must to understand that in natural landscapes the environmental impact is not absent, but is due to the populations of animals, especially large mammals.

Do not forget similar cases from our territory. For example, the elimination of herbivores (sheep, cows) considered as impacting by the managers of the Park of Monte Barro (Lecco) has triggered a change in the floristic composition of the pastures that, within a few years, would have brought to lose the peculiarity of the Park itself (one of the areas of greatest herbaceous biodiversity extent of Lombard Pre-Alps). Therefore, the animals have had to be reintroduced, though with controlled modes. However, the donkeys were introduced in place of sheep.

A similar shift away of flocks of sheep from the National Park of Abruzzo, in the 1970s, produced a thinning in the population of bears (*Ursus arctos marsicanus*), some of which have followed the herds outside the park, exposing themselves to poaching.

It is evident that, often, the fauna analysis requests from landscape ecology may concentrate on a series of careful observations on the presence or absence of species that have a strategic relationship with the structure of a landscape unit (LU) in examination, or on observations on the presence of impacts of animal origin in their natural habitat or in areas with little or no human influence. Always necessary, however, is to try to know the relationship of the animal species with the structure of the landscape, starting from their distribution in relation to the vital space per capita or standard habitat (SH, see Sect. 3.4.1), the distribution of ecotopes in the examined LU (at least in principle). In fact, the structure of the landscape acts in bi-univocal way on the animal populations in different forms, changing (or maintaining):

1. The heterogeneity of the landscape
2. Its carrying capacity
3. The conformation of patches and corridors
4. The model of metapopulation
5. The movements in the landscape
6. The natural habitat (NH) amount and distribution
7. The presence of man-made structures, etc.

For example, a study of landscape ecology of the nature reserve of Cornino in Friuli, designed for the protection of vultures (*Gyps fulvus*), showed that the perimeter of protection don't agree with the structural pattern of his LU, which must sheer suitable for nesting the whole range of rocks and the plateau behind, where several species of raptors (e.g. *Falco* spp.) go to find suitable prey. Moreover, according to data from Opdam et al. [4], we note that the probability of the presence of birds in an isolated grove passes from 76 % at a distance of 1 km from the forest, to 64 % at 18 km. You should also note that some animals (e.g. wolves, foxes, coyotes, lions, etc.) are also using artificial corridors, such as roads (the docks grassy and/or the tree-lined dirt roads), obviously if not very busy.

In Sect. 3.4.1, defining the concept of habitat standard per capita (SH) we pointed out that this concept is valid both for the study of animal populations with respect to the characteristics of their environment, as for human populations.

As already exposed in 3.4.1 in the case of the natural habitat (HN)

$$\begin{aligned} SH_N = {} & EXR + GEO + RSN + RSL + CON \\ & + PRT + PRD/n^\circ \text{ indiv.} \\ & \times Sp \left[m^2/\text{organism}\right]. \end{aligned}$$

The ungulates usually have different home ranges in different landscapes, but applying the SH function such measures are reduced to a unique thresholds value in relation to the SH quantity needed per landscape type, as explained by Ingegnoli and Giglio [5]. Note that this SH is referred to the landscape apparatuses which form a good bionomic state of the LU, going beyond the same trophic or climatic resources [6].

In Table 6.1 each landscape type is related to its average presence of NH (%) and the capacity to host the main ungulates of Central Europe. The SH per species is estimated in ha/organism: their differences are remarkable. The MVP (minimum viable population) is added, to complete the frame.

In suburban landscapes it is known that different species in recent years have become accustomed to live: just think to the foxes which are in search of food in the garbage, or to starlings (*Sturnus vulgaris*) which are known to occur in bulk, in Rome, being more protected spending the night on the trees of the city boulevards consequently, the starlings attracted the kestrel (*Falco tinnunculus*).

In Berlin it is quite common to meet a red squirrel (*Sciurus vulgaris*) or a pigeon (*Columba palumbus*) in central city parks; in Milan, in the public garden of the Palazzo Dugnani, one often sees a night heron (*Nycticorax nycticorax*) which has learned to exploit the presence of fish in the ponds of the garden (Fig. 6.4). Also in the garden, animals ever seen in big cities, like a king-fish (*Alcedo atthis*), often appear. Some birds, like the robin (*Erithacus rubecula*), have learned to sing at night, due to lighting and less traffic [8]. The most negative aspect of urban structures is linked to the introduction of exotic species: from the gray squirrel to the tortoises with cheeks black (water areas), to the others, to the shrimp of the lake (all mainly American) or parrots (e.g. some African species frequent in the gardens of Genoa).

6.1.2 Animal Populations and Landscape Dynamics

Permeant species of animal (i.e. multi-habitat vertebrates) implies a daily use of different habitats and, alternatively, different habitats

Table 6.1 Standard habitat (SH) referred to Central European ungulates and their relations with the main types of landscapes. In the first column, an average value of their natural habitat (NH)

NH %	Landscape types	BOAR	ROE	DEER	CHAM	STEINB
>90	Natural, Alpine	–	–	–	+++	+++
>90	Natural, forested	++	+++	+++	+++	++
90–75	Semi-natural, forested	+++	+++	+++	++	+
75–60	Agricultural and Forestry	+++	+++	++	+	–
60–35	Agricultural, protective	+++	++	+	–	–
35–10	Agricultural, productive	+	+	–	–	–
15–5	Suburban	+	–	–	–	–
<5	Urban	–	–	–	–	–
	SH (ha/animal)	3–4	4–5	15–16	5–6	5–7
	Survival number[a]	10	15	70	60	60

CHAM Chamois; *STEINB* Steinbock

[a]Data from Perco [7], but intended as MVP (minimum viable population) of sub-populations

Fig. 6.4 A red squirrel in a Berlin urban park (Charlottenbourg Garden) and a night heron often seen in the oldest urban park of Milan (Dugnani Garden, now Montanelli Park) right, from Figuier, 1926. Archive of the Author

may be used at different stages in the life cycle. Dynamics is the obliged consequence and in many cases, when these movements are repeated, the pattern of a landscape is structured with paths and places of standstill. Even if vegetation structure is a primary determinant of movement route, chemical signals, applied by animals to plants, soil and air play also an important role for many species (see also Sect. 3.5).

We may consider as "umbrella species" the species having the greatest ecological needs, hence being the first to decrease when their environment is altered. But generally these species are not truly permeant: they have distinct habitat exigencies, thus the umbrella effect is limited because they cannot properly be used to protect all the species living in a landscape.

It is better to refer to the concept of "focal species" [9] which concerns a small set of species representing the landscape heterogeneity of habitats, thus usable to protect the spatial and functional exigencies of a landscape. In the case of various species limited by the amount of resources, the less abundant one will be the focal species on which the landscape rehabilitation must be designed.

Transformation and disturbances of the landscape are very numerous and complex, so a wide

Fig. 6.5 Mediterranean Pine forest in the South of Portugal, semi-burnt, with sheep grazing on its edges. Before the fire, the sheep used to graze in clearings inside, but the strong dominance of daffodils after the fire has removed the animals from the forest

examination of these arguments in relation to animals in a book focused mainly on the theory and application of landscape bionomics is impossible. We have to limit necessarily to some spots.

The heterogeneity of the vegetated tesserae and ecotopes produced by the fires has an influence on wildlife resources that changed, both in quality and spatial arrangement. From studies carried out after the great fire of Yellowstone Park (USA) it was noted that the main factor for the survival of ungulates (e.g. *Cervus canadensis* and *Bison bison*) is the winter following the fire, when the food is lacking. If the winter is mild, an area of more than 60% of a LU has to burn to affect ungulates. Furthermore, as noted by Turner et al. [10], the spatial configuration of the residual stains not burned, of those semi-burnt and of those burned is important: this can create quite heterogeneous compensation sufficient to save the populations of animals.

If the fire is confined to a single tessera with a good resilience (e.g. a grass-grazing) inserted in an ecotissue in good condition, it has almost no influence on the invertebrate communities of the tessera itself, as noted by F. Bernini (personal communication) in a study for the Park of Monte Barro, especially for arachnids (the first re-colonisers).

In Mediterranean areas, the wildfires are usually more limited than those of the malicious and are notoriously recurrent disorders as underlined by Naveh and Liebererman [11]. However, even these facts create a special dynamic in relation to the pasture, since the strong regrowth of geophytes (e.g. *Asphodelus* spp.) prevents sheep grazing in forest clearings, as seen in Fig. 6.5, near Faro (Portugal), where the flock remains on the outside of a burned pine seeds, while before burning the sheep grazed in clearings inside the wood.

Even slow movements, e.g. *bradyseism*, can affect wildlife, as it is found in the lagoon of Venice. A subsidence (combined with slight eustatism) of about 2 mm/year is enough to change the shape of the salt marshes and to vary the halophytes, so in a few decades you may notice changes in the mosaic of marshes, with consequences both for nesting and for resources for certain types of birds, such as herons (e.g. *Egretta alba*), Terns (*Sterna caspia*), Shelducks (*Tadorna tadorna*), Oystercatchers (*Haematopus ostralegus*).

Climate change can occur in changing mosaic of wetlands in areas once cold-temperate (e.g. Germany) and allowing the introduction of species typical of Mediterranean countries, as evidenced by Ott [12]. Dragonfly species, such as *Crocothemis erytraea*, which 10 years ago had gone to the North, as well as 8–9 other species of dragonflies, in few years have become frequent in Germany. The movement of many species in a short time can cause effects on the landscape through changes in floristic composition and behaviour of animals in relation to the different regions. Four general reactions are predictable according to Graves and Reavey [13]: (1) change

Fig. 6.6 Fox in a meadow mowed just 15 m from the houses of the village of Panchià, Val di Fiemme. The deer often pass on this lawn and even in the garden of the house overlooking it

in the inhabited area, (2) a priori tolerance to new conditions, (3) adaptation to new conditions through microevolution, (4) increased mortality.

The *urbanisation* processes have led to the change of behaviour of some groups of gulls (e.g. *Larus ridibundus*) which appeared even in cities like Milan, in areas far from the sea or lakes, and contend with the crows (*Corvus cornix*) for the control of municipal waste areas.

Particularly interesting are the delays of extinction due to destructive changes of the agricultural landscape [14]. Some scholars have demonstrated in Britain that the spatial distribution of *Abax parallelepipedus* (a beetle) is more related to the connectivity of a network of Trees "bocager" of 1952 that to the present one, whose connectivity is decreased by 35 %. Den Boer [15] noted that the stable carabid habitat (with little power dispersion) remain at least 40 years after their fence was destroyed, while species with high dispersion only 10 years.

The abandonment of agricultural activities results in the birth of new vegetated tesserae, and new habitats available to wildlife. There is, however, a delay between the appearance of new habitats and the colonisation of animal species, except in the case of arachnids. Burel and Baudry [16] in a study on ground beetles in Normandy found that only the new vegetated tessera less than 20 m from a source (e.g. a lined corridor) are colonised relatively soon. Connectivity and circuitry are therefore essential for colonisation, as well as the area of the tesserae. Furthermore, it should be noted that the dynamics of the vegetation succession are not always correlated with the animal colonisation, both because such a sequence does not behave according to the traditional rules, as sustained by Ingegnoli [17], and because the polyphagia species are independent from this process.

It should be noted, moreover, that the ecological consequences of technical intensification within agricultural landscapes are not assessable staying at the scale of cultivated tessera. It is necessary to refer to the diversity of scale in the exploitation of resources by stenotopes and eurytopes animal groups, to the structural complexity of the landscape, and its τ (i.e. functional) diversity (see 3.7.1).

The disturbance to wildlife is evident in every aspect, and often these disturbances are not incorporated at LU scale. In studies on foxes (*Canis vulpis*) it was stated [18] that their dens remain at a minimum distance of 200 m from the populated areas, though the foxes -as we know- come to seek food in the suburbs of countries as in val di Fiemme (Fig. 6.6) or in the Abruzzo National Park (where we also met them in the

Table 6.2 Comparison of different forest management in two theoretical LU in the Alps, with forests of boreal type and feeding areas of ungulates. Notice of the diversity of landscape ecology characters in relation to feeding areas

Forest management	Age (years)[a]	Height (m)[a]	BTC (Mcal/m^2/year)	Feeding areas (%)[a]	Damages[b]
(a) Flush cut and reforestation	2–16	<8	2.0–3.5	30–100	3–5
	16–70	8–24	3.5–6.0	1–9	
	70–100	24–30	6.0–7.5	9–16	
(b) Natural renewal, patch cut	10–50	6–21	3.0–5.0	6–14	1–3
	50–80	21–27	5.0–7.0	14–30	
	80–120	27–36	7.0–8.5	30–100	

Order of magnitude, as a reference; [a]from Reimoser and Gossow [6], revised; [b]from Mustoni et al. [19], damage by stripping and barking

Table 6.3 Forest patches and use frequency by Capercaillies in landscape unit (LU) of Bavarian Alps (from Storch [20]; reworked)

Patch area ha	Patch presence (%)	Males (%)	Females (%)
<5	30	12	8.5
5–10	31	23.5	23
10–20	22	33	22.5
>20	17	31.5	46
–	100	100	100

The coverage of forest trees determines a further presence of grouse in reason of only 8 % for roofing <50 %, 70.5 % for roofing from 50 to 70 %, 21.5 % for roofing denser

centre of Civitella Alfedena). Consequently, the system of residual patches in good potential naturalness remains much reduced (see Sect. 9.2.2).

The *dynamics of forest management* (such as deforestation, reforestation, etc.) continuously change the local LU and affect animals in various ways. As can be seen in Table 6.2, comparing two different operations in theoretical LU of boreal alpine forest, (a) large flush cuts with reforestation and (b) cut in small pieces with natural renewal, you notice [6] that the feeding areas of ungulates are almost opposite, so different are the damages from barking and stripping [19]. Note also the different level of BTC reached in the two cases.

In Table 6.3 we note the presence of capercaille in a forest landscape in a LU of Bavarian Alps in function of the patch area of spruce forests. Changing their distribution (due to cutting) implies a change even in these big birds frequency.

In forest dynamics a heavy part is due to animals, especially insects. The argument is very broad, but at least we have to mention the role of spruce bark beetle (*Ips typographus* Coleoptera, Curculionodae, Scolytinae). This infesting animal can destroy in few month large surface of spruce forests, especially in the South-oriented forest edges: a predisposing "sun-effect" as exposed by Kautz et al. [21] studying the Bavarian Forest National Park. Very dangerous is also the most known pine processionary (*Thaumetopea pityocampa*).

The dynamics of the landscape lead to reducible spatial, transforming the relationship between the fauna. The changes in question are still to be explored; however, it must be said that some of these studies may fall within paragraph more properly structural. Dissection and fragmentation (and their inverses) are in every way the most important spatial processes in the landscape changes.

6.1.3 Main Effects of Fragmentation on Animals

As is found almost everywhere in Europe, the "dissection" sensu Forman [22] is the first and the most insidious process, which takes place (within certain limits) without, and against, any planning. It is the opening of a road in an agricultural or wooded landscape, followed by micro-interventions, often abusive, by construction and human activities in derivation from this line of traffic, some of which are enlarged and carry a disorder inwards, which recalls other minor road connections, in a very harmful process-to-back positive action.

The fragmentation (the breaking of a vast habitat into smaller pieces) is undoubtedly the most well-known process, though often ill-defined: it tends to include also the other spatial processes we have mentioned. The major effects of fragmentation on wildlife are, in descending order: the isolation, the increase in general and multi-species habitat, in edge species, in the nest predation, and the rate of extinction. Instead the dispersion of the internal species, of the species with large home ranges and of species richness of internal patches decrease.

In reality, the fragmentation tends to affect almost all major ecological processes and also genetic. For example, a growth of metapopulation dynamic and backcrossing are common. On the other hand, it decreases the area of the patches from disturbance. We must observe, also, that in some cases fragmentation can produce positive effects; it can be shown [17] that the general metastability of a natural landscape (LM) increases if we introduce a small proportion of human habitat (HH), never more to 3–10 %, and also biodiversity complex increases (up to HH = 35–45 %, See Sect. 3.7).

This is also demonstrated by the choices of habitat by many species of animals; it is known that the creation of ecotonal bands and new heterogeneous tesserae (also cultivate, if not intensive) are sought after by roe deer and wild boar. Even the deer will be attracted, but only if anthropic SH remains very high, that is, the population density remains low. Moreover, any naturalist can see by himself that the LU near the small villages of the Alps and the Apennines have a variety of bird species particularly high.

The effects of the opening of artificial corridors (roads, railways, canals, etc.) on the fauna are manifold: barrier, conduit, filter, source, sink. For example, for large mammals in forest landscapes the effect of a busy road on the surrounding tesserae can get from 90 to 200 m, and for the birds of the interior of the forest it is around 100 m, while in a prairie landscape that distance of disturbance can get to 1.5–2 km [22]. The streets are crossed rarely, even by animals that could do it easily during periods of low traffic, as large mammals such as caribou (*Rangifer tarandus*) or the mountain goat (*Oreamnus americanus*) in America or the red deer (*Cervus elaphus*) and roe deer (*Capreolus capreolus*) in Europe. Small mammals of the forest, such as red squirrels, generally do not cross roads greater than 12–15 m (including docks). Particularly frogs and toads have problems in crossing the roads, but they seems to better utilise artificial tunnels, as experienced in some places in Germany, Great Britain, Northern Italy and Poland.

It should be added that the killings of animals caused by the opening of roads (even with little traffic) are sometimes high, such as blacks bears and cougars in Florida, or the deer in val Pusteria or badges (*Meles meles*) in England. But a road can be dangerous even for birds. As a car roars towards birds standing on the asphalt, they do not check the driver's exact speed when judging how soon to flap out of the way [23]. Birds will take off further away from an approaching car on a faster road than on a slower road—regardless of the speed of the car. Researchers found that where there was a 50-km-per-h speed limit, birds on the road typically took off when the car was about 15 m away, whereas on a 110-km-per-h road, they took off when a car was nearer 75 m away. Birds did this even when faced with a car travelling faster on the slow road or slower on the fast road.

The opening of ski slopes may disturb the Alpine marmots, notwithstanding their hibernation period (October to April), because they live on pastures between 2,000–3,000 m, where the ski tourism may find snow from November to May (Fig. 6.7).

Fig. 6.7 The Alpine marmots have problems with the opening of ski slopes, especially today with the use of snow spray guns and their technical plants. This place is in Val di Fassa (Trentino). Note the technical hydrant at about 40–50 m from the marmot burrow

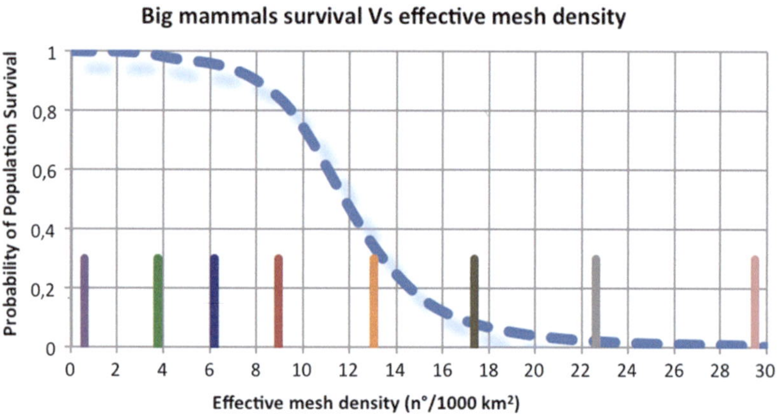

Fig. 6.8 A model proposed by the EUEA to indicate the survival probabilities of big mammals in the landscapes of the European countries: from *left* the colour bars represent the average mesh density in Sweden, UK, Austria, Italy, Switzerland, Poland, Czek and France

The rising of fragmentation in civilised countries was so strong that the European Environment Agency[1] [24] proposed a model (Fig. 6.8) to evaluate the probability of survival of big mammals in function of the effective mesh density (n° of mesh per Km^2). A comparison with the average values of mesh density per some European states is shown in this figure. The model[2] was deduced and reworked from Jaeger and Holderegger [25].

6.1.4 Towards the LaBiSF to Study the Fauna Component in the Landscape

In accordance with the previous for vegetation, a new method has been developing to study the

[1] The EUEA and the FOE (CH Federal Office for the Environment) published an interesting Study on Landscape Fragmentation (2011).

[2] The specific values of the thresholds depend on the particular species, traffic volumes on the roads, and the amount and quality of habitat present in the landscape (results of a computer simulation model). Once the threshold has been passed and the so-called 'point of no return' has been crossed, it will be impossible to rescue a declining population even with strong measures (EUEA).

Table 6.4 Evaluation of faunal sensitivity in a landscape unit

Faunal attributes	0	1	2	3	Score
Q. FAUNAL QUALITY (general characters)					
Q1- General rarity	Not Determin.	Rare, Risk	Vulnerab. (general)	Threaten.	Species rarity
Q2- Chorology	Cosmo-polites	Continent	Regional	Endemic	Areal
Q3- Critical state: local presence vs other areas	Other	Rare in district	Rare in region	Doubt in region	Presence in other areas
V. FAUNAL VULNERABILITY (local characters)					
V1- Abundance	Abundant >0.5	Common <0.5	Scarce <0.25	Rare <0.125	*i/p°*
V2- Habitat extension	Not selective	Rural Habitat	Ecotonal Habitat	Forest Habitat	Habitat sensitivity
V3- Fragility: living conditions	Any habitats	Not degraded	Colonial	Optimal habitat	Species needs vs. habitat
R. Results					
R1- Total scores	a	b	c	d	S(f)
Evaluation of faunal sensitivity				Total scores	
I	Critical S(f)			S(f) >72	
II	High S(f)			S(f) <72	
III	Relevant S(f)			S(f) <5 4	
IV	Mean S(f)			S(f) <36	
Level of attention					
V	Low S(f)			S(f) <18	
VI	Negligible S(f)			S(f) <9	

animal component that will be named "LaBiSF" (Landscape Bionomic Survey of Fauna), which currently sees as two cornerstones: the study of faunal sensitivity and the study for SH, similar to what is already in depth for human populations.

For the macrobenthic component relating to inland river functionality, please refer to the proper index (see Sect. 7.1.6). For the fish component there have not yet been conducted appropriate studies.

The method of fauna sensitivity presented has been developed with the collaboration of colleagues Renato Massa and Lorenzo Fornasari [26, 27] to measure the fauna sensitivity (FS) for a landscape unit (LU). This sensitivity is evaluated in two stages. The first consists on the evaluation expressed by the following equation:

$$FS = (Q1 + Q2 + Q3) \times (V1 + V2 + V3),$$

where FS is the sensitivity of a wildlife species, Q1 rarity, Q2 chorology, Q3 criticality, V1 abundance, V2 extension of the habitat, V3 the fragility (always of the same species).

For the quantification of the above equation, it is necessary to refer to the previous chart (Table 6.4), conceptually not too different from the standard form for vegetation assessment. The score values that are used are, however, in this case limited to a linear arithmetic sequence, not to a compound curve.

The final evaluation of the previous tessera must be compared with the second section of Table 6.4 which shows the fauna sensitivity classes in relation to the total score recorded by species. Obviously, if you do not have the time and the means to carry out a full investigation, it is important to make this assessment of key species, to be chosen earlier depending on the type of landscape units and the objective of the study. Bearing in mind the definition of umbrella species and focal species, we present the proposed index entries in more details:

Q1. Rarity

- *R. critical or high (3 points)*: The red list of endangered species become important both at a global scale or at a European (or in a Country) scale. Rare and vulnerable species for a

Country, or in the course of large decreases in Europe or species included in the Attached II EEC Directive 43/92. Species of birds with less than 25,000 breeding pairs in Europe.

- *R. average* (*2 points*): Species at indeterminate status for a Country, or that are included in the Attached IV EEC Directive 43/92 or in the lists of the Berne Convention with similar criteria, species with restricted areal and poor populations. Species of birds with less than 50,000 breeding pairs in Europe.
- *R. low* (*1 point*): Species not meeting the above criteria. Species of birds with less than 100,000 breeding pairs in Europe. Disperse species, typical of fragmented landscapes and colonial species; species excluded from the list of protected species in a Country.
- *R, none* (*0 points*): Species not meeting the above criteria, Species of birds with more than 100,000 breeding pairs in Europe. Common and widespread species.

Q2. Chorology

- *Reduced areal Chor.* (*3 points*): species or subspecies endemic in some regions in the entire country, particularly vulnerable to changes in the landscape.
- *Intermediate areal Chor.* (*2 points*): species with reduced areal (e.g. Mediterranean or infra-European).
- *Vast areal Chor.* (*1 point*): species with continental areal (e.g. European).
- *Wide areal Chor.*(*0 points*): species with intercontinental areal (e.g. Eurasian Palearctic) or cosmopolitan.

Q3. Criticity

- *Cr, high regional* (*3 points*): species not normally present in the region (dubious in the literature), or species present in less than 50 % of the region, but as elective area at national level.
- *Cr, low regional* (*2 points*): species present in less than 20 % of the region and in the study area with any density, or present in the area with higher density than the regional average.
- *Cr. local* (*1 point*): rare species at the regional level but relatively abundant locally.
- *Cr, not defined* (*0 points*): species do not meet the previous criteria.

V1. Abundancy or Frequency

- *A. rare* (*3 points*): rare species, e.g. for birds <0.125 individuals/point. For mammals and herpetofauna may take into account the frequency in the sensing unit (% area occupied), because little detectable species, e.g. sp. rare <12.5 %.
- *A. poor* (*2 points*): scarce species, e.g. for birds <0.25 individuals/point. Mammals and herprtofauna: sp. scarce <25 %.
- *A. common* (*1 point*): common species, e.g. for birds <0.50 individuals/point. Mammals and herpetofauna: sp. common <50 %.
- *A. abundant* (*0 points*): abundant species, e.g. for the birds >0.50 individuals/point. Mammals and herpetofauna: sp. abundant >50 %.

V2. Habitat Extension

- *E. forestry* (*3 points*): species with habitats restricted to forests or wooded spots with significant tree cover.
- *E. ecotone* (*2 points*): ecotone species (forest-campaign) less selective than the previous ones, or aquatic species in areas recently affected.
- *E. rural* (*1 point*): adapted species in agricultural landscapes also open.
- *E. non-selective* (*0 points*): species indifferent to habitat, capable of ageing in urban and suburban landscapes.

V3. Fragility

- *F. optimal status* (*3 points*): species related to characters of optimal habitat, Good BTC, no pollution, etc. and species with populations concentrated and isolated.
- *F. normal status* (*2 points*): species negatively affected by degradation, while not requiring optimal habitat; colonial species not as above, sp. sedentary or low reproduction rate.
- *F. precarious status* (*1 point*): species that are affected by environmental disturbances only if not incorporable and out of scale (i.e. destructive).

Table 6.5 Evaluation of the FS related to bird species in the Regional Park South Milan (1997–1998) LU of Chiaravalle-Selvanesco (Fornasari, Ingegnoli & Massa)

Species	Q1	Q2	Q3	V1	V2	V3	FS	Note
Knight of Italy, *Himantopus himantopus*	3	0	3	3	3	3	54	
Bittern, *Botaurus stellaris*	3	0	3	3	3	3	54	
Coot, *Fulica atra*	1	0	1	3	1	3	14	
Woodpecker May, *Picoides major*	2	0	0	3	3	3	18	
Blue Tit, *Parus coeruleus*	0	1	0	3	3	2	8	
Stiff neck, *Jynx torquilla*	2	0	0	3	3	3	18	
Nightingale, *Luscinia megarhincus*	0	1	0	0	2	1	3	
Buzzard, *Buteo buteo*	2	0	1	3	1	3	21	Nest
Stonechat, *Saxicola torquata*	1	1	0	3	1	2	12	
Shrike, *Lanius collurio*	3	0	1	3	2	2	28	
Screech owl, *Asio otus*	2	0	3	3	3	3	45	Roost
Hobby, *Falco subbuteo*	3	0	3	3	1	2	36	
Kestrel, *Falco tinnunculus*	2	0	0	3	1	2	12	Nest
Quail, *Coturnix coturnix*	2	0	1	3	1	2	18	
Lark, *Alauda arvensis*	2	0	0	0	1	2	6	
Swallow, *Hirundo rustica*	2	0	0	0	1	2	6	
White Wagtail, *Motacilla alba*	0	0	0	2	1	2	0	
Yellow Wagtail, *Motacilla flava*	0	0	3	1	1	2	12	
Average FS							19.28	

– *F. altered state* (*0 points*): species with elevate resilience and mobility, able to stay in altered environments.

The evaluation of Faunal Sensitivity is summarised in Table 6.4, below. An example of application of this method is reported in Table 6.5, related to a study in the suburban regional park of South Milan. As expressed in Table 6.5, the average Faunal Sensitivity was FS = 19.3, thus in the 4th class (medium FS).

6.1.5 Convergence Between Faunal and Vegetational Coenosis Evaluation in a Tessera

In many cases a traditional and simple method of animal detection can be used even in researches of landscape bionomics. For instance, in Table 6.6 we summarise the results of a survey to measure the validity of a LU restoration, in a mining area of Tavernola Bergamasca (Fig. 6.9). To check the level of ecological efficiency of the restored green tesserae (Ts), like Ts A in the figure, after 2–3 years of green planting, we compared these tesserae with natural ones (Ts N in the figure), detecting the presence of invertebrates, through pitfall traps as exposed in Ingegnoli and Giglio [5].

The most significant results can be seen evaluating the ecological indexes exposed in the table. The ratio between insects and invertebrates[3] in Ts A was only 82.5 % of Ts N, while the ratio invertebrates/plant biomass in Ts A is twice of Ts N. Note that the BTC resulted the exact inverse of the invert./plant biomass ratio, confirming the good significance of the experiment and denoting a phase of still not balanced state in artificial green Ts.

We have to underline with a particular stress the perfect convergence of a tessera (Ts) evaluation via animal parameters or via vegetation parameters. In fact, the best condition of a vegetation phytocoenosis signifies the best condition for the wildlife habitat thus for the insect community.

[3] We intend the invertebrates of the soil, as turbellarians, gasteropods, arachnids, crustaceans, millipedes.

Table 6.6 Comparison of two patches of bushy lawn area (about 500–600 m^2) in Tavernola Bergamasca (lake Sebino, North Italy). Detection of terrestrial invertebrates with pitfall traps, 1996–1997. Tessera A, artificial greening; tessera N, semi-natural grass (nearby)

Surveys	Measures	Tessera A	Tessera N
Vegetation characters of the tessera	Phanerophytae shrubs (%)	22	13
	Alien and cosmopolite (%)	14	8
	Above ground plant biomass (g/m^2 d.m.)	800	1.600
Invertebrate species (n°)	Detected	68	66
	Species in common	27	27
	Species insects	40	47
Captured Individuals (n°)	Total	302	312
	Insects	164	236
	Coleopteran	32	64
Biomass of invertebrates (g. d.m.)	Total (g)	36.77	35.68
	Coleopteran (g)	6.46	4.29
Ecological indexes	Sp. insect/Sp. invert. tot (%)	58.8	71.2
	Detected invert./plant biomass (n°/kg)	85.00	41.25
	BTC of the Ts Mcal/m^2/year	0.74	1.44

Fig. 6.9 The mining area of Tavernola Bergamasca (Lombardy) and the two sites of the present experiment

6.2 Analysis of Human Communities in the Landscape

6.2.1 Landscape and Human Population

From the dawn of the development of *Homo sapiens* (about 180,000 years b.p.) human populations have distinguished themselves in their ability to transform the environment even on a large scale, creating little by little man-made ecological systems, but also developing a mutualism with nature, as well recalled by Godron [28]. The ability to settle in areas appropriate for the safety and the exploitation of resources has therefore grown in a particular way, together with the capacity of villages planning and first agricultural areas.

However, only since the Classical period, especially in the Roman world, emerged the structuring capabilities of agricultural landscapes in a modern way, through the technique of *limitatio* (also called *centuriatio*). This technique was so effective that it has been echoed in the colonisation of the USA in sec. XVIII and XIX. Moreover, by checks carried out by Caravello and Ingegnoli [29], it takes into account different natural parameters and has not been applied in a destructive way, so that after 2,000 years, many landscapes in Europe and some tracks in North Africa still show their derivation from such a structure (Fig. 6.10). With regard to urbanised landscapes, you have to think that a urban ecosystem can not be formed in such a way in itself, being a type of heterotrophic landscape, but only in mutual and co-evolutionary way with other protective and productive ecosystems. The Romans had already guessed this process, since the "civitas" was intended as a system "urbs & territorium".

Fig. 6.10 Reconstruction on the basis of the Italian IGM of an area north of Parma in the first century. You can see the Flood plains (*blue*) and the grid of the land division (*lapis red*). From Ingegnoli [51]

Returning to the present, there are different studies aimed at understanding the relationship between man and the natural territory. It is known that Valerio Giacomini [30], promoter of the MAB (Man and Biosphere Program, UNESCO) in the 1970s had begun a study on the ecosystem of Rome. At the same time, research on human structures in agricultural landscapes were taken at the Polytechnic of Milan, with a collaboration between the Chair of the History of Architecture (Perogalli), the dean of geographers naturalists (Nangeroni) and other good teachers in the field of history and architectural restoration (Alpago-Novello, Langè). In this context, the historical and environmental evolution of the rural landscape, that is, the historical development of parallel adaptation to environmental conditions was studied by Ingegnoli (31–33). They were also dealt with the structural characters and the organisation of the agricultural landscape, the relationship between architecture and rural environment and between ecology and architecture; the ecological basis of the relationship between the villas, gardens and the environment; the ecological reasons of the development of the gardens and villas as control of urban ecosystems on the ecological resources of the territory.

Nature's ability to survive even in the cities has been analysed by Sukopp (University of Berlin) ever since the 1970s [34]. Studies on the city as an ecosystem have since that time very dilated. A "Green Paper on the Urban Environment" was published by the EEC in 1990. In June 1997, Ecology held in Leipzig organised by the German Ministry of Education, Research and Technology, which saw 380 participants from 48 countries.

It should also be pointed out that, in these years, the majority of human populations is starting to live in urbanised landscapes and/or suburban ones. The man does not live almost in the countryside or in semi-natural environments, and consequently the study of bionomics of the landscape on the human component take on increased importance.

6.2.2 Analysis of Urban Landscape Structure

As in the case of natural and semi-natural landscapes and habitats, even for cultural landscapes it is necessary primarily to recognise the structural pattern, which will be only partially related to the geological and geomorphological characteristics of the territory on which they insist, especially for those of more recent development.

6.2.2.1 Structure of Urban Landscapes

The city is the most typical of ecological systems formed by man, certainly one of the most recent in the history of the natural evolution of the biosphere. The city is the result of an integration of neighbourhoods with complementary functions of residence with gardens and protected orchards, market, management and public defence, religious centre, crossroad of transport, i.e. as a set of interacting human ecosystems. Then, you should not speak of "urban ecosystems", but more properly of "unity of the urban landscape", formed by a natural support due to the presence of units of agricultural landscape, necessarily complementary to urban areas. For these reasons, it is necessary to combine the traditional studies, which follow ecosystemic criteria, with studies of landscape ecology [11, 35–40, 17, 41, 5]. It is also necessary to remember once again that an urban landscape is also a specific level of biological organisation.

The origin of the structures of the urban landscape depends by the high specialisation of roles in the population, unthinkable in a village (where each family produces most of its needs). For these reasons, the city has a structure that is similar to an heterotopic organism, then just a landscape. In this complex system, the road network (which also includes the squares) is the most obvious aspect of the structure plan. In fact, the urban structures are part of the biological structures, as many scholars have understood since the early treatises of antiquity, from Vitruvius [42], to the Finnish Saarinen [43]. This architect suggested that the physical order of the urban communities was *organic* and used a section of tissue of mammals as a model to discuss the optimal circulation network for a city. Forman and Godron [35] also expressed a similar opinion and show that, in summary, urban structure is formed by two elements of the landscape: the *streets* and "*city blocks*" (i.e. the "insulae" in the Latin sense). The neighbourhoods are sets characteristic of the previous elements. The basic structural models for urban landscapes are, according to Forman and Godron, at least three:

a. *Concentric zone model*, where the neighbourhoods surrounding a centre of business in a similar manner for each direction
b. *Segments model*, in which a centre is surrounded by a halo of diverse neighbourhoods
c. *Model in multiple nuclei*, in which a mosaic of different districts asymmetric surrounds one or more central cores nuclear

In these models, however, you can add other structural patterns, as suggested by the history of town planning in the old world (Fig. 6.11), for example, we shall include at least the following:

a. The *market model and the citadel*, one of the oldest (early Iron Age), in which an "oppidum" or fortified hilltop is surrounded by neighbourhoods sprung up in the area around the market low;
b. The *model square mesh*, of Roman origin, in which the different neighbourhoods are set regularly around a main road junction (*cardus* and *decumanus*); model, among other things, often made in the USA;
c. The *model geometric*, symmetrical and complex geometry (e.g. on a hexagonal base) with walls particularly well developed;

To these types of structure must be added the great urban landscapes of today's "mega-cities", which include the suburban landscapes and to

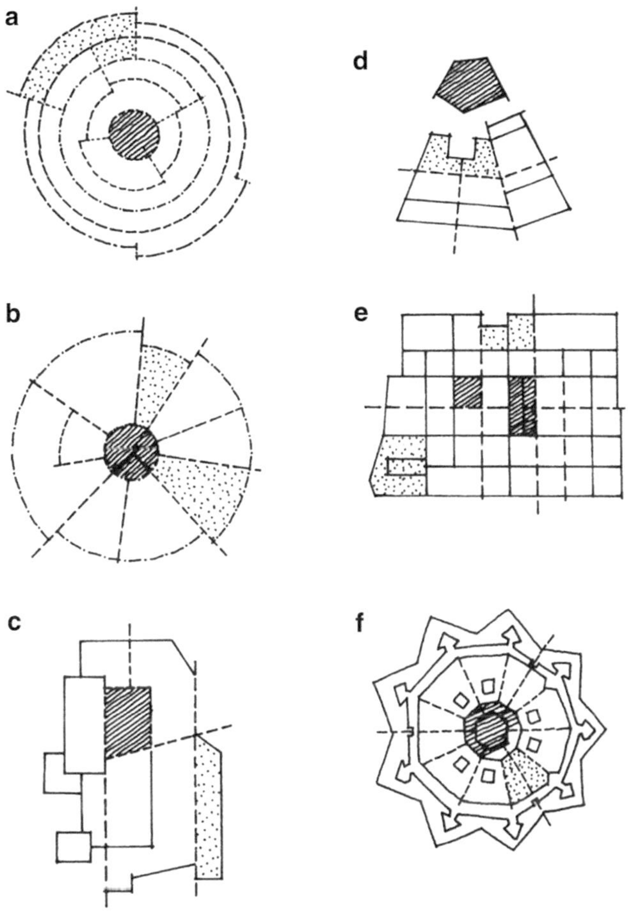

Fig. 6.11 Structural models of the city: (**a**) concentric zones, (**b**) segments, (**c**) multiple nuclei, (**d**) the citadel and the market (**e**) square mesh, (**f**) geometric

form ecological systems less stable and difficult to manage systems. Indeed, in ecology is known that the more you magnify the flow of input and output, storage and distribution, the less stable the system is. The structure of urban landscapes is still complex, although an Urban Region can be characterised with not many structural parameters, according to Richard Forman [44] (Fig. 6.12).

They are distinguished: (1) a Metropolitan Area, (2) one or more satellite cities, (3) an extensive network of mobility, (4) some area of natural reserves or biotypes, (5) a matrix of agricultural support. See also Sect. 3.4.2, on the functional bionomic study of an urban region. Forman makes comparisons among the 38 large and medium cities located on every continent, taking into account some concepts of landscape ecology (e.g. number of natural landscapes to urban region, urban parks, water resources, ecological networks). For the conurbation of Milan is estimated [41] an Urban Region of approximately 4.200 km^2 (see Fig. 3.12).

6.2.2.2 Structure of Suburban Landscapes

This type of landscape is that of the most recent development; it was originally a band ecotone between the urban and the surrounding agricultural landscape, and then, especially since the Roman Empire, has been expanding a lot, acquiring its own characteristics. Its structure is, however, less varied than one might think, being typically polynuclear, with alternation of industrial districts, dormitory ones and other more rural with the presence of vast tangles of technological networks. In recent decades many suburban landscapes have also altered, with syndromes not unlike those of the megalopolis.

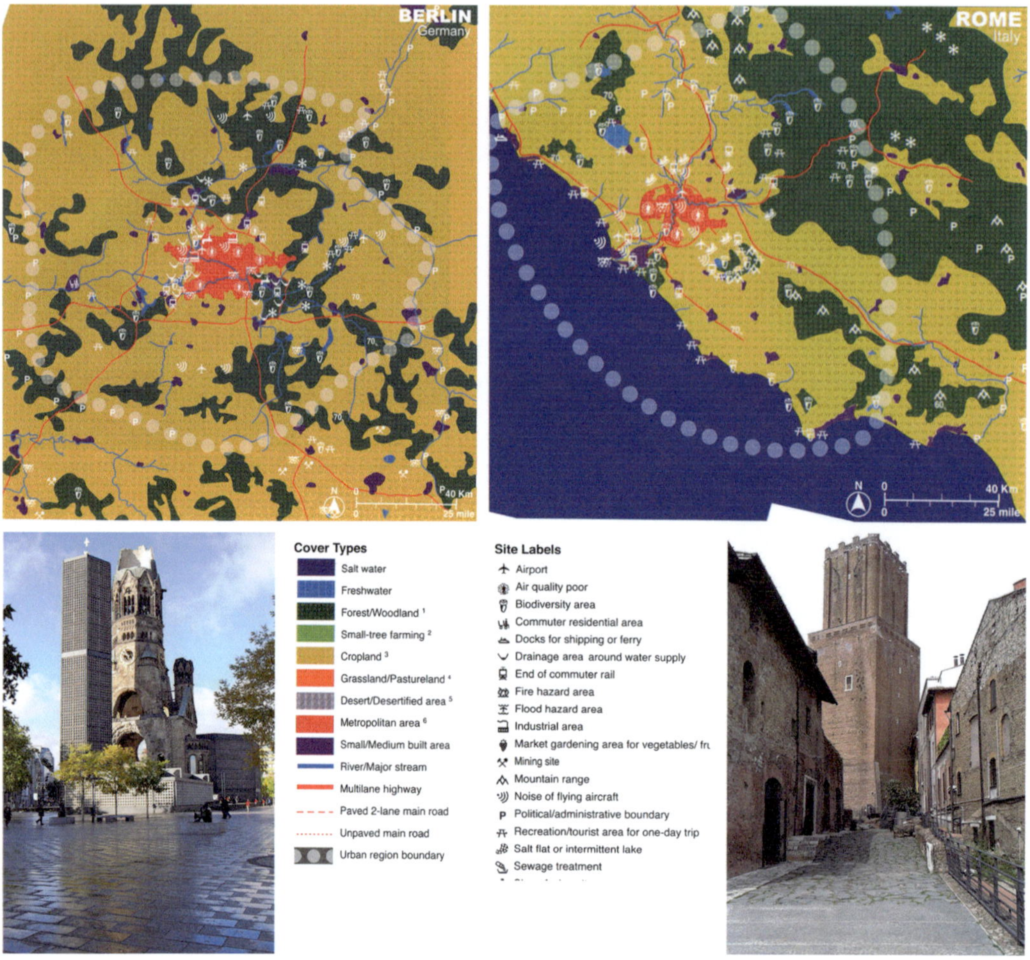

Fig. 6.12 Urban regions of Berlin (Germany) and Rome (Italy), from R.T.T. Forman [44] and views of the historical centre of both the cities

Oftentimes suburbia have been the glue for the formation of mega-cities themselves.

Especially in the USA, suburban landscapes are vast and dominated by residential areas of new development, in line with small cottage garden. In Europe, industrial presence is more frequent, as well as rural areas. In these structures are setting easily exotic species (*neophytes*), then suburban LU are sources of propagation of that species.

Even the study of architecture in relation to the landscape goes well beyond the traditional approach of the faculty of architecture (based on space matters, aesthetic, technical and functional arguments, but not natural ones) and must be placed within the bionomics. It should be therefore necessary to open a further chapter.

6.2.3 Agricultural Landscape Structures

In rural landscapes, as we have seen, several studies have been made by architects, naturalists, geographers [45] and ecologists, to verify the relationship between spontaneous architecture, villas, environment and landscape. As you can see from Fig. 6.13, a close relationship between

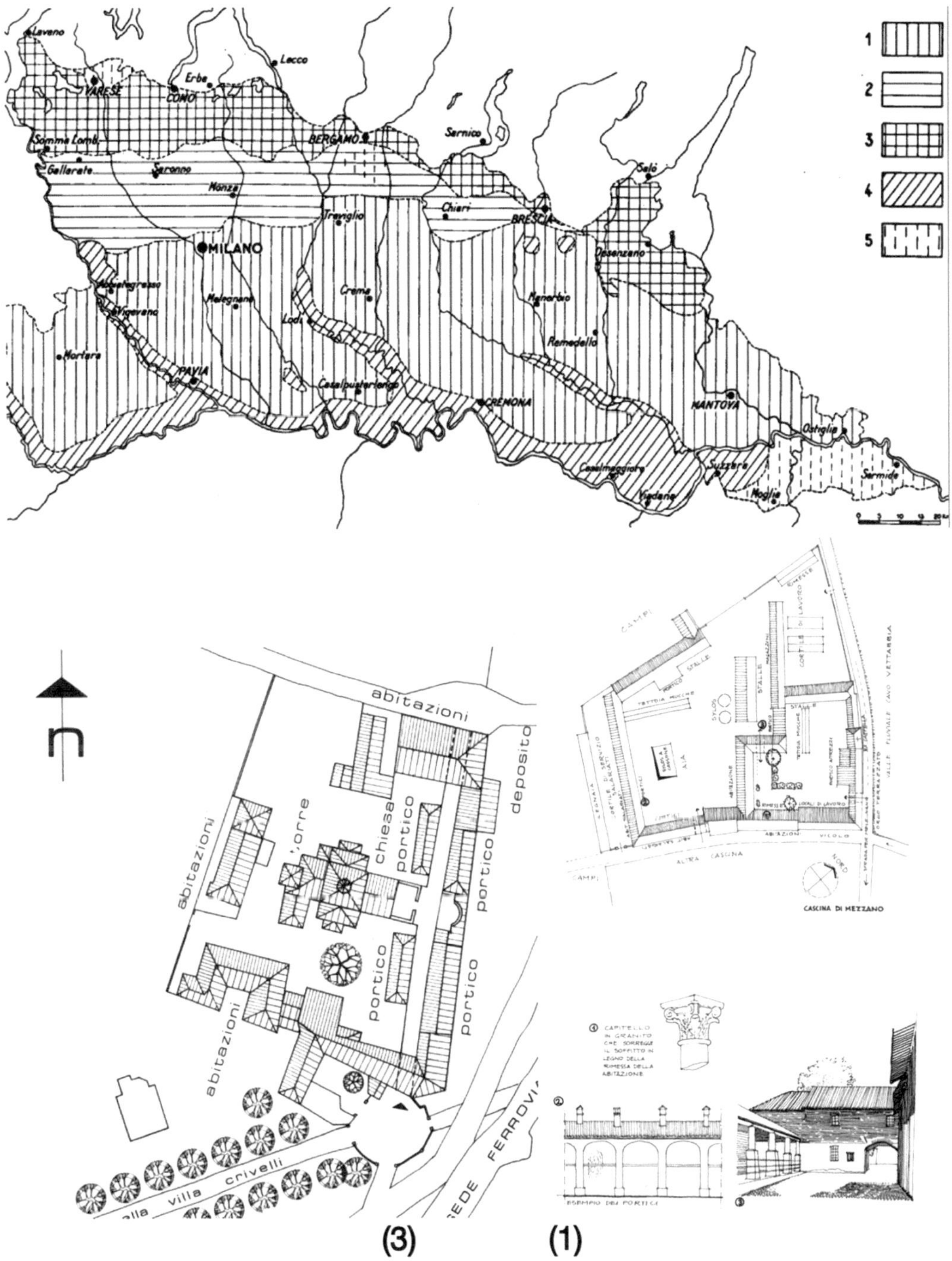

Fig. 6.13 Prevalent Landscape types: (1) wetted plain "Bassa Pianura", (2) dry terraced plain "Alta Pianura", (3) morenic hilly pre-alpine, (4 & 5) main flood plains corridors.

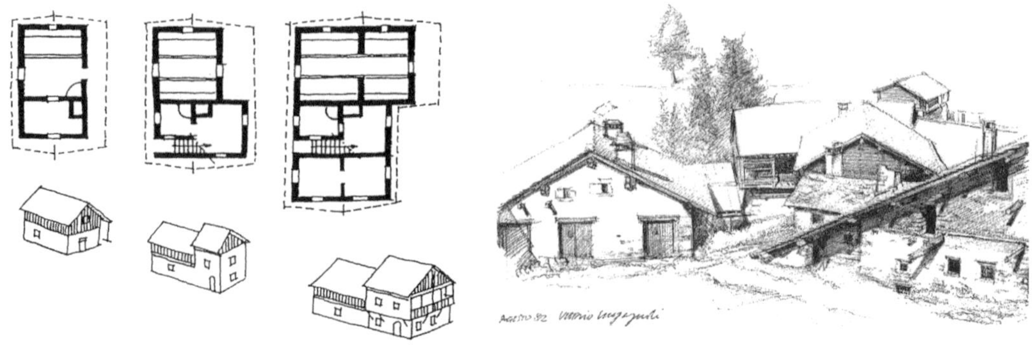

Fig. 6.14 An example of a vernacular architecture (map and section) from South Tyrol. Note the design method, planned step by step, to fit the house to the changing exigencies of the family. To the *right*, an old Alpine settlement: Mascognaz, Vallée d'Ayas (Italy) 1,900 m, fifteenth to sixteenth century. Note the localisation on a grassy side of the mountain, near the margin of a forest of spruce and larch, the exposure, the morphologic adaptation, the building typology and materials, the agriculture, the vegetable gardens, the sheep farming: all are perfectly assembled in this anthropic ecocoenotope, demonstrating a real "organic" structure (drawing by the author)

the types of rural architecture and the types of Lombard landscapes is evident.

It is noted that the materials, physically linked to the place, affect the structure and aesthetics of the buildings, the functions of use, ecological niches of the habitat; they shape the space and characters, the planning parameters, linked to the territory, influence types, routes, volumes and connections, the local culture, an expression of tradition and adaptations to the environment, interacts actively with the more academic culture.

The ecotopes formed by sets villa-park function as a contrast to the landscape matrix characterising the structure, while the flow of cultural information exerts control functions of the city on country LU. One can also observe a similarity between the mode of design and development of "spontaneous" or vernacular architectures and the evolutionary process of organisms. Vernacular design does not follow the method of rational design, rather the organic one. The similitude with the evolutionary criteria, through which an organism is designed is incredible. In fact, as pointed out by Konrad Lorenz [46], the structure of an organism can never be compared with that of a building designed by an experienced architect in a unitary project, but it is very similar to a country house built by a peasant.

The hut, built to protect from cold, wind and rain is enlarged (Fig. 6.14) proportionally to reflect the wealth and the family composition without destroying the old hut, rather changing it into a storeroom. In this way each room is enlarged and changed through time. The historical remnants are well recognisable and they are conserved because the building can only be restructured, not destroyed, since it is continuously inhabited. Similarly, it also happens to rural villages.

The historical villas often have obvious ecological roles [47], both in the countryside than in the city. In Fig. 6.15 a study carried out on some famous villas of Lake Como (Villa Carlotta and Villa Vigoni), is reported which shows the positive influence of the parks in the landscapes coasts of the lake [48]. Prior to the development of these parks, in fact, the average BTC of riparian LU, a rural area, was much less elevated than at present.

This complex relationship between ecological and cultural history of the local architecture has been seriously altered with the advent of the industrial age. It is purported to eliminate the characteristics considered "boundary" (see rationalism in architecture), generating the cult of the "ugliness" and "global", i.e. the indifference to the type of landscape.

Depersonalisation and energy deficiencies have been added and will still add to the degradation of

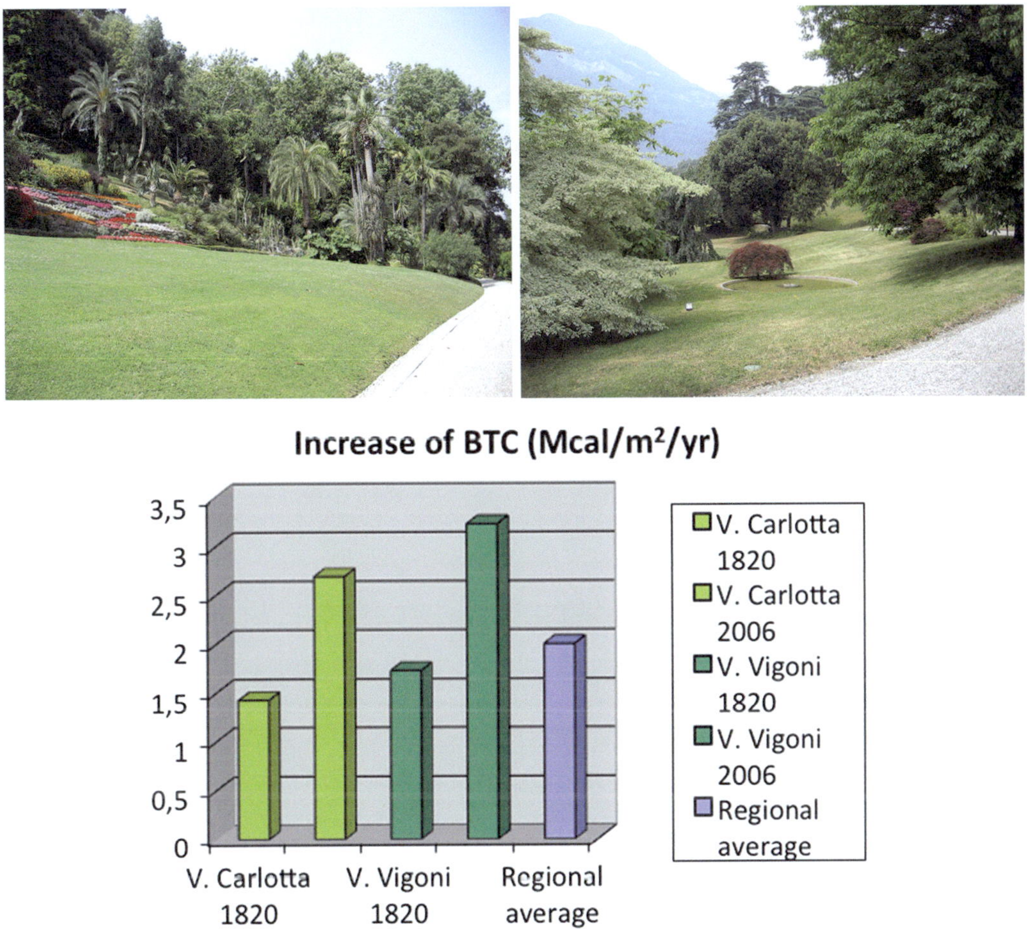

Fig. 6.15 Comparison between BTC values 1820–2006 related to the large parks of the Ville Carlotta (*left*) and Vigoni, on Lake Como. The high increase of BTC in the last two centuries shows that the landscape transformation of littoral LU from rural to park was positive

the landscape also understood in an ecological sense, not just aesthetic. The environmental ethic plays a great role in these cases: man want to impose his economic and rational view to the natural harmonies. It is a question of violence and justice, as explained in the Chap. 15.

The agricultural landscapes can be divided into at least four types: the *agricultural-forestry or agro-forestry-pastoral* landscapes, which are managed, while cultivated landscapes are *protective*, *productive* and *rural*. Figure 6.16 shows the difference between productive and protective agricultural landscapes.

The localisation of agricultural villages or large farms, in Europe, including the villas, the network of canals, the irrigation channels, the geometrisation of the fields are the elements that characterise the structures of agricultural landscapes. In Countries like Italy or France, with a high heterogeneity of environment, we can find nine types of agricultural landscapes:

1. *Mountain agricultural with huts*, in which still dominates the forest, with scattered meadows and small rural settlements, respecting geomorphological structures (e.g. in the Alps);
2. *Agricultural-rural valley floor*, hay meadows, fields and small countries, with remnant forest (alpine valleys), directed by rivers and alluvial fans and often characterised by tourism (e.g. alpine valleys);

Fig. 6.16 Comparison between two agricultural landscapes near Milan: productive (*above*, South to Abbiategrasso) and protective (*below*, N-E to Carate Brianza)

3. *Hillside vineyards and fields*, sometimes terraced, with countries and fields interspersed with craft areas, often with remnant forest on the top (e.g. in the Pre-Alps or Tuscany or Provence);
4. *Plain closed fields and farms*, structured by rows of trees and irrigation ditches, in open fields and farms complex (e.g. Po or Rhine Valley);
5. *Plain open fields,* dominated by monocultures in extensive fields, often with industrialised countries (e.g. the Po Valley);
6. *Mountain and open pastoral,* with vast grasslands and some open field and remnant forest with sparse rural settlements (e.g. in the Apennines)
7. *Hilly open fields*, ample arable land with rare appearances of trees and some vines and spaced countries (e.g. within the Sicily);
8. *Sub littoral with olive groves and vineyards*, dominated by olive groves and with often quite large villages and commercial areas (e.g. Mediterranean sub littoral bands);
9. *Sub littoral garden, the typical citrus gardens*, with small groves, vineyards and orchards, today surrounded by tourism (e.g. Mediterranean coasts).

6.2.4 LaBiSHH, New Method to Study the Human Habitat

Also for the study of the human habitat of various types of landscape we propose a code similar to what has already been done for the vegetation: LaBiSHH (Bionomic Landscape Survey of Human Habitats). In fact, the definition of

Table 6.7 Comparison of the interception values (%) of precipitations (rainfall of medium intensity) among different tesserae, with diverse percentage between vegetated and waterproof surfaces

Different tesserae	Nat	Ur-r	Ur-m	Ur-d	Imp
Vegetated patch	>80	25	10	<5	0
Semi-impervious surfaces	<15	10	10	<5	<5
Impermeable surfaces	<5	65	80	>90	>95
Processes					
Foliage interception	25–30	10	5	1	0
Water runoff	15–20	65	83	93	99
Soil absorption	60–50	25	12	6	1
Soil retention	15–10	5	2.5	1	1
Percolation by gravity	45–40	20	9.5	5	0
Ecological value	1.0	0.5	0.25	0.12	0.0

Nat natural, *Ur-r* rarely urbanisation, *Ur-m* urbanisation, *Ur-d* dense-urbanised, *Imp* impermeable

territory as landscape ruled by man, that is, as a multidimensional living entity co-evolved with man and nature, and the possibility of formalising the LU processes in an ecological sense, have helped to create some complex models that integrate more dependent variables, such as SH/SH* to HH (Fig. 3.10), g-LM/HH, considering HH as independent variable. It therefore becomes possible to evaluate the main functions of the human component in the sense of landscape bionomics, not in an architectural – urban one.

6.2.4.1 The Biotopflächenfaktor (BFF)

The analysis of built ecotopes and urbanised landscape units, even at a detailed scale, can be addressed to landscape bionimics through the measurement of a basic ecological factor, identified in the process of interception of rainfall. The alteration of this process is, in fact, one of the most serious ecological disturbances that produces a built tessera towards the environment. To understand this statement, it is sufficient to describe the situation of optimal normality natural vegetated tessera and place it in comparison with situations in urban tessera. In a patch of urbanisation the absorption of water in the soil passes from 55 % of a natural tessera to 12 %, while the interception of rain in anthropic vegetation decreases by over 80 %. This depends precisely on brick works constructed, by their relationship with the residual vegetated surfaces (usually in urban garden or vegetable garden) and their materials, especially as regards the floorings.

Some Germans ecologists [49] have proposed an index that measures the phenomena of interception of precipitation, giving the possibility to check the ecological status of urbanised ecotopes through this crucial process. This index was called *Biotopflächenfaktor* (BFF) translated into "Factor of naturalness of the urbanised tesserae". Tables 6.7 and 6.8 show the values to be applied to the analysis.

As you can see, the index ranges from 0 (total waterproofing) to 1 (natural ground vegetation). The use of this index is simple. We measure the components of the tessera in question with a breakdown by type of coverage, which is attributed to the relative value of BFF. In the case of intermediate types among those exposed you can help with the understanding of the processes involved. It is then measured in m^2 and the resulting value is related to the total area of the tessera.

6.2.4.2 Quantification of Human and Natural Habitats

To estimate the human habitat (HH) and natural (NH) is required the allocation of a class of primary importance, in proportion to the surface, relative to each type of element of the landscape, using a system similar to that of the detection of the vegetation in phytosociology, based on five classes of equal amplitude (20 %), of which the first and last can be divided for increased

Table 6.8 Attribution of values/square metres for measuring BFF (from Ermer et al. 1996)

Type of surface	Current components	BFF values
Total waterproof	Asphalt, concrete, house foundations	0.0
Almost waterproof	Paved grounds, flat stones	0.3
Semi-opened paved	Flat stones with greenery, or similar	0.5
Vegetated pavements	Garage greenery covers, etc.	0.5
Vegetated terraces	Green terraces, >0.8 m of soil	0.7
Garden vegetation	Vegetation on normal soil	1.0
Gravel areas	Water percolation areas with gravel/sand	0.2
Climbing vegetation	Walls covered by climbing greenery	0.5
Roof greenery	Roof gardens	0.7

Table 6.9 Survey classes related to HH and NH

HH classes	Amp. Y	N°	Types of tesserae and their characters
<5	5	1′	Ts of natural mature forest or semi-managed, Ts of natural vegetation, natural ponds, Ts of rock, sand, ice, etc.
5–20	15	1	Ts of coppice wood, managed shrub land, or natural pasture, Ts of riparian belt wood, etc.
20–40	20	2	Ts of semi-natural pasture, Ts of rural hedgerow with trees or shrubs, etc.
40–60	20	3	Semi-managed ponds and canals, semi-natural grasslands, etc.
60–80	20	4	Meadows, semi-natural arboriculture, hedgerows, biologic agriculture fields, semi-natural gardens (parks), semi-natural canals, etc.
80–95	15	5	Crop fields, technological arboriculture, urban gardens and parks, cottages with gardens, ruderal vegetation, artificial ponds and canals, etc.
>95	5	5′	Urban areas, industrial areas, technologic areas, paved roads and places, quarries and opencast mines, etc.

accuracy, as the following Table 6.9, valid for both HH and NH.

In practice, you have to detect the percentage of surface covered by elements referable e.g. to HH within the type of landscape element considered (e.g. the coppice typically has values between 10 and 15 %, depending on the presence and extension of the storage areas of the timber, access roads, of protection of woodcutters, etc. and the type of medium used for the removal of the cut material; agricultural fields oscillate between 80 and 90 %; sports and recreational areas have indicative values around 85 to 95 %).

After the first assignment of value classes, you must go on to assess the additional characters such CAH (to HH) or CAN (to NH) according to Tables 6.10 and 6.11. The additional characters are six and are estimated based on three pillars, of low, medium, high.

Table 6.10 Additional characters (CAH) for the human habitat (HH)

n°	Characters	Low	Medium	High
1	Intensity of man use	1	2	3
2	Change of the former structure	1	2–3	4
3	Presence of subsidiary energy	1	2	3
4	HH structure	1	2–3	4
5	Disturbances incorporation	3	2	1
6	Presence of NH components	4	3–2	1

For the estimation of HH, and similarly for NH, the score (p) maximum is: $p = 21$, while the minimum is: $p = 6$, then the amplitude of the usable values is equal to 15. Recalling that ‘y’ is the class interval of the previous table (Tables 6.10 and 6.11), the additional characters (or CAH or CAN) will be then:

Table 6.11 Additional characters (CAN) for the natural habitat (NH)

n°	Characters	Low	Medium	High
1	Presence of key species of vegetation	1	2–3	4
2	Presence of key species of fauna	1	2–3	4
3	Self-maintenance capacity	1	2	3
4	NH structure	1	2	3
5	Disturbances incorporation	1	2	3
6	Presence of HH components	4	3–2	1

$$CAH = [(p - 6)/15] * y.$$

In measuring HH is often necessary to clarify whether the result of the estimate is attributable to the entire geographical territory, or not. Except in the plain, in the hills and valleys, except the presence of important water bodies or lagoons, in the other areas there are often inaccessible areas or otherwise unusable for any type of human colonisation: e.g. rock walls, slopes >50°–60° or highly unstable, rivers with frequent flooding cycle, the presence of glaciers, etc. In such cases, HH remains in % of the entire LU, but in order to compare two different situations should be considered equivalent HHeq = HH/(LU−ti) × 100, where ti is the part of the territory unusable.

6.2.4.3 Functions Related to the Residential and Subsidiary Apparatuses

Based on experimental data reported in the analysis of tens of landscape units of North Italy at the municipal scale and on regional statistical data, it has come to establish a numerical relationship between the residential Standard Habitat (SHrsd) and the Human Habitat HH. This function varies linearly from minimum values in landscapes with high HH (for certain areas of Milan SHrsd = 69 m^2/inh, with HH 96.8 %), up to high values, even over 300 m^2/inh, for low HH (SHrsd of Petrignano (Lucca) = 306.5 m^2/inh).

In the model (Fig. 6.17) are taken into account, however, only the more balanced cases. When rural components prevail, HSrsd slows its descent for HH >65 %, up to values close to 90 m^2/inh, instead of the 55 m^2/inh of dense cities. For low values of HH (<20 %) we can have a lowering of HSrsd, even if present in the graph conventional values are considered, for the sake of optimality.

The equations of the model HSrsd, appear to be three: one for the first broken (by HH = 10 % to 20 %), shown as y_{RSD} (A) and valid for all the landscapes; one for the continuous line y_{RSD} (B) valid for all landscapes up to HH = 65 % and only for the urban ones later, the last for the broken line relative only to rural landscapes y_{RSD} (C).

$$y_{RSD}(A) = 10x + 100;$$
$$y_{RSD}(B) = -3,1(x - 20) + 300;$$
$$y_{RSD}(C) = -2,3\ (x - 65) + 160,5.$$

Concerning the subsidiary apparatus the variability of this area, that we remember correspond to industrial zones and large-scale commerce and major arterial roads, is significantly greater than the previous, but up to values of HH = 90 % there is a direct proportionality growth between HH and SHsbs (Fig. 6.17): with higher values of HH the descent is strong, since in dense city areas of technology are usually marginal (SHsbs of Milan m^2/inh = 15–20). The maximum values can also exceed 150 m^2/inh, but these values can not be considered to be optimal in any case. In addition HH = 65 % can, however, establish strong growth in the most urbanised landscapes.

When rural components prevail SHsbs usually do not arrive to exceed values of 100 m^2/inh, which, however, they are already significant. In the model, in fact these functions was taken into account with a certain width, given the tendency of their growth throughout the world. For the understanding of the relating equations note that there is a common trait that goes from HH = 10 to 65% to which will be indicated as y_{SBS} (A); this point there is a bifurcation: y_{SBS} (B) and y_{SBS} (C) correspond to the equations of cultural landscapes (the first until HH = 90 %, the second >90 % of HH); y_{SBS} (D) and y_{SBS} (E) correspond to the

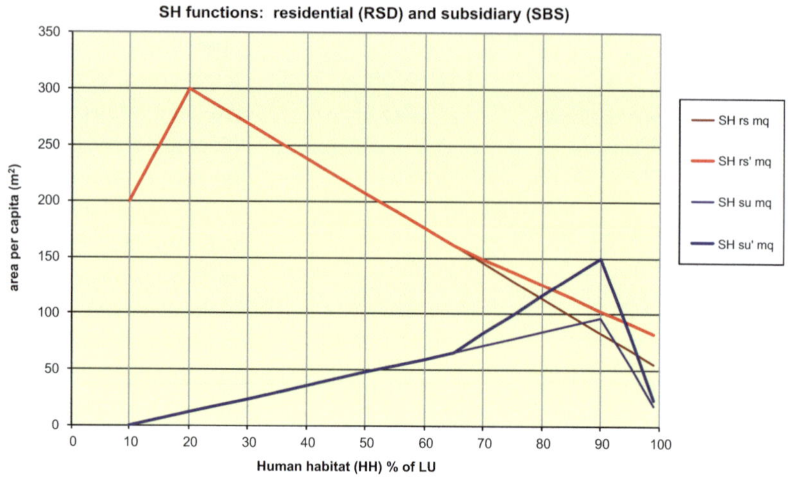

Fig. 6.17 Models of the anthropogenic functions RSD (residential) and SBS (subsidiary). In abscissa the percentage of HH per unit of landscape; in ordinate the residential SH and the subsidiary SH. The forks over HU = 65 % indicate both rural and urban transformations

equations of rural landscapes (the first to HH = 90 % is the actual continuation of y_{SBS} (a), the second for the higher values is different).

$$y_{SBS}(A) = 1.2\ (x - 10);$$
$$y_{SBS}(B) = 3.36\ (x - 65) + 66;$$
$$y_{SBS}(C) = -14\ (x - 90) + 150;$$
$$y_{SBS}(D) = 1.2\ (x - 10);$$
$$y_{SBS}(E) = -8.6(x - 90) + 96.$$

Also for the study of the subsidiary apparatus, evaluation boards for ecological industrial areas have been proposed by Elena Giglio [50] using concepts and principles of integrated biological and landscape ecology. It can thus approach the study of the EIA (Environmental Impact Assessment) and SEA (Strategic Environmental Assessment) in a way quite different from the usual, for more effective environmental protection.

6.2.4.4 Functions Related to the Other Apparatuses

Functions of the Human Habitat included in Natural Habitat (HHnh). From the experimental data the average amount of HH present in the landscape elements of NH usually ranges in Europe between 5–10%, albeit with a few exceptions (e. g. 3 % in Bialowieza, Poland). In the construction of this model we take into account a conventional value of 8.5 %, which is quite abundant, but which reflects the situation of the current trend. Proportioning the average value for each NH complementary to each HH considered, the equation HHnh of the model is to contribute to the total SH through an hyperbolic curve, decreasing with HU. The equation is the following:

$$y_{HHnh} = 0.085\ (100 - x)\ 1/x.$$

So for HH = 20 %, HHnh = 0.34 Σ (SHrsd, SHsbs, SHprd, SHprt), then we will have a contribution to the SH total of approximately 25 %, which however drops quickly to 3.5 % with HH = 80 %. Note that in the system of equations to obtain the total SH_{HH}, the equation y_{HHnh} must be considered as a multiplier to the sum of the other four components.

When we pass to the productive apparatus, experimental data on agriculture and regional statistics allow us to consider the variation of SHprd compared to HH in descending order, starting with values of SHprd = 4,100 m^2/inh for HH = 10 %. If you continue to have a rural character of the landscape, there is an inflection at HH = 85 %; is then guaranteed a SH/SH* (theoretical) productive apparatus of 2. Even with high HH, SHprd remains about 2,000 m^2/inh. When urban technological components prevail, the development follows a bifurcation for HH >65 %, with the next inflection, which leads to very low values of SHprd in landscapes with dense urbanisation (e.g. 227 m^2/

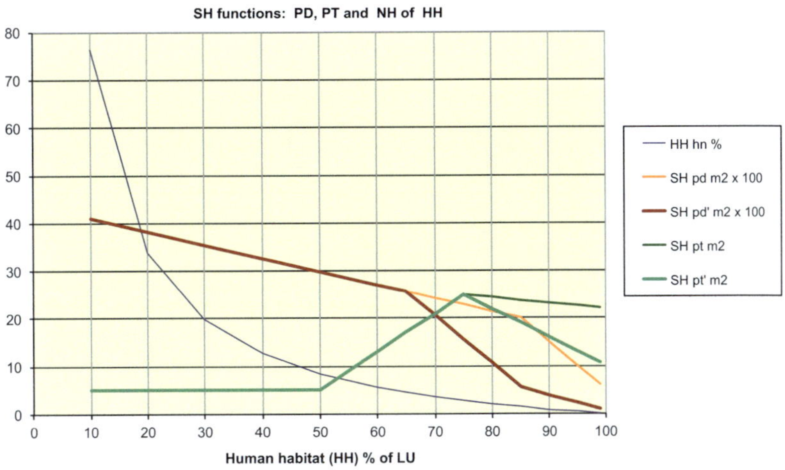

Fig. 6.18 Models of the functions of the natural part of the human habitat (HHnh), the production areas (SH_{PRD}), protective areas (SH_{PRT}), always in relation to HH (%), in m^2/inh

inh for HH = 95 %) (Fig. 6.18). Consequently, the equations of the model SHprd are as follows:

$$y_{PRD}(A) = -28\,(x-10)+4100$$
$$(\text{for all landscapes} < HH = 65\%);$$

$$y_{PRD}(B) = -28(x-10)+4100$$
$$(\text{the same line, for rural landscapes to } HH = 85\%);$$

$$y_{PRD}(C) = -100(x-85)+2000$$
$$(\text{for rural landscapes to } HH > 85\%);$$

$$y_{PRD}(D) = -100(x-65)+2560$$
$$(\text{for urban landscapes}, HH = 65 \text{ to } 85\%);$$

$$y_{PRD} = -35(x-85)+560$$
$$(\text{for urban landscapes}, HH > 85\%).$$

Functions related with the protective apparatus (PRT). In the protective apparatus experimental and statistical data can be used only partially, since today almost anywhere we found a heavy alteration of the protective components, especially in the agricultural landscape [14]. Therefore, it is necessary to calculate the theoretical protective Standard Habitat. It is based on the equations of the present model for the determination of the values of SHrsd, SHsbs, SHprd and for measuring the relative deficit of transformation (always referring to HH, as the independent variable).

As you recall (see Sect. 4.1) this measure needs an average BTC of reference, which on a large scale should be the BTC of the region, but at the municipal scale that would be difficult, as the higher levels of human activity (HH >75 %) would lead to values of SHprt of about 300 m^2/inh (Fig. 4.2). So, the green lines in Fig. 6.18 are only indicative. To value the SHprt we must follow what expressed in Sect. 4.1.

References

1. Johnston CA (1995) Effects of animals on landscape pattern. In: Hansson L, Fahrig L, Merriam G (eds) Mosaic of landscape and ecological processes. Chapman & Hall, London, pp 57–58
2. Pavan M (1981) Significance of ants of the Formica rufa group in Italy in ecological forestry regulation. Bull SROP 161–169
3. Figuier L (1926) Gli uccelli. F.lli Treves Editori, Milano
4. Opdam P, Van Apeldoorn R, Shotman A, Kalhoven J (1992) Population responses to landscape fragmentation. In: Vos CC, Opdam P (eds) Landscape ecology of a stressed environment. Chapman & Hall, London, pp 147–171

5. Ingegnoli V, Giglio E (2005) Ecologia del Paesaggio: manuale per conservare, gestire e pianificare l'ambiente. Sistemi editoriali SE, Napoli
6. Reimoser F, Gossow H (1996) Impact of ungulates on forest vegetation and its dependence on the silvicultural system. For Ecol Manage 88
7. Perco F (1987) Ungulati. Lorenzini Ed. Udine
8. Petretti F (2003) Gestione della fauna. Il management delle popolazioni animali negli ambienti naturali, agricoli e urbanizzati. Edagricole, Bologna
9. Lambeck RJ (1997) Focal species: a multi-species umbrella for nature conservation. Conserv Biol 11:849–856
10. Turner MG, Wu Y, Pearson SM, Romme WH, Wallace LL (1992) Landscape-level interactions among ungulates, vegetation and large-scale fires in Nothern Yellowstone National Park. In: Plumb GE, Harlow HJ (eds) 16th annual report. University of Wyoming national Park Service Researche Center, Laramie, pp 206–211
11. Naveh Z, Lieberman A (1984) Landscape ecology: theory and application. Springer, New York
12. Ott J (2001) Expansion of Mediterranean Odonata in Germany and Europe- consequences of climatic changes. In: Walter GR, Burga CA, Edwards PJ (eds) Fingerprints of climate change, adapted behaviour and shifting speciues ranges. Kluwer, London, pp 89–111
13. Graves J, Reavey D (1996) Global environmental change. Plants, animals and communities. Addison Wesley, Reading, MA
14. Burel F, Baudry J (1999) Écologie du paysage: concepts, méthodes et applications. Technique & Documentation Ed, Paris
15. Den Boer PJ (1985) Fluctation of density and survival of carabid populations. Ecologia 67:322–330
16. Burel F, Baudry J (1994) Reaction of ground beetles to vegetation changes following grassland derelictions. Acta Oecol 15:401–415
17. Ingegnoli V (2002) Landscape ecology: a widening foundation. Springer, Berlin
18. Storm GL, Andrews RD, Philips RL, Bishop RA, Sniff DB, Tester JR (1976) Morphology, reproduction, dispersal and mortality of mid-western red fox populations. Wildlife Monogr 49:5–82
19. Mustoni A, Pedrotti L, Zanon E, Tosi G (2002) Ungulati delle Alpi. Nitida Immagine Ed, Cles, Trento
20. Storch I (1997) The importance of scale in habitat conservation for an endangered species: the capercaille in central Europe. In: Bissonette JA (ed) Wildlife and landscape Ecology. Springer, New York, pp 310–330
21. Kautz M, Schopf R, Ohser J (2013) The "sun-effect"—microclimatic alterations predispose forest edges to bark beetle infestations. Eur J For Res 132:453–465. doi:10.1007/s 10342-013-0685-2
22. Forman RTT (1995) Land Mosaics: the ecology of landscapes and regions. Cambridge University Press, Cambridge
23. Legagneaux P, Ducatez S (2013) European birds adjust their flight initiation distance to road speed limits. Biology Letters, vol 9 no 5
24. EUEA and FOE (2011) Study on landscape fragmentation. EU edition
25. Jaeger JA, Holderegger GR (2005) Schwellenwerte der Landschaftszerschneidung (Translation: thresholds of landscape fragmentation). GAIA 14:113–118
26. Fornasari L (1997) I rapporti tra i vertebrati e il paesaggio: teoria ed esempi. In: Ingegnoli V (ed) Esercizi di ecologia del paesaggio. CittàStudi-Utet, Milano, pp 131–168
27. Massa R, Ingegnoli V (eds) (1999) Biodiversità, estinzione e conservazione: fondamenti di conservazione biologica. Utet Libreria, Torino
28. Godron M (1984) Ecologie de la végétation terrestre. Masson, Paris
29. Caravello G, Ingegnoli V (1991) Landscape ecology in the Roman world. IALE World Congress of Landscape Ecology. Carleton University, Ottawa, ON, p 37
30. Giacomini V (1980) Qualifying aspects of MAB Project 11 applied to the city of Rome: Provisional draft report. Italian Committee for the Programme on Man and the Biosphere
31. Ingegnoli V, Roncai L (1975) Ambiente e architettura rurale. In: Perogalli C, Alpago-Novello A (eds) Cascine del territorio di Milano. Milani Editrice, Milano, pp 23–42
32. Ingegnoli V (1977) Il territorio di Melegnano e le architetture rurali. In: Perogalli C (ed) L'arte nel territorio di Melegnano. Nuove Edizioni, Milano, pp 16–56
33. Ingegnoli V (1981) Organizzazione agricola e casa rurale. In: Pirovano C (ed) Lombardia: il territorio, l'ambiente, il paesaggio. Electa, Milano, pp 27–64
34. Sukopp H, Hejny S (eds) (1990) Urban ecology. SPB Academic, The Hague
35. Forman RTT, Godron M (1986) Landscape ecology. Wiley, New York
36. Zonneveld IS (1995) Land ecology. SPB Academic, Amsterdam
37. Farina A (1998) Principles and methods in landscape ecology. Chapman & Hall, London
38. Farina A, Belgrano A (2005) The eco-field hypothesis: toward a cognitive landscape. Landsc Ecol. doi:10.1007/s10980-005-7755-x
39. Ingegnoli V (1980) Ecologia e progettazione. Cusl, Milano
40. Ingegnoli V (1993) Fondamenti di Ecologia del paesaggio. CittàStudi, Milano
41. Ingegnoli V (2011) Bionomia del paesaggio. Springer, Milano
42. Vitruvio Pollione M (I. century) De Architectura. (many editions)
43. Saarinen A (1962) Eero Saarinen on his work. Yale University Press, New Heven
44. Forman RTT (2008) Urban regions, ecology and planning beyond the city. Cambridge University Press, Cambridge

45. Saibene C (1955, 1980) La casa rurale nella pianura e nella collina Lombarda, (CNR vol 15) L Olschki, Firenze
46. Lorenz K (1978) Vergleichende Verhaltensforschung: Grundlagen der Ethologie. Springer, Berlin
47. Ingegnoli V (1989) Basi ecologiche del rapporto ville-ambiente. In: Brusa C (ed) Ville e territorio. Lativa Edizioni, Varese, pp 17–24
48. Ingegnoli V (2007) Significato ecologico dei giardini nei paesaggi delle riviere lacustri: villa Vigoni e villa Carlotta. In: Lodari R (ed) Il giardino e il lago, specchi d'acqua fra illusione e realtà. Conoscenza e valorizzazione del paesaggio lacustre in Italia e in Europa. Gangemi Editore, Roma, pp 119–125
49. Ermer K, Hoff R, Mohrmann R (1996) Landsschaftsplanung in der Stadt. Ulmer, Stuttgart
50. Giglio E (2005) Proposta di valutazione ecologica di un lotto industriale o commerciale e delle sue eventuali trasformazioni. Alcuni esempi reali dalla Provincia di Lucca. Estimo e Territorio, 4
51. Ingegnoli V (1992) Valutazione dello stato ecologico del territorio di Torrile e Colorno. In: Cattini M (ed) Torrile alla ricerca di una identità. Regione Emilia Romagna, Bologna, pp 109–130

7 General and Bionomic Analysis of the Landscape

7.1 General Landscape Analysis

7.1.1 Structural Pattern Analysis of a Landscape Unit

These analyses depend to a large extent on the methods of geomorphology, with contributions from geo-pedology and geo-tectonic. It is not the case to expose these methods in this volume, both for their ponderousness and because we do not want to replace experts that are part of the Departments of Earth Sciences. What we will just emphasise is the need to use these analyses as part of the criteria that are at the basis of landscape bionomics. In fact, as also in geography, even if people talk a lot about landscape, it is usually considered in a different way from what we have exposed in previous chapters. This is not an obstacle in itself, because the intentions of Earth Sciences are different from those of the Science of Nature, for obvious reasons. But we must keep in mind these differences of opinion, to not deviate from the already difficult task of integration of the data according to a single criterion for defining the landscape.

The analysis of the structural pattern is for a bionomist not so different from the analysis of the skeleton for a student of anatomy. A naturalist has the opportunity to outline the structures and processes related to the examined territory, possibly with the help of specialised publications. Fitting together the geological and geomorphological data, however, can give results different from the traditional maps of Earth Sciences for the need of studying a landscape, because the need for data integration of that sector with the more specific ones of landscape bionomics can lead to charts specifically designed for the case. As an example, we report processing made for a map of the Alto Garda Bresciano, where, in Fig. 7.1 you can see the summary of a morphological map prepared by the method of "maxiclive" [1].

The geological and geomorphological data, however, which constitute the basis for all further and more specific processing in the study of a landscape, are the following:

(a) *Topographic data*: level lines, units, hydrography (e.g. maps IGM[1]).
(b) *Lithological data*: information divided into surface (native and non-native) and geological formations (intrusive, extrusive, metamorphic, limestone, sandstone, etc.).
(c) *Structural data*: attitude layers, discontinuities, faults, etc.
(d) *Morphogenetic data*: types of processes and their distribution, and, in addition, data relating to them.
(e) *Endogenous processes*: volcanic forms and their derivatives.
(f) *Littoral processes*: the actions of the sea and on the shores of the lakes, the formation of the beaches, etc.

[1] IGM means Geographic Military Institute of Italy (analogous to other European Countries ones).

V. Ingegnoli, *Landscape Bionomics Biological-Integrated Landscape Ecology*,
DOI 10.1007/978-88-470-5226-0_7,

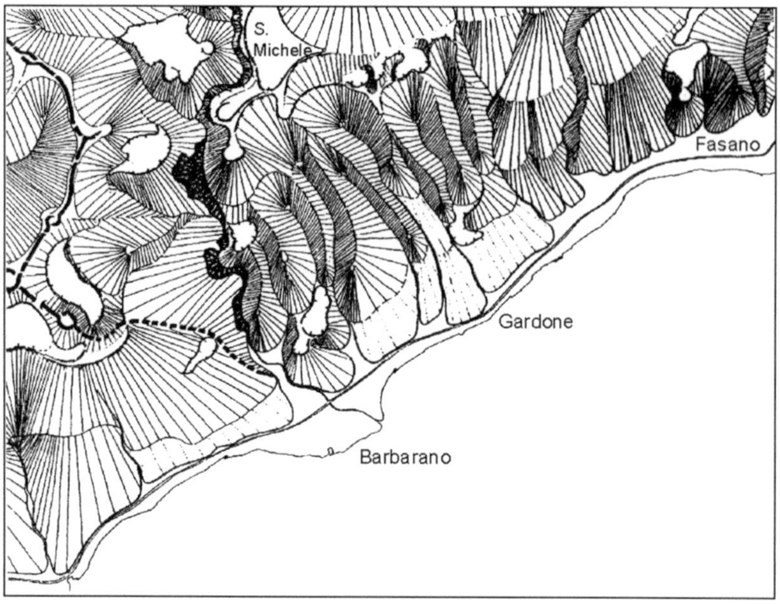

Fig. 7.1 Morphologic map of the landscape following a method of "maxiclive" by Valerio Romani [1]. This territory is in the West part of the Lake Garda

(g) *Fluvial processes*: erosion, meanders, bars, suspensions, braids, etc.
(h) *Karst processes*: sinkholes, rutted fields, trays of corrosion, etc.
(i) *Wind processes*: dune, deposition, erosion, etc.
(j) *Glacial processes*: glaciers, moraines, cirques, boulders, etc.
(k) *Cryo-nival processes*: accumulation of loess, ice-flow, etc.
(l) *Slope processes*: ravine, landslides, debris flows, etc.
(m) *Anthropic processes*: reservoirs, canals, changes pads, roads, etc.
(n) *Polygenetic forms*: forms generated by the competition of many processes, etc.

7.1.2 Climatic Analysis

For the analysis of climatic aspects of a landscape unit (LU), it's difficult to have the values of "normal" temperature or rainfall to which referring and checking the consistency of those measured. With appropriate triangulations, or suitable transects, we are able to highlight phenomena (e.g. heat islands of the city) and the distance at which they are affected by these effects. The first step of the evaluation is to define the climate at different spatial scales, (macroclimate, mesoclimate, microclimate) sensu Venanzoni and Pedrotti [2], where the synthetic one expresses the environmental constraints on the types of vegetation present in a given place. More detailed scales allow operating further subdivisions in the study area and/or may help to explain any differences in local habitat areas apparently homogeneous. The scales of detail (from meso- to- micro-climate) assume considerable importance in areas where morphology is not flat—for the complex circulation of air established in these areas—as well as within, and in the surrounding of, large conurbations (*heat islands*, etc.).

For a proper evaluation, the averages of the data should refer at least to 30 years, so that the regressions and trends were noticeable and should not miss, where data were available, even small rainfall stations only. Moreover, as we shall see, it is appropriate to consider not so much the climate, but to prefer the phytoclimate, that is more directly related to aspects of the vegetation cover.

After having thoroughly identified the stations, both inside and outside of the LU in question, in place and in an adequate numbers to understand the phenomenon at different scales, for each of them it is necessary to draw a thermo-precipitation diagram, according to the diagram in Pignatti [3], or alternatively, with simplified thermo-precipitation diagrams.

The raw data of temperature and precipitation should be developed with a series of *bioclimatic indexes* (Rivas Martinez [4] and subsequent processing; Ingegnoli and Giglio [5]). The importance of such climatic indexes is high, because they allow considerations that relate to climate and plant cover (which as we know is a structuring factor of the ecotissue), but considerable attention is required in their use, as they allow a bioclimatic classification from one side, but have well-defined limits and are not always generalisable on the other side: they are valid only within the limits for which they had been created for. We report some index only for the purpose of example.

1. *Stress indices of monthly aridity* of Mitrakos: (MDS) = 2 (50-p) where p defines the monthly rainfall (mm). MDS defines the intensity and duration of dryness monthly, 0 being the minimum value that corresponds to the absence of aridity.
2. *Monthly stress index from cold* of Mitrakos (MCS) – 8 (10-t) with t the monthly average minimum temperature (°C). Expresses the duration and intensity of the monthly stress from cold, 0 being the minimum value equal to the absence of cold.
3. *Stress index by summer aridity (SDS)* of Rivas-Martinez as the sum of June MDS + July MDS + August MDS.
4. *Winter cold stress index* (WCS) of Rivas-Martinez as the sum of MCS in December + MCS January + MCS February.
5. *Simple warmth index (En)* = $(T + m + M) \times 10$ *and Index of warmth and plywood (ITC)* = En + C (for values of En > 18) of Rivas-Martinez. It is an index that relates the quotient of Emberger with annual temperature. It is used to determine the bands and bioclimatic horizons and stresses the importance of the minimum temperature of the coldest month as one of the limiting factors for vegetation.
6. *Simple continentality index* (Ic) = T_{max}–T_{min} of Rivas-Martinez, with T_{max} average temperature of the warmest month and T_{min} the average temperature of the coldest month (°C). The limits are: 0 < Ic < 21 = oceanic climate (divided into hyper-oceanic between 0 and 10; EU-oceanic between 10 and 15; semi-oceanic 15–21), 21 < Ic < = 65 continental climate (divided between the semi-continental 21–27; eu-continental between 27 and 46; hyper-continental between 46 and 65).
7. *Aridity index* (AI) = P/(T + 10) of De Martonne. Edit the simple relationship (P/T) of Lang. The ecological values of AI are: 5 = desert vegetation; 5–10 = steppe grassland, 10–20 = prairie, >20 forest.
8. *Ombro-thermic annual index* (Io) = Pp/Tp, Rivas-Martinez, with: Pp sum of the average rainfall (mm) of the months with temperatures >0 °C, Tp sum of the average monthly temperature > 0 °C. Proposed to determine the climatic regions states that: 0.1 < I < 1.5 = Mediterranean climate; 2 < I < 3.8 = temperate climate. For areas of transition between temperate and Mediterranean situations (1.5 < I < 2) an Ombro-thermic summer index (IOV) is added referring to the months of June, July and August, and possibly an Ombro-thermic summer plywood (IOS4) index with the addition of May.

Please note that, processing data with climate indices such as those listed above, can allow to the identification of geographic and climatic areas: macroclimate, bioclimates, climate horizons or thermotypes (mesoclimate), ombrotypes (topo-climates), sub-horizons [5].

7.1.3 General Ecological Indexes

7.1.3.1 Gravity Model

A configuration of patches in a landscape sometimes needs to be evaluated in prevision of a possible transformation. In cases like these, it could be useful to understand the effect of the totality of interactions (unknown process) among the patches or between them in relation to another element which remains unchanged. It is possible to utilise a neutral model, the *gravity model*, which is derived from classical physics and based on the gravity between two masses:

$$I_{ij} = K\left(P_i \times P_j\right)/d^2$$

where: I_{ij} is the interaction measure; P_i the attributes of the patch *i*; *d* the distance between two patches and *K* is a constant.

It is necessary to appropriately choose the attributes of the patches and the correct setting of the *K* index. For instance, *K* can be calibrated on the interference of barriers and/or possible disturbances in the area.

7.1.3.2 Diversity and Dominance

The following step represents the investigation through ecological indexes, many traditional ones of which may be utilised even in landscape ecology, with some adaptations. In general ecology, the indexes are essentially based on species, while we have to use landscape elements. We describe here only the most important indexes, specifying that the term "habitat" means "landscape element."

$$\text{Relative richness} \quad R = (S/S_{\max}) \times 100$$

where: S is the number of habitat types and $S_{\max}$ the maximum possible number of habitat types [6].

$$\text{Diversity} \quad H = -\textstyle\sum_{k=1}^{s} p_k \ln p_k$$

where: s is the number of habitat types and p_k the proportion of area in habitat k [6]. It derives from information theory.

$$\text{Dominance} \quad D = \ln s + \textstyle\sum_{k=1}^{s} p_k \ln p_k$$

where: s is the number of habitat types and p_k the proportion of area in habitat k [6]. Note that *ln s* corresponds to the maximum diversity ($H_{\max}$).

$$\text{Relative evenness } E = \left(H_j/H_{\max}\right) \times 100$$

($H_{\max}$) is reached when each component has the same probability; H_j is the diversity of the studied LU [7].

$$\text{Contagion } C = 2s \log s + \textstyle\sum_{i=1}^{s} \sum_{j=1}^{s} q_{ij} \log qij$$

where: s is the number of habitat types and q_{ij} the probability of habitat i to be adjacent to habitat j [6, 7].

$$\text{Dispersion of patches } \mathrm{R_c} = 2\mathrm{d_c}(\lambda/\pi)$$

where: d_c is the average distance from a patch (its centre) to its nearest neighbouring patch and λ the average density of patches. If $\mathrm{R_c} = 1$, patches are randomly distributed; if $\mathrm{R_c} < 1$, patches are aggregated; if $\mathrm{R_c} > 1$ (max 2.149), patches are regularly distributed. The equation is also a measure of aggregation [8].

7.1.3.3 Grain and Contrast

The grain is the extent and distribution of the elements in a LU. The grain is usually measured with respect to the tessera or patches of tesserae (always specify!) that make up the LU, but can be useful, or necessary, to measure the grain of ecotopes of the same, or different, LU that compose a certain landscape. Interesting is, for example, the variation in size of the agricultural fields in time or of different vegetated tesserae in a forest, while in synthesis we can use the patches of tesserae of analogues forest types.

Very useful is, also, the examination of the contrast between the tesserae present in a LU, which indicates the degree of difference and the speed of transition between two adjacent areas according to the type and quality of landscape elements [5]: it is much greater the stronger are the differences between the elements, and the narrower the transition zones, i.e. the margins between a type and the other, although it is not easy to encode a high and low contrast. The contrast is more simple when, instead to refer to the characteristics of the individual tessera, one considers their belonging to a landscape apparatus (considered on the basis of the function most characterising the tessera in question): in this regard, we provide a table which can be taken as indicative of situations of contrasts of different entities (Table 7.1).

Table 7.1 Estimation of the contrast among the most important different landscape apparatuses, from Ingegnoli and Giglio [5]

	RNT	CON	EXR	SOUR	RES	CHAN	DIST	PRT	PRD	RSD	SBS
RNT	–	–	–	–	+	++	+++	++	++	+++	++++
CON		–	–	–	– +	– +	++	+	–	– +	++
EXR			–	–	–	– +	++	++	+	++	+++
SOUR				–	–	– +	++	++	+	+++	++++
RES					–	–	–	– +	–	++	+++
CHAN						–	–	– +	– +	++	++
DIST							–	–	– +	+	++
PRT								–	– +	–	–
PRD									–	+++	++++
RSD										–	+++
SBS											–

Note: ++++ = very high; +++ = high; ++ = medium; + = medium-low; – + = low; – = null
RNT resistant, *CON* connective, *EXR* excretory, *SOUR* source, *RES* resilient, *CHAN* change, *DIST* disturbance, *PRT* protective, *PRD* productive, *RSD* residential, *SBS* subsidiary

7.1.4 Connection Analysis

The analysis of a landscape network may be quite complex and may range from application of the autocatalytic system criteria to more simple empirical considerations. Useful suggestions come from the application of the topological *theory of graphs* [6, 7]. A graph is a geometrical entity applicable to make explicit the relations among objects in a general way, reducing the elements to points (vertices or nodes, V) and their relationships to arcs (linkages, L). A graph is important for finding paths among its nodes: a path may have a determined length or be closed, therefore forming a cycle. A graph is connected if its nodes are all linked by one or more paths. A complete graph with K nodes has by definition a number of arcs equivalent to:

$$L = [K\ (K - 1)]/2$$

L depends on a binomial coefficient (K choices 2): it describes the situation in which all the couples are in relation; the result is that the number of linkages is $< K^2$. The degree of a node is the number of arcs connected to it. Graphs derivable from maps are called planar graphs.

Thus, a complete graph is no longer planar for $K = 5$. If a graph accepts the case of arcs departing and returning in the same node (loop) it is called multi-graph. A tree is a connected graph without cycles. A simple graph is called Hamiltonian if it has a cycle of Hamilton, which is a closed path crossing all the nodes only once each.

From this theoretical framework (it is much more complex) we can derive two useful indexes: for a network connectivity (γ) and for a network circuitry (α).

$$\gamma = L/3\ (V - 2); \quad \alpha = (L - V + 1)/2V - 5$$

To identify key gaps and to apply the connectivity and circuitry indexes, the conversion of a LU basic mosaic into a network is needed. In general, it is necessary to put in evidence the vegetation patches and the main species sources as nodes and the existing corridors and sporadic routes as linkages.

The connection of the analysed landscape structure is always monitored by comparison with its past configurations. Frequently, these data can be used to design a restoration plan for natural conservation purpose or for planning a new suburb. In cases like these, we may observe that the mentioned γ and α indexes do not always have a clear significance, because they are not related to the types of studied landscape.

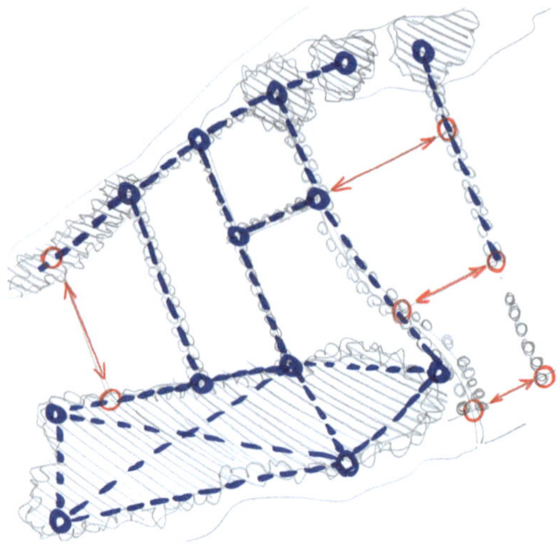

Fig. 7.2 Example of analysis of the connection of an ecotope using a planar graph

A drawing (Fig. 7.2) may indicate how you can apply the theory of planar graphs to a map of a LU. As you notice, in a LU in appropriate scale the following elements of an agricultural matrix are present: (a) forest areas, (b) rows of trees, (c) tree-lined small spots. The nodes are placed at the ends of a corridor, in the centre of a small spot or at the top of a forested area.

If two or more vegetated elements are neighbours (distance < height) they can be considered as a corridor made of "*stepping stones*," quite not in presence of barriers representing an incorporable disturbance. In large forest patches, the wings of margin function as corridors; areas of interior, though clearly present and without barriers, indicate the need for further links, often crossed, between the corner nodes of the patch itself: for corner nodes we mean a change of direction in the margin with angle >30°. The relationship between bonds and vertices (L/V) is an interesting number, as it gives a first idea of the state of connection of the system: for L/V < 1, the circuitry is zero (or negative). After applying the indices to a state of fact, we can propose a new network structure designing new elements to bond and comparing the results of the new indices, up, for example, to have an acceptable range. In this regard, we should recall the ecological law of the "appropriate amount" (not too much, not much, but enough). To get the references, you may use the values of L/V most frequent for each type of landscape: 0.5–0.75 for urban landscapes, from 0.6 to 0.85 for those suburban; 0.75–1.15 for those agricultural; >1–1.2 for those agricultural-forestry.

7.1.5 Fractal Analysis of the Landscape

Many characters of a system may change with scale, but the intrinsic part of them does not. This is in accordance with the description of the landscape given in Chap. 1, as a specific level of life organisation and as a self-organising system. A similar process which links form and function is verified also in geographical aspects, such as the coast line: changing the scale, the line changes, but something similar remains and may be measured.

In mathematics it is possible to describe a process like this, applying the concept of homology, in particular the invariant of homothety. This is what Mandelbrot [9] did in the theory of fractals. A fractal is a geometric not Euclidean dimension which is not an integer one but a fractionary one (from Latin *fractus*). A fractal dimension expresses a rule between form and functions.

The concept of homology is part of projective geometry and it is defined as a bi-univocal correspondence among Euclidean elements (e.g. lines and points) maintaining their bi-relations; homothety is a homology with an improper axis. Thus, a homology is characterised by a centre, an axis and a pair of corresponding points. In a homothety the relation r between two corresponding segments is constant (homothetic invariant); looking for instance at Fig. 7.3 [10]:

$$r = \left(OR_{\infty}AA'\right) = \left(AA'O\,R_{\infty}\right) = \left(AA'O\right) = AO/A'O$$

In the case of a parallelepiped (Euclidean dimension = 2) the homothetic relation r,

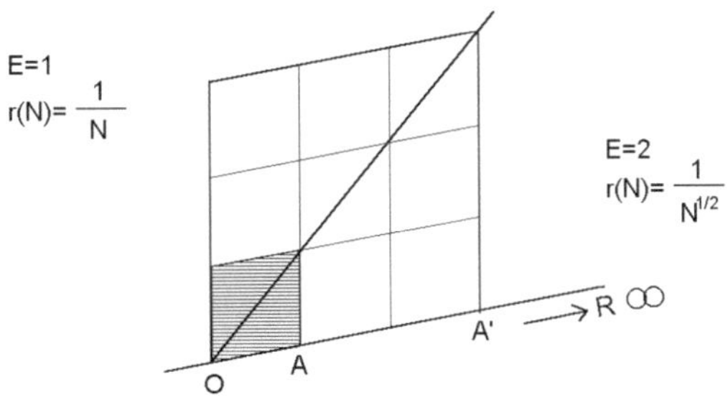

Fig. 7.3 Homothety on a plane, with an improper axis. See text for explanation (From Ingegnoli [9])

referred to the N parts in which a figure is related to the others, is:

$r\ (N) = 1\ /\ N^{1/2}$ from which $\log r\ (N) = \log (1\ /\ N^{1/D}) = -\ (\log N)\ /\ D$ therefore:

$$D = -\log N/\ \log r\ (N) = \log N/\ \log(1/r)$$

This is the most well-known Mandelbrot equation, in which D is the fractal dimension. Thus, D is calculated as a relationship between two logarithms quite easy to be identified.

Note that the rule between form and functions may be constant or variable. The fractals can be exact when exhibiting a very regular structure because they are formed on the basis of a simple and constant rule, like snow crystals; or they may be irregular when formed on the basis of variable rules, like a coast line or vegetated patches. For instance, the rule may be a landslide process on a mountain chain, thus the fractals may have also a statistical nature.

The meaning of D may be represented in comparison to Euclidean dimensions. For example, $D = 0$ is the point, therefore $D = 0.55$ is a set of points along a line; $D = 1$ is a line, $D = 1.29$ is a curve with fluctuations at all scales; $D = 2$ is a plane, $D = 2.4$ is a rough surface, like a folded sheet.

On the other hand, observing self-similar figures, it is possible to generalise the fractal dimension. A segment is a mono-dimensional self-similar figure, composed of two parts measuring one half. A bi-dimensional figure is a square, composed of four parts of one half. A three-dimensional figure is a cube, composed of eight parts of one half. Thus, a self-similar figure of dimension D is formed by n^D parts of magnitude $1/n$. So, the fractal dimension D may be interpreted as a relation between a quantity Q and the scale of magnitude L at which Q is measured: $Q\ (L) = L^{Dq}$. This is applicable also to irregular fractals. Even the relation area/perimeter of patches (A/P) can be measured with fractal dimensions [11], with β, η as constants:

$$A = \beta L^{Da} \quad P = \eta L^{Dp}; \text{since } D = Da/Dp,$$
$$\text{it is possible to write} : \log A \cong D \log P$$

Voss [12] presented a fractal as a set of S points in a spacc of dimension d, for instance (d — 2) in a map. The spatial dependence among the points was studied by measuring the probability $P(m,L)$ of m points observed with a window of size L^2 centred on individual points of the set S (e.g. a landscape cover type). The measure of the mass $M^q(L)$ of the S points at a given L is:

$$M^q(L) = \textstyle\sum_{m=1}^{N} m^q\, P(m, L)$$

where q is an index of the moment of the fractal distribution on the map and $N(L)$ is the number of different values of m observed for a given L. This formula has the significance of statistical moments. When $Q = 1$, $Q(L)$ indicates the means of the points found in a window of size L; when $Q = 2$, $Q(L)$ indicates the variance of points.

One of the most important utilisations of fractals in landscape ecology is the evaluation of the irregular dispersion of patches in a territory, measured by counting the number of grid cells occupied by the mosaic, from low to high resolution. Given a certain patch distribution on a map, e.g. forested tesserae in a landscape unit, we have to overlap a grid of an appropriate size able to contain the mosaic. This grid has low resolution and is proportional to the scale.

Each cell measures L per side, and L will be divided hierarchically; for instance $r\ (N) = 1/2$, 1/4, 1/8, 1/16, 1/32, 1/64. Thus, each grid cell is divided in N parts: 4, 16, 64, 256, 1,024, 4,096. Stating a value of significance (e.g. > 25 %) of the attribute (e.g. vegetation) per cell, we calculate per each passage of scale the number of N parts occupied by the patches.

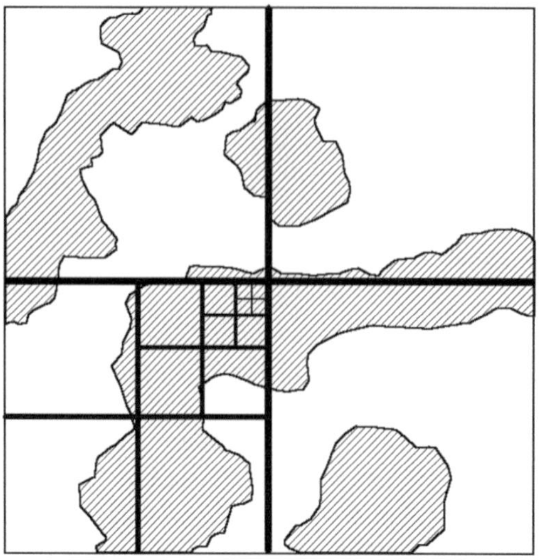

Fig. 7.4 Example of the distribution of vegetated patches and the grid for fractal analysis. The value of significance per cell was 30 %

Table 7.2 Example of calculation of the fractal dimension *D* related to Fig. 7.4

$1/r$	$\log 1/r$	N'	$\log N'$	D'
2	0.301	3	0.4771	1.585
4	0.602	11	1.0414	1.729
8	0.903	35	1.544	1.709
16	1.204	115	2.06	1.711
32	1.505	396	2.598	1.726
64	1.806	1604	3.205	1.7774

N' Number of the N parts occupied by the vegetation, D' Fractal dimension per each N'

Then, we have to apply the Mandelbrot equation: $D = \log N/\log (1/r)$. A simple example (Fig. 7.4; Table 7.2) is presented.

The linear regression of D' leads to the equation: $y = 0.089x + 1.6119$, from which $D = 1.701$ (when $x = 1$).

The meaning of the use of *D* in ecological systems investigation does not refer to the simple measure per se, but to its variation in time. Given a case study, maintaining the same grid for the fractal analysis and the same scale, but referring the measures to different periods, thus to different configurations of landscape elements, the calculation of the fractal dimensions could be useful to check the trend of transformation, even when difficult to be detected by direct visual observation.

7.1.6 River Functionality Index

The importance of the fresh water network in landscape ecology was already emphasised in relation to the excretory apparatus. The analyses of the components, the rivers, have long been considered to be a competence of limnology, which is only interested in the water body and the riverbed: however, in limnology none of the principle of landscape ecology has been taken into consideration. But about 20 years ago a few scientists noted the necessity to study the river functionality in a more complete way, considering the entire river corridor: the vegetation conditions of the riparian belt and the landscape characters in which the river flows, the morphological structure of the riverbed and the biological condition of the macro-benthos.

From the studies of Robert C. Petersen of the Limnologic Institute of Lund, Sweden [13], continued by Siligardi and Maiolini [14] of the Istituto Superiore di Ricerche Agrarie of San Michele all'Adige (Trento), Italy, an ecological index of river functionality, the *Indice di funzionalità fluviale* (IFF), was presented (Table 7.3) [15].

Note that two categories (2 and 12) present two alternative versions: only one of these should be considered. Moreover, the distinction between lotic and lentic water flow is done looking at the water surface: if it is rippled or flat. The score values of the river functionality index (IFF) can be ordered in five-nine levels of functionality (Table 7.4).

7.1.7 Gradient Analysis using Segmented Lines

It is a type of analysis very useful in the study of the gradients at the level of LU or ecotope, or in situations in which the basic ecomosaic proves a high heterogeneity or it's apparently difficult to recognise the regularity or even to identify the

Table 7.3 Survey schedule for the estimation of the river functionality index (IFF) (from ANPA [14], modified)

Characters of the river corridor	LB		RB
River:			
Site:			
Tract:			
Altitude:			
1. *Main composition of the surrounding landscape*			
a. Forests and woods ecotopes	25		25
b. Meadows and pastures	20		20
c. Cultivated ecotopes and/or scattered urbanisation	5		5
d. Urbanised ecotopes and industrial suburban areas	1		1
2a. *Vegetation of the natural peri-fluvial belt*			
a. Arboreal riparian plant formations (dominant)	30		30
b. Shrubby riparian plant formations and/or reed thickets	25		25
c. Non-riparian arboreal formations	10		10
d. Non-riparian shrubby or herbaceous plant formations (or absent)	1		1
2b. *Vegetation within the artificial river-banks*			
a. Arboreal riparian plant formations (dominant)	20		20
b. Shrubby riparian plant formations and/or reeds thickets	15		15
c. Non-riparian arboreal formations	5		5
d. Non-riparian shrubby or herbaceous plant formations (or absent)	1		1
3. *Breadth of the vegetation peri-fluvial belt (trees and/or shrubs)*			
a. Vegetation belt >30 m	20		20
b. Vegetation belt 5–30 m	15		15
c. Vegetation belt 1–5 m	5		5
d. Absence of vegetation belt	1		1
4. *Continuity of the vegetation peri-fluvial belt (trees and/or shrubs)*			
a. Without interruptions	20		20
b. With interruptions	10		10
c. Frequent interruptions or only herbaceous	5		5
d. Bare soil or rare herbaceous	1		1
5. *Water conditions of the riverbed*			
a. Moderate flow width < triple of the wetted riverbed		20	
b. Moderate flow width > triple of the wetted riverbed (seasonal)		15	
c. Moderate flow width > triple of the wetted riverbed (frequent)		5	
d. Wetted riverbed very reduced or absent (or impermeablised)		1	
6. *Riparian morphology*			
a. Presence of arboreal vegetation and/or rocks	25		25
b. Presence of shrubs and herbs	15		15
c. With thin herbaceous strate	5		5
d. Bare shores	1		1
7. *Trophic input retention structures*			
a. Riverbed with rocks and/or embanked trunks or reeds and hydrophytes		25	
b. Rocks or branches with sediment or reeds and hydrophytes		15	
c. Retention structures dependent only on floods (or reeds absent)		5	
d. Sand sediments without algae		1	
8. *Erosions*			
a. Non-evident and non-relevant	20		20
b. Only in curves and narrows	15		15
c. Frequent, with riparian excavation	5		5
d. Very evident, with excavations and landslides	1		1

(continued)

Table 7.3 (continued)

Characters of the river corridor	LB		RB
River:			
Site:			
Tract:			
Altitude:			
9. *Transversal section*			
a. Natural		15	
b. Semi-natural (few artefacts)		10	
c. Semi-artificial (few natural remnants)		5	
d. Artificial		1	
10. *Structure of the bottom of the riverbed*			
a. Stable and diversified		25	
b. Partially movable		15	
c. Easily movable		5	
d. Artificial or with concrete		1	
11. *Scrapings, puddles or meanders*			
a. Clearly distinguished and recurrent		25	
b. With irregular succession		20	
c. Few meanders, or long puddles and short scrapings		5	
d. Absence of meanders, scrapings and puddles (canalised)		1	
12a. *Lotic waters vegetation*			
a. Very scarce periphyton and macrophytes		15	
b. Some presence of periphyton and macrophytes		10	
c. Discrete periphyton and high presence of macrophytes		5	
d. High presence of periphyton and macrophytes		1	
12b. *Lentic waters vegetation*			
a. Very scarce periphyton and tolerant macrophytes		15	
b. Some presence of periphyton and tolerant macrophytes		10	
c. Discrete periphyton and high presence of tolerant macrophytes		5	
d. High presence of periphyton and tolerant macrophytes		1	
13. *Debris*			
a. Vegetal fragments (recognisable and fibrous)		15	
b. Vegetal fragments (fibrous and pulpous)		10	
c. Pulpous fragments		5	
d. Anaerobic debris		1	
14. *Macrobenthonic community*			
a. Well-structured and diversified, in accordance with the river type		20	
b. Quite diversified but with altered structure		10	
c. Badly balanced community with pollution tolerant taxa		5	
d. Absence of community, few pollution tolerant taxa		1	

transition between two ecotopes or two LU: this analysis is an alternative to the simple relative frequency percentage of the element itself, because it allows to observe the tesserae or the contiguity between different elements. In addition, this method is also useful to quantify the possible loss of information in response to a drastic transformation provided in the same LU (Fig. 7.5).

Depending on the purpose, it can be done, therefore, at different resolutions (usually using

Table 7.4 Evaluation and levels of functionality of the IFF

RFI values	Functionality levels	Sense of functionality	Colour
261–300	I	High	Blue
251–260	I–II	High–good	Blue–green
201–250	II	Good	Green
181–200	II–III	Good–mediocre	Green–yellow
121–180	III	Mediocre	Yellow
101–120	III–IV	Mediocre–low	Yellow–orange
61–100	IV	Low	Orange
51–60	IV–V	Low–bad	Orange–red
14–50	V	Bad	Red

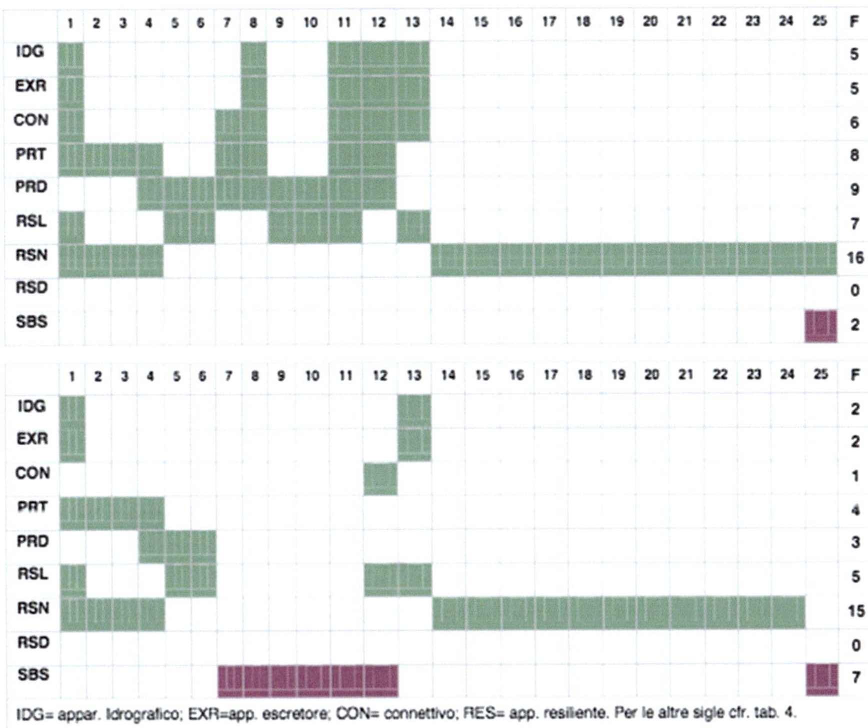

Fig. 7.5 Matrices detection of structural scheme of LU in examination, see Osio-Brembo (13.2.1) according to the method of segmented lines (Forman and Godron [6]): (above) ex ante, (below) ex post. Note the heavy alteration ex post, corresponding to an information decrease of about −24%

cards at a scale of 1:25,000 or 1:10,000, but you can go into greater detail or summarised with a scale of 1:50,000). The relief lines are drawn in parallel so as to cross the most characteristic environmental gradients and then divided into segments of equal length: each segment is characterised by the elements included in it. The result is therefore a matrix having rows for the number and types of elements in the LU in examination and columns for the number of segments. In each column the presence (with a x) and the absence (with a dot (.)) of the various elements are marked. The result is a drawn of the typical configuration of the examined LU.

If it were necessary to have a general design of more detail, i.e. higher resolution, it could be divided into two segments before being considered. The comparison between the different matrices obtained from each reading level will highlight what is the reading resolution best suited to the purposes. This means that is able to describe in a more complete way the reality, optimising the synthesis of the content information without thereby losing important data. Just to learn more about the information content I of the i-th element at every level of reading an equation was developed [7]:

$$I_i = \log_2 S!/F!(S - F)!$$

where the logarithm is in base 2, S represents the total number of segments that make up the line in question, and F is the number of segments in which the i-th element is found (i.e. the absolute frequency of the i-th). Therefore, it is possible to calculate which is the resolution that provides the highest content of global information also numerically. As reported by Forman [7] if, increasing the level of synthesis of the information, it tends to the unit we are in the presence of macro-heterogeneity, otherwise of micro-heterogeneity. It is also possible to find the optimal location of the boundary between two groups of landscape elements [16, 17] or the beginning of a significant presence of another type of element

$$I_d = I_i - \log_2(S - D)!/(F - 1)!(S - D - F + 1)!$$

with D = number of the segment in which the element in question "appears"; or the end of the presence of another element

$$I_f = I_i - \log_2(E - 1)!/(F - 1)!(E - F)!$$

through the number of the segment in which the element in question "disappears" (further applications of these calculations are cited in Forman and Godron [7]).

7.2 Bionomic Analysis of the Landscape

7.2.1 Influence Fields of Landscape Elements

Prominent patches or corridors generally acquire the role of attractors and direct many landscape patterns. These functions may be put in evidence by the emergence of *influence fields*: the intensity of this influence varies with the distance from the corridor and it is proportional to the degree of its role. As we know, a vegetated corridor crossing cultivated fields acquires a prominent natural role. The measure of its influence field requires a complex analysis regarding:

- Microclimate influence, where a distance is given over which a variable differs significantly from conditions in the open [7, 8].
- Soil erosion and control, depending on prevalent winds, type of soil, porosity of the hedgerow and aridity.
- Structure, form, porosity, width and height of the corridor.
- Home range of key species using the corridor as their own habitat.
- Heavy seeds dispersal of ecologically indicative species.
- Number of species in the corridor vs. number of field species.
- Landscape contrast among the surrounding elements.
- The average biological territorial capacity of the landscape.
- The presence of a canal within the corridor.
- The presence of a range of local disturbances.

At landscape scale it is often impossible to measure this influence field, therefore it is useful to propose a synthetic index. This can be done referring to the potential capacity of ecological re-balancing of the corridor vs. the landscape matrix and to its main microclimate influence.

The BTC of the corridor is compared with the average BTC of the landscape (or even the region) and a balance is made: the BTC deficit of the adjacent fields and roads, such as in

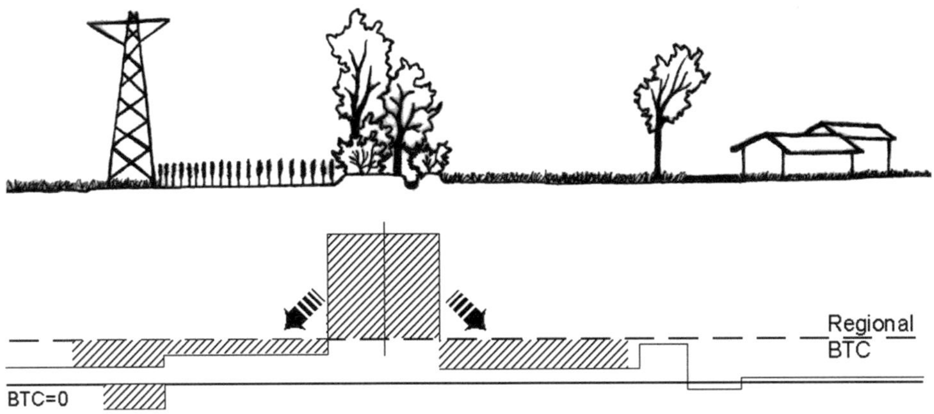

Fig. 7.6 Example of the measure of the ecological field of influence of a hedgerow in a rural landscape unit

Fig. 7.6, results in a measure of distance from the boundary of the corridor. Another distance is given by the microclimate influence, referred to the height of the corridor and its porosity. Thus, we measure two factors:

$$Lb = [BTC(+) \times Wc/2]/BTC(-); \; Lm = 6H$$

where *Lb* is the length of BTC balance, *Wc* the width of the corridor, *Lm* the length of the microclimate's normal influence, *H* the height of the corridor trees, *BTC*(+) the BTC exceeding the regional or landscape media, *BTC*(−) the BTC deficit of the adjacent landscape elements. Then we consider the mean of the two factors multiplied by the eventual presence of singular elements of disturbances crossing the matrix near the corridor (field roads, power-lines, small buildings) [6]:

$$CISF = 1/2\,(Lb + Lm) \times 0.81/d \quad (\mathrm{m})$$

CISF is the corridor synthetic influence field, *d* the number of singular disturbances (see Fig. 7.6). This ecological index is particularly useful in the control of complex networks, in which both natural (or semi-natural) and human (technical) corridors are present.

Note that the concept of influence field of the corridors can be used also for the technical elements. In these cases it is necessary to refer to other specific measures. For instance, from 2 to 4 μT (micro Tesla = 0.796 A/m) for electrical power lines of high tension (e.g. 300 kV), corresponding to a distance of about 25–100 m from the line. For the roads, the influence field it is more variable, from 40 to 320 m (see Sect. 11.3).

7.2.2 Analysis of the Landscape Apparatuses

Please note that the apparatuses of the landscape are made by multifunctional tesserae or elements. It is, therefore, necessary to list the elements or components for landscape dominant function, but then estimate the minor functions for each one. This estimation is done by giving a conventional weight to each class of functions: for example, the dominant ones may have a function with weight f = 0.8–0.9, while those with secondary functions a weight f = 0.3–0.4 and minor f = 0.1.

Note that these values may change from case to case, since it must be estimated on the spot. Obviously, as happens in the sizes of components of the ecotissue, the sum of weights for conventional ecological function may be different from 100 (usually higher), so it is necessary to normalise the results returning them proportional to 100, so that they are comparable to each reference storm.

LANDSCAPE ELEMENTS	1900	1950	2000	RSD	SBS	PRD	PRT	RNT	STB	ETN	RSL	CON	EXR	GEO
Residential (RSD)														
Urbanized	0,6	3,3	4,9	0,9	0,2		0,1							
Subsidiary (SBS)														
Industrial	0,5	0,6	1,4	0,1	0,9									
Roads	0,1	1,4	1,5		0,9							0,1		
Airport	0	0,02	0,3		0,9									
Canals	2,75	3,42	4,95		0,9							0,1	0,4	0,2
Productive (PRD)														
Cropland	13,3	19,9	22,8			0,9					0,1			
Woody agrarian	9,5	7,8	0,4			0,9	0,1							
Fish ponds	0,1	0,3	0,5			0,9				0,2				
Protective (PRT)														
Urban green	0,1	0,3	2,1	0,1			0,9							
Camping green	0	0,4	0,3	0,2			0,9	0,1						
Resistant (RNT)														
Woods	0,6	0,2	0,2				0,4	0,9						
Filling in vegetation	0	0	0,8				0,2	0,9	0,1	0,1	0,1			
Stabilizing (STB)														
Barene (sandbanks)	13,8	9,05	5,68				0,4	0,1	0,9		0,4	0,2		0,1
Submerged prairies	5	5	5				0,2	0,1	0,9					0,1
Ecotonale (ETN)														
Marshes	4,8	0,03	0,02				0,2			0,9				0,1
Resilient (RSL)														
Shrubland	0	0	0,1				0,2		0,1		0,9			
Sand vegetation	0,3	0,2	0,1				0,1		0,4	0,1	0,9			
Connective (CON)														
Banks & hedgerows	1,2	1	0,4				0,2					0,9		
Excretory (EXR)														
Wide channels	2,75	2,28	1,65										0,9	0,4
Reed thickets	0,3	0,2	0,1				0,2		0,1		0,1		0,9	
Rivers	0,5	0,6	1,4										0,9	0,4
Skeleton (GEO)														
Shallows	24,8	26,9	28,3							0,1				0,9
Sea littoral	17,7	16,5	16,5							0,1				0,9
Seashores	1,3	0,6	0,6							0,2		0,1		0,9

Fig. 7.7 Analysis of the landscape apparatuses of the Venice Lagoon in the last century

An example of evaluation of the apparatuses in the landscape of the Venice lagoon is given (Fig. 7.7). The landscape components (type of elements) are listed in the first column: the yellow columns list the percentage values of their surface extension in the Venice lagoon in the three considered temporal thresholds; the eleven columns on the right express how much each element fulfils the function related to a specific Landscape apparatus (0.9 means 90 %; 0.1 or 0.4 means 10 or 40 %): these coefficients are necessary for the measurement of the functions, dominant and secondary, of each element, as explained by the theory, to catch the real weight of each apparatus.

From this table we derive the feedback overall percentages for each apparatus and for each time reference (Fig. 7.7, values in italics); for example, for the residential apparatus (RSD) in the year 2000 you get:

$$\begin{aligned} \text{RSD (2000)} &= (4.9 \times 0.9) + (1.4 \times 0.1) \\ &\quad + (2.1 \times 0.1) + (0.3 \times 0.2) \\ &= 4.82 \end{aligned}$$

The results of estimation in the three periods considered (1900, 1950, 2000) give respectively: 121.315, 115.689, 113.345 %, which need to be then normalised in order to compare the periods and appropriately evaluate the dynamics. The results are reported in Fig. 7.8.

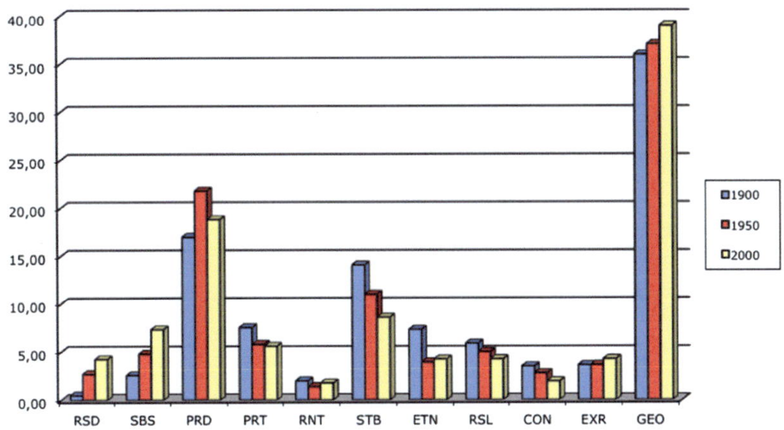

Fig. 7.8 Control of the dynamics of the main landscape apparatuses of the system under test. Among L. apparatuses the first two, RSD (residential) and SBS (subsidiary), are growing rapidly, the productive (PRD) less, the protective (PRT) is in slight decline. Among NH you can see the proportions and the growth of the geomorphological (GEO); severely under necessity the resistant (RNT) and connective (CON). In sharp drop the stabilising apparatus (STB) and the ecotonal one (ETN). The resilient (RSL) is also decreasing, the excretory (EXR) increased slightly

7.2.3 The Function HH/BTC

As we have shown in Sect. 3.3.1, the biological capacity of the territory, or BTC sensu Ingegnoli [18, 19], is a quantity related to the metastability of the vegetated tesserae, the value of which is measured in Mcal/m^2/year and is calculated taking into account: (a) the strategy of resistance of the most representative plants formations of the biosphere, (b) their metabolic data (biomass, gross primary productivity, respiration) and (c) their level of organisation. The BTC measures the flow of energy that a system of vegetation must dissipate to maintain its level of organisation and metastability. The BTC is an indispensable magnitude for the study of various aspects of a LU, such as:

(1) Identification of thresholds for transformation of a LU.
(2) Control of the distribution of structural and functional components.
(3) Control of the metastability of the landscape.
(4) Analysis and evaluation of biodiversity eco-landscape.

After having studied the ecological status of 35 LU first, and then, with further additions, of 48 LU (on average at the municipal level), most in Northern Italy, of which 10 were urban and suburban-technological, 22 were of rural-suburban, agricultural and agriculture-protective type, 16 pertain to agricultural-forestry and forestry landscape type, it was found that the correlation between HH and BTC, deduced from these experimental data, was very close ($r^2 = 0.95$).

These data represent a current ecological state, as only few of the considered LU were altered, but even few of them were in optimal state, while most of them represented an average state. Not taking into account the worst cases, this function has been improved, so it can now be considered as a function of normality. The mathematical equation that describes the behaviour of this function was found to be:

$$y = 0.0007\, x^2 - 0.1518\, x + 8.85$$

As you can see from Fig. 7.9, the function allows you to define the most frequent thresholds of characterisation of the main types of landscape: forest-natural, forest semi-natural, forest agriculture, agricultural, rural/suburban, suburban/urban rarely, dense urban. The BTC interval between the two LU with extreme values of HH (HH = 0 and HH = 100) is the following: 8.85–0.67 = 8.18 Mcal/m^2/year.

In case of presence of a barren and rough component on a LU in question, we must refer to a different function from that just presented.

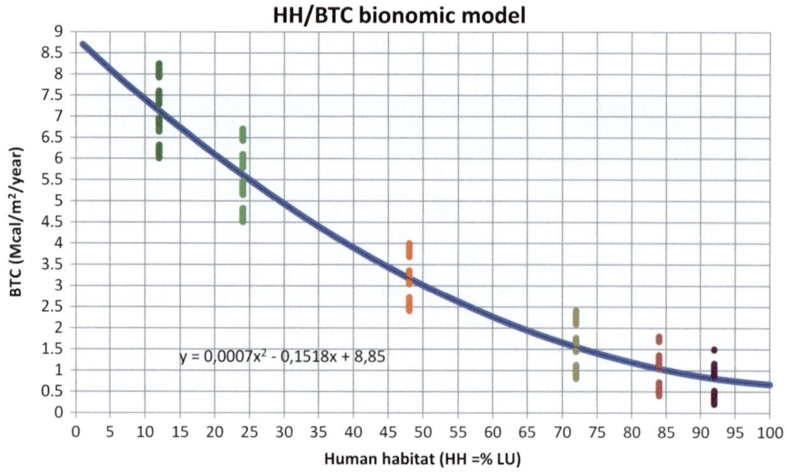

Fig. 7.9 Model HH/BTC (from Ingegnoli and Giglio [19]) indicating the curve of normality useful for comparing the ecological status of landscape units (more or less populated) and consequently have a clinical picture of the territory in question. The vertical lines divide the types of landscape (*from left*): natural forest, semi-natural forest, forest-agriculture, agricultural, rural, suburban (or urban rarely), dense urban

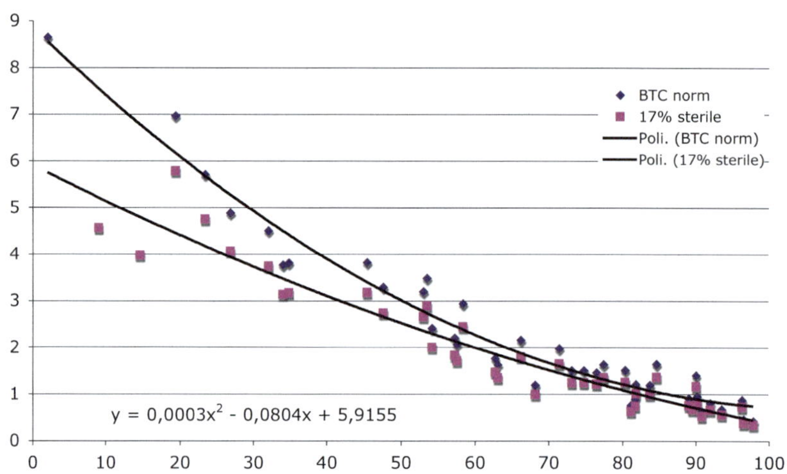

Fig. 7.10 Model HH/BTC and its transformation in the case of the territory partially unusable (17 % sterile). Note the lowering of the values, especially for HH not high (<50 %)

For example, if the impervious area is 17 %, e.g. the case of the Alpine municipality of Tesero (Trento), the function that serves as a reference model acquires a lesser curvature which, for medium-low values of HH, obviously provides less high values of BTC (Fig. 7.10), while the differences with the higher HH values (> 55 %) are minimal.

Being able to refer to the function HH/BTC as a model of normality is undoubtedly useful, because if the LU in question diverged appreciably from that model, then you can be sure that there is some alteration or pathological form (see Sect. 4.5).

7.2.4 Standard Classes of BTC

The analysis (and diagnosis) through BTC needs to have a standard subdivision of the BTC values into classes, through which it is possible to compare the studied cases of landscape evaluation. As mentioned in Sect. 3.3.1 the proposed standard is composed of nine BTC classes (Mcal/m²/

Table 7.5 The standard classes of BTC and their range of values

Biological territorial capacity	BTC classes	Mcal/m²/year		Average value
Semi-desert, scrubs	I	0.00	0.40	*0.20*
Grass, fields and urban green	II	0.40	1.20	*0.80*
Vines, shrubs, plantations, gardens	III	1.20	2.40	*1.80*
Young forest, temperate coppice, urban parks	IV	2.40	4.00	*3.20*
Near adult forest, young "macchia" woods	V	4.00	6.00	*5.00*
Adult forest, temperate woodland	VI	6.00	8.40	*7.20*
Mature forests, adult "macchia" woods	VII	8.40	11.20	*9.80*
Dense Schlerophyll forests, tropical forest	VIII	11.20	14.40	*12.80*
Dense tropical forests, tall pluvial forest,	IX	14.40	18.00	*16.20*

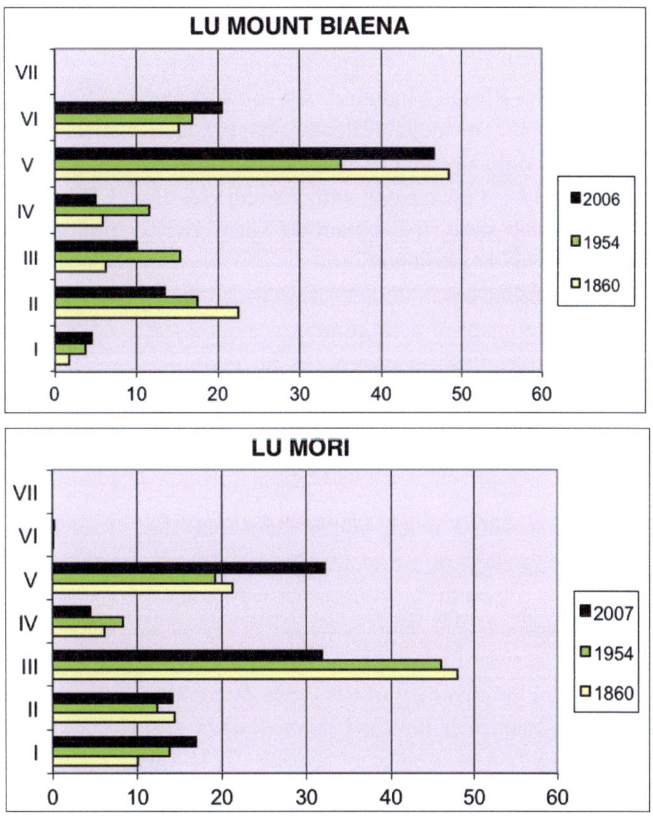

Fig. 7.11 Differences between the distribution of BTC classes in LU n°4 (*above*) and n°1 (*below*) in the municipality of Mori (Trentino)

year), because of their good significance (Table 7.5).

These classes are in geometric proportion. Given that LU of tropical surroundings have not yet been studied, the last two classes (VIII and IX) are only hinted at, waiting a financial support to carry out research in these areas.

The BTC classes are distributed in different ways in all types of LU and the analysis of this distribution is necessary to evaluate the bionomic state of the LU under examination.

It is reported, for example, a comparison between two LU of the town of Mori (Pre-Alps of Trento, Italy) (Fig. 7.11): LU1 Mori Centre

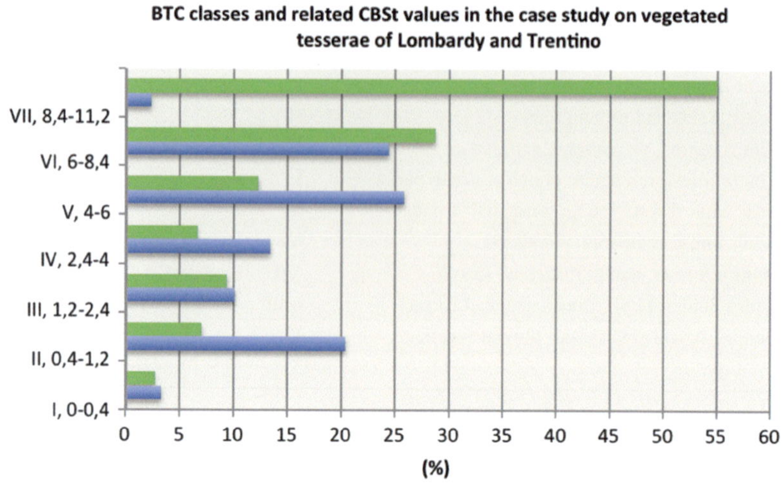

Fig. 7.12 The 430 tesserae (Ts) of the case study (Lombardy + Trentino), divided in the seven more significant BTC classes, present strong differences among the CBSt values (*green*) associated with the percentage of presence per class (*blue*). Five classes are coupled with a CBSt < 20, remaining in a scarce bionomic state

and LU4 the Biaena Mountain. As you can see, the bar graphs are quite different, having LU1 practically only the first 5 classes with a wide class I, and LU4 six classes with a small class I. You can also study the dynamics, since 1860, which is also very different.

The distributions in object can in theory have an infinite number of configurations, even if the three most significant ones are: (1) the equiprobable, that is maximum entropy, minimum information; (2) the bell-shaped trend, having a maximum in the centre, which presents greater dominance and a minor presence in each class; (3) the one with a much higher class than the others, up to verge on zero value in one of them, which has maximum dominance, maximum information.

But remember that in a hyper-complex adaptive system the diversity of the types of components must consider both the heterogeneity and the information as important aspects. It is for these reasons that the index of functional diversity (see Sect. 3.7.4) of the landscape is $\tau = H \times (3 + D)$.

7.2.5 Concise Bionomic State and BTC Classes

In Fig. 7.12, the 430 tesserae (Ts) of a case study (Lombardy + Trentino), divided into the seven more significant BTC classes, present strong differences among the CBSt values (green) associated with the percentage of presence per class (blue). Only two classes are coupled with a CBSt > 20, 24.47 % reaching a normal bionomic state, only 2.35 % a good one. The other 73.18 % remain in a weak or precarious state. A weight media of CBSt of our case study results in CBSt = 14.86 (weak state of Table 5.15).

The problem is that the real ecological situation of these Alpine Regions is still more weak! In fact, the distribution of the BTC classes in Lombardy and Trentino presents much more huge percentages in the first two classes: class I = 19.59 % (vs. 3.29 of the sample in Fig. 7.12), class II = 39.16 % (vs. 20.47 of the sample). The weight media decrease abruptly to a CBSt = 10.61.

Another observation is the confirmation that the vegetation communities with lower resistance strategy (hence lower BTC) reach a lower bionomic state. Only forest coenosis can go over the normal state, arriving to measure a CBSt from 40 to 65 % (good state). This seems to be impossible under BTC < 7.5 Mcal/m^2/year.

Forest reach the highest values of CBSt, even if only few of the surveyed tesserae show an optimal state (CBSt > 65 %). Note that surveys like these have just the role to testify a recognised optimum state of Central European forest vegetation, mainly resulting from the Bialowieza Natural Park, *Querco-Carpinetum* and *Q-Tilietum* [20]. Therefore, only 5-6 of the 430 tesserae (all

Table 7.6 Range of LTpE related to the main landscape types

Landscape type	Min	Max	Average value
Dense urban	0.30	3.00	1.68
Rare urban	2.70	6.30	4.50
Sub-urban rural	5.70	11.00	8.35
Agricultural	10.20	24.20	17.20
AGR-protective	22.00	40.00	31.00
AGR-forestal-tour	34.00	55.00	44.50
AGR-forestal	45.00	88.00	66.50
FOR-tour	65.00	140.00	102.50
Forestal semi-natural	100.00	245.00	172.50
Forestal natural	170.00	500.00	335.00

from Lombardy and Trentino) have been found in optimum bionomic state (see Sect. 5.3.1).

7.2.6 The Landscape Type Evaluation (LTpE)

The assessment of the Landscape Type Evaluation (LTpE), which classifies the types of landscape in relation to HH as the independent variable, is based on two indicators:

(a) NH_d/HH, where: NH_d = NH – [NH × (SH/SH*deficit)] and SH/SH*dcit is the amount ot heterotrophy of a landscape when SH/SH* < 1.
(b) BTC × (A + F) = the product of BTC by the presence of major ad hoc components: the agricultural (A) and the forest (F) measured as % of PRD and RNT elements.

The following index is:

$$\text{LTpE} = 10\left[(\text{NH}_d/\text{HH}) + (\text{BTC} \times (\text{A} + \text{F}))\right]$$

In Table 7.6 we reported the range of values of the LTpE index, related to the main types of landscapes in the temperate regions. These values form an exponential curve, going from 1.68 to 335.

As we will see forward, the diagnostic evaluation of a LU depends on the correct individuation of the type of landscape under examination. The dynamic of transformation also depends on the type of landscape. We have seen that in some cases an indication on the landscape type could be done, but only in general, e.g. recalling Fig. 7.9 on correlation HH/BTC. But it is indispensable to have much more precision in the identification of the landscape type. Therefore, the use of this LTpE index becomes necessary.

For example, coming back to Sect. 2.5.1 and Fig. 2.12, we had to classify the 55 municipalities of the province of Monza to show the structure of the territory, part of which is comprised in the metropolitan area of Milan.

We created a model placing in sequence for each of the 55 municipalities in question, the components FOR, URB, AGR and then the known parameters HH, BTC, SH/SH* (i.e. σ), NH, deficit σ and the two indicators *NHd/HH* and *BTC* (*A* + *F*), arriving to measure their LTpE. Each column of this matrix was evaluated for intervals of the measured parameters in relation to the normal values, with three levels of esteem, equal to 1 (yellow), 2 (hazel), 3 (violet). The values of the index LTpE, which in this case ranged from 0.49 to 16.90, allow to distinguish the types of landscape, with reference to Fig. 2.12 and Table 2.5. An example of this, 24 on 55 municipalities, is reported in Fig. 7.13.

The municipalities of the territory of Monza-Brianza fall into the following four types of landscape:

(1) LTpE [0.30–3.00] = dense urban, No. 14 municipalities.
(2) LTpE [3.00–6.00] = rare urban, No. 13 municipalities.
(3) LTpE [6.0–10.0] = suburban rural, No. 14 municipalities.
(4) LTpE [10.0–25.0] = agricultural productive, No. 14 municipalities.

It is noted that in the territory of Monza the classification of landscapes is equally divided into three groups of 14 and a group of 13 municipalities, however, mainly grouped properly with each other, as exhibited in Fig. 2.12. Although other investigations will be needed for a scientific demonstration of belonging to the metropolitan area of Milan sensu Forman [21], it's already evident from this figure that the incorporation of the common type of dense urban and normal urban represents the direct extension of the Metropolitan Area of Milan towards North-East.

Note that traditional methods to distinguish the landscape characters of the municipalities

Municipality	% FOR	% URB	% AGR	HH (%)	BTC	SH/SH*	NH	deficit σ	*NHd/HH*	*BTC A+F*	LTpE
TRIUGGIO	28,86	29,83	41,09	66,46	2,06	0,49	33,54	0,51	0,25	1,44	16,90
BRIOSCO	24,16	32,40	43,18	69,93	1,82	0,56	30,07	0,44	0,24	1,23	14,70
BELLUSCO	8,04	9,82	82,15	77,89	1,21	0,57	22,11	0,43	0,16	1,09	12,55
VEDUGGIO-COLZANO	19,86	39,07	40,90	73,71	1,58	0,41	26,29	0,59	0,15	0,96	11,05
MISINTO	16,97	38,30	44,73	75,61	1,45	0,58	24,39	0,42	0,19	0,89	10,80
CAMPARADA	16,19	37,13	46,68	75,98	1,42	0,49	24,02	0,51	0,15	0,89	10,47
LAZZATE	15,33	37,27	47,41	76,58	1,38	0,42	23,42	0,58	0,13	0,86	9,92
USMATE VELATE	12,14	36,66	50,49	78,37	1,23	0,59	21,63	0,41	0,16	0,77	9,32
AICURZIO	8,03	30,15	61,61	80,46	1,08	0,68	19,54	0,32	0,16	0,75	9,17
CARATE BRIANZA	13,44	48,73	34,73	78,11	1,19	0,33	21,89	0,67	0,09	0,57	6,66
BURAGO-MOLGORA	6,70	41,36	51,92	82,88	0,95	0,47	17,12	0,53	0,10	0,55	6,51
VIMERCATE	3,34	34,60	61,40	84,02	0,83	0,46	15,98	0,54	0,09	0,54	6,23
CAVENAGO BRIANZA	4,75	36,56	52,49	81,15	0,84	0,40	18,85	0,60	0,09	0,48	5,72
BARLASSINA	16,24	65,50	18,25	79,63	1,24	0,25	20,37	0,75	0,06	0,43	4,91
MEDA	19,04	66,90	12,77	77,46	1,35	0,21	22,54	0,79	0,06	0,43	4,89
ALBIATE	4,69	50,03	45,28	85,36	0,80	0,32	14,64	0,68	0,06	0,40	4,52
CONCOREZZO	1,62	44,11	53,82	86,48	0,69	0,35	13,52	0,65	0,05	0,38	4,35
BIASSONO	6,09	62,61	30,66	85,74	0,77	0,25	14,26	0,75	0,04	0,28	3,26
GIUSSANO	5,72	67,51	26,49	86,84	0,73	0,28	13,16	0,72	0,04	0,23	2,76
BRUGHERIO	1,72	61,42	33,38	87,22	0,55	0,19	12,78	0,81	0,03	0,19	2,23
VERANO BRIANZA	7,12	72,81	16,15	85,25	0,73	0,23	14,75	0,77	0,04	0,17	2,10
NOVA MILANESE	0,50	63,91	29,48	87,34	0,46	0,17	12,66	0,83	0,03	0,14	1,64
MONZA	0,92	80,02	18,42	91,54	0,42	0,17	8,46	0,83	0,02	0,08	0,97
VEDANO AL LAMBRO	0,49	89,17	10,29	93,24	0,34	0,16	6,76	0,84	0,01	0,04	0,49

Fig. 7.13 Example of LTpE (landscape type evaluation) of 24 on 55 municipalities of the province of Monza; from above: agricultural productive LT, suburban-rural LT, rare urban LT, dense urban LT

forming a territory (e.g. a province) are often misleading: in this case, if we limit ourselves to consider the geographical density of population plus the ratio of urbanised/agricultural and the presence of forest on the 55 municipalities of the province of Monza, we would arrive to an error of 21.8 % (unacceptable!) with respect to the method of LTpE. This is because we must not forget that we are studying a complex system and that to have reliable results we should seek a diagnostic index.

7.3 Drawing Analysis of the Landscape

7.3.1 Natural Sciences and Artistic Formation

The writer's passion for nature was transferred to him by his father, a painter. He harshly taught his art and, with it, a strong passion for nature which, to be studied, should be observed through a depth approach. A confirmation of the importance of this method derived, later, by the great naturalist Konrad Lorenz, in his book "The ethology: fundamentals and methods" [22]. He stressed that the approach of a researcher who wants to analyse an "all organic" must be just like that of a painter: "This procedure, starting from the system to its parts, is mandatory in biology."

It is, therefore, only right, as well as profitable, try to understand more fully the importance of an artistic education also for those who study natural sciences.

It is said that art is a subjective expression, so it would be incompatible with a scientific background. However, it is still Lorenz to say that the process of knowledge and the object of knowledge cannot legitimately be separated from each other; in addition, it is simply *not true* that the subjective experience relates exclusively to the private sphere of the individual.

Max Hartmann [23] has defined a logical relationship between the physiological processes and internal processes, but the correlation between them is so secure that a subjective phenomenon, for example the perception of a complementary colour, can be used, in the phenomenon of contrast, as a reliable indicator of the occurrence of the parallel physiological event. Indeed, the overcoming of scientism (see Chap. 15) allows us to understand that the more strictly human knowledge is defined as expressed on the verbal, the more it becomes clear that a large number of primary phenomena do not allow themselves to express anything through words, or, if you prefer, by means of rational logic. The good, the beautiful, the sensitivity to the harmonies, the meaning of life or the ultimate truth certainly exist, but they are unspeakable. Yet we can also understand and express the unspeakable, as every true artist, every poet knows. This is also important in nature, as in the case of beauty.

The *beauty* exists in nature, manifests itself in many forms, and is also used by living systems as an attraction towards integration purposes less immediate, or rather more complex. For example, it is known that the flowers are beautiful and colourful when they have to call the pollinators, but if the pollen is spread by the wind that beauty is not present. The hyper-complexity of living systems on a large scale, such as really harmonic landscape, affects not only men, as shown by Jane Goodall [24, 25] studying the emotions of chimpanzees in front of the perception of landscapes of exceptional beauty.

Lorenz again wrote that science is now able to show that matter and energy, corpuscular radiation and electromagnetic waves, physiological and emotional events are part of the same real process, which we undoubtedly experience, but through two ways to know that are immeasurable. Yet there is a principle accessible to both ways of knowing, because at the base of knowledge itself: the truth. Overcoming the conflict between rationality and inner experience is possible through the truth, which is in itself one, unique, invariant, only if the state of an object is not considered independent from the way in which the possible sources of information about the same object are altered by the observation, following the words of Wolfgang Pauli [26].

Talking about epistemology (Chaps. 1 and 15), we saw that there is a precise way to approach the truth and this road is a correspondence of harmonies that allows understanding the manifold in the unit. It is yet another aspect of the binomial indistinguishable *truth/non-violence*. The art and the science have in common the same quest for unity in variety: the beauty and emotion inherent in a creative act reveal this essential correspondence, in a poem as in a theorem, which allows the generation of thought and identifies itself with it. "The best feeling is the mysterious side of life, -writes Albert Einstein [27]— it's the deep feeling that is always in the cradle of art and pure science."

Art and science are not opposed, but they are complementary activities and so people dedicating to research shouldn't and couldn't radically separate them: on the contrary, researchers must cultivate both them, because only through this way they can deepen their knowledge.

We observe, incidentally, that the drawing may be more than the requirement of logical sequence of writing and this is important to understand what Lorenz said about the method of the painters in the study of nature: this reference is fundamental not only for the method, but also for the development of sensitivity to differences in value among not rational processes. The sensitivity for the "harmonies," which are the processes of integration of hyper-complex systems, depends in large part on the ability of our sense organs and our brain structures —thanks to their special type of organisation— to "perceive the shapes."

One of the most important functions of the perception of forms is to allow us to *distinguish* what is *healthy* and what is *sick*. Lorenz wrote that it is essential for physicians, scholars of animals and landscape ecologists. Essential, because the "*clinical eye*" can never be replaced by a collection of quantitative data, although computerised, for the impossibility to express, in rational sense, many of the essential functions of a biological system. An artistic education based on the design can develop *clinical eye* particularly well with regard to the state of environmental system: so, for example, the Field Studies Council in London have been forming

Fig. 7.14 Alterations both in the art (painters) and in the environment (urban ecology) during the last century, which expressed the most tragic period of the history (sketch of the Author)

the naturalists also through drawing watercolour since 1946. At this point, the problem is: should a good art education be possible today? If we look at Fig. 7.14 it does not appear so simple: the degradation of the environment seems to have the same ultimate reasons of the alteration of the art. It is not a case if these degradations arose in the most tragic and violent period of the History, well reported by Woody Allen: "God is dead, Marx is dead, and I'm feeling not so good. . . ." A cynic attack to the heart of man is really dense of consequences.

If framed as a field of research or as a language to express arguments otherwise untreatable or as a new decorative language in technology and architecture, very different from the tradition, then, and only then, even the so-called "avant-garde" is acceptable. However, if one attempts to set himself out of research,[2] if one expresses any topic with the same semantics, if one considers surpassed the "figurative" field so as to eliminate it from any Art Centre, or worse, if one consider for art what it is not art ("Artist's Shit," cuts on the canvas, etc.,), then it is a moral obligation to be opposed.

The artistic expression of today seems to be oriented on the novelty of signs and forms rather than on the content, as Marcus Tullio Cicero [28] already warned:

> Sed adsiduitate cotidiana et consuetudine oculorum adsuescunt animi neque admirantur neque requirunt rationes earum rerum, quas semper vident, proinde quasi novitas nos magis quam magnitudo rerum debeat ad exquirendas causas excitare.

("But daily recurrence and habit familiarise our minds with the sight, and we feel no surprise or curiosity as to the reasons for things that we see always; just as if it were the novelty and not rather the importance of phenomena that ought to arouse us to inquire into their causes"). Here again another reason for having criticised the definition of landscape as the "visual perception of a territory" (Chap. 1).

[2] Together with Mallarmé and Picasso, the "avant-gardists" sustained that the forms of Creation have to be destroyed and that "nous ne cherchons pas, nos trouvons" (=we do not try, we find). This is quite presumptuous!

7.3.2 The Contribution of Drawing in Landscape Bionomics

After having defined the landscape in a scientifically correct way, we need to expose how the discipline of landscape bionomics considers aesthetic and semantic features information on the systems in question, pointing out that it does not intend to refuse such information, but treat them less impromptu.

Coming in mind, in summary, the main fields of important use of design and colour in the natural sciences, with particular reference to the bionomics of the landscape, we can see that these are six fields: (1) the scientific illustration, (2) the descriptive representation of complex territorial systems, (3) the scientific relief of the physiognomy of landscape components, (4) the reconstruction of landscapes (or their components) in the past, (5) the eco-design of territorial systems or their components and (6) the environmental design and control of technical interventions in the environment.

1. *The scientific illustration.* The importance of scientific illustration is perhaps the most well-known ability to combine art with science. This illustration covers fields as diverse as zoology, botany, comparative anatomy, geomorphology and, lasts a long time in spite of the advent of photography (the second half of the nineteenth century), also in the clinical-diagnostic field, for the unsurpassed ability of interpretation connected to it.
2. *The descriptive representation of complex territorial systems.* The use of photography often is not enough to describe a hyper-complex system as the landscape, because this medium allows only snapshots representations, without being able to eliminate redundancies and noise to the images, process necessary for a great number of considerations about the ecological environment. This is even more relevant when we consider partially humanised landscapes and relationships between natural elements and historic buildings.
3. *The scientific relief of the physiognomy of the landscape components.* The study of physiognomy of systems composing a landscape also demands the contribution of the drawing. Take, for example, the need to detect vegetation transects along critical gradients, ecotonal edges, or to represent a summary of the differences between plant formations similar to each other for the species involved, but with locally different characters.
4. *The reconstruction of landscapes (or their components) in the past.* The need for reconstruction of landscapes in the past requires, almost always, the ability to provide scientific representations through the design and colours, at least of most of the parts that have changed with respect to the current time of the study. In certain cases, these reconstructions should be integrated based on several images more or less partial and approximate, from cadastral maps as from both literary and iconographic material.
5. *The eco-design of territorial systems, or parts of them.* To understand that no design can avoid sketches and drawings to create and control the necessary components, whether they are natural or anthropogenic parts, one should remember that drawing in itself also means project. The project interventions on hyper-complex systems (e.g. ecotopes and parts thereof) requires integration among the purposes of environmental remediation, the indications on the ecological parameters involved coming from the diagnosis and the choices of ad hoc intervention.
6. *Environmental design and control.* It differs from the previous point not only for the scale, usually of greater detail, but also for the need to integrate technological devices and bioengineering with natural needs of subsystems. Often this work needs to study natural parts as a reference model. The aesthetic control has to be understood here, as always in the case of bio-ecological studies, in the sense of semiotics. One then adds the naturalistic control, that is based on the diagnostic evaluation, but also has the need to control Gestalt.

Fig. 7.15 Rural landscape near Fort Collins, Colorado, USA. Two old Poplars, protecting the farm, indicate the wind direction. Drawn by Vittorio Ingegnoli

Fig. 7.16 Study of a natural Alpine ecotope. Watercolour representing the "blue lake" in the upper valley of Ayas (Aosta) at an altitude of 2,120 m, at the foot of the glacier of Verra, coming down from the peaks of Breithorn (4,168). Painted by V. Ingegnoli

7.3.2.1 Some Examples

We will start showing a field study for the detection of an agricultural ecotope (Fig. 7.15). It is a typical farm near Fort Collins, Colorado, USA. Two old big Poplars emerge as landmarks: their protection of the farm is evident. Note also the precise direction of the main wind. Light conditions were difficult for a photograph, but not for a sketch.

Figure 7.16 is a study of a stretch of natural landscape in the Alps, where the characterisation of the arrangement and wrinkling of the rocks is predominant. It is the ecotope of the "Blue Lake" in the upper valley of Ayas (Aosta Valley,

Fig. 7.17 A characteristic farm in a forest landscape in Rajasthan, India. Pen and watercolour with only complementary colours: indigo and burnt sienna. Drawn by Vittorio Ingegnoli, 1985

Fig. 7.18 Rural landscape with an old cemetery in South Ireland, 1989. Note the uncommon field depth. Watercolour painted by Vittorio Ingegnoli

altitude 2,120 m) at the foot of the Verra glacier which falls from the Breithorn (4,168 m). The colonisation of the banks by the larch is clearly documented. Even in this case, the drawing and the watercolour were also required for reasons of difficulty in the use of photography, as well as to better understand the complex structure of the landscape.

Figure 7.17 in a forest landscape in Rajasthan, India. This typical farm presents an elevated porch of entrance and a robust roof to protect from heavy rains. Local people prohibited to photograph, but not do draw.

Figure 7.18 is a watercolour of an Irish rural landscape. The weather was continuously variable: when the foreground was clear, the background was covered and vice-versa. With patience, watercolour was able to understand the whole landscape with its exceptional depth of field.

Figure 7.19 is a pen drawing that represents a traditional settlement and fortified walls still

Fig. 7.19 Study of a LU referred to a very old town. Pen drawing, which represents the walled village in relation to production areas (PRD) and protective (PRT) surrounding. Glorenz-Glurns, Trentino-Alto Adige (South Tyrol) (From [6]). A rare case of non-expanded town in the last century

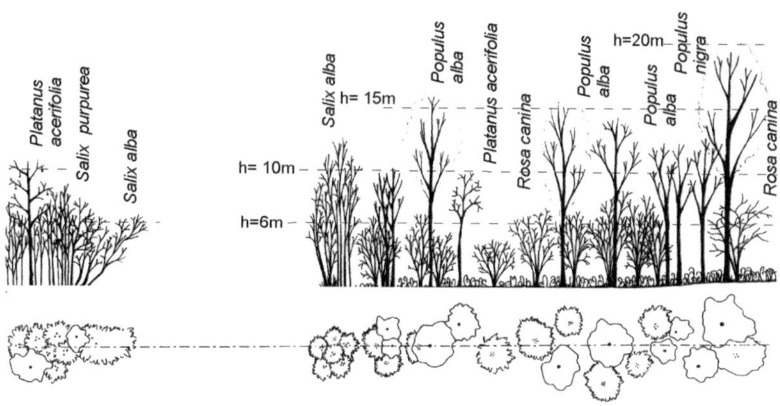

Fig. 7.20 Drawing a Transect of riparian forest in the Ticino Regional Park "Bosco Vedro." Note that the physiognomy of the vegetation does not permit shortcuts through geometric informatics reductions of the plants present. From [6]

intact in relation to the landscape of the valley. This is the town of Glorenza-Glurns, in the Trentino-Alto Adige (South Tyrol), in Val Venosta. As the city works similarly to a cell, the landscape elements productive (PRD) and protective (PRT) are located to its surroundings and based primarily on detritus on his shoulders.

Moving on to a more technical and descriptive use of the drawing, we show a transect of forest vegetation of the "Bosco Vedro," in the Regional Park of the Ticino River, Lombardy (Fig. 7.20).

This study, which shows in plan the strip of vegetation considered and raised in the physiognomic characteristics of the plants, was carried out in a suitable scale (1:200) and plotted for the heights to 20 m. Note that this design could not be geometrically stylised by computer, without losing all the characters of its ecological real status.

Another use of the technical drawing is presented in Fig. 7.21: the photo to the left shows the environmental status (1991) of the slope next to the town of Tavernola Bergamasca, Iseo Lake (Lombardy), where there is a large open pit mining of cement marl, which invades the middle of the side of the mountain behind it. After the Region have imposed an ecological

Fig. 7.21 Studies for the environmental rehabilitation of the opencast mine of cement marl in Tavernola Bergamasca (Lake Iseo). *To the left*, the picture of the environmental status in 1991. *At the right*, a tempera drawing that shows the forecasts of ecological restoration project

Fig. 7.22 The pen sketches show one of the ways of pattern matching semiotic purposes. The clinical eye of a ecoiatra allows you to trace the ecological alterations (margin, materials, percolation of water, the vegetation barrier) due to the introduction of the rigid road and guess (with subjective projection) the type of distortion of the ancient village, before you even get there. *To the right* of the photo: church of Camuzzago (Bellusco) has been converted into residential area

sanitation of the territory, the studies carried out by Vittorio Ingegnoli showed a prediction of the effects of interventions in which the morphological reconstruction was performed by computer program, but then the distribution of patches of reforestation was done by drawing and tempera colours, as can be seen.

In the last example, a vision aesthetically negative, we expose a significant case in Bellusco (Monza-Brianza). The pen sketches in Fig. 7.22 show one of the ways of "pattern matching" for semeiotic purposes, based on a comparison with a not altered situation, easily imaginable by an expert. The "clinical eye" of an *ecoiatra* allows to connect the introduction of the rigid road to many ecological alterations (e.g. margin, materials, percolation of water, barriers in the vegetation) and to assume, with subjective projection, the type of distortion of the ancient village, before one even get there. It joins, by the way, a photo of the Parish of Camuzzago (Bellusco) now converted into a residential neighbourhood.

References

1. Romani V (1988) Il paesaggio dell'Alto Garda bresciano: studio per un piano paesistico. Grafo ed, Brescia
2. Venanzoni R, Pedrotti F (1995) Il clima. In: Pignatti S (ed) Ecologia vegetale. Utet, Torino, pp 7–24
3. Pignatti S (1995) Ecologia vegetale. UTET, Torino
4. Rivas-Martinez S (1995) Clasificación bioclimática de la Tierra. Folia Bot Matritensis 16:1–27
5. Ingegnoli V, Giglio E (2005) Ecologia del Paesaggio: manuale per conservare, gestire e pianificare l'ambiente. Sistemi editoriali SE, Napoli
6. Ingegnoli V (2002) Landscape ecology: a widening foundation. Springer, Berlin
7. Forman RTT, Godron M (1986) Landscape ecology. Wiley, New York
8. Forman RTT (1995) Land Mosaics: the ecology of landscapes and regions. Cambridge University Press, Cambridge
9. Mandelbrot BB (1975) Les objets fractals: forme, hasard et dimension. Flammarion, Paris
10. Ingegnoli V (1993) Fondamenti di Ecologia del paesaggio. CittàStudi, Milano
11. Milne BT (1991) Lessons from applying fractal models to landscape patterns. In: Turner MG, Gardner RH (eds) Quantitative methods in landscape ecology. Springer, Berlin
12. Voss RF (1988) Fractals in nature: from characterization to simulation. In: Peitgen HO, Saupe D (eds) The science of fractal images. Springer, Berlin
13. Petersen RC (1991) The RCE: a riparian, channel and environmental inventory for small streams in agricultural landscape. Freshwat Biol 27:295–306
14. Siligardi M, Maiolini B (1993) L'inventario delle caratteristiche ambientali dei corsi d'acqua alpini: guida all'uso della scheda Rce-2. Biol Ambient VII (30):18–24
15. ANPA (2000) I.F.F. indice di funzionalità fluviale. Manuale ANPA, Roma
16. Godron M (1972) Echantillonage linéaire et cartographie. Investigacion Pesquera 36:171–174
17. Godron M, Bacou AM (1975) Sur les limites "optimales" séparant deux parties d'une biocénose hétérogène. Ann Univ Abidjan, Série E 8(1): 317–324
18. Ingegnoli V (1991) Human influences in landscape change: thresholds of metastability. In: Ravera O (ed) Terrestrial and aquatic ecosystems: perturbation and recovery. Ellis Horwood, Chichester, UK, pp 303–309
19. Ingegnoli V (1999) Definition and evaluation of the BTC (biological territorial capacity) as an indicator for landscape ecological studies on vegetation. In: Windhorst W, Enckell PH (eds) Sustainable landuse management: the challenge of ecosystem protection. EcoSys: Beitrage zur Oekosystemforschung, Suppl Bd 28, pp 109–118
20. Falinski JB (ed) (1994) Vegetation under the diverse anthropogenic impact as object of basic phytosociological map: result of the international cartographical experiment organized in Białowieża Forest. Supplementum Cartographiae Geobotanicae 4. Phytocoenosis 6 (NS)
21. Forman RTT (2008) Urban regions, ecology and planning beyond the city. Cambridge University Press, Cambridge, UK
22. Lorenz K (1978) Vergleichende Verhaltensforschung: Grundlagen der Ethologie. Springer, Wien
23. Hartmann N (1964) Der Aufbau der realen Welt. De Gruyter, Berlin
24. Goodall J, Berman P (1999) Reason for hope. Waren Book, New York.
25. Goodall J (2000) 40 years at Gombe. Stewart, Tabori, and Chang, New York
26. Pauli W (1959) Aufsätze Vortäge uber Physik und Erkenntnistheorie. Tr. It. Fisica e conoscenza. Boringhieri, 1964
27. Einstein A (1965) Pensieri degli anni difficili. Bollati Boringhieri, Torino
28. Cicero MT (I sec.) De Natura Deorum, II, XXXVIII, 96 (many editions)

8 Historical Evaluation of the Landscape

8.1 The Importance of History in Landscape Bionomics

8.1.1 Towards a New Definition of History

In Sect. 1.3.9 we pointed out that, following the non-equilibrium thermodynamics, history is an intrinsic concept in all complex systems. This means that history has to deal not only with human affairs, but also with the natural and ecological ones, as the landscape. Unfortunately, some Author still insist to regard history as referring only to man, thereby reducing the landscape history to the anthropogenic signs "that have formed it". With a so limiting setting, the concept of natural landscape would lose its meaning: pay attention that those human signs are largely addressed and supported by natural factors, as today confirmed by scientific topic. Note that the persistence of this ultra-anthropocentric vision of history is also ethically questionable, as it places man as a potential tyrant detached from nature: this preconceptions is at the base of all kinds of degradation.

It is definitely wrong also the opposite approach, namely the attempt to define the landscape as exclusively natural: "the unitary and dynamic totality of the world in which we live". Even this view is brutal and very risky because it "tends to exclude everything that does not fit into a kind of repeated scientism tended to detect the law of development of this universe", as written by Langé [1]. Refusing scientist epistemology, for the reasons explained in the first chapter, please note that the development of a landscape as a living entity cannot be reduced to a simple occurrence. As a result, both these extremist views, scientism and fanatic anthropocentrism, are issues to be overcome, if we want to seriously study a complex system like the landscape. This system, in fact, exceeds

(a) The deterministic occurrence, as it is a level of biological organisation, then a hyper-complex adaptive system manufactured by unpredictable events and that develops in the sense "of event", of which man is an integral part being himself a biological organism
(b) The design and the anthropic product, as, in fact, it is itself a living entity, never entirely reducible to the will of one of its components, even the most gifted: the whole is always greater than the sum of its parts, according to the principle of emergent properties

Landscape bionomics requires, therefore, to re-examine the meaning of history.

In common parlance the term history is used to characterise the temporal changes of a system or, more generally, of an entity, while the scholars refer to the greek-latin etymology of the term *historia* that really means "research, investigation, cognition", derived from an Indo-European root, hence the greek 'ιστωρ is also "one who knows" [2].

V. Ingegnoli, *Landscape Bionomics Biological-Integrated Landscape Ecology*,
DOI 10.1007/978-88-470-5226-0_8, © Springer-Verlag Italia 2015

Following the more thrust anthropocentric view, the story is the knowledge of the human past, a definition that does not indicate a simple work of narration of the past, neither a study or a research of elements ending in themselves. The recognition that the landscape acquires its characterisation as a result of the changes inflicted by man over time allows, however, that the landscape, like all human endeavours of the past, can become the object of operations and business of the historian. From the orderly arrangement of the exhibits, documents and historical events, and taking into account the intentions of those who produced them (the general guidelines of the company and of cultural expressions), the landscape is essentially and exclusively seen as a "cultural product".

Following the scientist and reductionist vision, which presupposes to be able to deduct certain times events from typing, in accordance with certain laws, history is the description of natural dynamics, encoded for use by academic and educational field. A landscape is then studied according to its "natural history", that is considering a territory (or part thereof) as a system that contains within itself, at the same time, once and for all, each one of its own determining factors. For this reason, the natural transformations are reflected in patterns of evolution abstractly formalised in accordance with the reductionist rules. Some authors believe, therefore, to be able to evaluate the environmental status of a landscape basing on the "distance" of existing vegetation from potential vegetation composed by the final stages of appropriate series of vegetation in the absence of disturbances.

Just as expressly provided above, it is clear that both these definitions of history forget considering the more advanced paradigms of current knowledge in practice and can lead to serious methodological prejudices. In fact, these visions are essentially due to the epistemological limitations of knowledge and research. As we wrote, a new definition of complex adaptive systems, of life, of evolution, of time, of the physical and mathematical principles, as well as greater attention to the truth, unequivocally lead to a very different definition of story, closer to the original etymology of *historia*: the story is the research on the evolution occurred in natural systems, that is "on the event" of the phenomena in the elapsed time.

So, once again, it must be emphasised that all of reality is transformed following an arrow of time in not reductionist evolution, but in total creative freedom: for this, without history is quite impossible to understand the procedure and the meaning of events, both natural and human. The irreversibility of time leads to a succession of processes: the flow of time is not homogeneous, and then the phenomena occur at different times and are scalars. Therefore, it is theoretically possible to recognise two behaviours in relation to the arrow of time, either as a variation or as a stasis. The clearance of such behaviour constitutes real systems, making it impossible a tetragon determinism, which moreover is not able to give an account to the creativity of living systems.

8.1.2 The Concept of Environmental History

It must be understood that only a conception "of event" of nature is capable of overcoming the dichotomy between an ultra-anthropic sense and a scientist sense of history, being able to develop correct historical studies on complex systems such as landscape. Following that concept, you can avoid the occurrence of distortions leading:

(a) To the arbitrary fragmentation of landscapes as if they were indifferent geometric portions of space hired to support events and choices of politicians and administrative men, more or less opportunistic or chronicle; or, on the contrary
(b) To the deterministic conception of the landscape, according to a descriptive meaning of the succession of typing, abstractly formalised in accordance with certain rules that are presumed to be valid in the absence of human interference

As written by Zanzi [3], it is right today to speak of "environmental" history. This well-known historian, who has not surprisingly been in contact with Ilya Prigogine, says that this concept of environmental history is to be understood properly as local history, that is, as the history of a place (e.g. a landscape unit) considered in its autonomous and differential individuality from its environmental consistency. The "natural history" must be changed in the "history of nature", first of all attentive to the process of forming and continuously metamorphosing of individual "local landscapes". Each place has its own story, completely unique, which can be reconstructed according to the traces that the events of territorial transformation have originated in the place itself.

The history of an evolutionary entity, e.g. a landscape unit (LU), consists mainly in the narrative of the events through which it was formed. The retrospective considerations of an evolutionary process may aim to differentiate stages of development, turning points, stopping, breaking, thresholds of progression or regression, but introducing factors of interpretation and stressing the flow of events in one or more episodes particularly charismatic. All this completes a use of individual data and facts obtained from the "archives of history" for a theorisation. The history must, therefore, propose to itself the "reconstruction" of the facts starting from their residual traces, without uprooting them from the frame of "local" events where these facts have occurred, indeed, "contextualising" them in their sole and exclusive relevance of "space-time" on that system within which continuity and complexity have been produced.

As we know, each landscape unit must be always contextualised in many different ways, given that different theoretical questions configure different perimeters, related to scale, or to morphology, or to texture, etc. Orogenic aspects, geomorphology, vegetation, evolution of ecological, cultural transformation, etc. refer to different fields of "event history". Thus, it can reaffirm the importance of a new structural model of landscape according to the concept of multidimensional ecotissue (See Sect. 2.3).

8.1.3 The Historical Interpretation of a Landscape

After having defined history as the "research on the evolutionary events of nature in the past (including human being)", you must keep in mind that, for the accomplishment of this aim, often difficult, in-depth studies both in humanistic and in scientific disciplines are required. Since the contextualisation needs specific studies to arrive to correct purposes (e.g. to get to the perimeter of an ecotope or a LU in landscape bionomics), it must be recognised that the interpretation of the events that have marked the evolution of a landscape requires ad hoc studies.

For these reasons, the presentation of the principles of landscape ecology (today bionomics) is accompanied [4–6] by a summary of the foundations of systems theory (Sect. 1.3), which can help in the more complex interpretations, in which human and natural components are integrated in various ways. Knowledge of archaeology and history of nations, history of local populations, agricultural history, architectural history, history of technology are of undoubted importance in the case of cultural landscapes too. Ecology, botany, zoology, geomorphology, pedology, climatology, current and referred to the quaternary age, are indispensable in every case study on past landscapes.

It should be noted, incidentally, that incomplete interpretations, or distorted by a limiting definition of history, on the development of major components of the landscape are more common than one might believe. An emblematic superficial interpretation was reported by Ingegnoli [5, 7] about the history of the parks and gardens between the eighteenth and nineteenth centuries. The official historiography speaks of transformation by formal gardens ("French") to near-natural ("English") ones, starting around the mid-eighteenth century and reaching almost the end of the following century: the reasons given for this profound change are purely cultural or humanistic, such as the thought

Fig. 8.1 View of the castle of Racconigi by its "English" park. This vast system has a rather complex story of transformations both anthropogenic and natural and can serve as an example to remind you that the story of a landscape unit must be designed according to the "event history".

of philosophers and poets, the suggestion of landscape painters, the influence of oriental garden. But such historical interpretations are only apparent: the main reason for change came from the environmental crisis of that time due to the strong development of industrialisation. As it is well known, the industrialisation was born in England, spread primarily in rural and soon after in urban landscapes, performing the first extensive degradation of the environment, the first strong urbanisations and starting to change the way of life of the society. It is, therefore, logical that the parallel reaction—mostly unconscious—in such a state of crisis produced the first public parks in the city and the first "near-natural" gardens, which will then be exported to all Europe and even in America, during the second half of the nineteenth century and beyond. Only the landscape bionomics today is able to demonstrate that, despite the great changes of the Era, many European landscapes have managed to maintain their average BTC thanks to the new way of setting up the parks (see Sect. 4.2.2).

We report a view of the castle of Racconigi (Piemonte, Italy) with its well-known park (Fig. 8.1), the history of which has relations with numerous anthropogenic and natural events of transformation linked to both the rural area and the village. The geometric park designed by the famous French landscape architect André Le Notre in the seventeenth century was replaced, in the nineteenth century, by the "English-natural" park created for the king Carlo Alberto by Xavier Kurten (a German architect). The whole, now partly to be restored, is home to several pairs of storks. The LU formed by the above system can be a prime example of the need to study history in an environmental sense.

The study of the development of each complex adaptive system and the study of the history of a landscape unit (LU) too is concerned with very different time fields, for example relating to processes of short, medium, long and very long term: this often makes difficult the interpretation of an historical event. Recalling the hierarchical systems (see Sect. 1.3.5), it should be noted that the speed of processes and the space-time scale are not always immediately connected.

One should keep in mind that we can get in front of not imaginable system behaviours that are the result of past data, maybe secondary or unconscious, but that have evolved affecting other components, which in turn have proceeded self-delaying or vice versa self-exciting. Recall that some times a complex system reacts very quickly to an input, perhaps with violence, while in other cases the robustness of a system, that is its ability to disturbances incorporation, allows it to absorb for a long time a significant amount of disturbances and of stimuli without anything happening, but shortly after, suddenly, to change its structure in disruptive way, and/or its organisation and its behaviour. Precisely because of the complexity of these interactions, often a system can lose the direct causal correlation with the input and shows the effects of very different time horizons too. It follows that also the output loses the temporal correlation with the direct input. The historical study of the evolution of the landscapes is thus assuming the importance of an anamnesis, because the appearance of certain symptoms, due to the occurrence of delayed effects, could indicate that there is no more time to remedy, at least in the short term.

8.1.4 The Historical Sources and the Landscape

In the historical reconstruction of a landscape an undoubted importance is linked to sources of information, to how to find them, to their critical discussion, to the interpretation of the data. In the light of what previously expressed, it is essential to look for the sources of information suitable for our purposes and then to read and to interpret the data properly. Our assumption—environmental history—is not always easy to achieve, because generally today too many people think to history in a purely humanistic sense, so historical archives are structured on the basis of these humanistic aspects, with rare exceptions. On the other hand, the natural history museums present certain processes over time, but limited to areas of very large time scale (geological and biotic evolution), or otherwise set mainly on the taxonomy of the species and on species ecology. It follows that finding the data required to reconstruct the landscapes in the recent past become, more often, a not easy process.

As it is known, the main sources of information about the past of human and natural components are five complementary sets, all to be investigated, and then related, in an iterative manner. Each nation may have developed these sets in a different way and degree of deepness and the aims of this book are not concerned with them. These sets are [6]:

(I) Bibliographic sources
(II) Cartographic sources
(III) Iconographic sources
(IV) Direct real sources
(V) Real indirect sources

(I) *Bibliographic Sources.* The written sources, especially books, are diverse and varied, but much more difficult to find, if we exclude the summaries of registers, which are always found in the same archives where the maps are. We are educationally inclined to distinguish some group of publications of interest to the historical studies on landscapes, more than anything else to highlight areas of interest where to search for texts suitable for our studies.
   (a) The *European Union Institute of Statistics* (EUROSTAT), with offices in major cities, publishes data useful for our purposes in the Statistical Yearbooks [8]. They can be useful on the following topics: climate, forests, agriculture, hunting and fishing, population, type of employment, consumption of energy, roads and railways, industries, etc.
   (b) *Other publications of statistics.* In libraries one can find regional or provincial yearbooks since the nineteenth century. For Italy, the De Agostini [9] has been publishing concise geographical-statistical calendars since the beginning of 1900 (national and regional).
   (c) *Climate history*: few scientifically correct publications exist. An interesting example for the study of the landscape is the text of Le Roy Ladurie [10], based among other things on the beginning of the grape harvest. The general data of the IPCC [11] (watch out for errors!) and those opposing of the N-IPCC [12] (Non-Governmental IPCC) are also to be pondered without controversy.
   (d) *History of the landscape*: there are not many publications. Among the best known, and among the first, we should mention the history of the English landscape by Hoskins [13] and that of the Italian agricultural landscape by Sereni [14] on the changes mainly due to

cultural, social and economic reasons. Books about the history of the landscape of Lombardy were published by Electa [15], with texts by geographers (Turri), historians (Perogalli), scholars of the landscape (Ingegnoli).

(e) *The history of forest landscapes*. Some news about the history of forest landscapes at the Italian regional level can be found at the National Academy of Forestry Sciences of Florence and/or faculties of this discipline. But it is easier to find individual studies on the state of the local forests at the end of the nineteenth century or the first half of the twentieth century, published by Doctors of Forestry. A well-known story of the woods was written by Kuster [16].

(f) *Historical and archaeological studies*: from the Chalcolithic Age and especially from the Iron Age, they may be useful to some reconstructions of local landscapes or province. Interesting, for example, the publications of the Center for the Study Camuni [17, 18].

(g) *History of landscape architecture*: not many texts are reliable for our purposes, since the humanistic criterion followed by many authors. The most recommendable for Italy is the text of Maniglio Calcagno [19], while at a wider level you have to cite the book of Pregill and Volkman [20].

(h) *History of rural architecture*: important attempts to relate these forms to the characteristics of the environment. Interesting studies were published by geographers [21] and architects on Lombardy [22, 23].

(i) *History of the villas and castles*: studies on the villas and castles are very numerous, but not many of them talk about basic relationship with the land with competence. For Northern Italy, interesting books are concerned with villas [24, 25] or castles [26].

(j) *History of urbanism and territory*. It's an argument analogous to the previous point. In this case one can cite, for example, books of Sica "History of Town Planning" in the three volumes on the eighteenth, nineteenth and twentieth centuries [27].

(k) *Stories of local territories*. There are many cases to choose from, depending on the location of interest, often with iconographic sources. Interesting publications are those of Italian Touring Club (TCI) divided by regions, documenting historical monuments, historic centres and landscapes from the years 1925–1930 [28].

(II) *Cartographic sources*. The cartographic references remain the essential fact, in some ways the primary for each study on the historic landscape: they have three main sources in European countries and others secondary. We will quote briefly from the recent to the distant past: *Regions Cartographic Services*; *Military Cartographic Institutes*; *Historical Archives of State,*[1] *Private Archives and various other sources*, such as Ecclesiastical Archives, Archives of prints, Archives Noble or local institutions, Archives of trade associations, etc.

Figure 8.2 shows the ancient Map of Villa Carlotta (Lake Como) in the eighteenth century, before the formation of the great gardens of the family Sachsen-Meiningen, however, very rich of any kind of data on the Villa and the park. From the map you can see that the current "Valley of the Ferns" was then a tract of land still natural.

(III) *Iconographic sources*. The historical artefacts covering every type of representation can transmit information, which in many cases are very useful: they form the iconographic sources. They are graffiti, copper prints, lithographs, drawings, oil paintings, watercolours, etc. Many of these types can extend over a very long time, dating at least from Roman times to today.

[1] For example, for Lombardy we have the Cadastre of Charles VI (1721–1723), the updating of Maria Theresa (1770–1780) measured in trebuchets (scale of about 1: 2,800), the maps of the IGM Imperial Royal Government of Austria-Hungary (1840–1850) (K.u.K Militar Geographischen Institute) at 1:25.000. In analogy, the Belgium maps of Ferraris (end of eighteenth century) were made by the same Austro-Hungaric administration.

Fig. 8.2 Map of Villa Carlotta (Lake Como) in the eighteenth century, before the formation of the great gardens of the family Sachsen-Meiningen (courtesy of archive Ente Villa Carlotta)

More recently, from the late nineteenth century to the present, you can also use old postcards, old photographs, old newspapers or magazines. In Fig. 8.3 we can see the landscape of Loveno Menaggio (Lake Como) at the beginning of the nineteenth century and note that, prior to the formation of the large garden of Villa Mylius-Vigoni, the area was agricultural with spots on the tree-lined hill behind: a question that proved history to be important in the study of the ecological status of the Villa.

The iconographic sources can be found both in the archives of the prints and in libraries, or in the photographic archives (e.g. Alinari in Florence), while on drawings and paintings there are many publications, including monographs, concerning Italian or European painters that are useful when dealing with subjects that affect our research in the localities in question.

(IV) *Direct real sources.* Among the most crucial data sets of historical character of the landscape and its components there are information "deducted on the field", derived more or less directly, from inspections on the territory in question. In this case it is necessary to know how to observe in great detail, and through a scientific method, the components that form a part of the landscape, at different scales, to be able to extract the necessary information.

These sources revealed to be simply indispensable for human components, because it is not certain at all that a settlement, or fields, apparently remained unchanged for a couple of centuries on the maps, had actually still its old structure. Then the verification is mandatory.

Equally essential are these direct sources for natural components, because it is not said, of course, that a forest marked on a map (without additional attributes) was today what could have been yesterday. The most common case may concern a forest that was once coppiced but not today, which may weigh on the type of organisation of the LU in question.

(V) *Real indirect sources.* Although usually less pregnant, a set of information can be derived from the local reality in an indirect way, that is mediated by a controlled program of interviews with some representatives of the native population.

Fig. 8.3 View of the landscape of Loveno Menaggio (Lake Como) at the beginning of the nineteenth century: note the Villa Mylius Vigoni (behind the church), still without its large park, and the hill behind, with the presence of few patches of forest

We recall, for example, older people who played a particular role (forest rangers, farmers, teachers, etc.) and who can provide information sometimes very useful in the absence of other documentary sources. Of course, such investigations must be carried out in a scientific way and should be checked by comparison with other available sources for careful and logical analysis. Other types of information can be derived from surveys calibrated to know the meaning of perception of the landscape felt by the local population and which may contain useful data on the historical memory of the most significant components of a landscape unit.

8.1.5 Historical Methodology and the Landscape

It must be emphasised that the critical reading- integrated, compared and reasoned- of the information derived from the sources necessary for the study in question, reading targeted to the main objectives of the research, is a basic methodological scheme, which remains in force. Please note that, to study the dynamics of transformation of a landscape unit, at least two reconstructions are required and this indication is essential for guiding the historiographical analysis. In fact, at least three points are needed to recognise the dynamic of a curve: if you want to avoid reducing the movement to a straight line, you should consider at least one point between the two extremes (past and present) to see if it is a concave or convex curve. The distance from the current state of the two reference periods in the past is quite variable and depends on two factors:

(a) The main scale of interest and
(b) The availability of sources in gathering the data

It is known that the more the spatial scale of a LU is widest, the more it's necessary to withdraw the historical reference periods. In addition a general method is not sufficient to address our theme: it may be necessary to have an additional track from the main types of sources. Therefore, we regard, always in summary, the sources cited in the previous paragraph, to give some hint of method of reading the information on the past to the historical reconstruction of a landscape unit.

(a) *References*. Being usually very difficult to find exhaustive sources referred to the territory in question, the critical comparison between the various issues addressed in the literature becomes crucial. The most important question is, therefore, the need of constant comparison between the literature sources. It is also essential:

1. To compare bibliographic information to map data with method and patience, keeping in mind the need to work at different spatial and temporal scales. It also suggests you to read texts not only directly describing the components of a landscape, but to expand on topics of culture and customs, cultivation techniques, in order to understand how this landscape could be perceived and managed, ensuring better relationship between the local population and its territory
2. To note that the statistical information require a reworking of control. In fact, the themes affecting the ecology of the landscape are not taken into account by National Institutes of Statistics, which maintain a not-ecological setting on the territory and environment sectors. In some cases, we must go back to the past to reconstruct the most recent data, such as the distinction between unproductive urbanised and natural theme that, until the beginning of the twentieth century, was detected in a distinct manner and that by mid-century was inexplicably unified. It is, therefore, essential to use interpolation methods, remembering also in this case that three reference points are always necessary

(b) *Maps*. There is no doubt that for the historical reconstruction of a landscape units, the local geographical and topographical maps are undoubtedly the primary source of information. Let's recall some basic concepts.

1. The critical reading of the cards must always start from the present situation, in order to fully understand the mode of transformation from the past and to interpret correctly the types of landscape elements that can be recognised, by reference to a series of lines and areas that characterise the locality.
2. The categories of relief mapped in the past must be clarified at the best that it's possible, to catch the correct ecological information and to be able to compare this information with that of today. Perhaps this is the most difficult phase, as in the past in the "legend" we don't find descriptive nomenclatures that were taken for granted, but that today are almost completely changed. For example, in the precise cadastral maps of the Lombardo-Veneto or of Belgium (Habsburg), one can read conditions such as "coppice, sweet forest, strong forest", as it can be seen in the reconstruction of the area where the Park of Monza was born [29], in the years 1721–1722 (Fig. 8.4). Coppice it still says today, "sweet" and "strong" are terms instead referred to the dominant species of temperate softwood or hardwood, namely: poplar-willow or alder (soft wood) and maple or oak-hornbeam-linden (hard wood). On a larger scale synthesis (e.g. 1: 25,000) it is more difficult to get information about forests.
3. The old maps show only the use of land (often synthetic), while the reconstruction of the ecotissue of a LU cannot be limited to this first type of information. As a result, the main issues of land use (e.g. forests, meadows, fields, orchards, urban) must be specified with deeper distinctions for ecological characteristics, indicating at least the main faces of forests, meadows, cultivated fields, etc. Note that some maps were quite detailed even in the past, as we can see from the Habsburg map of South Tirol (Fig. 8.5). Otherwise assessment with ecological indices becomes too rough. Sometimes it's necessary to refer to literature and iconography.
4. The measures of area and perimeter of old maps had been taken over by hand and their geo-referencing in relation to national or international networks topography is almost non-existent. So it is

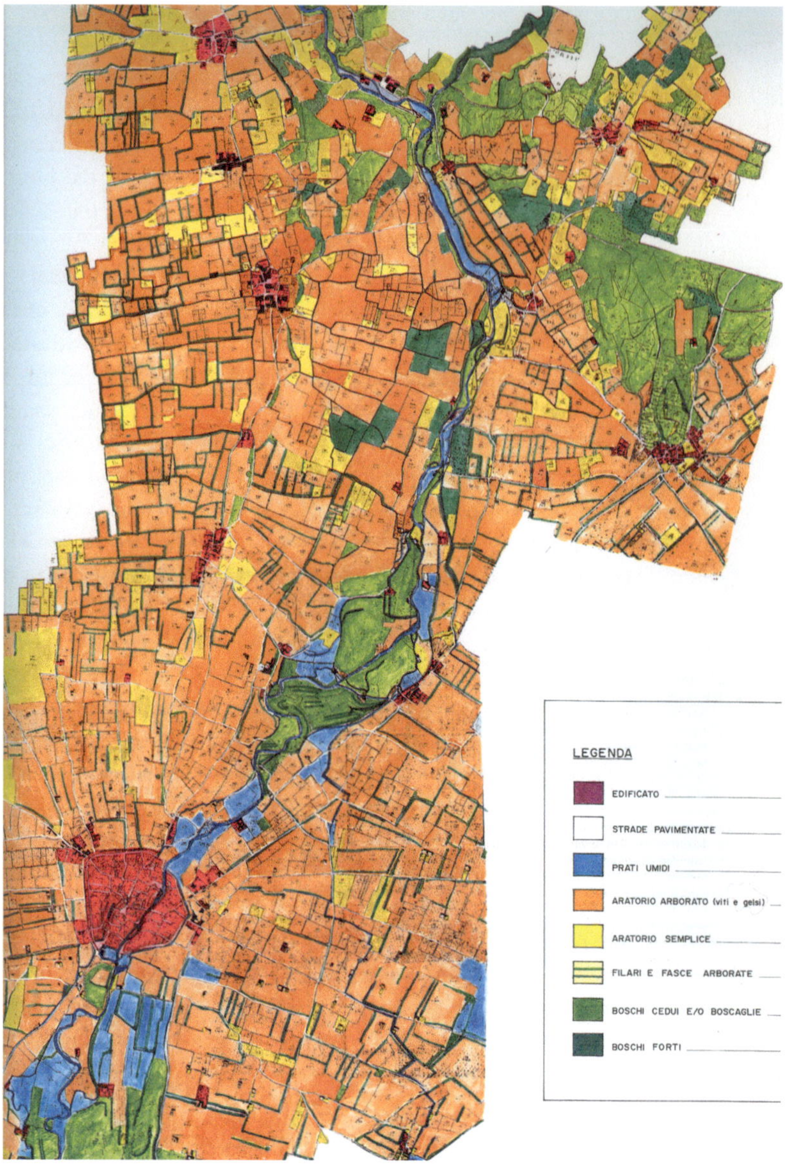

Fig. 8.4 The territory north of Monza, on which was afterwards built the famous Park, has been reconstructed using the maps of the Cadastre of Charles VI of Habsburg (1721). It can be seen that the spots residual "strong forest" (*dark green*) correspond to the location of the so-called medieval Bosco Bello (Beautiful Wood), with dominant sessile oak and pedunculate oak

always necessary to compare the current data with the old ones, from the general to the particular ones, and to update proportion with georeferenced data based on geophysical structures less susceptible to transformation in recent centuries.

(c) *Iconography*. Usually we tend to overlook these sources of representation, but it is a blunder. The information can be used to better specify iconographic elements detected in the maps in generic title. For example, the use of the terms "cultivated land "or "farmland" does not say anything about the type of cultivation (cereals, vegetables, legumes, fodder, fruit, vineyards, etc.) nor on the presence of rows or hedges between fields; it is essential, therefore, to make some attempt to try to catch more information from the representation of a print, a drawing or a painting of the same places. Even more important is to

Fig. 8.5 An example of old map and its information on land use. This is an Habsburg map (1860) of South Tirol, the municipality of Mori, scale 1:25,000

try to better understand the dominant species in the forest vegetation, or the type and historical-architectural value of buildings marked on the map only as a generic space.

It highlights the need to understand what the position of the perspective point of view (even for prospects rough or otherwise) was and possibly to make a comparison with the current situation, taking into account the landmarks still present today (see Fig. 8.11).

The iconographic information may serve also for reasons of symptomatology, i.e. to understand the symptoms of the state of ecological alteration of the landscape in question. On the contrary, only the visual perception of a landscape is often considered, but this is an issue decidedly formalistic and often useless.

Caution: You must always take precautions to figure out if the representation reproduces the reality or instead if it is only symbolic. For this you need to know the purpose of iconography, the author and the traditions of the major iconographic schools, which sometimes requires the opinion of an expert in archaeology or art history.

(d) *Field observations*. To catch information through this way is less developed, adducing lack of objectivity and difficulties of interpretation. The reality is quite different, because the information derived from the surveys is always of the utmost importance, even if only as a confirmation of the data tracking down through other methods and from other sources.

1. Regarding the *human components*, inspections are always required to verify the maps (which are unlikely to be updated to the current time of the study) and to understand the real situation, the mode of the changes occurring in recent years. More detailed analysis of ancient artefacts and of their architectural details can figure out how much of old, and sometimes even why, is still remaining. Note that usually a settlement or a road that is maintained as the same as a drawing on the maps so far have not been left in its original state and, indeed, several modifications or alterations on the same grounds may have distorted the pre-existing structures.
2. Regarding the *natural components*, especially the vegetation cover, it often lacks the discovery of descriptive studies on forest a few decades ago or even more than a century ago. Essential are, therefore, the accurate field surveys, to make observations on the past: if to understand the structural state and the indicator species is not sufficient, to infer what that forest tessera could have been a century earlier, you will need to carefully analyse the stands, mapping both the cut stumps and the stems in place (with precise position and diameter) plus their canopies, and observing age distribution, the presence of non-native species, type of environment, role of water, any findings of former human use, etc. You can supplement this information with more sophisticated methods of investigation, such as those of dendrology and dendro-ecology with Pressler gimlet or on the rhizomes or other parts of plants (see Schweingruber [30]). It can also be useful to analyse whether the state of vegetation is changing (recreation dynamic or just regenerative, or degradation) or if there were traces of ancient coppice management. Even the soil surveys can be useful, for the presence of paleo-soils, or abnormal change in the characteristic horizons. In some cases, even archaeological excavations can give very useful information for the study of landscape units.

By the way, see Fig. 8.6, which shows a view of the mountain spruce forest of Ronzo-Chienis (TN), altitude 1,150 m a.s.l. The presence of a good age distribution, with trees up to about 35–40 m in height and 75–85 cm in diameter, indicate a respectable age of the tessera of forest, approximately around two centuries. Do you see 3–4 dead trees for barking from *Ips tipographus* (dangerous fir bark beetle), still standing, a sign that the damage occurred not more than 4–7 years ago? As a result the forest is evidently stressed.

Figure 8.7 shows the reconstruction of the dynamics of a recent tessera of *Querco-Castanetum insubricum*, analysed by Ingegnoli

Fig. 8.6 Mountain spruce forest near Ronzo-Chienis (Trento), altitude 1,150 m a.s.l. See the 3–4 dead trees for barking from *Ips tipographus* (fir bark beetle), still standing. The forest as a result is evidently stressed

Fig. 8.7 Recent dynamics of a wild forest tessera. From the map you can see two larger chestnut trees (Cs1 and Cs2, 115 and 120 years old), all other individuals present, *Fraxinus excelsior* (Fx), *Ostrya carpinifolia* (Oc), *Quercus robur* (Qr) are characteristic species the *Querco-Castanetum insubricum*, but with age between 20 and 35 years. *Above*, a plant of *Polygonatum multiflorum* (Pm)

and Giglio [6]. The mapping show the presence of two larger chestnut trees (CS1 and CS2, 115 and 120 years old), and of other individuals, that are *Fraxinus excelsior* (Fx), *Ostrya carpinifolia* (Oc), *Quercus robur* (Qr), all characteristic species of *Querco-Castanetum insubricum*, but with age between 20 and 35 years; the same species are found as *plantulae*, flanked by plants of *Polygonatum multiflorum* (Pm, whose age is established from the number of thicker areas of the rhizome). There are some cut stumps of chestnut (of which you have to observe the condition of the wood to try to estimate the age of cut). We are in the presence of an old chestnut cutting, abandoned 30–35 years ago, which allows the entrance of the fittest species typical of the vegetation of the area (Canton Ticino, Switzerland); 20 years ago it was used as pasture, until 5 or 6 years before this study: in fact, this appears to be the age of *Polygonatum,* that in the presence of grazing would have been eliminated. This is the reason for the current lack of young trees under the age of 20 years.

8.2 Ancient Landscapes Characters

8.2.1 Birth of Agricultural Landscapes in Europe

The domestication of plants and animals (Neolithic Revolution) occurred in the Near East, probably in Anatolia, about 8,000–8,500 years BC, apparently by proto-Indo-European people. These people, expanding, brought language, agrarian technology and stable settlement towards

Fig. 8.8 View of the old agricultural landscape of Camonica Valley (Central Alps, Lombardy) and below one of the incised maps of Bronze-Iron Age representing this landscape, which still shows its ancient structure. See also Fig. 1.5

Europe, but during 3,000 years they arrived only to Greece and South Italy, a distance of about 1,000 km. Then the speed of expansion doubled and, in other 3,000 years those novelties arrived to the north of Europe, that is the southern part of Scandinavia and Finland. In England, near Dartmoor, the oldest recognisable corn fields, small irregular plots of ground associated with hut-circles, date possibly from the Early Bronze Age (1900–1400 BC) as written by Hoskins [13].

The landscape transformations, advanced with the speed of about 35 then 70 km/century, were important, a true symbiosis with nature resources, but the sharp change of the environment followed thousands years later, with the urban revolution. It was a drastic revolution, a sort of fruit of the agricultural symbiosis, which began in the tenth to eleventh century BC in Greece and was completed by Romans in Britannia in the fourth century AC.

Therefore, from the Bronze Age and the late Iron Age *transitional* landscapes, based upon farming and hunting/gathering, became typical in Europe. An example of these landscapes should be done in Fig. 8.8 in which, still today, the rural landscape of Capodiponte (Camonica Valley, Lombardy) is structured following the Bronze-Iron Age criteria, as it is proved by the map incised on a smooth rock just over the slope of the picture.

Even when the full agricultural landscapes were developed, many of them in Europe maintained their first structure, accurately impressed by old populations. Many examples derive by the Roman "*Centuriatio*", very frequent in Italy, France, Spain, but also in U.K.[2], in the South of Germany and in the old Illiria region. The major traditional agrarian landscapes, of old origin and still imprinting (at least partially) the European landscapes [31], are the following:

1. Mediterranean agriculture
2. Three-field farming
3. Hardscrabble herding-farming

8.2.1.1 Mediterranean Agriculture

It was indeed the most ancient one, coming directly from Neolithic culture of Anatolia without changes (or very few) because of the similar climatic zone and vegetation. This environment allowed a good adaptation of field agriculture, horticulture and pasture. The field, devoted to the winter cultivations, does not need irrigation, being the Mediterranean winter very rainy.

Mediterranean cropland is formed by wheat and barley (40–50 % in peninsular Italy and Greece in 1950s). Orchards (and Olive trees) and vineyards (Greek cultivation), composed by perennial plants species able to withstand the summer dry season, have also a great importance. Often typically horticulture and field agriculture have intertillage, so that wheat and barely were sown among the large spaced orchard trees. Consequently, these Mediterranean landscapes reach a very high biodiversity. Pasture, on the contrary, remains confined on the mountains with herds of sheep, goats and swine. In the winter, herds migrate to marshy lowlands through a "transhumance".

8.2.1.2 Three-Field Farming

North of the Mediterranean, in fertile lands beyond the mountains with humid continental or sub-continental climate, a second type of agricultural landscape prevails. It was developed since the Roman times in Gallic regions (both *Gallia Cisalpina* and *Transalpina*) in Britannia, in South and Western Germany, in Pannonia and Tracia, but it was extended by other German peoples and also Slavic. Grains were raised on a three-field rotation, theorised also by Columella (first century AC). A close relationship between crops and livestock characterised these landscapes, in which rye (*Secale cereale*) was preferred to wheat (*Triticum aestivum*) for ecological reasons: the *Triticum durum* is available only in Mediterranean climates or similar. On the ills, especially of Piedmont and Champagne, vineyards was common, but with the so-called Etruscan cultivation.[3] Livestock furnished the manure to fertilise the fields and each farm needs to cultivate also wide meadows for cattle.

8.2.1.3 Hardscrabble Herder-Farmers

Landscapes with broken or sterile soil, within the mountains or in excessive clouded climates, developed an agriculture in which cattle were the favoured animals, pastures a wide presence and crop only a marginal function. This is typical of Walser or Ladin landscapes in Alpine valleys.

Many characters of these three types of old agricultural landscapes, as underlined, subsisted till now. But another criterion differentiated all these landscape: along the Western tradition the practice of dividing the land equally among offspring resulted in high fragmentation over time, while the North-Eastern tradition of the primogeniture dominated and avoided the division of the fields.

8.2.2 Rebirth of Agricultural Landscapes in Europe

After the fall of Roman Empire (fifth to sixth century AC) many of the agricultural landscapes were abandoned and partially re-natured. The decrease of population, began after the third century, brought the about 70–75 million inhabitants

[2] Landscapes formed by *centuriatio* were found in Kent, Sussex, Hertfordshire, Essex and Middlesex.

[3] The "Greek cultivation" of vineyards is directly on the soil surface, without grass, adapted to Mediterranean climate, while the "Etruscan cultivation" is on pergolas, upon a grass, adapted to Central European climate.

Fig. 8.9 The gothic Abbey of Morimondo (1134) founded by *Cistercenses* near Abbiategrasso (Milan)

of Roman Europe to 27–30 million in seventh to eighth century, even because the barbarian invasions added very few people. At the beginning of Medieval Age Flanders, Brabant and Lombardy were the most populated regions of Europe. The re-growth started after the eleventh/ twelfth century, with the help of new agrarian techniques or derived by the classic agrarian volumes of Varrone and Columella. The use of horse gave a great impulse to crop production and also the extension of irrigation and "*marcitae*" (water meadows). In this rebirth of agricultural landscapes the abbeys had a crucial importance (Fig. 8.9).

The beginning started from the Benedictine reform of Cluny (948–981) which founded 1,300 monasteries in France and 250 in Europe (especially U.K. and North Italy). But the most part of the work on agricultural landscapes was done by the second reform of Citeaux (1098), due to Robert de Molesme. Soon after 1112 the nobleman Bernard de Fontaines with his friends founded four base monasteries (*abatiae primigeniae*): la Ferté (1113), Pontigny and Morimond (1114) and Clairvaux (1115). Morimond in Champagne-Ardenne founded founded in 1134 Morimondo (Fig. 8.9) in Lombardy and many other. Especially Bernard de Clairevaux gave the bigger impulse to found affiliated monasteries in Europe: Tiglieto (1120) in Liguria, Kamp (1123) in Rhineland, Chiaravalle (1135) in Lombardy and other 160 all over Europe. About 600 *Cistercenses* Abbeys were founded in summary. Each Abbey was organised in many "granges",[4] at least 3–7, through which the territory was transformed and improved as agrarian landscape, especially with field reshaping and water regulation.

8.2.3 The Reconstruction of Ancient Landscapes

By way of example, it is assumed to have to reconstruct past states of a landscape of Trentino (in the municipality of Mori, near the Garda lake), going back at least to two historical thresholds from the current period, considered around the year 2007–2008. This requires the formation of two maps of the landscape concerned at the historical thresholds of 50–60 and 120–150 years from now. Those thresholds were not chosen at random: the major changes in the area in fact occurred in the last World War, while, in the nineteenth century, there had been still a situation roughly valid since the beginning of the century (cities included).

[4] Granges (from French "grange" = barn) typical rural farm of Benedictine reform.

1. First of all you will have to perform the completion of the current map, around the year 2007. We must often speak of "reconstructions" also for the cards today, as it is difficult to find the update of all the main themes (geographical forms, urbanisation, mobility, hydrography, etc.) for the same date. Also to check through the multispectral satellite images is a must, as well as a georeferencing control, after shipping on a computer. It should be noted that the computer support is not always valid for studies on the landscape, as computer screens are still too small to make observations on the set of components of landscapes that can reach even to the extent of hundreds of km^2; therefore it is also need to work on paper and switch between phases of work.
2. The second phase involves tracking the perimeter of study of the landscape in question (see Sect. 2.2.2), or of each one of its units (four in the example of Mori, see Fig. 3.16): perimeter which will be later considered, except for some possible local change, even in the reconstruction of past states.
3. The map of the historical mosaic of patches of tesserae and ecotopes, dating from the mid-1900s, can be reconstructed using the IGM maps (1: 25,000) detected in the early 1950, and with the help of aerial photographs of 1954. The Habsburg map from 1860 has to be verified at the borders, since at that time the perimeters of municipalities could have been different from those of today, often gathered in a larger territory.
4. Caution: the reconstructions cannot be considered as the representations of two instants or 2 years: in fact, the available materials do not allow it (some exceptions) and the situation in an instant cannot be considered representative of the average conditions of a period for a complex system like a landscape. We want, however, to frame those average characteristics of the land structure that have been found in the transition between two centuries and in the middle of the last century, to be able to calculate indicative values taken by some ecological indices and highlight the trends up to the present period.

Figure 8.10 shows the reconstruction of the territory of Mori (TN) in 1860 using Habsburg maps 1:25,000 (above) and in 2007 (down). Historical centres (brown) are linear at the foot of the mountain. Greenland was not wide, while cropland and vines were very developed on the bottom of the valley.

Figure 8.11 reports a significant comparison between the valley of Loppio at the beginning of the twentieth century (before the War 1915–1918) (top) and today (2007). Notice the contrast and complementarity: (a) the campaign is now almost a treeless plain, while a century ago it was full of plants and fruit rows of protection; (b) the woods on the lower slopes of the mountains, at one time exploited and sparse, are today more dense and continuous.

8.2.4 Estimation of the Ecological Parameters in the Past

Unfortunately, it frequently happens that you cannot quantify all the themes of landscape components present in a time series. There are many reasons: mode of merging different data per period, lack (or loss) of data on some issue and even lack of data or their elaboration by the government of its territory, but which gets to be necessary for the study of a LU. This also applies to the current state, or in any case for very recent times.

8.2.4.1 Structural and Functional Configurations

Continuing with the same examples of Mori, the territory that is taken as a reference in 1860 (Fig. 8.10 above), as we have seen, does not entirely coincide with the current borders (Fig. 8.10 down).

This occurs in relation to a different division of the management area, a time longer tied to features and local resources. Mori was divided into smaller municipalities, more similar, although not entirely coincident, with the 4 LU

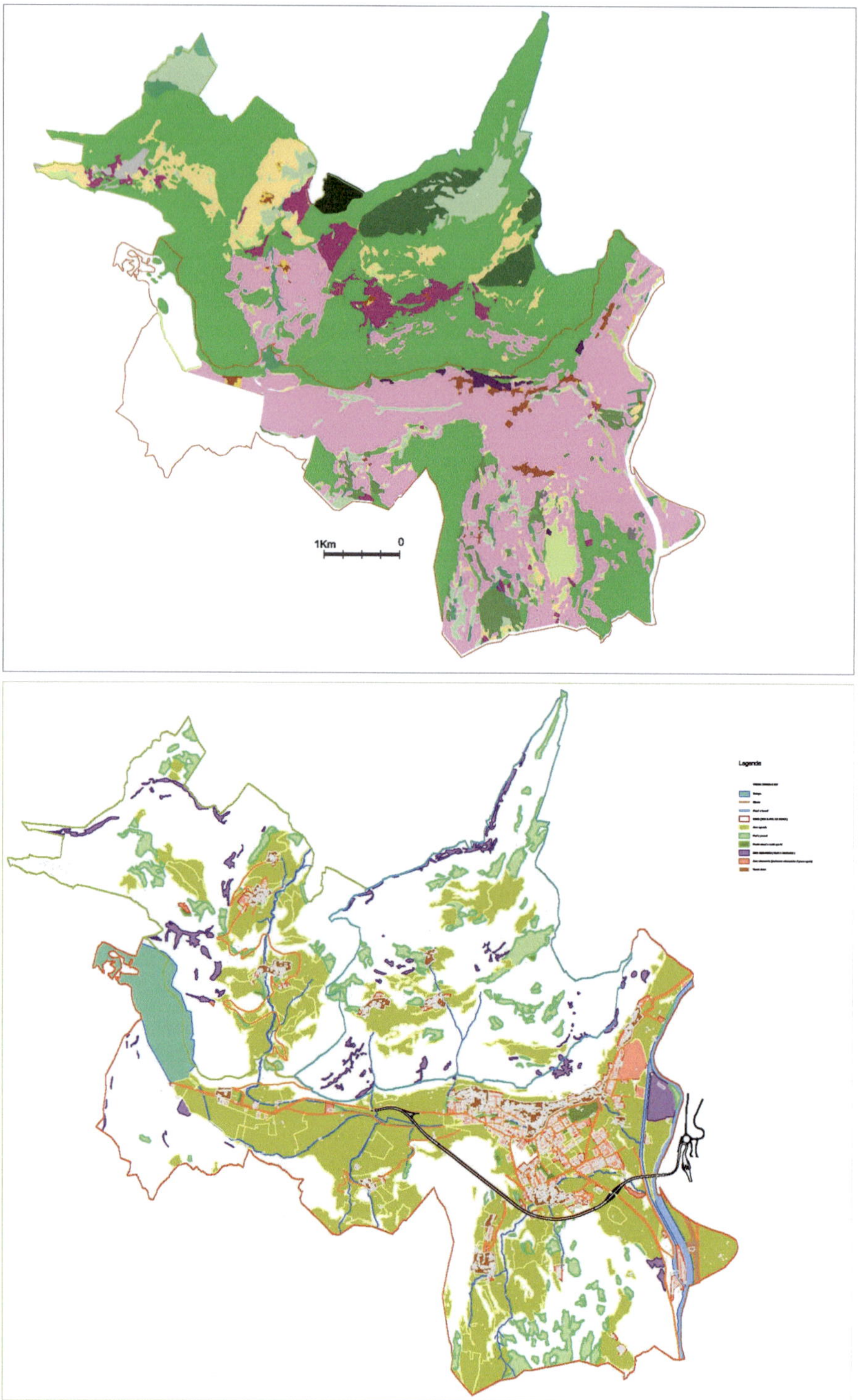

Fig. 8.10 Reconstruction of the territory of Mori (Trento) in 1860 (*above*), using maps Habsburg 1:25,000 and in 2007 (*down*) through modern administrative maps. In 1860 historic centres were in *brown*; cultivated vines in *pink*, coppices in *light green*; coniferous forest in *dark green*, pastures in *yellow* and beech (also coppice) in *light green*. Note that the area south of Lake Loppio is not coloured, because it is not included in the maps of Mori and the Val di Gresta. In 2007 forest are in white, cultivated vines (and agricultural fields) are in yellow-green while urbanised is in grey+pink with roads in red+black. Rocks are violet

Fig. 8.11 Comparison between the valley of Loppio at the beginning of the twentieth century (before the War 15–18) (*top*) and today (2007). Notice the contrast and complementarity

identified according to the principles of the bionomics of the landscape (remember Fig. 3.16).

In cases like these you have to adjust the scheme of "structural pattern" landscape unit, being careful to highlight the parts that have changed. Later, you have to assess whether the boundaries recognisable at the present time are still valid in the light of the findings. If they are found not to be valid, there are three options:

1. To re-draw the boundaries of the current LU on the basis of the new look (and then correct all the indexes of the current situation and the measures that had already been made);
2. To also consider in the past the same current borders (with some minimal adjustment to the limit not exceeding 10 % of the total surface area under examination) specifying that it is, in this case, an operational LU;
3. To analyse the two LU independently, if instead, today and in the past, there are ecologically well-defined boundaries but distinctly different; then to bring back all the considerations made and the reference values (e.g. in the mosaic of conversion or other descriptions) directly in percentage (without omitting, however, the two surfaces of the two LU in hectares, so as to facilitate reflection of what is observed).

Be careful!! What certainly need to be redrawn are the apparatuses of landscape, as it is necessary to observe if they have changed over time:

1. As the amount of covered area, with absolute percentage value;
2. As the mutual relation of importance (i.e. what is the degree of importance of each of them at the different hierarchical historical thresholds examined);
3. As the number of apparatuses present at different historical thresholds;
4. As the distribution on the territory, through the calculation of the variation of their fractal dimension (See Sect. 7.1.5) at the various historical thresholds for the apparatuses considered more important or characterising, modified or more, and/or through the use of the index of dispersion/aggregation of the tesserae or components of the same apparatus as the level of contagion and contrast between the apparatuses themselves;
5. As the ratio HH/NH to the different historical thresholds.

We report a study carried out always on the landscape of Mori, and referred to the LU 1 (Mori valley). From Table 8.1 we can see the dual comparison between the components in each apparatus and the complete calculation of the apparatuses themselves (see Sect. 7.2.2), while the Fig. 8.12 shows the bar graph of direct comparison between apparatuses.

Warning: it's always important to consider if substantial changes have occurred in the shape, size and characteristics of ecotones, both within the ecomosaic (i.e. between ecotopes or apparatuses) and on the edge of LU, and outside, that is if, in one of the historical thresholds examined, the whole LU—or any substantial part of it—has played the role of ecotone and possibly why (to inquire) it had acquired or lost it.

Table 8.1 Differences between landscape apparatuses of LU1 of Mori municipality in the period 1860–2007. Note also the diverse measures between element x apparatus and their value of relative multifunctionality

	Element x apparatus		Multifunction apparatus	
Mori, LU1	1860 (%)	2007 (%)	1860 (%)	2007 (%)
Stabilising woods	2	7.95		
STB	2	7.95	5.93	9.55
Temperate woods	20.52	22.68		
Chesnut	4	2.92		
Other woods	2	3.31		
RNT	26.52	28.91	23.4	26.66
Rocks and gravel	2	2		
Water network	3.66	3.54		
HYG	5.66	5.54	4.88	5.08
Pastures	6.59	0.82		
RSL	6.59	0.82	8.12	2.77
Meadows	5.84	3.3		
Arable	0.95	4.22		
Modern vineyards	0	21.08		
Traditional vineyards	45.28	9.84		
PRD	52.07	38.43	47.3	34.7
Hedgerows with trees	1	0		
Urban gardens	0.74	0.94		
PRT	1.74	0.94	5.29	4.84
Historical centres	1.92	4.56		
New urbanisation	1	5.91		
RSD	2.92	10.46	2.88	9.54
Industries	0.5	3.62		
Roads	2	3.33		
SBS	2.5	6.94	2.21	6.88

8.2.4.2 Time Series of Data

Let us consider another example: the study of a series of historical data relating to the LU of the Regional Natural Park of Monte Barro (Lecco, Italy) (Fig. 8.13), data collected on the basis of computerised mapping (1: 10,000), the reference years being 1888, 1959, 1995. The data were incomplete (Tables 8.2), so they were not usable for the reconstruction of the dynamics of the landscape ecological state. We had to try to make the extra processing based in part on maps even at synthetic scale, partly on logical analysis. Of course, it was necessary to have first made a series of careful inspections.

(a) We started by the most recent data (1995). Compared to the past, data on "rocky outcrops" are almost non-existent, so we can assume that they were grouped together in the closest index entry (which is excessive, compared with field testing): that of the "shrubs". The opposite was the case for 1888, where the voice "shrub" is absent. The correct reference is, therefore, the figure for 1959 (the quarries were absent) and the

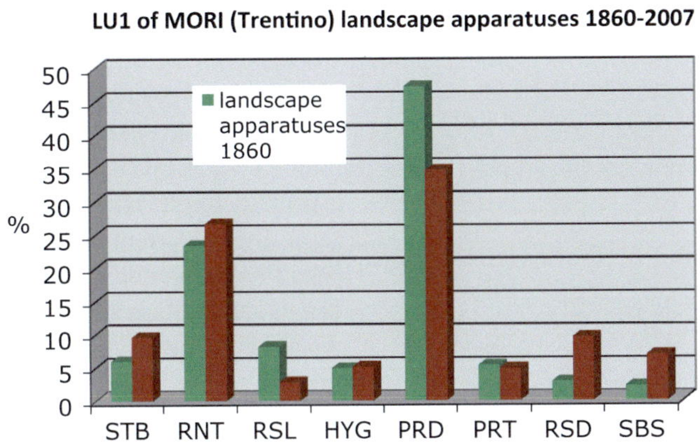

Fig. 8.12 The dynamics of main landscape apparatuses in the LU1 of Mori (Trentino), period 1860–2007

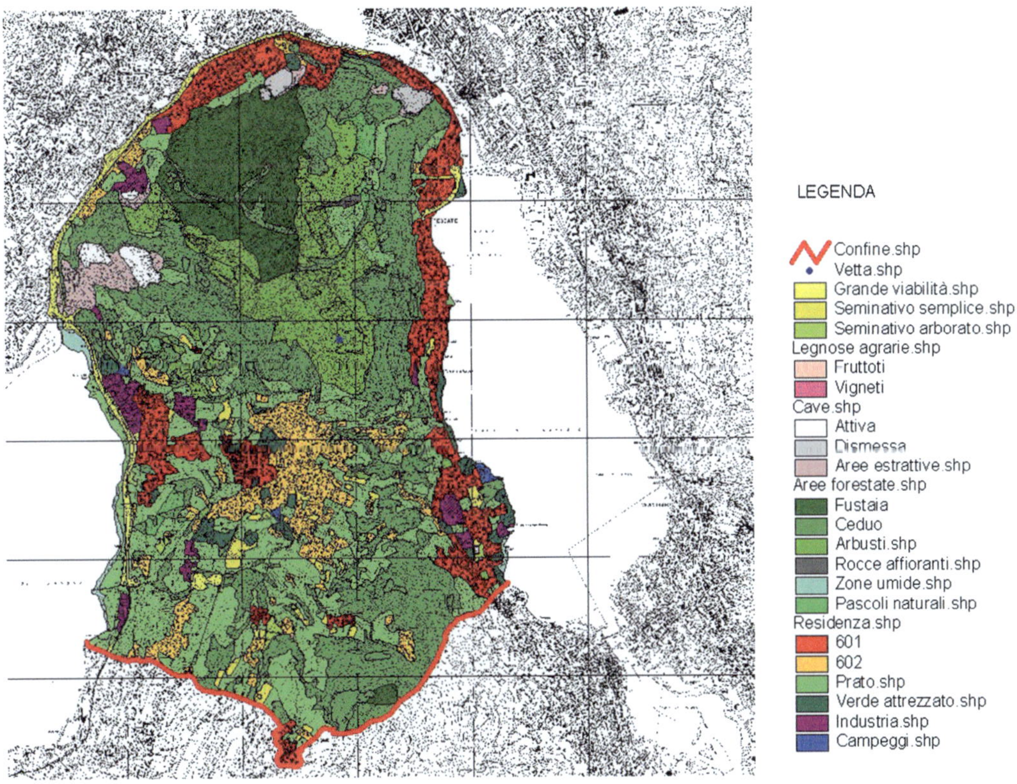

Fig. 8.13 The reconstruction of the ecotissue of the Mount Barro Regional Park in 2004

adjustment is to accept as constant value for rocks the one reported in 1959.

(b) The portion missing for recent data (South-East) can be estimated in its component elements by referring to a regional map 1:25,000, with its distinct issues: forests (55 %), urbanised (20 %), agriculture (15 %), industrial (10 %). The pastures, the shrubs and reeds do not appear in that quadrant. Proportioning these values in Table 8.2 issues you can get the adjustments. Estimates are acceptable, as the main voice (forests)

Table 8.2 Measure of the surfaces of main types of landscape elements in the LU of the park of Monte Barro (Lecco)

	1888	1959	1995
Types of elements	ha	ha	ha
RSD	24.4	68.3	113.1 Urban. dispersed
			186.8 Urban. dense
SBS	23.4	70.3	16.4 Industrial etc.
			36.3 Roads etc.
PRD, Meadows	35.9	35	372.1
PRD, Arable land	328.7	506.1	22.8
PRD, Arable with trees	–	–	5.3
PRD, Wooded agrarian	608.3	396.4	2
PRT, Urban green	–	–	31.6
PRT, Camping green	–	–	3.3
RNT	506.7	504.3	175 Forests
STN			133.7 Coppices
STN. Grassland and pastures	97.4	14.2	13.5
RSL, Shrub land	–	59.3	134.4
EXR, Reed thickets	4.1	–	8.1
GEO	55.2	31.0	2 Rocks
			49.0 Mines and caves
Total	1,684.1	1,684.1	1,335.4

Present data do not consider the South-East part of the territory (348.9 ha, 20.6 % LU)
Data from the Polytechnic University of Lecco (C.Farina, G. Gaudenzi, S. Isacco, G. LoTennero, L. Maffescioni, P. Pileci)

come back to about the same order of magnitude data of 1888–1959 (these elements have remained constant obviously).

(c) Other adjustments (minor) were made for reeds (1959) and for shrubs (1888). In addition, the summary of the woods had been divided on the basis of the current ratio, confirmed by surveys. Even the urbanised was divided, but in proportion to the information derived from historical maps and surveys.

Results of all this kind of proceeding are shown in Table 8.3.

8.2.4.3 Estimation of the BTC Map-Based

Once completed the time series of data on a LU, you can begin to develop the ecological reconstruction of the states through the use of appropriate ecological indices. For some of them, especially for the BTC, it is necessary to compare carefully the current measures to the states of the past; this estimation is not easy because the map data would require additional information on the types of elements considered, in particular for forests, which are the components with greater BTC. Unfortunately, it is rare to find suitable sources for this information, apart from the guide values obtained in the field, according to the criteria already mentioned.

The detection of the BTC of the current state has to be programmed also taking account of what expressed above: so after the study of the cartographic data and appropriate inspections, they can be conducted at various levels of depth, depending on the structure of LU in examination and objectives of the study. If it is a preliminary research or applications to comprehensive planning (synthesis scale) excessive precision can be avoided. If, instead, the work needs a scale of greater detail, so that it requires accuracy, you will have to resort to a statistical sampling, with a percentage error, for example, less than 12 %.

The more detailed information, with regard to the forest tesserae and in part to the shrub one, (for example, if you want to know what could have been the BTC of a forest tessera 40 or 80 years before) need to rebuild the model development of that type of forest (see Fig. 5.4). Basically, the procedure is as follows:

1. According to the present surveys (estimated age) you will have to determine the position of the tested tessera on the model reported in Fig 5.4; then
2. You must mark the BTC “Y”, corresponding to the estimated current age, on the curve of the theoretical model; after
3. You have to go back on the model until you find the point “x”, corresponding to the age that the tessera had at the historical threshold that interests you; finally,

Table 8.3 Controlled adjustment. Estimation of the surfaces of the main types of landscape elements of the LU of the Park of Monte Barro (Lecco)

	1888		1959		1995	
Types of elements	ha	%	ha	%	ha	%
RSD, Urban. rare	**4.9**	0.2	**17.2**	1.1	139.6	8.3[a]
RSD, Urban. dense	**19.5**	1.2	**51.2**	4.1	230.1	13.7[a]
SBS, Industrial etc.	23.4	1.4	70.3	4.2	33.7	2.0[a]
SBS, Roads etc.					53.8	3.2[a]
PRD, Meadows	35.9	2.1	35	2.1	420	24.9[a]
PRD, Arable land	328.7	19.5	*498.9*	*29.6*	25.8	1.5[a]
PRD, Arable with trees	–	–	–	–	6.3	0.4[a]
PRD, Wooded agrarian	608.3	36.1	396.4	23.5	2.5	0.1[a]
PRT, Urban green	–	–	–	–	33.0	2.0[a]
PRT, Camping green	–	–	–	–	3.3	0.2[a]
RNT, forests	**202.7**	12.0	**201.7**	12.0	200.3	11.9[a]
STN, Coppices	**304.0**	18.1	**302.6**	18.0	300.4	17.8[a]
STN, Grassland and pastures	97.4	5.8	14.2	0.8	13.5	0.8
RSL, Shrub land	*24.0*	*1.4*	59.3	3.5	*134.4*	*8.0*
EXR, Reed thickets	4.1	0.2	*6.5*	*0.4*	8.1	0.5
GEO, Rocks	*31.2*	*1.9*	31.0	1.9	*31.4*	*1.9*
GEO, Mines and caves	–	–	–	–	49.0	2.9
TOTAL	**1,684.1**	100	**1,684.1**	100	**1,684.1**	100

Data proportional to the historical maps controlled by surveys are given in bold
In italics: data deducible from other voices
[a]adjustment with the previous lack of 349 ha, following the Regional maps

4. You must calculate the variation (percentage) of BTC (y/Y) between this second point and the first and apply the same change to the BTC surveyed in reality.

Note: if the estimated age of the card exceeds 220 years, consider that the curve proceeds after that time threshold, in an asymptotic way, thereby maintaining the same inclination.

The above operations are fine if the forest is regrown after an abandonment of cultivation or after clear-cutting of a tessera or after a destructive disorder. The cases are perhaps very frequent, even if quite different, as it comes from assessing the woods with long persistence. Usually, however, these forest formations are managed:

(a) If high forest, with a thinning restricted to mature trees (with the classic method of the forest "hammered");
(b) If coppice, with cuts cyclic frequency varying from 8–10 to 18–20 years.

The result is that, usually, a persistence media of the value of BTC is obtained, which consequently remains constant or differs little compared to today. However, in several cases, we must take account of local history and perform the estimates by appropriate coefficients, in relation to disturbances that can be found and after a careful check on the field, as is shown in the same example of LU of the Natural Regional Park of Barro (Table 8.4).

8.2.4.4 Theoretical standard habitat per capita SH*

From the second to the fourth Century, 1 heredium = 0.506 ha was divided in two jugera =2,530 m^2 and this quantity was intended as the minimum area to sustain a person. In the high Medieval period this quantity was not more sufficient, because of the decrease in agricultural technology: we have to consider at least 1.5 times this area (i.e. 3.750 m^2/person).

After the agricultural revolution, beginning in Central Europe in the second half of eighteenth century (from Lombardy to Brabant to England), the carrying capacity of the landscapes became

Table 8.4 Estimation of the BTC, average per element of the LU under examination. Note that the elements presenting higher BTC were surveyed on the field, while the others were evaluated from analogous cases

	Surveyed data		Adjusting coeff.			BTC estimation		
Element	BTC	Average	88	59	95	88	59	95
Coppice W.	4.67 + 5.14	4.90	0.85	0.85	0.95	4.20	4.17	4.65
Wood	6.25 + 7.00	6.62	0.95	0.85	1.0	6.30	5.63	6.62
Shrubland	1.73 + 1.50	1.61	1.0	0.95	1.0	1.61	1.53	1.61
Grassland	0.9 + 0.77 + 0.86	0.84	1.05	0.95	1.0	0.88	0.80	0.84
Urban park	4.39 + 2.50	3.45	–	–	0.95	–	–	3.27
Arable	0.9–1.2	1.05	1.05	1.0	0.95	1.10	1.05	1.00
Meadows	0.65–0.7	0.67	1.05	1.05	1.0	0.70	0.70	0.67
W. agrarian	1.5–1.7	1.60	1.05	1.05	1.0	1.70	1.70	1.60
Reeds	1.0–1.3	1.15	0.95	0.95	1.05	1.10	1.10	1.20
Rare urban	0.4–0.7	0.55	1.0	1.0	1.0	0.55	0.55	0.55
BTC_{UdP}						2.38	2.22	2.06

Adjusting coefficients are proportioned to the historical periods, e.g. lack of technology in agriculture before 1900, the destruction of the war, etc. BTC in Mcal/m^2/year

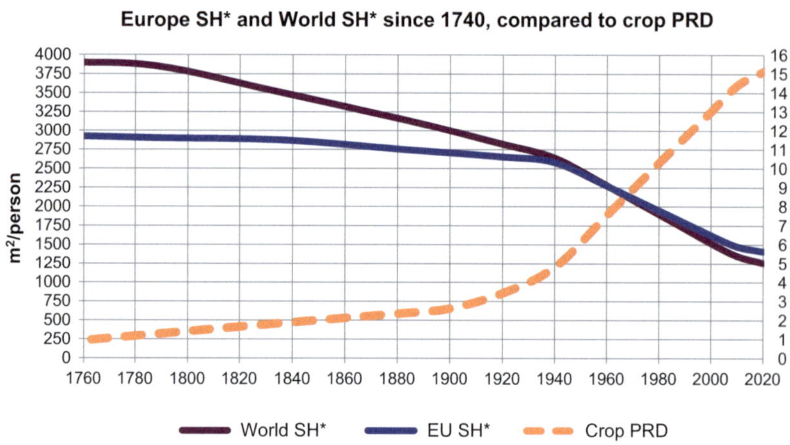

Fig. 8.14 We can see the growing of crop production, that has been passing from 1 to 15 times since 1760. The delay between Europe and the entire World may be considered of about 1.5–1.6 century

growing and in 1750 the productive standard habitat per capita could be considered below 3,000 m^2/person. At World scale, these quantities were to be enlarged of about 600–700 m^2/person. The trend of crop production increased from 1 to about 3 times in the period 1780–1880 and again three times or more from 1940 to 2010.

In Fig. 8.14 we propose an estimation of the changes in SH* in the period 1750–2010, in which the trend of the European situation is compared with the World one. In some advanced European regions, as Lombardy or England, we may consider still contained SH* (about 10–20 % less then in Europe). Note that the measures of SH* decrease in inverse proportion of the crop production, but not through a mirror proportion, due to the interference of diverse security factors and the presence also of non-agricultural areas.

8.2.5 The Mosaic of Conversion

In Table 8.3 we observed that the amount of coppice and high forest in the LU of Monte Barro seems to have remained unchanged over time. However, you should be really sure:

1. That all the coppice we see today were the same of 1888; or
2. that they were former coppice or cultivated field or abandoned ones;

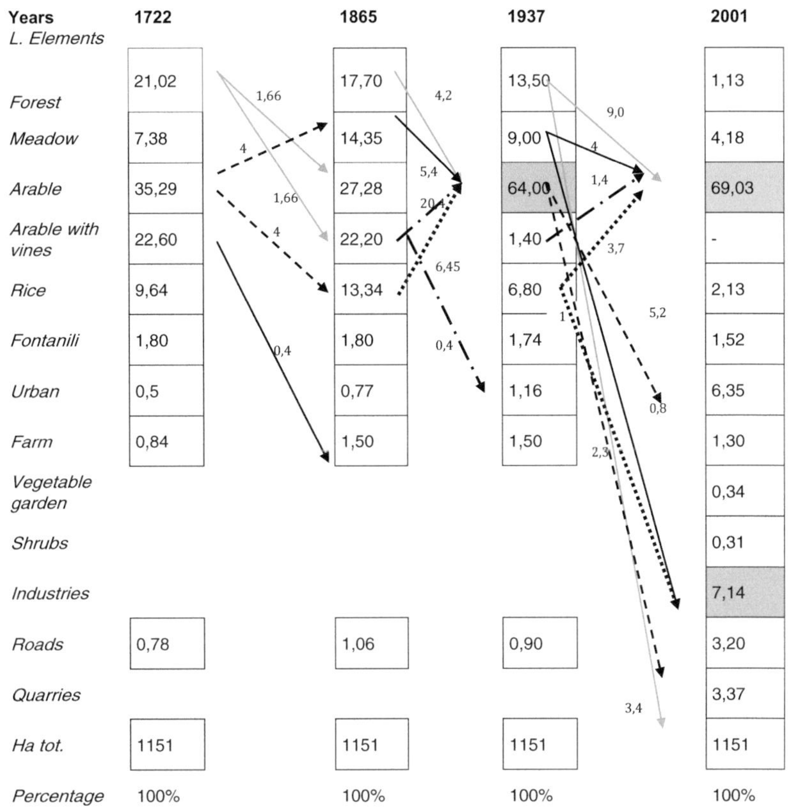

Fig. 8.15 Mosaic of conversion explaining the transformations of the basic ecomosaic incurred by the Municipality of Cusago (near Milan) over time (LU operational). The *arrows* indicate the amount, expressed in percentage points of the conversion, from one type to another. The *grey cells* indicate the types to which we find the convergence of two or more arrows [6]. Fontanili, or Risorgive, are water springs with following small streams typical of the Po Plain when passing from the permeable to the not-permeable plain

3. That any tessera of current high forest were not an ex-coppice and some present coppice were not an ex-high forest then cut

So very useful is another instrument, the "mosaic of conversion" or shifting mosaic sensu Forman and Godron [32]. It consists of a matrix, where the columns gives the years of detection (in this case also understood as representative of the situation of a period and not as the year actually quoted) and in rows the amount (in hectares or percentage) of each of the items corresponding to the elements of the landscape detected (as mentioned in the previous paragraphs). This tool is used to understand the "transformation rates cross", that is to test the qualitative changes in the mutual relations of the mosaic of ecotopes, even when the same type of item remains quantitatively constant over time, so it does not seem to vary (Fig. 8.15).

For the measurement of eco-mosaics at the different historical thresholds previously rebuilt you can use a planimeter, unless also historical maps have been georeferenced and processed in digital format, in which case only the possess of the correct computer tools available serves. Here is an example of a mosaic of conversion (Fig. 8.15) reported to the municipality of Cusago (MI) with the following series: 1722, 1865, 1937, 2001.

For the building of the mosaic of conversion hectares, corresponding to each element of the landscape at every historical threshold previously considered, must be detected (and eventually calculated as the percentage of the total). Then:

1. Transcribe the measured data in the matrix
2. Compare the corresponding eco-mosaics to each historical threshold and then, starting now from the oldest one, highlight graphically the areas that, in the transition to the next threshold, have not changed the type of landscape element from those which have become something other

3. For each transformed landscape element, measure the type of source element and the type of final element plus what was the amount, in hectares or percentage, of such change (e.g. from a deciduous forest like an oak-hornbeam type, a number of hectares would have become arable, another quantity lawn mowed and another quantity would have been used for urbanisation; in turn, a certain amount of arable land would have been abandoned and transformed in a woods, another would have been subtracted to build factories, etc.)
4. Via an appropriate system of arrows, highlight the relations between the different elements of change, bringing above each arrow the magnitude of the change itself (in percentage or in hectares)
5. Whatever type of measurement had been carried out, sign in the last row of the matrix the amount of hectares of the observed area. In fact, subject to the limits of LU studied, there may have been some adjustments in hectares between different periods

Note: an additional element may be constituted by a table, which shows the "rate of change" between the various elements in time, to highlight if the processes are still in place or are now in a phase of consolidation or progressive conclusion.

Although not frequently used, the mosaic of conversion that has been described may be formed recurring to landscape apparatuses, to be performed with the same procedures as shown above, based not on the elements per apparatus, but on multifunctional apparatuses.

References

1. Langé S (2003) Presentazione del Master Analisi e Gestione del Patrimonio Paesistico, Politecnico di Milano, Polo Regionale di Lecco
2. Duro A et alii (1986) Vocabolario della Lingua Italiana. Istituto della Enciclopedia Italiana G. Treccani, Roma
3. Zanzi L (1995) Viatico per una avventura nella storia della Val Grande, In Val Grande, storia di una foresta, a c. di Alberti L, Rainaldi G, Rizzi E, Anzola d'Ossola, Fondazione Arch. Enrico Monti
4. Naveh Z, Lieberman A (1984) Landscape ecology: theory and application. Springer, New York
5. Ingegnoli V (2002) Landscape ecology: a widening foundation. Springer, Berlin
6. Ingegnoli V, Giglio E (2005) Ecologia del Paesaggio: manuale per conservare, gestire e pianificare l'ambiente. Sistemi editoriali SE, Napoli
7. Ingegnoli V (1993) Fondamenti di Ecologia del paesaggio. CittàStudi, Milano
8. EUROSTAT (2013) The grange. Pulloxhill, Bedfordshire
9. Boroli P (2013) Calendario Atlante 2013 + Deawing. Istituto Geografico De Agostini, Novara
10. Le Roy Ladurie E (1967) Histoire du climat depuis l'an mil. Flammarion, Paris
11. IPCC (2013) Climate change 2013, the physical science basis. IPCC Secretariat, Geneva
12. NIPCC (2013) Climate change reconsidered II: physical science. Center for the Study of Carbon Dioxide and Global Change, Tempe, AZ
13. Hoskins WG (1971) The making of the English landscape. Penguin, Harmondsworth
14. Sereni E (1972) Storia del paesaggio agrario italiano. Laterza, Bari
15. Ingegnoli V (1981) Organizzazione agricola e casa rurale. In: Pirovano C (ed) Lombardia: il territorio, l'ambiente, il paesaggio. Electa, Milano, pp 27–64
16. Kuster H (2003) Geschicte des Waldes. Von der Urzeit bis zur Gegenwart. CH Beck, Muenchen. Tr. It. Storia dei boschi, Bollati Boringhieri, 2009
17. Anati E (1985) I Camuni. Jaca Book, Milano
18. Sansoni U, Gavaldo S (2009) Lucus Rupestris, Sei millenni d'arte rupestre a Campanine di Cimbergo. Edizioni del Centro
19. Calcagno Maniglio A (1983) Architettura del Paesaggio. Calderini, Bologna
20. Pregill P, Volkman N (1999) Landscapes in history, 2nd edn. Wiley, New York
21. Saibene C (1955, 1980) La casa rurale nella pianura e nella collina Lombarda, (CNR vol 15) L. Olschki, Firenze
22. Perogalli C (ed) (1975) Cascine del territorio di Milano. Milani sas, Milano
23. Perogalli C (1981) L'architettura fortificata lombarda. In: Pirovano C (ed) Lombardia, il territorio, l'ambiente, il paesaggio. Electa, Milano, pp 65–108
24. Langé S (1972, 1980), Ville della provincia di Milano. Sisar, Milano
25. Langé S (1989) L'eredità romanica. La casa europea in pietra, Milano
26. Tabarelli GM (1981) Castelli del Trentino. De Agostini, serie Gorlich, Novara
27. Sica P (1976-1980) Storia dell'urbanistica. (Il Settecento. L'Ottocento. Il Novecento.) Laterza, Bari
28. TCI (1930-1950) Attraverso l'Italia: illustrazione delle Regioni Italiane. Touring Club Italiano, Milano

29. Ingegnoli V, Langé S, Süss M (1987) Le ville storiche nel territorio di Monza. Pro Monza, Monza
30. Schweingruber F (1996) Tree rings and environment. Paul Haupt Bern (Available at Swiss Federal Institute for Forest, Snow and Landscape Research, CH-8903 Birmensdorf)
31. Murphy AB, Jordan-Bychov TG, Bycova-Jordan B (2009) The European Culture Area, a systematic geography. Rowman and Littelfield, London
32. Forman RTT, Godron M (1986) Landscape ecology. Wiley, New York

9 Diagnostic Evaluation of the Landscape

9.1 The Bionomic Evaluation of Vegetation

9.1.1 Some General Premises

The evaluation of plant components in a landscape unit (LU) is performed by the analysis presented in Chap. 5, according to the principles and methods of landscape bionomics. We know, however, that the evaluation criteria go beyond the phases of analysis, because the data retrieval often does not exhaust the critical reading of the phenomenon in question and, therefore, it is necessary to give some guidance on assessing.

We always keep in mind (see Sect. 4.5.2) that the diagnostic evaluations depend on the comparison between the conditions of the ecological system under consideration and those of a state regarded as "normal". In other words, is the relationship between "pathology" and "physiology" of the system that allows a diagnosis in the clinical sense of the landscape in question. You have to understand how the system moves from the state of normality due to pathogenic stimuli and, with a projection of the information, to assess where the damage could rise to the structure and functions at a time congruent. We expose briefly some longer operational consideration.

As expected from the survey forms of LaBiSV (Landscape Bionomic Survey of Vegetation) the main groups of parameters (see Table 5.1) are four (T, F, E, U) for a total of 28 ecological characteristics (in case of forest), divided as follows: 6 parameters of tessera (T), 3 parameters of aboveground phytomass (F), 10 parameters of ecocoenotope (E), 9 parameters of landscape units (U). Evaluating the data collected in relation to the *maximum potential* for each set of parameters the results are measured as ecological quality percentages QT (%), QF (%) QE (%), QU (%) (e.g. Fig. 9.1 and 9.6). These results are very significant, especially as they can distinguish comparatively the vegetation characters of forests in examination, even when the traditional ecological data are equal (e.g. syntaxonomy, dominant species, average height, phytomass, etc.).

In the mentioned method (LaBiSV), we reported the equations for estimating the biological territorial capacity of vegetation (BTC) for the vegetation types in question (see Table 5.6). Applying these equations to the results of surveys of various vegetated tesserae, we obtain the estimation results in Mcal/m^2/year. This makes it possible to measure, in a whole unit of the landscape, its average BTC. It is necessary, of course, to refer to the set of all the formations of vegetation after analysing, for each type to estimee the surface and the percentage of presence with respect to the LU territory, in order to assess the average weighted BTC.

V. Ingegnoli, *Landscape Bionomics Biological-Integrated Landscape Ecology*,
DOI 10.1007/978-88-470-5226-0_9, © Springer-Verlag Italia 2015

PLACE	TREE COMPOSITION	Q.T	Q.F	Q.E	Q.U	Hc	v PB	BTC	bQ	M-Th	CBSt
Tessera (Ts)	*Dominant Species (%)*	%	%	%	%	m	m^3/ha	$Mcal/m^2/yr$	%		
Poland, Bialowieza, Ts3	*Q. robur 85, Tilia c. 10, Carp.bet. 5*	100	84,9	81,8	94,6	29,4	768	**10,34**	90,10	82,51	**74,34**
Poland, Polana, Ts1	*Pinus syl. 85, Picea abies. 9, Q.robur 6*	64,7	58,7	65,6	57,8	24,0	257	**5,89**	62,1	34,74	**21,58**
Belgium, Asse, Ts2	*Acer pspl. 30, Prunus av. 22, Fraxinus ex. 17, Alnus gl. 8, Q. rob.4*	72	84,9	55,9	70,7	26,9	584	**7,60**	67,2	55,112	**37,04**
Austria, Galtur, TS1	*Picea abies* 100	72	70,7	78,8	81,3	26,1	838	**8,31**	77,30	64,94	**50,20**
Trentino, Paneveggio, Ts4	*Picea abies. 100*	72	70,7	76,8	95,1	32,4	992	**9,11**	80,92	76,99	**62,30**
Lombardy, V. Paghera, Ts 1	*Picea abies 90, Larix d.10*	79,3	70,7	78,8	70,7	26,8	635	**7,78**	75,43	57,53	**43,39**
Lombardy, Groane, Ts 2	*Q.robur 48, Q.petraea 6 P.syl. 18,5, Acer pspl. 15, Robinia. psa 7,*	73,3	74,2	64,6	69,7	25,8	348	**7,58**	69,10	47,39	**32,75**
Lombardy, Colli Briantei Ts 3	*Q.robur 60, Q.petraea 25, Plat. hy 13, Robinia psa 2*	77,3	69,7	57,3	57,6	30,1	551	**7,64**	68,2	55,64	**37,95**
Tuscany, TOS2 (Cala Violina)	*Q ilex 74, Frax.or. 14, Q.suber 8, Q.pub. 5*	60,9	30,4	58,3	59,4	14,7	260	**6,26**	56,2	31,34	**17,60**
Sardinia, Marganai, Ts.2	*Q. ilex, 100*	47,8	56,5	68,7	76,8	18	347	**7,47**	65,5	42,18	**27,61**
Sicily, SIC 1 (Ficuzza)	*Q.gussonei 90, Q.pub. 10*	37	41,5	63,1	57,8	13,9	204	**5,73**	53,4	27,07	**14,47**

Latitudes: Bialowieza 52°40'N-23°50'E; Polana 52°20'N-23°25'E; Asse 50°50'N-4°21'E; Galtur 47°08'N-10°31'E; Paneveggio 46°25'N-11°37'E; Val Paghera 46°15'N-10°23'E; Groane 45°40'N-9°05'E; Colli Briantei 45°41'N-9°20'E; Cala Violina 42°54'N-10°51'E; Marganai 39°19'N-8°32'E; Ficuzza 37°29'N-13°18'E.

Hc = heght of the canopy, v PB = volume of plant biomass, the other signs have already been explained in the text.

Fig. 9.1 Transect of Europe, from Sicily to Poland, comparing some examples of forested tesserae of different types: temperate deciduous (*green*), boreal coniferous (*blue*) and Mediterranean (*orange*). Hc = height of canopy, v PB = volume of plant biomass, CBSt = Concise Bionomic State (see Sect. 5.3.1), M-Th = Maturity Threshold (see Sect. 5.1.2.2. and fig. 5.6)

9.1.2 Evaluation of the Bionomic State of Vegetation

We wrote, in Chap. 5, that an integrated evaluation is essential to check landscape pathology and that the BTC function is very useful, but in some analysis we need to consider both the metastability and the efficiency of vegetation. Therefore, we highlighted the necessity to refer to a new bionomic index, the "concise bionomic state" (CBSt) of vegetation, linked with the state of a vegetation patch, relating to the maturity level of a vegetation phytocoenosis (MtL) and its bionomic quality (bQ).

As exposed in Sect. 5.3.1, the CBSt of a vegetation tessera should be reached considering: (1) the significance of the surveyed BTC of the patch in relation with the "maturity level" (MtL) of its vegetation coenosis and (2) its bionomic quality (bQ) always resulted from the parametric survey; thus

$$CBSt = (MtL \times bQ)/100.$$

After having applied the CBSt [1] to a sufficiently wide pool of surveys, the CBSt correlations with the BTC of each tessera (Ts) have been investigated. Let us show two examples.

The first regards data derived from the Central European Regions of Lombardy and Trentino-S. Tirol, in which 430–440 vegetated tesserae of about 1.5–3.5 ha have been examined in the last 10 years: 272 of natural and semi-natural vegetation (18 formations) and 168 of human vegetation[1] (12 formations). These data are synthesised in the 30 formations ranked in Table 9.1; they are useful to be compared with the surveyed tesserae, giving a first diagnostic base.

In Fig. 3.6 the CBSt from this research has been showed; natural formations were plotted in green, while the human formations in red. A wide variability of CBSt appears at low BTC values (<3 Mcal/m²/year), less in medium and high values. The correlation with BTC is good, having a $R^2 = 0.8953$ for forested tesserae.

The second anticipates a detail of the synthesis of a research carried out with University KU-Leuven (Belgium) on the comparison of the bionomic state of two rural-suburban landscapes in Brabant and Brianza, both at the Northern periphery of the cities of Brussels and Milan. It shows the evaluation of the forest patches characterising the municipalities of Asse (Brabant) and Bollate (Brianza).

The smaller areas delimited in the plot of Fig. 9.2 by the green curves of Bollate in both BTC and CBSt measures give an immediate perception of the best ecological conditions of the forest tesserae of Asse. Note that point 7 represents the average values of the two sets. The use of the CBSt index is in this case very symptomatic, because the ratio BTC Asse/BTC Bollate is limited to 129 %, while the ratio CBSt Asse/CBSt Bollate arrives to 185 %.

[1] Here we intend for human vegetation or human formations the following: cropland, hedgerows, meadows, vineyards, gardens, urban parks, poplar cultivations, etc.

9.1.3 Other Evaluation Criteria and Methods

9.1.3.1 Summary Tables of Evaluation

After having surveyed the various vegetation tesserae composing a LU, following the LaBiSV method, it could be useful to create a summary table, in which the most characteristic analysis related with the percentage of territory covered by each type of tessera will be expressed. As an example, we propose the case study of the municipality of Ronzo-Chienis, near the Trentino Lake Garda, about 1,000 m a.s.l., 13.76 km².

In this territory, 53.29 % forested with six types of woods, we made 18 surveys of about 1.5–3.0 ha each, thus a sampling covering about 6–7 % of the area. The nine tesserae nearest to the average values of BTC and representing the six types of local forest are ranked in Table 9.2. Other tesserae have been added, representing scrubby areas, agricultural lands and urbanised, following the same criteria. The estimation of the average BTC is made with weighted method resulting in 3.84 Mcal/m²/year. The differences between the forest and non-forest contribution to the average BTC may be important in the bionomic evaluation of the LU. Comparing these two values we can see that the average forest BTC = 6.12 Mcal/m²/year contributes for only 53.29 % while the other 46.71 % of the area of the LU brings, in total, only 0.38 Mcal/m²/year.

Anyway, the bQ averages are not good, because the difference between the forest bQ and the human bQ is too small: 59.4 vs. 56.3 %. This fact means that the forest cover is too exploited, confirmed by the presence of few tesserae of good value, as the Ts 7 Mountain Spruce forest, 1,270 m (North slope of mount Biaena) presenting a BTC = 8.76 Mcal/m²/year and bQ = 84.60 %.

A second example, relating with forest tesserae assessed along a transect from North to South Europe (Fig. 9.1).

This transect is quite wide for Europe, going from Bialowieza, Poland (52°40′N-23°50′E) to Ficuzza, Sicily (37°29′N-13°18′E) a latitude gap of more that 15°N. Five tesserae are of Central

Table 9.1 CBSt and BTC (average), with their standard deviations. From the survey of vegetation in Lombardy and Trentino (430–440 Ts)

	CBSt	± SD (CBSt)	BTC	± SD (BTC)
Human formations				
Urban plot + green	2.60	± 2.50	0.33	± 0.23
Poplar plantations	4.85	± 4.20	3.30	± 1.20
Meadows	6.17	± 4.07	0.59	± 0.13
Tree-lined roads	4.38	± 1.92	1.69	± 0.38
Tree-corridors	7.50	± 5.00	3.40	± 0.82
Exotic conifers	5.65	± 2.70	3.58	± 0.69
Fields	6.20	± 2.76	0.79	± 0.13
Parks & gardens	7.05	± 3.42	3.08	± 0.56
Vineyards	8.10	± 4.10	1.75	± 0.33
Robinia woods	10.08	± 5.66	4.59	± 1.18
Oak + exotics	12.00	± 5.50	5.09	± 0.96
Olive groves	17.70	± 5.30	2.40	± 0.34
Natural formations				
Mountain spruce forest	35.62	± 10.76	7.19	± 0.80
Alpine spruce forest	40.11	± 14.84	7.53	± 1.08
Spruce and fir forest	48.52	± 17.35	8.09	± 1.08
Alnus Glutinosa Wood	28.00	± 10.50	7.00	± 1.29
Beech forest	29.30	± 9.40	7.27	± 0.65
Ilex wood	19.50	± 5.00	6.44	± 0.68
Querco-Carpinetum forest	27.90	± 10.5	7.05	± 1.04
Larch forests	21.60	± 8.00	5.86	± 0.93
Pinus sylvestris wood	19.10	± 6.70	5.72	± 0.77
Orno-ostryetum wood	13.70	± 4.10	5.26	± 0.67
Pinus nigra wood	13.74	± 5.04	5.05	± 0.68
Reeds	15.68	± 7.40	1.48	± 0.27
Poplar woods	11.90	± 4.30	5.08	± 0.70
Chesnut woods	11.90	± 6.20	5.13	± 1.11
Beech coppice	16.30	± 3.90	5.77	± 0.54
Willow groves	7.60	± 2.90	4.32	± 0.70
Scrubs	9.62	± 5.94	0.38	± 0.13
Grass & pastures	7.48	± 5.97	0.66	± 0.16

SD = standard deviation

Europe (Austria, Trentino and Lombardy, from about 47.5° to 45.5°N). The differences in BTC are large, from 5.73 to 10.34 but not as the differences in CBSt, from 14.47 to 74.34.

The forest typologies are mainly three: (1) boreal conifer, in Poland and on the Alpine regions, (2) temperate deciduous, in Poland and Lombardy (the two extreme of the habitat of *Quercus-Carpinetum*) and (3) Mediterranean schlerophyllous, from Tuscany to Sardinia and Sicily.

In the example two tesserae have a CBSt >60, but this is far from the European average, is just to show two excellences useful to compare the other surveys. Climatic and human disturbances determine the lower values of the hot Mediterranean forest of *Quercus gussoney* (Ficuzza) and the cold Boreal forest of *Pinus sylvestris, Picea abies* and *Quercus robur* (Polana), about BTC = 5.73 and 5.89 Mcal/m^2/year and CBSt = 14.47 and 21.58, which are very scarce.

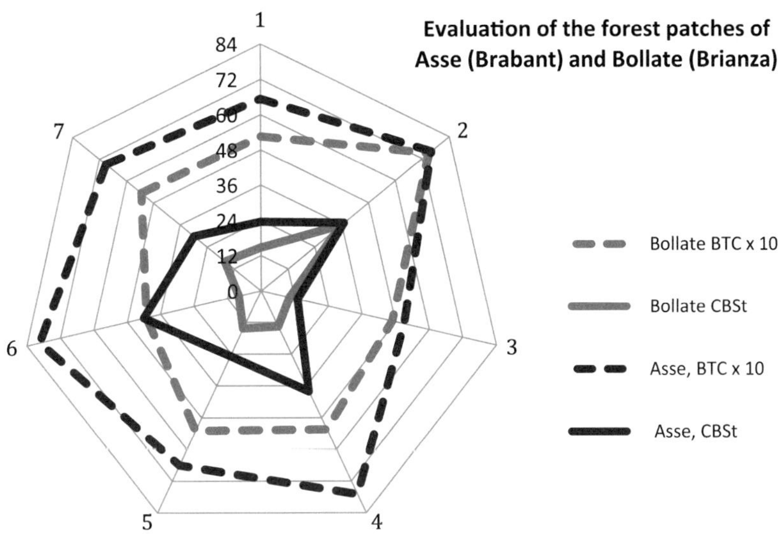

Fig. 9.2 Comparison between six forest tesserae for each municipality plus their average values (7), after the evaluation of their BTC (×10 in the plot) and CBSt. Note the best state of the forests of Asse: e.g. in only one case CBSt >36.0 in Bollate, Vs. 3 cases in Asse

Table 9.2 Example of summary table for the evaluation of the average BTC in a small Landscape Unit, the municipality of Ronzo-Chienis in Trentino near the Lake Garda (1000 m a.s.l.)

Tesserae	Dominant species	Height (m)	bQ (%)	BTC (Mcal/ m^2/year)	% LU	Weighted average BTC
Mixed oaks wood, Ts 6	*Quercus pubescens, Fraxinus ornus*	5.9	46.4	4.44	1.19	0.0528
Beech coppice Ts 2	*Fagus sylvatica*	16.5	49.6	5.46	9.3	0.5078
Beech coppice Ts 3	*Fagus sylvatica*	13.7	53.4	5.44	9	0.4896
Pine wood, Ts 5	*Pinus sylvestris, Fagus sylvatica*	15.2	49.9	5.13	1.2	0.0616
Pine wood, Ts 6	*Pinus nigra*	20.5	56	5.82	1.5	0.0873
Larch mixed wood, Ts 11	*Larix decidua, Fagus sylvatica*	17.5	59.6	6.59	7.4	0.4877
Spruce mixed forest, Ts 15	*Picea abies, Fagus sylvatica*	23.6	70.3	7.17	2.2	0.1577
Spruce mixed forest, Ts 17	*Picea abies, Fagus sylvatica*	28.0	70.4	7.21	10.7	0.7715
Spruce forest Ts 14	*Picea abies, Larix decidua*	26.0	78.6	7.80	10.8	0.8424
Forest, average			*59.36*	*6.12*	*53.29*	*3.46*
Tall shrubs, Ts 3	*Alnus viridis, Pinus mugo*	2.8	72	2.10	8.5	0.1785
Rocky patches, Ts 2	*scattered herbaceous*	0.3	67	0.12	8.9	0.0107
Meadow, Ts 3	*Arrhenatherion sp.*	0.75	52	0.59	5.4	0.0319
Field of cabbages, Ts 4	*Brassica oleracea*	0.5	47	0.66	8.3	0.0548
Field of potatoes	*Solanum tuberosa*	0.5	51	0.76	7	0.0532
Field of cabbages, Ts 7	*Brassica oleracea*	0.5	60	0.87	5	0.0435
Urban with gardens	gardens, vegetables	4.5	45	0.2	2.7	0.0054
industrial and roads				0	0.9	0.0000
Human and rocky, average			*56.29*	*0.66*	*46.71*	*0.38*
Total BTC of LU					100	3.84

Table 9.3 Evaluation of the real ecological character of some tesserae (TS) of Spruce forest in the municipality of Mori, Trentino . CaSp = species character (see text).

L.U.	mB	mB	mB	mB	mB	mB	mB	mB	mB	vG	vG
Survey N°	1	12	21	2	14	11	16	10	17	18	23
area TS (ha)	1.1	2.2	1.5	1	2	1.5	1	1.5	2.8	2	2
SP/TS	60	41	64	32	41	44	41	36	72	50	49
PICEA ABIES (%)	**92**	**63**	**59**	**79**	**63**	**54**	**46**	**62**	**42**	**54**	**57**
PINUS SYLV. (%)	0.1	5	1.2	15	0.7	8	22	5	22	33	31
FAGUS SYLV. (%)	7	20	26	5	25	25	14	33	15	10	5
LARIX DECIDUA (%)	0.8	10	12	5	8	12	14	8	20	0.1	6
PB (m^3/ha)	542.1	736.7	687.9	387.5	534	531.3	343.3	564.1	606.2	497.1	573.6
H canopy (m)	27.5	27.9	27.7	26.6	26.2	27.4	21.9	25.7	29.2	23.6	27.1
BTC (Mcal/m^2/yr)	6.91	7.91	7.89	5.34	7.52	7.35	7.19	7.48	8.35	7.17	7.00
bQ	68	72	72	55	72	71	74	71	82	70	67
M-Th	0.74	0.82	0.82	0.55	0.78	0.76	0.73	0.78	0.84	0.74	0.73
SP Pine (Ps)	2	6	5	3	5	5	4	3	9	6	4
SP/SP* (Ps)	0.17	0.50	0.42	0.25	0.42	0.42	0.33	0.25	0.75	0.50	0.33
CaSp (Ps)	8.51	94.05	48.71	67.82	40.70	91.67	102.74	47.02	231.17	176.41	115.18
CaSp/CaSp*(Ps)	1.82	20.10	10.41	14.49	8.70	19.59	21.95	10.05	49.39	37.70	24.61
SP Spruce (Pa)	11	10	11	6	10	14	10	8	15	4	2
SP/SP* (Pa)	0.50	0.45	0.50	0.27	0.45	0.64	0.45	0.36	0.68	0.18	0.09
CaSp (Pa)	248.29	198.95	214.11	128.73	198.95	264.58	179.15	158.32	260.70	75.60	38.49
CaSp/CaSp* (Pa)	54.22	43.45	46.76	28.11	43.45	57.78	39.12	34.57	56.93	16.51	8.40
SP Beech (Fs)	13	8	9	12	14	14	14	15	19	13	11
SP/SP* (Fs)	0.35	0.22	0.24	0.32	0.38	0.38	0.38	0.41	0.51	0.35	0.30
CaSP (Fs)	73.93	64.56	79.27	61.00	121.70	121.70	100.31	143.04	139.31	83.27	55.92
CaSp/CaSp* (Fs)	15.76	13.76	16.89	13.00	25.94	25.94	21.38	30.49	29.69	17.75	11.92
CaSp/CaSp*(Ps)%	2.53	26.00	14.05	26.06	11.14	18.96	26.62	13.38	**36.31**	**52.39**	**54.77**
CaSp/CaSp*(Pa)%	**75.52**	**56.21**	**63.14**	**50.56**	**55.64**	**55.93**	**47.45**	**46.03**	**41.86**	22.94	18.70
CaSp/CaSp* (Fs)%	21.95	17.80	22.81	23.38	33.22	25.11	25.93	**40.59**	21.83	24.66	26.52

Remember that a synthetic view of the variability of ecological parameters at regional or continental scale represents a good help in the diagnostic evaluation of any biological system.

9.1.3.2 Real Ecological Character

In frequent cases we may have to evaluate some tesserae characterised by plant species dominant as physiognomy, but with an ecological character different from what appears, both for reasons of forest management and for natural local transformation processes. In such cases, what is important is the identification of the real ecological state, i.e. not concerning the appearance of physiognomy, but the systemic features.

In Table 9.3, for example, the comparison among the three possible phytocoenosis (Alliances: *Erico-Pinion, Piceion abietis, Fagion*) in the evaluation of 11 tesserae of Spruce forest in the municipality of Mori is highlighted.

Ecological data, measured using the LaBiSV method on 9 tesserae in the LU-4 (mount Biaena) and 2 in the LU-2 (Gresta Valley), are summarised in the first part of the table. In the second part are ranked the parameters needed to assess the specific characters of Alliance (*sensu* phytosociology) then:

SP = characteristic species found in the Ts for a certain Alliance,

SP/SP * = ratio (%) with SP potential of the same Alliance

CaSp = specifics of the formation in question (in this case following Pignatti [28])

CaSp/CaSp * = ratio (%) with CaSp * (i.e. the theoretical maximum)

The CASP* are marked in the Table 9.3, while the evaluation of Vegetation Formations (S) was obtained according to the following formula:

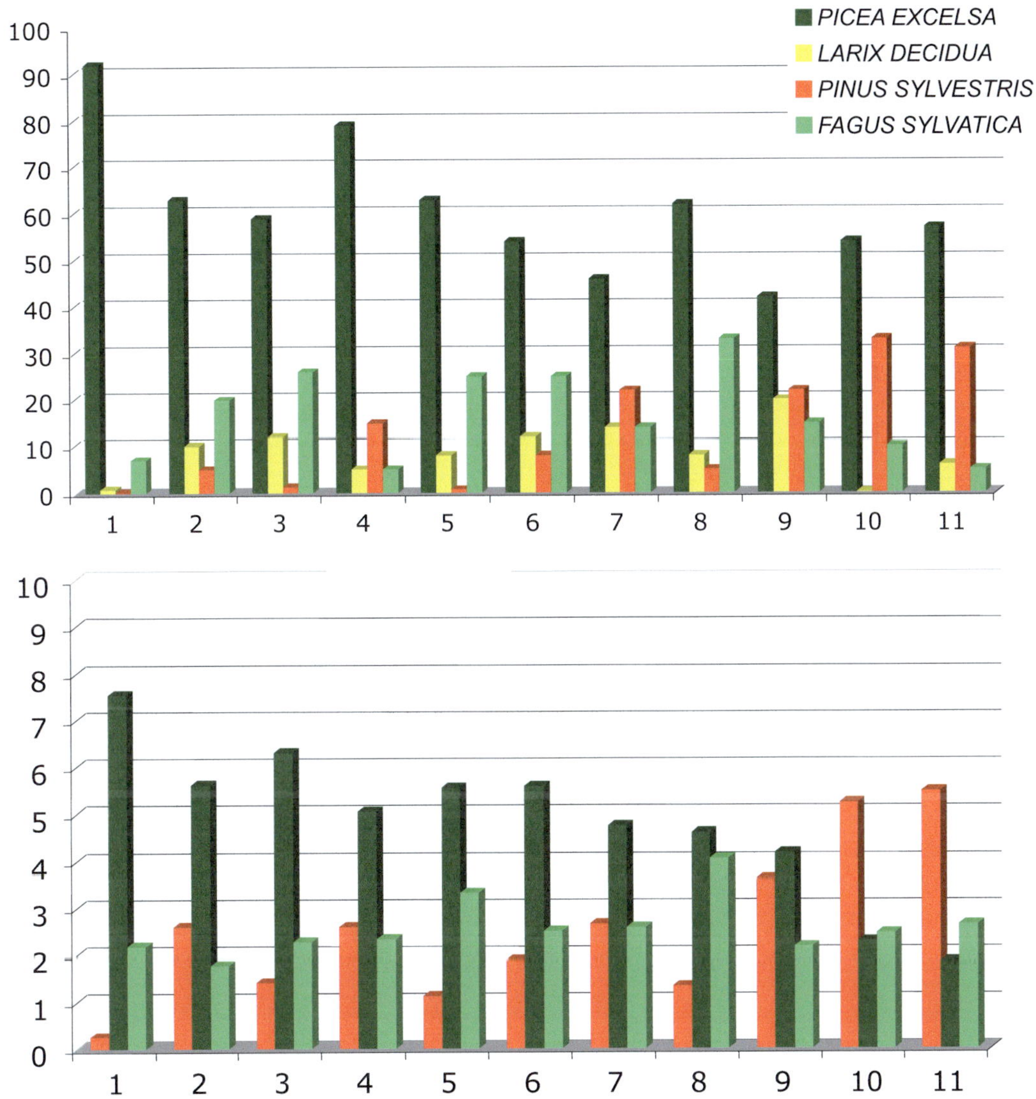

Fig. 9.3 Comparison among 11 Ts of forest dominated by *Picea abies* in Mori (Trentino). *Above*, the % species per Ts, indicating the present Spruce physiognomy; *below*, the systemic characters of the same Ts, after the elaboration of the proposed Eq. (9.1)

$$CASP = [k\,SP/SP^*] \times DM^{1/3}, \qquad (9.1)$$

where:

k = coefficient to take account of companions and/or transgressive SP (e.g. k = 1.1),

DM = dominant in proportion to the vPB (volume of phytomass), high 1/3

In the example, each Tessera (Ts) shows the values of CaSp/CaSp* to assess the specific characters of the three Alliances, and it is observed that, for each Ts, at least one of these values is considerably higher than the others; comparing these values to their sum (see the last three rows of the table) there is a real assessment of the weight of the distinguishing characters that allow us to go beyond appearances. So, as can be seen in Fig. 9.3, comparing these 11 tesserae all dominated by *Picea abies* in percentage, we don't find 11 spruce forests, but only 7 (!), other two being mixed forest of Spruce, Pine and Beech and the last 2 Pine forests.

9.1.3.3 Characterising Transects

The ecological evaluation frequently is called to understand the structural differences of forests (or other formations). For this purpose you use the *transects*, i.e. the sections measured along lines of gradient typical of the woods in question. The survey should be carried out considering a range of a few metres wide (in proportion to the structure of the forest itself) and for a length of at least 2.5–5 times the average height of the canopy. As an example, see Fig. 7.20.

Obviously, this method can also be used for the study of herbaceous communities, as shown in Fig. 9.4, which concerns the halophytic vegetation of the salt marshes in the lagoon of Venice.

The transects were detected on a sandbar to the East of the island of New Lazaretto in the Venice Lagoon. It is observed (a) the presence of an *Artemisietum* (Ac) on the right and the presence of *Spartina* in *Limonietum* (Sm), perhaps due to subsidence. In the second transect (b), we note an *Artrocnemetum glaucum* on a small hillock, but also radiates at the margin with *Limonietum*, which in turn form a band with an ecotonal *Spartinetum*. The synthetic transect added is taken from Pignatti [2] and shows a side of our saltmarsh. We note, however, 50

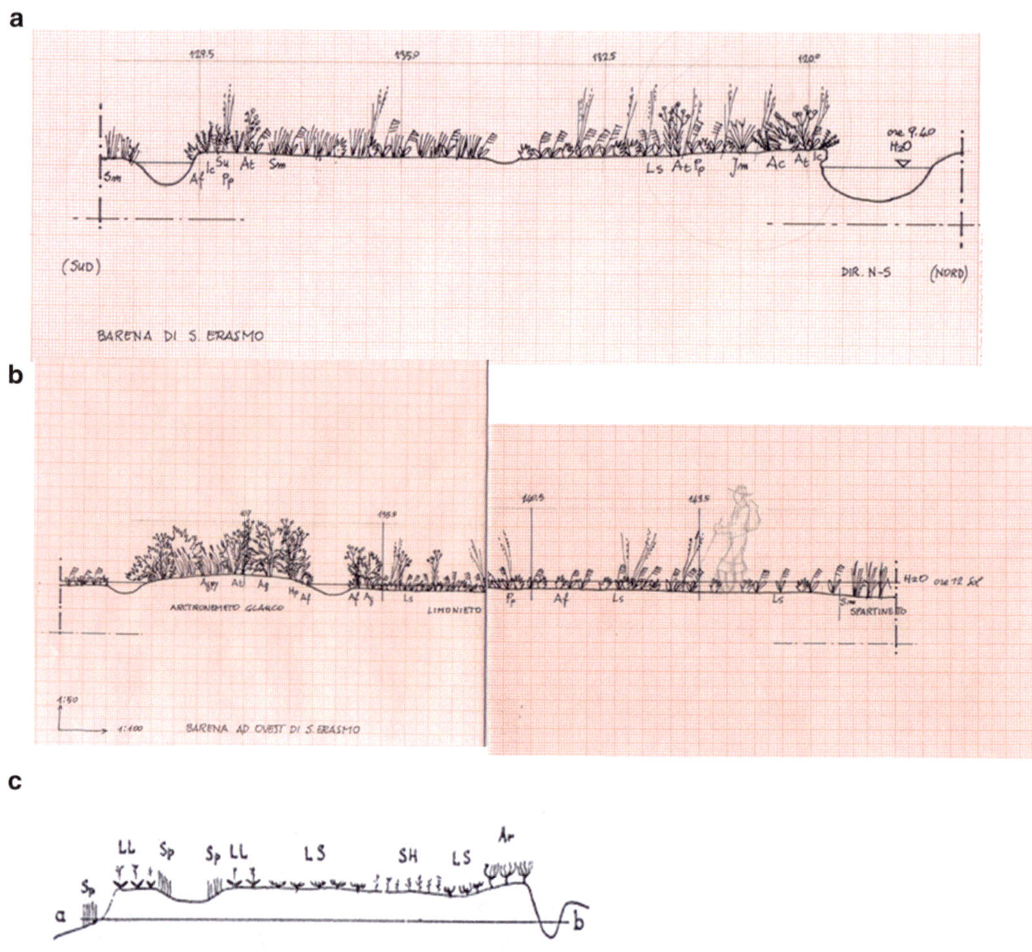

Fig. 9.4 (**a**) Transect on a sandbar near New Lazaretto (Venice). Note the presence of an *Artemisietum* (Ac); (**b**) Transect in the same salt marsh. We note an *Artrocnemetum glaucum* on a small hump (*left*); (**c**) The synthetic transept added is from Pignatti [2]

years ago, sharper distinctions between the types of associations present (compare a/c cases).

9.1.3.4 Control Diagrams

Radar graph could be useful and utilised in different ways: for example to check the ecological qualities of a group of forest types in a landscape unit (LU). In Fig. 9.5 we present the case study of Robecco, in the Ticino River Regional Park, near Milan. The formations concern (a) human vegetation (Arboriculture and Poplar cultivation), (b) allocthonous wood (*Robinia* and Mixed Oak), (c) natural forest (*Quercus-Carpinetum* and *Alnetum glutinosae*). None of the forests reach a good bionomic state, notwithstanding the conservation area (Regional Park).

Figure 9.6 shows a similar graph, for a comparative assessment of human vegetation tesserae, as cultivated corn in the Po plain. This graph is different from the previous one, because on the spokes are placed ecological parameters and the lines show the three LU of reference: Oltrepo, Besate, Adda Sud. This is necessary when normal values are not the same for all the measured parameters. It is noted, however, that even in the case of common cultivation such as corn, there may be unexpected differences between the various LU.

9.1.3.5 Evaluation Parameters of Urban Green

The types of urban green areas are varied, much more than is commonly believed. The principal can be summarised in a dozen types: urban gardens, cultivated sub-urban, private gardens, sporting green, tree-lined avenues, green squares, public gardens, small parks, large parks with semi-natural patches, urban forests, patches of residual forest. For each of these types of green, you can highlight the ecological values able to characterise them (Table 9.4).

The main parameters are as follows:

(a) Surfaces constructed or built-total [%].
(b) Factor of water percolation (BFF),
(c) non-native or exotic plant species [%],
(d) general tree cover [%],

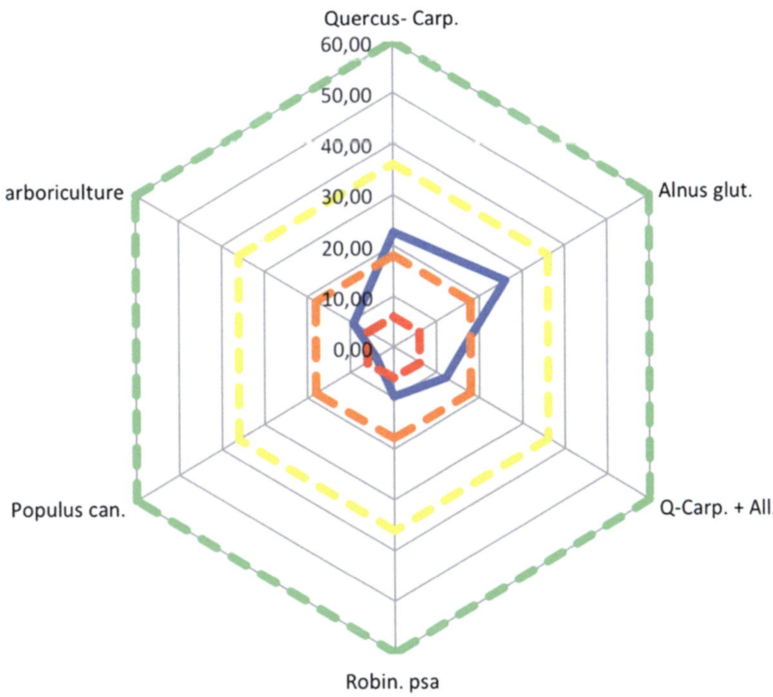

Fig. 9.5 A control diagram to evaluate six types of tree formations in the municipality of Robecco (Ticino River Park) through the ecological index CBSt. The green line is the threshold of the optimum state; the red one is the worst (see. Table 5.15)

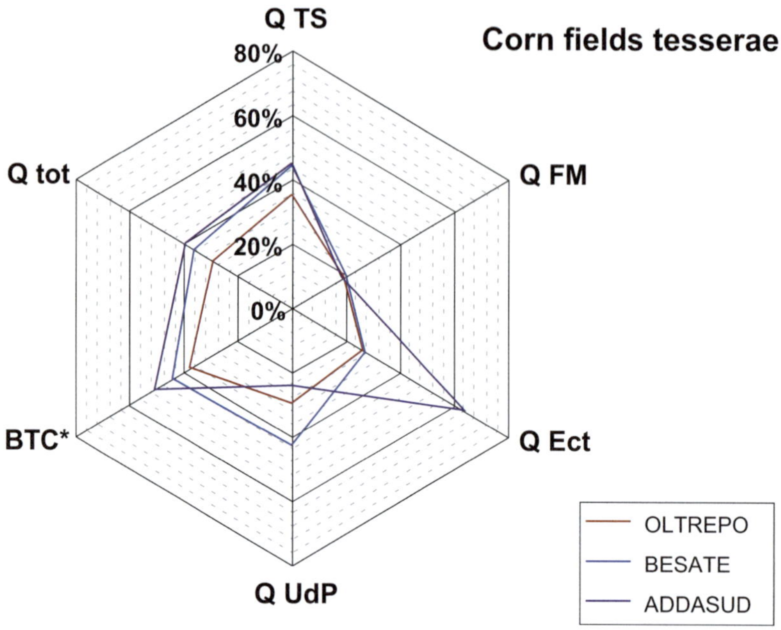

Fig. 9.6 Assessment of ecological status of corn fields in three landscape units (LU) in Lombardy Oltrepo (Pinarolo Po), Besate Ticino, Adda Sud: the main ecological quality (*sensu* LaBiSV) and the relationship of maturity are given. FM = phytomass (i.e. plant biomass), UdP = LU (from Giglio [3])

Table 9.4 Ecological parameters characterising the main types of urban green (UG)

Types of urban green	Ecological parameters Built (%)	Allocthonous plants (%)	Soil over built (%)	Tree cover (%)	Permeability BFF	BTC Mcal/ m^2/year
1 Sport green	65	35	50	10–20	0.30	0.8–1.5
2 Road green	30	30	50	5–20	0.50	0.4–1.5
3 Square green	75	40	60	10–20	0.40	0.4–2.0
4 Vegetable garden	15	20	20	0–10	0.85	0.9–1.3
5 Remnant field	5	15	–	0–10	0.95	0.7–1.0
6 Private garden	20	35	40	20–45	0.80	1.1–3.0
7 Public garden	30	30	50	25–35	0.70	1.3–2.0
8 Urban park	10	20	15	30–40	0.85	2.0–3.0
9 Park with natural core green	2	5	–	50–70	0.95	3.5–4.5
10 Urban forest biotope	0.5	–	–	>60	1	>6.00

BFF index for soil permeability (Ermer et al. [4])

(e) Average (BTC) [Mcal/m^2/year].

To these parameters can be added: (f) the stratification of vegetation and (g) the biodiversity of landscape tesserae. It will be understood, therefore, that the range of values for the urban green is truly remarkable. These parameters are required for both the evaluation and the design of parks and gardens, as you can easily understand.

Note that without comparison with normal ecological states, also in this field it would be impossible to make critical assessments of any kind. An example of application will be illustrated in Chap. 13.

9.2 Human Habitat Evaluation

9.2.1 Evaluation Criteria of Human Habitat

In 6.2 we have already outlined the information that could be used also for the evaluation

Fig. 9.7 The small village of Resy, Ayas Valley (m 2070), in the group of Mount Rosa (4,635 m) of which two huts have been turned into mountain hut for tourism

of human habitat. However, these human components of a landscape have a particular complexity, because they follow ecological laws by focusing on the cultural mediation. For example, farmhouses, villas, castles or churches and monasteries, which are often incorporated into a variety of landscapes, even in semi-natural areas, should be evaluated as architectural organisms with cultural, technical, and especially historical values as well as for their important ecological functions. It is, therefore, necessary to summarise some of the more general evaluation criteria, to be complemented and integrated with the more directly ecological criteria shown in 6.2.4.

9.2.1.1 Consideration on Architectural Nuclei in the Landscape

Farms and rural complexes. These buildings maintain an evident ecological sense when they are still included in agricultural landscapes or at least suburban-rural ones. In fact, there is a clear complementarity between the natural resources and control centres of human management.[2]

[2] This role has been known since the Roman age, as you can read in the volumes of Vitruvius and Columella.

In cases (unfortunately frequent) of encapsulation in metropolitan ecotissues, the evaluation of such items also decreases greatly and change their land uses. Their protection may still be of some importance, especially if there are architectural expressions of historical and aesthetic value.

Even in the mountains, after the abandonment of agriculture or sheep-farming, we find small rural villages that maintain a high environmental value also ecologically (Fig. 9.7), as we can see in the example of the small village composed of the Walser Huts of Resy (Ayas Valley). Their tourist conversion can avoid destruction, even if sometimes modern, out of scale, buildings could be added. In some cases, however, we must recognise that it would be necessary to do everything possible to subsidise farm families to continue to reside on the ground and control the environment.

9.2.1.2 Ville (Historical) and Castles

As already mentioned (see Sect. 6.2.3) the villas have a clear ecological role, with several current relapses, because:

- Represent an inspection by the city on agricultural landscapes complement to the latter, and irreplaceable

Fig. 9.8 The feudal castle of Tolcinasco, province of Milan. Note the deep integration with the agriculture: a fortified farm, annexing the farmer residences and a small church. At the centre is the castle (*left*). Sketch of the Autor

- Can reconvert into recreational ecotopes, when they are incorporated in the city, thanks to their large gardens
- Maintain the role of increasing BTC centres, in landscapes that need intense spots of vegetation (although partly exotic), as in the landscapes of lacustrine or marine coasts
- Manage to combine elements of cultural, historical and architectural value, with a shrewd placement in the environment, usually not degrading

The castles of the villas have less potential, but often represent important structural elements of agricultural landscapes for centuries, with management action similar to that of the villas, but less tied to urban control. A peculiar example of the integration between a castle and the agricultural landscape is shown in Fig. 9.8 in the province of Milan, North Italy.

9.2.2 Evaluation of the Undisturbed Areas of Strategic Importance

Crucial is the following observation: the structuring of suburban landscapes, due to the transformation of agricultural landscapes as a result of continued fragmentation and alteration by humans, usually leaves a mosaic of remnant patches or residual spots [5] of different magnitude and importance. Pay attention to the fact that the *remnant patches* are not residues of woods scattered here and there in the countryside. On the contrary, all those areas are (remnant) patches of the type of the current pre-existing landscape that are located at a suitable distance from the main sources of disturbance (chronic) —even in the case of some old agricultural fallow-distance sufficient to avoid the disturbances.

Furthermore, the term patch can indicate also a band with a direction of the prevailing development. Therefore, they are perhaps best defined as "undisturbed areas". The mosaic of remnant patches should be carefully evaluated by means of the principles of landscape bionomics, if you want to prevent an ever more chaotic landscape itself and if you want to take action to bring greater order, going towards a good level of environmental remediation. The complete and accurate assessment of the ecological mosaic can turn the patches into "undisturbed areas of strategic importance" (or "undisturbed areas of strategic rehabilitation").

The structural pattern of remnant patches is composed of many kinds of elements, characterised by their size and shape [6], a role into the ecotissue and biological territorial capacity of vegetation [7–9], the ability to disturbances

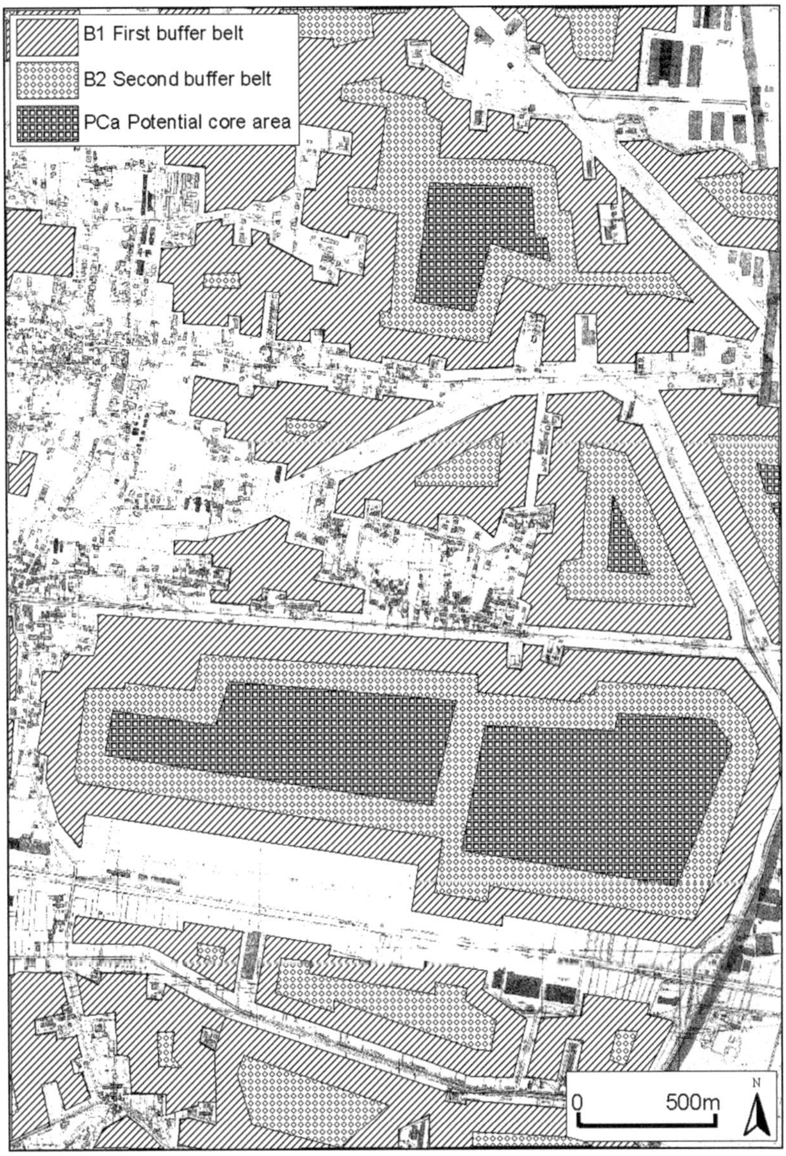

Fig. 9.9 An example of the application of a set of belts to measure the gradient of disturbances exerted on the remnant vegetated patches by their surroundings in the territory of Capannori (Lucca). The potential core area is darker

incorporation, etc. Consequently, the residual patches cannot be considered as indifferent to human activities, but on the contrary their possible compatibility profile and their suitability to the planning and "biotope-und Artenshutz" should be evaluated [4].

This can be done by mapping a defined set of belts, the width of which depends on the characters of the landscape unit; for example, there will be a difference between an agricultural landscape unit and a suburban or open-forest one. In fact, one of the principal functions of a landscape element is its capacity to incorporate a disturbance. So, the width of a considered belt will be minor in the presence of a landscape element with a high BTC value and larger when the landscape element, on which the disturbance is exerted, has a lower BTC value (thus enhancing the extreme difficulty to incorporate the disturbance). In the case of the study for the municipality of Capannori (Lucca) (Fig. 9.9; Table 9.5), a set of belts, moving from the outside to the inside, has been defined as:

Table 9.5 Example of a set of belts for the evaluation of remnant patches studied for the municipality of Capannori (Lucca). Note that the width of the belts depends on the BTC value of the landscape element through which the belt passes

Type of disturbance	BTC value of the landscape element close to the disturbance source	Belt type	Distance from the source of disturbance (m)
Residence buildings, small railroads, small roads, small industries, etc.	Low BTC	Db	20
	High BTC	Db	10
State roads, large industrial plants, main railroads, quarries, etc.	Low BTC	Db	40
	High BTC	Db	25
Airports, highways, dangerous industrial plants, etc.	Low BTC	Db	100
	High BTC	Db	60
Buffer zone perimeters.	Low BTC	B1 and B2	100
	High BTC	B1	60
		B2	80

Table 9.6 Evaluation levels of a mosaic of remnant patches characterised by a five belt set

Number of belts	i < 20 %	i > 20 %	BTCp > BTCm	SR
2 (Db + B1)	A	A	C	D
3 (Db + B1 + B2)	A	B	C	D
4 (Db + B1 + B2 + PCa)	B	C	D	D
5 (Db + B1 + B2 + PCa + Pci)	C	C	D	D

i is the most inner belt mapped, *BTCp* the BTC value of the remnant patch, *BTCm* the mean value of the BTC of the landscape unit, *SR* the strategic role of the patch within the landscape unit; *A, B, C, D* evaluation levels from worst to best

- Disturbance belt (Db), 10–100 m in width
- First buffer belt (B1), 60–100 m in width
- Second buffer belt (B2), 80–100 m in width
- Potential core area (PCa), 100–120 m in width
- Potential inner core area (PCi), >25 m in width

The resulting bounded area inside the third belt can be called the "potential core area". Thus, a remnant patch, even a large one, may lack a potential core area due to its shape or to the presence of disturbances that are too serious. Inside the "potential core area", it is sometimes possible to define a "potential inner core area" with better restoration possibilities.

The second step will be to give to each remnant patch a specific quality value (Table 9.6): the presence of a potential natural or semi-natural core area, a strategic linking capacity to connect other remnant patches, a good level of BTC of its vegetation, will lead to a high value of the patch. For example, two close "core areas" can have different values of BTC, thereby needing different types of evaluation. Analogously, a patch without intrinsic value could be very important as a stepping stone for an ecological network, so assuming a strategic functional role (SR).

Obviously, the strategic roles of remnant patches have to be clearly expressed after a specific study of their distribution in a landscape unit. For example:

- Stepping-stone function for connection
- Meso-climatic mitigation of an intensively urbanised area
- Island for birds temporary refuge (e.g. along a flyway)
- Protection for hydrologic processes (e.g. a spring or resurgence)
- Necessity for maintaining good porosity of the landscape matrix

Depending on their evaluation, remnant patches present different constraints which are reflected in the different planning capacities. To measure these capacities, the main transformation processes need to be checked. For instance

- N: Each transformation improving the naturalness of the patch (e.g. reforestation)
- M: Each transformation bringing the patch back to the original landscape matrix condition (e.g. from abandoned fields to cultivation)
- RG: The creation of recreational green areas (e.g. public gardens)
- HS1: The building of light human structures (e.g. rural roads or canals)
- HS2: The building of heavy human structures (e.g. industrial roads or new technical plants)
- TU: The urbanisation of the entire patch

Referring to the last two Tables 9.6 and 9.7 note that the patch values (A, B, C, D) can be obtained in different ways: for instance, a 3D value can be obtained starting from a 3A or a 3C; analogously, a 4D value can derive from a 4B or it can be a simple 4D. Therefore, the value degrees are generally more than four and they could be indicated, respectively, as 3D(A), 3D (C), 4D(B) or 4D(D).

These evaluation degrees give the possibility to map with different colours the system of patches of a landscape (Fig. 9.10). A map like this can help to study an ecological network and to compare some case studies of transformation proposed by the urban planners of the municipality. Moreover, it can be used to restore the landscape structure and to divide the landscape in different planning subunits.

In conclusion, it should be noted that the preservation of nature need to protect the most important parts of the system of residual patches of the agricultural landscape, to stop the fragmentation and arrive to an environmental rehabilitation of the territory. The methodology presented here is not alone sufficient to reach this purpose, but it certainly can be of considerable utility, together with the principles of landscape bionomics already known in the literature. It should be noted that the system of residual patches should be considered in conjunction with:

(a) The distribution of cumulative impacts, e.g. the dangerous pollution patches around chemical industries or toxic waste dumps;
(b) The system of the river network, to which you usually give the appropriate values for the bands of protection.

Table 9.7 Different planning capacities of the remnant patches

Patch values	N	M	RG	HS1	HS2	TU
A	+	+	+	+	+	+
B	+	+	+	+	(+)	[+/−]
C	+	+	(+)	(+)	[+/−]	−
D	+	(+)	(+)	[+/−]	−	−

Where: + Freedom of transformation; (+) transformation possibility only marginal and only with mitigation projects; [+/−] transformation possibility only marginal and only with mitigation and compensation projects; (−) no transformation possibility

In the study of an ecological network for a municipal or provincial territory remains essential to carry out an evaluation of the system of residual patches.

9.3 Evaluation of the Ecological State of a Region

9.3.1 Evaluation Criteria

Mindful of the words on the importance, in a study of landscape bionomics, to keep in mind the Theory of Hierarchical Systems, which leads us to consider a suitable level of interest for the purpose of our research and, at least, two levels of scale directly related to it —that is, a lower one to analyse the components and explain the status of the level of interest and an upper one to analyse the constraints and understand the meaning of the level we are interested in— in order to analyse the ecological state of a landscape unit (LU) it is essential to compare this state with the higher scales.

For this purpose it is necessary to have an idea of the state and trends of the landscape apparatuses, especially the productive (PRD), of the BTC and of other ecological indicators, such as HH or the SH/SH*, at regional level. On this depends the subsequent stage of processing therapeutic design and the control over the sizing of the tesserae to reforest or on the corridors, in areas of possible expansion, etc.

It is our duty to clarify that in fact an ecological region, or ecoregion, and an administrative region almost never coincide. In the case of the Lombardy

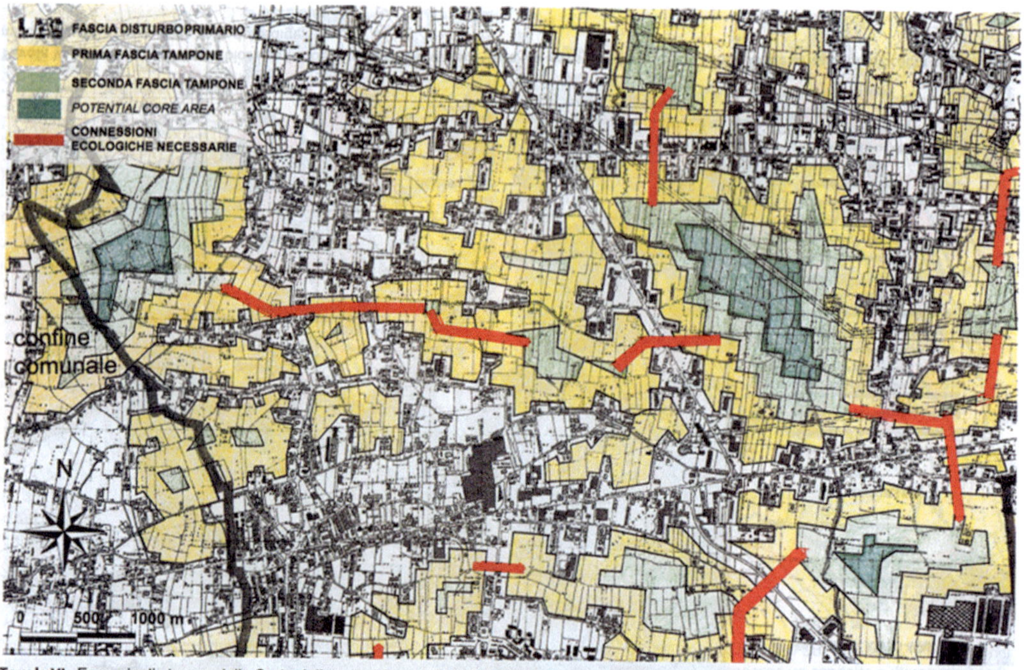

Tavola VI Esempio di stesura della Carta delle macchie residuali (Stralcio dalla Carta omonima del Comune di Capannori, Lu) con evidenziate le connessioni ecologiche necessarie per la creazione di una rete ecologica. Talune macchie presentano forma e dimensioni tali da impedire la

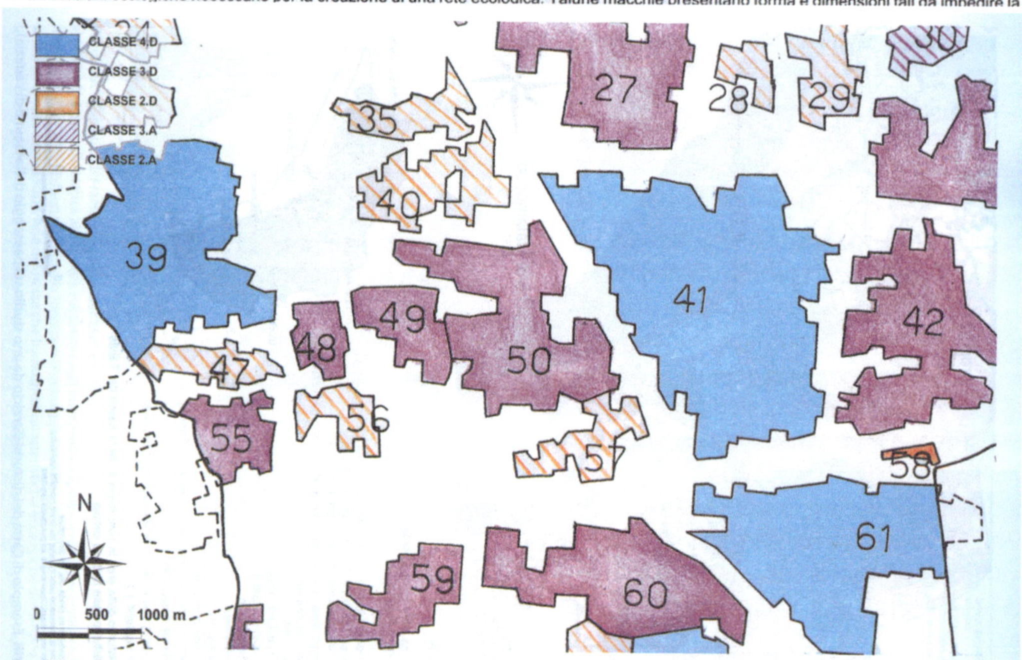

Tavola VII Esempio di Carta di valutazione delle macchie residuali riportate nella tavola VI (Stralcio dalla Carta omonima del Comune di Capannori, Lu). Si noti come, nella valutazione di ciascuna macchia, entrino in gioco non solo il numero di fasce ma anche la qualità della vegetazione presenti, il suo valore di BTC e la posizione strategica.

Fig. 9.10 Example of application of the method of undisturbed strategic patches in the municipality of Capannori. In the *upper part* of the figure, in *dark green* the potential core areas; in the *lower part* of the figure, the spots are coloured according to the assessment: those with greater strategic value are *blue*, followed by *purple* and *orange* ones full (from Ingegnoli and Giglio [9])

LOMBARDY (2008)	Eurostat 09 Km^2	DUSAF, 2010 ha	%	HH	HH	BTC	BTC
Arable land	7.862	740.779	31,03	96	29,79	0,77	0,24
Rice fields		116.310	4,87	95	4,58	1,10	0,05
Permanent crops		36.830	1,54	94	1,42	2,10	0,03
Other cult. & plantation		37.771	1,58	88	1,34	3,10	0,05
Pastures	3.993	112.500	4,71	86	4,05	0,59	0,03
Natural grassland		146.000	6,11	25	1,22	0,63	0,04
Shrubs & herbaceous	406	85.600	3,59	15	0,36	2,80	0,10
Boreal Coniferous		169.700	7,11	4,5	0,28	6,59	0,47
Temperate, broad-lived	6.543	408.502	17,11	9	1,49	5,46	0,93
Mixed forest		2.950	0,12	7	0,01	6,15	0,01
Urban green & sport		31.008	1,30	92	1,14	1,50	0,02
Urbanized	3.039	180.350	7,55	100	7,55	0,10	0,01
Mine, dump, etc.		18.957	0,79	100	0,78	0,10	0,00
Transport		14.447	0,61	100	0,61	0,00	0,00
Industrial etc.		92.302	3,87	100	3,87	0,00	0,00
Inland water & glaciers	1.272	86.896	3,64	10	0,36	0,05	0,00
Rocks, gravel and sand	761	106.815	4,47	10	0,22	0,05	0,00
SURFACE TOT. Ha	23.876	2.387.600	100,00		**60,01**		**1,98**
Population tot.		9.750.000					
HH, total (ha)		1.434.947,6					
SH m^2/inhab.		1.469,5					
SH prd m^2/inhab.		1.071					
SH/SH* [SH* = 1.435 m^2/inhab.]		**1,02**					

Eurostat 2009, based on Corine Land Cover; DUSAF 2010 (Lombardy Region), based on 2007-2009 surveys.
Blue numbers: CRA Forest Inventory, 2007-08; Red numbers: BTC measured with experimental data; orange HH and BTC mean weighted average

Fig. 9.11 Example of evaluation methodology to measure the components of a regional landscape system. Eurostat 2009, based on Corine Land Cover; DUSAF 2010 (Lombardy Region), based on 2007–2009 surveys. Blue numbers: CRA Forest Inventory, 2007–2008; Red numbers: BTC measured with experimental data. Orange: HH and BTC mean weighted average (Attention: commas and dots in the numbers follow Italian way)

region, for example, it includes the ecoregion "oceanic division" (240 Bailey, [10]) and part of the division "Mountains oceanic regime" (M240), both into the Dominion "humid temperate" (200). It is true, however, that the set of Lombard landscapes belong to the above ecoregions and the Lombardy region is configured as peculiar space within the ecoregions. The same scale of space-time reference (Lombardy: 25,000 km^2 and several thousand years) is fully within the meso-macro ecological scales of ecological regions and/or their significant parts. In add, it is also undeniable that the statistics are strictly referred to the administrative regions.

On the other hand, the European subdivision in Regions has been recognised as the best way to measure environmental characters even by the European Administration, as Eurostat, because of the exceptional variety of European landscape systems. Note that small nations can be considered as regions, ecologically speaking. Figure 9.11 give an example of the evaluation methodology to measure the components of a regional landscape system.

When possible, the normal way to measure land use and cover is to refer to the official "Corine Land Cover System", if the territory surface is above 10–15,000 km^2. The differences between Eurostat and DUSAF[3] [11, 12] are in some cases (e.g. forests) very important: the

[3] DUSAF is the database of landuse and landcover in the Lombardy region.

Table 9.8 Synthesis of the bionomic state of the landscape systems of Central European Regions across the Alps (45°to 50°N of latitude)

Central Europe (Alps)	Bayern, 2009		N-Tirol, 2008		Trentino/S-Tirol		Lombardy, 2008	
Landscape elements	km^2	%	km^2	%	km^2	%	km^2	%
Cropland	22,200.0	31.55	110.4	0.87	1,335.0	9.81	8,570.9	35.90
Wood agrarian	723.0	1.03	71	0.56	453.0	3.33	746.0	3.12
Grassland	12,040.0	17.11	3,523.2	27.86	2,030.0	14.91	2,585.0	10.83
Shrubs & herbaceous	470.0	0.67	615	4.86	1,028.0	7.55	856.0	3.59
Boreal Coniferous	16,794.0	23.87	4,337.0	34.30	5,727.0	42.06	1,697.0	7.11
Temperate, broad-lived	7,830.0	11.13	1,084.0	8.57	1,584.0	11.63	4,114.0	17.23
Urbanised, RSD	5,147.0	7.32	412	3.26	387.0	2.84	2,110.0	8.84
Urbanised, SBS	3,377.0	4.80	279	2.21	197.0	1.45	1,257.1	5.27
Inland water & glaciers	1,407.0	2.00	770.4	6.09	416.0	3.06	1,230.0	5.15
Rocks, gravel and sand	370.0	0.53	1,442.0	11.40	458.0	3.36	710.0	2.97
SURFACE TOT. Ha	70,358.0	100.00	12,644.0	100.00	13,615.0	100.00	23,876.0	100.00
Population (million)	13.40		0.95		1.25		9.75	
Human Habitat (HH)	**57.72**		**21.38**		**28.68**		**59.08**	
Ecol. density inhab/km^2	330		351		320		674	
SH m^2/inhab	3,008.4		2,845.3		3,123.7		1,470	
SH prd m^2/inhab	2,312.7		2,675.7		2,518.5		1,071.0	
Carrying capacity SH/SH*	1.81		1.71		1.88		1.02	
BTC (Mcal/m^2/year)	**2.64**		**3.11**		**3.91**		**1.98**	
Latitude of capital city	*48°-15′N*	*11°-30′E*	*47°-15′N*	*11°-25′E*	*46°-10′N*	*11°-05′E*	*45°-30′N*	*9°-15′E*

accurate Forest Inventory by CRA [13] (Italian Centre for Agro-forest Research) is indeed more correct that the Eurostat value. The HH estimations per landscape element are made in analogy to experienced case studies and literature, while the BTC evaluation is referred to personal surveys, for Lombardy and Trentino/ South Tirol.

Elaborating the main landscape elements exposed in the table of Fig. 9.11, we put in evidence these bionomic indicators:

1. The human habitat (HH) in % of the total area of the region,
2. The ecological density of population, referred to HH,
3. The biological territorial capacity of vegetation (BTC), as weighted average,
4. The total HH as measure of the surface (ha),
5. The standard habitat per capita (SH), m^2/ inhabitant,
6. The productive standard habitat per capita (SHprd), m^2/inhabitant,
7. The carrying capacity SH/SH*, where SH* is dependant on the climatic belts.

9.3.2 Case Studies in Europe

Following a similar method, we studied 8 European Regions in a sort of transect from North to South, considering only countries personally visited and at least surveyed in some patches of forest. These Regions are: Denmark, Belgium, Bayern, North Tirol, Trentino/South Tirol, Lombardy, Latium and Sicily [11–17]

If measuring at least 17–18 landscape elements per region (see Fig. 9.11) is necessary to reach a sufficient precision in the values of HH, BTC, SH and carrying capacity, to have a synthetic view of every studied region we reduced these values to 10–11, even because of the differences among the eight groups of data, ranging from 12 to 24. Here we present two Tables 9.8 and 9.9, grouping (1) the Central European-Alpine Regions, from about 50°N to 45° of latitude and 8°30′ to 14°30′ E of longitude representing the Alpine transect: Bayern-N/S Tirol-Lombardy; (2) the Northern and Southern regions, two of the North, from about 49°30′N to

Table 9.9 Synthesis of the bionomic state of the landscape systems of North (50°to 58°N of latitude) and South (36°to 43°N of latitude) European Regions

North to South Europe	Denmark, 2006		Belgium, 2008		Latium, 2009		Sicily, 2009	
Landscape elements	km^2	%	km^2	%	km^2	%	km^2	%
Cropland	27,859.0	63.29	5,020.0	16.44	3,820.0	22.22	7,780.0	30.25
Wood agrarian	4,598.0	10.45	414.0	1.36	1,360.0	7.91	4,075.0	15.85
Grassland	844.0	1.92	10,250.0	33.58	3,094.0	17.99	6,184.0	24.05
Shrubs & herbaceous	1,313.0	2.98	229.0	0.75	820.0	4.77	1,870.0	7.27
Boreal Coniferous	1,801.7	4.09	3,371.8	11.04	3.9	0.02	71.0	0.28
Temperate, broad-lived	2,089.7	4.75	3,827.2	12.54	4,010.0	23.32	1,382.0	5.37
Mediter. schlerophyllous	0.0	0.00	0.0	0.00	2,053.0	11.94	1,746.0	6.79
Urbanised, RSD	2,769.0	6.29	4,229.0	13.85	863.1	5.02	989.0	3.85
Urbanised, SBS	469.2	1.07	2,385.0	7.81	705.0	4.10	717.0	2.79
Inland water & lagoon	1,372.5	3.12	466.0	1.53	293.0	1.70	420.0	1.63
Rocks, gravel and sand	902.3	2.05	336.0	1.10	173.0	1.01	482.0	1.87
SURFACE TOT. Ha	44,018.4	100.00	30,528.0	100.00	17,195.0	100.00	25,716.0	100.00
Population (million)	5.75		10.75		5.90		5.20	
Human Habitat (HH)	**80.93**		**63.22**		**55.59**		**69.84**	
Ecol. density inhab/km^2	161		557		617		290	
HS m^2/inhab	6,195.5		1,795.3		1,620.0		3,453.7	
HSprd m^2/inhab	5,743.0		1,144.3		1,229.3		3,076.6	
Carrying capacity HS/HS*	3.73		1.20		1.16		2.50	
BTC (Mcal/m^2/year)	**1.54**		**1.89**		**2.77**		**1.70**	
Latitude of capital city	55°-41′N	12°35′E	50°-50′N	04°-21′E	41°-53′N	12°-28′E	38°-06′N	13°-21′E

57°30′N and 4° to 15°E (Belgium and Denmark) and two of the South, from 36°30′ to 42°30′N and from 11°30′ to 15°30′E (Latium and Sicily).

Remember that in Europe the climatic gradient is double, not only North to South: a cold triangle having its Southern vertex into the Alps between Aosta-Chamonix is present, from which depart one side to Sweden and another to East Hungary. Around this triangle are disposed the climatic gradient, so that the North-West portion of Europe is similar to the Southern Rhone and Po plains to Croatia.

Table 9.8 presents an evident symmetry. The two piedmont regions of Bayern and Lombardy have a similar HH, respectively 57.7 and 59.0 % of their territory and a BTC lower than the strictly Alpine regions of Tirol: 2.6 and 2.0 Vs. 3.1–3.9. The population density is very high even in N-S Tirol, because of the low HH (21 to 28 %), about the same of Bayern: from 320 to 350 inhabitants/km^2, while in Lombardy the density reaches 674 inhabitants/km^2 especially due to the big metropolitan area of Milan, near the same of Berlin (3.5 to four million people in only 960 km^2). As we can note from productive SH, Lombardy is near the limit of agricultural subsistence, consequently its carrying capacity is about 1.

Table 9.9 presents two regions of the near North and two of the near South. Denmark is one of the most anthropised countries having a HH = 81 %, but only 161 inhabitants/km^2, favourable to export. Its BTC is too low, because of less than 9 % of forest. Belgium is very similar to Lombardy (Table 9.8): same population size, similar ecological density (560 vs. 670), similar BTC (1.9 vs. 2.0) and similar SHprd = 1,144 vs. 1,071. Even Latium has a high ecological density (620), due to the presence of Rome, but a good BTC value (2.77 Mcal/m^2/year). The Mediterranean forests represent about ½ of the temperate deciduous, while in Sicily it is the opposite (large part of shrubs is “macchia wood”). Sicily shows a good SHprd (3,080 m^2/inhab), exporting fruit all over Europe.

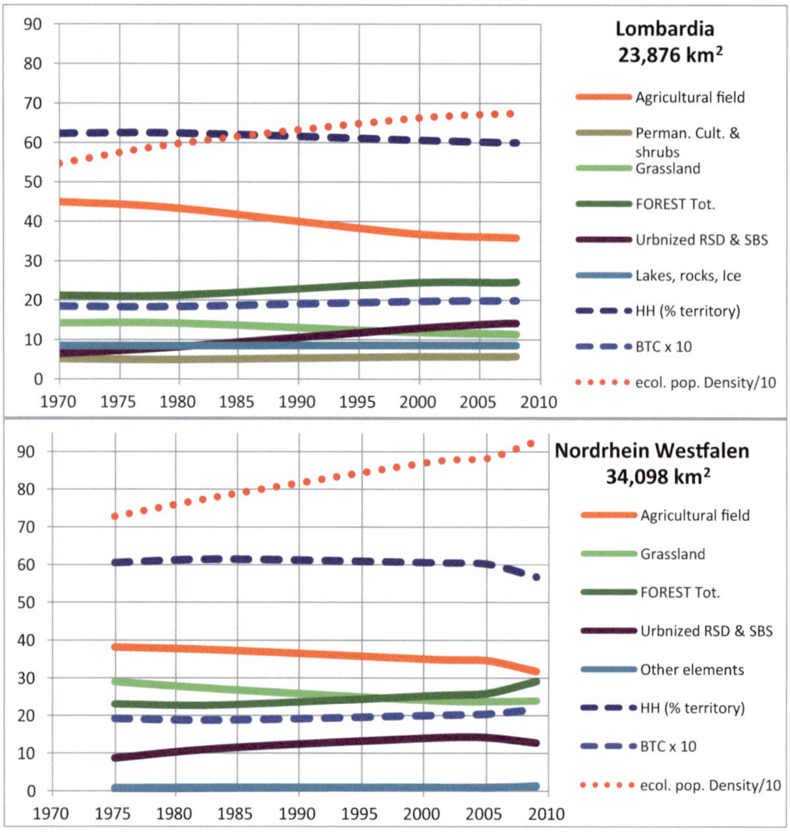

Fig. 9.12 The transformation dynamics of two landscape systems of high regional population density in Europe: the Lombardy Region and the North Rhine Westphalia Region, in the last 33–38 years

9.3.3 Region Transformation Dynamics in Detail

In Figs. 4.7 and 4.8 we exposed the transformation dynamics of landscape systems from a World scale to a local one, since 1880 (a period of 130 years). Now, we have to pass from general processes to more detailed, in order to use these data in diagnostic evaluations.

Figure 9.12 shows the dynamics of the last 33–38 years, comparing two of the most overcrowding regions of Europe, Lombardy and North Rhine-Westphalia. The scope of this study is to consent a critical frame at the upper scale, when operating at local scale, that is at LU level. Remembering that a superior level of system organisation imposes environmental constraints, it results evident that the transformations in these two regions impose exceptionally careful interventions. These regions present many similar processes:

(a) The increasing trend of the ecological density of population, passing since 1975 from 575 to 675 inhabitants/km^2 (Lombardy) and from 720 to 920 (NR-Westphalia)
(b) A quite constant human habitat (HH = 61 %) practically the same in both regions, with a decreasing trend in the last 10 years, more accentuated in NR-Westphalia
(c) A quite constant BTC level, in a pale increasing trend in the last 15–20 years and with the same value: BTC = 1.98–2.02 Mcal/m^2/year
(d) A similar decreasing trend for agricultural fields, which arrive about 35–37 % in 2005, with a recent major decrease in NR-Westphalia

Table 9.10 Synthesis of bionomic data in the dynamics of landscape systems of two European Regions

Lombardy (23.876 km²)	**1970**	**1980**	**1999**	**2008**
Population × 1000	8,140	8,900	9,400	9,750
Agricultural fields (%)	44.90	43.25	36.96	35.80
Grassland (%)	14.16	14.10	11.88	11.30
Forests (%)	21.13	21.20	24.38	24.60
Urbanised RSD + SBS (%)	6.40	8.14	12.65	14.10
Other (%)	13.40	13.30	14.10	14.20
HH (% territory)	***62.30***	***62.40***	***60.69***	***60.01***
BTC (Mcal/m²/year)	***1.84***	***1.83***	***1.96***	***1.98***
Ecological pop. Density (inhab/km²)	*546*	*597*	*660*	*674*
SH (m²/inhab.)	*1,827.36*	*1,674.00*	*1,541.53*	*1,469.54*
SH regional EU (m²/inhab.)*	*1,860.4*	*1,748.40*	*1,535.60*	*1,434.8 0*
Carrying capacity SH/SH*	***0.98***	***0.96***	***1.00***	***1.02***
NR-Westphalia (34.084 km²)	**1975**	**1984**	**2001**	**2009**
population × 1000	15,000	16,400	18,010	17,930
Agricultural fields (%)	38.06	37.30	34.82	31.72
Grassland (%)	28.94	26.94	23.87	*23.87*
Forests (%)	23.00	22.78	25.33	*29.08*
Urbanised RSD + SBS (%)	8.68	11.31	14.08	12.71
Other (%)	1.32	1.67	1.90	2.62
HH (% territory)	***60.51***	***61.47***	***60.47***	**56.70**
BTC (Mcal/m²/year)	***1.91***	***1.87***	**2.00**	**2.16**
Ecological pop. Density (inhab/km²)	*727*	*783*	*874*	*928*
SH (m²/inhab.)	*1,374.95*	*1,277.53*	*1,144.46*	*1,077.66*
SH regional EU (m²/inhab.)*	*1,804.40*	*1,703.60*	*1,513.20*	*1,423.60*
Carrying capacity SH/SH*	***0.76***	***0.75***	***0.76***	***0.76***

(e) A continuous increase of forest lands, with an acceleration in the last 5 years in NR-Westphalia

The increase of urbanisation (RSD + SBS landscape apparatuses) is also similar, from 8 to 14 % (1975–2008) in Lombardy and from 9 to 14 % (1975–2001) in NR-Westphalia, but in the last 8–9 years the tendency results inverted in the German region (12.7), probably due to the very high pressure of ecological people density: 928 inhab./km² in 2009!

A more detailed synthesis of these dynamics is shown in Table 9.10 which permits the measure of the carrying capacities of the two regions. Remember (see Sects. 3.4.1 and 4.4.1) that the theoretical SH* is continuously changing, especially after the "Green revolution", with a decreasing trend of about 11.2 m²/inhabitant per year, so that since 1975 we have been having a decrease in the minimum vital space per capita of about −21 %, vs. an increase of population of about +16.5 %. To these values we have to add the decrease of agricultural fields (about−16.7 to −18.8 %) in the same period, therefore the carrying capacities of these overcrowded regions remained near constant. But the capacity to reduce the vital space is a process without any guaranty and in any case it reduces the natural structure and the bionomic state of the rural landscapes.[4]

Therefore, the main bionomic problem is the fragility of these systems, especially in the case of NR-Westphalia, because a value SH/SH* = 0.76 indicates a heterotrophic landscape system, bypassing even a security coefficient of 0.20. On the other hand, it should be interesting

[4] Remember the possibility to measure these alterations of the rural landscape applying the principles of Landscape Bionomics, as the CBSt index etc.

to observe that the strong increase of forests, due to the agricultural land decreasing, brought to the grow of BTC which, in both regions, tends to reach and remain not far from the average BTC of the Emerged Continents, about BTC = 2.10 Mcal/m^2/year (Tables 4.2 and Fig. 10.11).

As underlined in Sect. 4.2 (cybernetic control of landscape system), the human culture is guided by Nature to maintain the best possible living capacity of a biological system: this fact is another confirm. Moreover, the problem of soil consumption has to be (partially) reduced, because in a widening bionomic balance other priorities must be followed.

9.4 Diagnostic Evaluation of a Landscape Unit

9.4.1 Diagnostic Evaluation Criteria and the Concept of "Normal state"

We must remember and again emphasise that it is the relationship between "pathology" and "physiology" of the systems that allows a diagnosis in the clinical sense of the landscape in question. You have to understand how the system moves from the *state of normality* due to pathogenic stimuli and, with a projection of the information, assess where the damage could rise to the structure and functions at a congruent time. Recall also that the detection of symptoms requires a good medical history, given the importance of history even in the natural processes.

Note that the concept of "norm" has its place in relation to the body —the vital norms of temperature, of blood pressure, of heartbeat and the like— perhaps of anatomy and physiology more generally: being the landscape a living entity, it is impossible to avoid the concept of normality in any diagnostic evaluation.

It is not possible to solve diagnostic problems focusing the interest only on the individual components of a system under test; it is necessary to jump from one side to the other, keeping in mind the entire set. It is recalled, in fact, that a rigid and linear logic is incompatible with complex systems. For this reason we reiterate the importance of the "Method of Painters" [18]: "a correct methodology must proceed from the system as a unit toward its parts".

These methods are defined as "Gestalt", because they are based on gestalt perception "a German word which means form and form-ness, pattern and pattern-ness" [19]. Furthermore, in complex systems, the observation also of one or two symptoms (not repeated over time) may acquire a diagnostic significance, excluding in those cases a statistical approach. Iterative and Gestalt methods like these are based on the sequence:

(a) Observation of the case
(b) Identification of the main components
(c) Construction of a model of our system
(d) Comparison with normal behaviour
(e) Return to observation with renewed criteria

For a diagnosis of the ecological state of a landscape unit (LU) we must base on the comparison of the ranges of *normal*ity for each parametric analysis carried out according to the classic clinical-diagnostic method also used in medicine. It establishes, first of all, what is the type of landscape that is found in the LU in examination, to identify the normal ranges for the parameters analysed.

9.4.1.1 Characters of Normality

To find the *normal* conditions for the most significant ecological parameters, studies have been made on correlations between the parameters of the surveys carried out on 232 forest tesserae in Central Europe (Austria, Belgium, Lombardy, Trentino, Switzerland) according to the principles of the bionomics of the landscape (LaBiSV). This study indicates that the best correlation for forests parameters ($R^2 = 0.899$) is due by BTC in function of bionomics qualities bQ (see Sect. 5.3.1 and Fig. 9.13).

We can see from Fig. 9.13 the limits to a good diagnostic evaluation. In fact the average values of this set are: BTC = 6.26 ± 1.40 and bQ = 59.05 ± 11.99. The dispersion of BTC is ± 22.36 %, therefore for many cases unacceptable. That is why we introduced the CBSt index to measure the bionomic efficiency of a forest.

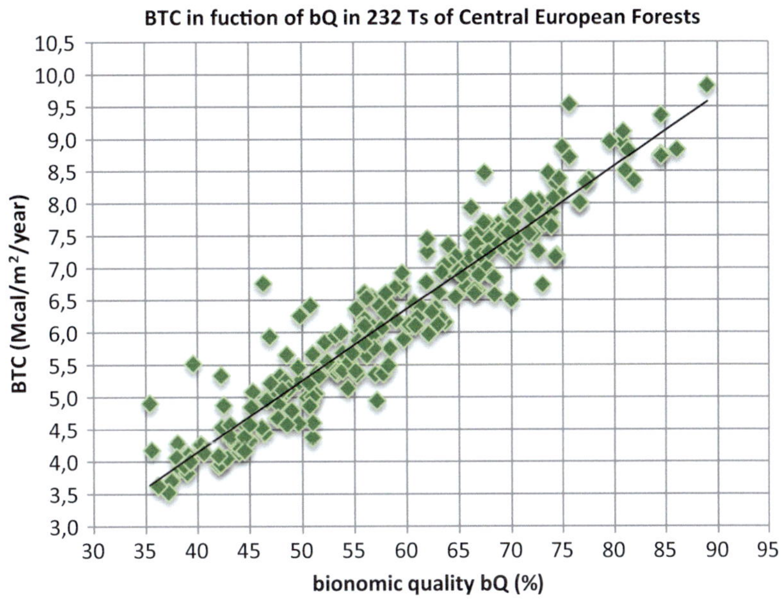

Fig. 9.13 Correlation between BTC and bQ in 232 Tesserae (TS) of Central European Forests (Austria, Belgium, Lombardy, Trentino, Switzerland). $R^2 = 0.899$

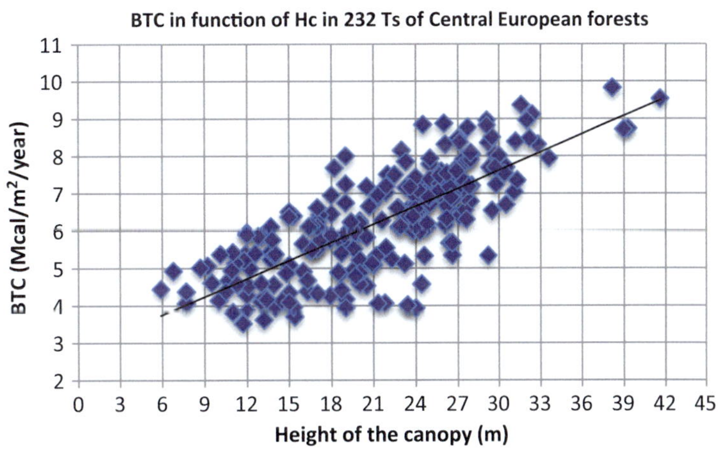

Fig. 9.14 Correlation between BTC and Hc (height of the canopy) of 232 Tesserae (Ts) of Central European Forests (Austria, Belgium, Lombardy, Trentino, Switzerland). $R^2 = 0.581$

Anyway, we may extract some indications, e.g. a forest tessera with BTC = 6.00 cannot reach bQ >64 %: it represents a modest value.

Wider limitations are present correlating BTC and Hc (the height of the canopy), as we can see in Fig. 9.14 ($R^2 = 0.581$). The average values of this set are: BTC = 6.26 ± 1.40 and Hc = 21.52 ± 6.61, so that the dispersion of the BTC remains 22.36 % but the dispersion of Hc reaches 30.72 %. Anyway even in this case we may extract some indications, e.g. a forest tessera with Hc = 16.5 m, cannot reach a BTC value >6.80 Mcal/m^2/year: it represents a modest value.

The averages resulting from the main recorded data on the types of natural vegetation (about 240 tesserae) and para-anthropogenic (about 190 tesserae) in northern Italy have been tabulated in Table 9.1.

To get an idea of the normal range of the most used type of landscape ecological parameters in temperate climates, we need other two tables. The first (Table 9.11) covers the 5 types of predominantly anthropogenic landscapes: *URB* Urban, *SUB.T* suburban-technological, *SUB.R* Suburban-rural, *AGR* Agricultural, *AGR.PRT* protective-agricultural.

Table 9.11 Ranges of bionomic *normality* in temperate landscape types in which human activities prevail

Bionomic parameters	URB	SUB.T	SUB.R	AGR	AGR. PRT
RNT, forest area (%)	4.0–8.0	5.0–12.0	8.0–25	10.0–30	20–55
CBSt of forest	16.5–22.5	18.5–25.5	22.5–30.5	23.5–32.0	27.5–37.5
Allocthonous forest formations (%)	2–5.0	1–4.0	1–2.0	1–2.0	0–1.5
pCA, potential core area (%)	5–10.0	10.0–20	20–35	35–50	50–65
CON, connection ($\alpha + \gamma$)	0.50–0.65	0.55–0.70	0.60–0.75	0.65–0.85	0.70–0.90
Network efficiency	0.6–0.8	0.6–0.8	0.7–0.9	0.8–1.2	1.0–1.8
PRD, agricultural areas (%)	5.0–15	10.0–35	30–55	45–75	35–60
RSD, urbanised areas (%)	25–50	10.0–25	8.0–15	4–9.0	2–6.0
SBS, industrial & transport areas (%)	10.0–20	15–50	7.0–12	4–7.0	2–4.0
HH, human habitat (%)	85–95	80–92	70–86	60–82	42–68
SH/SH* carrying capacity	0.1–0.5	0.4–1.0	0.6–3.0	1.5–9.0	3.0–12.0
BTC, LU (Mcal/m^2/a)	0.40–0.90	0.60–1.20	0.80–1.60	1.00–2.00	1.70–3.50
CBSt, of entire territory	4.0–12.0	6.0–16	8.0–20	10.0–24	12.0–30
g-LM, gen. landscape metastability	3.5–6.0	5.5–8.0	7.5–15	14–24	20–30
LTpE = 10 (HNd/HU) + BTC_{A+F}	0.35–3.0	2.7–6.3	5.7–11.0	10.2–24	22–40
IFF, river functionality	120–180	120–180	180–200	180–200	200–250
Low noise areas (<60 dB) %	40–50	35—45	45–55	50–60	55–70

Landscape types: *URB* urban, *SUB-T* suburban technologic, *SUB-R* suburban rural, *AGR* agricultural, *AGR-PRT* protective agricultural

Table 9.12 Ranges of bionomic *normality* in temperate landscape types in which prevail natural processes

Bionomic parameters	AGR-SYL-TOUR	AGR.FOR	FOR-TOUR	FOR.SN	FOR.NAT
RNT, forest area (%)	25–65	40–75	45–80	60–85	65–95
CBSt of forest	28–38	29.5–40.5	30–41	33.5–45.5	>39.5
Allocthonous forest formations (%)	0–1	0–1	< 0.5	0	0
pCA, potential core area (%)	45–70	60–75	65–80	75–85	>85
CON, connection ($\alpha + \gamma$)	0.65–0.95	0.75–1.00	0.7–1.0	0.8–1.1	0.8–1.1
Network efficiency	1.1–2.0	1.5–2.0	1.2–2.1	1.6–2.2	>1.8
PRD, agricultural areas (%)	15–40	20–50	10.0–25.0	5.0–20	3.0–6.0
RSD, urbanised areas (%)	2–7.0	1–3.0	0.5–2.0	<1.5	<1
SBS, industrial & transport areas (%)	1.5–3.0	1–2.0	<1	<1	<0.5
HH, human habitat (%)	20–40	22–48	10.0–25	5.0–20	0–12
SH/SH* carrying capacity	1.2–12.0	3.5–15	1.5–15	4.0–18	>4.5
BTC, UdP (Mcal/m^2/a)	1.8–4.5	3.00–5.00	3.5–6.0	4.50–7.00	6.0–8.5
CBSt, of entire territory	14–34	16–38	18–42	20–46	>45
g-LM, gen. landscape metastability	27–37	32–43	40–47	48–52	52–55
LTpE = 10 (HNd/HU) + BTC_{A+F}	34–55	45–88	65–140	100–245	>170
IFF, river functionality	200–270	250–270	250–300	261–300	261–300
Low noise areas (<60 dB) %	70–80	65–75	75–85	80–90	>90

Landscape types: Agr-Syl-Tour = agro-sylvo-touristic, AGR-FOR = agro-forest, For-Tour = Forest-touristic, FOR-SN = forest-seminatural, FOR-NAT = forest-natural

The second (Table 9.12) relates to other 5 types of landscape, predominantly semi-natural: *AGR-SYL-TOUR*. Agro-forestry-touristic; *AGR. FOR* Agricultural-Forestry; *FOR-TOUR*. Forestry-touristic; *FOR.SN* Forest semi-natural; *FOR.NAT* natural Forest. The 17 ecological parameters taken into account (from Ingegnoli [20] updated) are listed as follows:

1. RNT, forest area (%), comprise remnant patches in rural areas;
2. CBSt of forest, the average of surveyed tesserae of forest;
3. Allocthonous forest formations (% on forest area);
4. pCA, potential core area (%), see undisturbed areas of strategic importance;
5. CON, connection ($\alpha + \gamma$), see Sect. 7.1.4;
6. Network efficiency, (CON x DCV)—INTF, where: DCV = vegetated corridors, INTF = interference (see Sect. 11.2.3.1);
7. PRD, agricultural areas (%), arable land, permanent crop, pasture, etc.;
8. RSD, urbanised areas (%), residential elements;
9. SBS, industrial & transport areas (%), subsidiary elements;
10. HH, human habitat (%), see Sect. 6.2.4.2;
11. SH/SH* carrying capacity;
12. BTC, LU (Mcal/m^2/a), the weighted average of the entire components of the LU;
13. CBSt, bionomic state of vegetation, referred to the entire territory;
14. g-LM, general landscape metastability;
15. LTpE = 10 (HNd/HU) + BTC_{A+F}, see Sect. 7.2.6;
16. IFF, river functionality, see Sect. 7.1.6;
17. Low noise areas (<60 dB) %, sustainable daily noises;

To these parameters you may add the value of the areas of biological conservation (e.g. SIC, ZPS, local biotopes) or the deviations from the range of normality given by bionomic models, as HH/BTC or g-LM.

It must be recalled that the values that define the normal ranges are not to be considered rigid, in the sense that they, although resulting from experimental cases and theoretical cross controls, may require small local adjustments, given that each case has its own characters. These performance measures, however, can give a pretty good idea of the differences between the types of landscapes most frequently found in our temperate zones (compare Tables 9.11 and 9.12).

9.4.2 Bionomic Diagnostic Indexes

After establishing accurately the type of landscape in examination, and after having processed all the reliefs and the data concerning the set of ecological parameters needed for a diagnostic evaluation of the health status of LU in object, it must subsequently quantify the "distance" of each ecological parameter detected by the standard, giving an appropriate score according to the offset values (in percentage) from the threshold of normality. This means using a diagnostic index (DI).

The verification of a diagnostic index (DI) requires a method based on the "weight" difference percentages from the normal ranges for each parameter found in the various cases under consideration, scraps that are considered in absolute value. These weights can be measured both for appropriate intervals that in a continuous way, as we shall see.

Table 9.13 reports a case of diagnostic evaluation for the changes which occurred in the town of LU1 Mori (Trentino), in which 20 parameters are used for comparison with normal ranges for a rural-suburban landscape.

The reference intervals for the evaluation of the scores are:

Evaluation Scores: 0–10 = 2; 10–30 = 1; 30–60 = 0.5; >60 = 0.

The diagnostic index DI is given by the total score divided by the number of estimated parameters multiplied by the maximum score of 2. In the example of LU.1 of Mori, we see that the ecological status of the landscape was outside the normality already in 1860 (DI = 31/38 × 100 = 81.58), even if only slightly (81.58 < 85). Today DI/2007 is dropped to the value of 66.25, losing almost 19 % of DI/1860.

Wanting instead to use a valuation method of the continuous type, keeping waste to nonexistent ($\Delta = 0$) the value 2, you calculate the values defined by the following linear equations:

- $y = -0.0667 + \Delta\ 2$ if the differences are within the range 0–20 %.
- $y = -0.0083 + 0.83\ \Delta$ if the differences are in the range 20–100 %.

Table 9.13 Diagnostic evaluation of the LU.1 of the municipality of Mori (1860 al 2007)

	Sub–rur	1860	score	1954	score	2007	Score
RNT forest areas	8–25 %	27.6	**2**	27.7	**2**	33.6	**2**
BTC of forest areas (Mcal/m^2/a)	6.00–6.50	4.75	**1**	4.5	**1**	4.91	**1**
Allocthonous forest areas	2–5 %	4.5	**2**	5.4	**2**	6.75	**0.5**
pCA, potential core area	20–35 %	>30	**2**	>25	**2**	20	**2**
CON (α + γ), connection	0.5–0.75	0.46	**2**	0.4	**1**	0.254	**0.5**
Network efficiency	1.2–1.8	0.9 ca	**1**	0.7 ca	**1**	0.435	**0.5**
HH, PRT (HH protective)	2–6 %	0.74	**0**	0.65	**0**	1.45	**1**
PRD (agricultural areas)	30–55 %	53.07	**2**	49.98	**2**	41.19	**2**
RSD, urbanised	8–15 %	2.92	**2**	7.1	**2**	10.47	**2**
SBS transportation & industry	7–12 %	2.5	**2**	3.66	**2**	6.95	**2**
HH human habitat	65–80	58.64	**2**	60.47	**2**	58	**2**
SH/SH* carrying capacity	0.7–2.8	1.17	**2**	0.75	**2**	0.6	**1**
BTC (Mcal/m^2/a) of LU	1.19–1.94	2.27	**2**	2.05	**2**	2.3	**2**
BTC (model offset)	<5 %	3.7	**2**	8.1	**0.5**	3	**2**
LM (lands. metastability)	8.5–13	11.1	**2**	10.26	**2**	11.8	**2**
LM (model offset)	<10	29.5	**1**	32.1	**0.5**	27.2	**1**
LTpE (10 NHd/HH + BTC_{A+F})	1.5–5.3	4.52	**2**	2.53	**2**	2.41	**2**
River corridors artificiality	0–50 %	65	**1**	75	**0.5**	85	**0**
Conservation areas	15–25 %					12.4	**1**
IFF (rivers functionality)	180–200	150 ?	**1**	150 ?	**1**	48	**0**
Landscape type/Total Score		Sub-rur	31	Sub-rur	27.5	Sub-rur	26.5
Diagnostic Index (DI)	**85–100**		**81.58**		**72.37**		**66.25**
Bionomic status	Normal		Altered		Altered		Very altered

Evaluation Scores: ***0–10*** *=* ***2****;* ***10–30*** *=* ***1****;* ***30–60*** *=* ***0.5****;* *>****60*** *=* ***0****.*

- Remains a value 0.00 if the differences are higher than 100 %.

In summary, recalling the physiology/pathology ratio, the main process allowing the diagnosis, we must understand the movement of the system shifting from the normal state. The comparison with normal bionomic parameters, and the evaluation of a diagnostic index (DI), just based on the difference (%) between surveyed parameters and their normal values, will follow. A historical perspective is here indispensable to have a dynamic response. Moreover, the landscape bionomics theory shows the correlations among classes of diagnostic index and levels of LU pathology (Table 9.14). It is necessary referring to five ranges of pathological levels (which, in case, may be divided into two subsets):

I. DI = 1.00–0.85 indicates a *normal* state of landscape health, in which the system remains into a normal homeostatic plateau, thus it may be a bit frail, but substantially it needs only actions of prevention;
II. DI = 0.85–0.65 indicates a state of *alteration*, in which compensation is needed because of instable health, thus interventions and adequate therapies are inevitable;
III. DI = 0.65–0.40 indicates a state of *disorder* both in structure and functions, presenting some physiological damages, thus strong therapies, wide and complex interventions, are needed;
IV. DI = 0.40–0.10 indicates a state of *high disorder* in structure and functions, bringing to harmful effects, thus therapies and interventions must be very strong but even so difficult that the prognosis should be insecure;
V. DI < 0.10 indicates a phase of *extinction* of the L.U., characterised by irreversible and multiple damages, thus degenerative transformations are inevitable.

Table 9.14 Diagnostic evaluation of a landscape after the measure of the D.I

Pathology levels	Diagnostic index (DI)	Diagnostic evaluation	Physiological-pathological notes	Ecological health & interventions
I	0.85–1.00	Normal	Homeostatic plateau	Quite good health, only prevention
II	0.65–0.85	Alteration	Compensation needed	Instable health, some therapies & intervention
III	0.40–0.65	Dysfunction	Some physiological damages	Dysfunction, wide intervention needed
IV	0.10–0.40	Severe dysfunction	Harmful effects	High dysfunctions, difficult intervention
V	<0.10	Extinction	Irreversible damages	Degenerative transformations

Table 9.15 Bionomic "Medical Record" referred to the Landscape of the Venice Lagoon

Basic Territorial Data	• Landscape of the Venice Lagoon and its margins, pertaining to the landscape system of the Po-Adige Plain. • Area: ha 107,974.6 (45.7% of the water basin), population 580,574 (2/3 residents); Littoral marine belt 16.5 %; Lagoon & canals 36.3 %; Terrestrial elements 47.2 %
Preliminary Study on the Ecologic State of the Landscape of the Venice Lagoon (2003-04)	
Syndromes found	• Functional and structural complex alteration with inhomogeneous LU, ψ constant but reduced τ as evident diagnostic sign. • Decrease of the ratio barren/tidal area 0.285 to 0.097 in only one century and wide extension of water surfaces with increase of erosion of bottom and barrens; • Lost of 0.1%/year of vegetated tesserae, decrease of BTC, but already altered since 1900 (-23.5 %); • Apparatuses: (RNT) constant, decreasing ecotonal (ETN) with insufficient values (< 5%), reduced connective (CON); residential (RSD) and subsidiary (SBS) passed from 3.1 to 11.6% in 100 year; productive (PRD) too simplified, some human disturbances not incorporable (e.g. boats and pollution). • Undisturbed strategic patches with non correct distribution and too small < 47% (P.C.A < 24 %).
Diagnosis	• This landscape changed from rural to suburban (HH from 45 to 70%, SH/SH*– 0.5). These syndromes are very serious.
Prognosis	• This trend could bring in 20 years to HH=78-79%, BTC=0.23-0.24 Mcal/m^2/year, SH/SH*=0.41-0.42, going towards a full urban landscape but without an appropriate structure. • The ratio barren/tidal-area should be reduced to 0.1 producing more erosion even in the case of tide control (MOSE), because of wind and boat traffic. • The complex alteration of the system could go out of control. • NB. Therapeutic prescriptions will be resumed in another following table.

9.4.3 The Bionomic "Medical Record" for a Landscape Unit

It stated, lastly, the reasons for the altered parameters in a real *"medical record"*. It becomes crucial, here too, the calibration of the *normal range*, derived from the study of both the models and the observations in the field. Please note that each LU has its own characteristics that must be respected. We will report (Table 9.15) an excerpt from the medical records of the state of the ecological landscape of the Venice Lagoon, drawn up after research commissioned by the Consorzio Venezia Nuova for the Water Authority of Venice in 2003–2004 [21].

In fact, a complete bionomic medical record that can summarise the conditions of the landscape, the analyses carried out, the prognosis and the first therapeutic indications should be divided into the following sections:

(a) *Ecoiatra* (Physician of living systems from the landscape level up to the whole Earth),

i.e. Professional Office that have examined the "sick" landscape (or LU)
(b) *Spatial data*: the classification of the type of landscape and geographical data
(c) *Client and motivations*: the responsible Land Managers, who have entrusted the task to the consultation
(d) *Data provided*: general information and maps provided by the competent authorities
(e) *Symptoms*: alterations identified by the competent authorities and their ecological-technical staff
(f) *Analyses*: the list of all the clinical tests performed by doctors, ecologists, after being charged for the consultation
(g) *History*: reconstruction of past states and study of the historical dynamics
(h) *Diagnostic evaluation*: control and processing, assessment of key ecological parameters, finding anomalies in place
(i) *Syndromes encountered*: a description of malfunctions detected and clinical-pathological classification
(j) *Prognosis*: trends, dangers and damages to the structures and functions of the landscape in question, predictable recovery times

In conclusion of the diagnostic assessment of the lagoon landscape, it was possible to give "Therapeutic indications" after a first detection of the "movement" of the landscape of the Venice Lagoon. It should be noted (see Fig 1.13) that it passed from 28.5 to out of the range of normality: note also the projection of the movement in the two intervention hypotheses (a = red) and (b = blue) and the different distances from the field of normality. Remember that the biological capacity of the territory (BTC) measures the level of organisation of the components of landscape vegetation, while the ratio of salt marshes/tidal area (B/T) allows us to understand in a nutshell, the structural alteration of the landscape in question.

9.5 Evaluation of a Technical Infrastructure in a LU

9.5.1 Comparison Between Impact Vs. Bionomic Criteria

After the landscape bionomic evaluation at a regional and at a landscape unit scales, we have to descent to a more detailed scale. A very frequent case study consists in the need to evaluate a technical infrastructure in a LU. Today, this evaluation is generally elaborated following the concept of "environmental impact assessment" (EIA), therefore we have to understand the differences between two criteria, based (1) on the concept of impact or (2) on the landscape bionomics principles (Table 9.16).

At the light of the theory of landscape bionomics, the interpretation of the table is clear: the traditional impact criterion could bring to ecological errors. But it is important to deeply understand why to go over the evaluation once based on impact.

The definition of "impact", according to the largest Italian vocabulary [22], is "shock", both in the real sense and figuratively. According to the English dictionary [23], the meaning of impact is similar: the effect or the action of one object hitting another. This term derives from the Latin (impactus) and implies a lot of force. Despite today's environmental assessment is still entirely guided by this criterion of impact, it should be noted that, for the most advanced science paradigms, this criterion is considered to be fully exceeded [24, 25]. The revision of the ecosystem concept, expressed by O'Neill et al. [26], already posed some premises to do so, when they defined "disturbance", including anthropogenic—commonly regarded as negative- as the need for structuring and processing of ecological systems, except in cases of destructive (out of scale) disturbances.

Table 9.16 Comparison between the two main criteria of environmental evaluation

	Traditional "IMPACT"	Bionomic COEVOLUTION
Man/nature relationships (**epistemology**)	Selective conflict, therefore, ***impact***	Integration, Cooperation with nature, hence, ***sharing***
Landscape (**definition**)	Aspect of a territory, formed by natural and human components	Biological organisation which characterises a territory, living entity.
Plan/project evaluation	**Measure of the impacts on landscape components**	**Control of the bionomic state of a LU through systemic procedures**
Primary objective	Variation of components before/after an operation	Control of *disturbances incorporation* ex ante, ex post
Methodology	Technical screening of *thematic* components and their factors	Clinical *diagnosis* of bio-systemic units and of their subsystems
Spatial framework	Arbitrary	Following landscape *anatomy*, through identification of the *ecotopes*
Temporal dynamic	Often *not required* or limited to recent past	*Anamnesis* on 2–3 historical periods to understand *system dynamics*
Example: environmental evaluation for a small hydroelectric plant in the Alps		
Analysis	Parameters of the basin, fluvial parameters, quality of river landscape, protection level	Ecological parameters of LU and ecotope, from general to specific and biophysical-semeiotic control
Evaluation	Screening for thematic and cumulative impacts	Diagnosis with integrated ecological indicators (bionomic) for LU and ecotopes
Main discriminant	Technical limits for impact (thematic and cumulative)	Terms of normality for type of landscape and its subsystems
Judgment	Technical and perceptive level of disturb	Bio-system pathology induced by the operation
Purposes	Ecological Balance	Metastable level and health protection of bio-ecological system (and human)
Possible improvement	Proposed mitigation and compensation of the impact	Rehabilitative or therapeutic interventions on the LU or its portion

We observe that the importance given to the old concept of "impact" comes from errors in the hypothesis of Darwinian evolution, in which selection is mainly based on the concept of conflict (impact). On the contrary, today we know that evolution is driven by more complex processes, and that the copy (variation and selection) is to be attached at least to the coding process (rules and meanings), as noted in Sect. 4.3. The conflict is exceeded and regulated by the ability of expression.

It is noted, therefore, the increasing evolutionary importance, for every biological system, of the concepts of cooperation, integration, sharing. Please note that the landscape participates in the process of evolution. And also in this light human action (if not drastically out of scale) should be read as order, but in cooperative sense, in the sense of stimulus to the system. It follows the conceptual error of the other two terms "mitigation and compensation", linked to an exchange of territory = static sheet of paper, "impact", then distortion with shock (pain) to compensate and mitigate; it is necessary to refer instead to "strategic intervention to rehabilitation and therapy", which is performed on a living entity, so dynamic.

What stated above is therefore applied to the landscape, as a direct consequence of the proof that the landscape must be defined in science "as a specific level of biological organisation that characterises an area". This reinforces the need to abandon the old concept of impact, not suitable for checking the status of processing of a given ecological system, such as a landscape unit or an ecotope in the sense of biologic-integrated landscape ecology [27, 8, 9]. The only acceptable meaning of the term impact in landscape bionomics is related to destruction out of range, or "zero event". But in an eco-design the zero event is

Fig. 9.15 The profile of the green mountain at the end of the famous touristic town of Canazei (Fassa Valley, 1430 m) along which will be built a new cable car from Alba-Penia (1500 m) to Rosch Col (2350 m)

precisely what must be avoided. Note, also, that in the design and planning, the abandonment of the old concept of impact produces positive results because the proposed interventions, if properly conducted, can be a stimulus to an improvement of the ecological landscape unit or ecotopes involved. Although preliminary studies, if they are set according to the principles of bionomics of the landscape, they can provide a check-up of the state of ecological health of ecotopes involved and not only an assessment relating the element at play (e.g. a stretch of stream).

Therefore, we insisted that the evaluation of a hyper-complex living system needs the comparison with its field of existence, checking the normal parameters chosen ad hoc for each case, which allow you to express and monitor the behaviour of the system. In the bionomics of the landscape is recalled that, except for a few necessarily descriptive, all other parameters are considered "systemic parameters" already designed as complex interrelations of variables. These parameters have to indicate the real behaviour of a system through ecological functions such as space-time, in which the value expressed by the parameter is, in itself, comparable with that of the higher scales.

9.5.2 Suitability Assessment of a Technical Infrastructure in a LU

Given a case study on the environmental assessment of a new technological project in a LU, e.g. a design of a new cable car in an Alpine LU (Canazei, Fassa Valley, Fig. 9.15), it is necessary to perform a diagnostic assessment of the transformation of the LU in question, to verify the status and capacity of bionomic incorporation of disturbances, following what previously expressed in this chapter.

Established the absence of disease taking place in the case study of Canazei, you can proceed to the *second level* of bionomic control for the suitability of the work, not less important than the first, although necessarily consequent. It is a new diagnostic framework, however, centred primarily on the interference that the proposed work could lead to the adaptive complex system examined. We briefly summarise the parameters considered as the most significant and their values (Table 9.17).

1. Corridor of forest fragmentation, due to the passage of the cable car in the forest, estimated at 28 % of the affected section.

Table 9.17 Environmental assessment of the new cable car of Canazei (funivia Alba-Col di Rosch)

B	EA, parameters	Normal	Prev. 2015	Deviation %	Score
b1	Corridor forest fragmentation (%)	<30	28	0	2
b2	Wildlife disturbance (1/3 fragm.)	<10	9.3	0	2
b3	Infringement SIC-ZPS (%)	<10	0	0	2
b4	Geologic risk (classes)	0.2–0.5	0.93	86	0
b5	Derived building increase (%)	<20	0.18	0	2
b6	Ecological network barriers (%)	<40	60	50	0.5
b7	Vehicular traffic increase (%)	<10	5	0	2
b8	Distance from archaeological area (m)	>50	55	0	2
b9	Distance from historical buildings (m)	>250	670	0	2
b10	Visual disturbances (Visual p.ts)	<35	47.1	34.6	0.5
b11	Ecotope BTC change (%)	<2.0	0.74	0	2
b12	HH change in the LU (%)	<3.0	0.63	0	2
b13	Change of SH/SH* (%)	<5.0	1	0	2
b14	Traffic decrease in neighbour LU (%)	>20	28	0	2
b15	g-LM/g-LM*	0.95–1.05	1.038	0	2
b16	Bionomic status of the LU	0.85–1.00	0.90	0	2
b17	Remnant disturbances incorporation	>0.025	0.075	0	2
b18	Disorders due to the construction site	05–15	8	0	2
B19	**Diagnostic index (DI)**				**86.11**
Evaluation Scores: 0–10 = 2; 10–30 = 1; 30–60 = 0.5; >60 = 0					

2. Disturbance to wildlife communities, following the work, estimated as a percentage of previous disturbance (at least 1/3 of the stretch of fragmentation).
3. Infringement SIC-ZPS, in our case non-existent.
4. Geological hazard, estimated by class of risk, according to provincial legislation (very high here).
5. Increase of derived infrastructure, estimated on the rise in the parking lot of a conventional 15–20 %.
6. Closure of the ecological network connections, which today consists of the corridor "stream with ripe shrub, tree-lined headband of Spruces" estimate in anticipation of the destruction of at least 60 % of the same (transverse section).
7. Increase in vehicular traffic, estimated at around 5 %, as the positive effect of Canazei.
8. Distance from archaeological sites, estimated at >50 m.
9. Distance from the historical and cultural buildings, estimated at >250 m.
10. Visual perceptual disorder, estimated at a higher optimum of 35 %, although not in an unacceptable way.
11. Change of the ecotope BTC, estimated to be around −0.74 %, negligible.
12. Change of HH in the LU, estimated to be around −0.63 %, negligible.
13. Changes in the carrying capacity of the population (SH/SH*), estimated to be around −1 %, negligible.
14. Decrease in neighbouring LU traffic, estimated at around 28 %.
15. Ratio of tolerance of general metastability g-LM/g-LM*, where * is the g-value of the LM curve: good.
16. Bionomic state of LU, deduced from the table above, equal to 0.9 (very good).
17. Residual disturbances incorporation, equal to 7.5 %.

Table 9.18 Eligibility of technological network in a LU, following landscape bionomics principles

Classes	Diagnostic index	Eligibility of technological work	Terms of adaptation	*Post operam* monitoring
I	0.9–1.0	Executable	Congruence of the project	Minimum (routine)
II	0.8–0.9	*Sub conditione* executable	Mitigation and compensation	Only main parameters
III	0.7–0.8	*Sub conditione* executable	Finishing touches to the project + mitigation and compensation	All environmental parameters
IV	0.6–0.7	Postponement of adjustment	Remaking of the project + mitigation and compensation	All environmental parameters
V	<0.6	Prohibition of execution	Ban design	Nonsense

18. Disorders of running the site, only partial and transient, limited to the support, interest only 8–10 % of the way.

The diagnostic values of the index (DI) are obtained, as already seen in the previous Sect. (9.4.2), but the opinion of assessment change: while in the case of control of the pathological state of a LU was referred to an apposite table, in the case of the control of the eligibility of a work of technological intervention (buildings, equipment, infrastructure, transport, etc.), values are easily deductible from Table 9.18.

As can be seen from Table 9.17, the diagnostic index is here equal to DI = 86.1 %, so that comparing with the eligibility requirements (Table 9.18) shows that we are in class II, that is executable with mitigation and compensation, except for contingent geological or technical hazards.

References

1. Ingegnoli V (2013) Concise evaluation of the bionomic state of natural and human vegetation elements in a landscape. Rend Fis Acc Lincei. doi:10.1007/s12210-013-0252-2
2. Pignatti S (1953) Introduzione allo studio fitosociologico della Pianura Veneta Orientale. Arch Bot 28–29
3. Giglio E (2002) Metodologie di valutazione dello stato ecologico dei paesaggi agricoli. In: Gibelli G, Padoa Schioppa E (ed) Aspetti applicativi dell'Ecologia del Paesaggio: conservazione, pianificazione, valutazione ambientale strategica. Atti Congr. Naz. SIEP-IALE, Università degli Studi di Milano Bicocca, Milano
4. Ermer K, Hoff R, Mohrmann R (1996) Landsschaftsplanung in der Stadt. Ulmer, Stuttgart
5. Forman RTT, Godron M (1986) Landscape ecology. Wiley, New York
6. Forman RTT (1995) Land mosaics: the ecology of landscapes and regions. Cambridge University Press, Cambridge
7. Ingegnoli V (1991) Human influences in landscape change: thresholds of metastability. In: Ravera O (ed) Terrestrial and aquatic ecosystems: perturbation and recovery. Ellis Horwood, Chichester, pp 303–309
8. Ingegnoli V (2002) Landscape ecology: A Widening foundation. Springer, Berlin
9. Ingegnoli V, Giglio E (2005) Ecologia del Paesaggio: manuale per conservare, gestire e pianificare l'ambiente. Sistemi editoriali SE, Napoli
10. Bailey RG (1996) Ecosystem geography. Springer, New York
11. Eurostat (2013) Eurostat regional yearbook. doi:10.2785/44451
12. DUSAF (2009) Uso del suolo. Regione Lombardia, Milano
13. CRA-CFS (2007) Inventario Nazionale delle foreste 2005. Roma
14. UBA [Hrsg.] (2004) Workshop CORINE Land Cover 2000 in Germany and Europe and its use for environmental applications. Berlin, 20–21 January 2004, UBA Texte 04/04, ISSN 0722-186X
15. EEA (2012) The European environment: state and outlook 2012. OPOCE (Office for Official Publications of the European Communities)
16. EEA (2010) Corine Landcover 1990-2000 changes. OPOCE (Office for Official Publications of the European Communities)
17. EEA (2006) Towards integrated land and ecosystem accounting. OPOCE (Office for Official Publications of the European Communities)
18. Lorenz K (1978) Vergleichende Verhaltensforschung: Grundlagen der Ethologie. Springer, Berlin
19. Naveh Z, Lieberman A (1994) Landscape ecology: theory and application. Springer, New York
20. Ingegnoli V (2011) Bionomia del paesaggio. L'ecologia del paesaggio biologico-integrata per la

formazione di un "medico" dei sistemi ecologici. Springer, Milano
21. Ingegnoli V (2006) Sintesi dell'esame preliminare del paesaggio della Laguna di Venezia: Cartella Clinica e terapie proponibili, vol 9. VA Valutazione Ambientale, pp 10–18
22. Duro A et al (eds) (1986) Vocabolario della lingua italiana. Itsituto della Enciclopedia Italiana Treccani, Roma
23. Sinclair J et alii (eds) (1988) Collins Cobuild English language dictionary. Collins, London
24. Ingegnoli V (1993) Fondamenti di ecologia del paesaggio: Studio dei sistemi di ecosistemi. CittàStudi (poi UTET-Cittàstudi), Milano
25. Ingegnoli V, Pignatti S (eds) (1996) L'ecologia del paesaggio in Italia. UTET-Città Studi, Milano
26. O'Neill RV, De Angelis DL, Waide JB, Allen TFH (1986) A hierarchical concept of ecosyestems. Princeton Univ. press, Princeton, NY
27. Zonneveld IS (1995) Land ecology. SPB Academic, Amsterdam
28. Pignatti S (1998) I boschi d'Italia: sinecologia e biodiversità. UTET, Torino

10 Landscape Therapy and Territorial Planning

10.1 Landscape Therapy and Ecological Design

10.1.1 Therapeutic Prescriptions

We will begin to emphasise that protection and environmental remediation laws must be applied not only in the light of the needs and current uses of the area, but especially in relation to the different trends that historical events have driven in the current state of the system under consideration, some of which have produced profound changes. So, to stop the degradation of the landscape, it is not sufficient to go directly to the elimination of the causes that have resulted in changing its ecological balance, i.e. a pathological cause, without choosing the search for a new metastable equilibrium. This new balance has to be capable of greater integration between ecotopes and between the types and functions of the components of the landscape, never forgetting the significance of the relationship between man and nature. Man must obey the laws of nature that govern all biological systems and their components (including humans) and must also preside over and safeguard the organisation of life that allows him to live correctly.

Returning to the example of the Venice Lagoon we will note, in this regard, that the road taken by the Venetians since the Middle Ages is correct and is to be continued. Coevolution man-lagoon has created a landscape in a continuous dynamics directed to maintain a substantially stable state, confirming the management role of man. To continue on this road means to retrieve all the planning capacity and protection for integrating and harmonising the activities and the man-made structures with the evolution of the lagoon, but also to use new criteria and media for a new environmental structure.

It is not possible, therefore, to consider the landscape as a "museum," intervening only with protection constraints, with reconstruction of old structures because they were in the past, with prohibitions to build new facilities, bans of using new materials, merely some maintenance. A landscape is a living system! As landscape bionomics has proven time and again, you need a type of action of nature conservation in the active sense, capable of reactivating dynamics of ecological balance in the metastable sense.

This means being able to modify the structure and dynamics of the landscape to bring it to overcome its current dysfunction, because the concept of metastability implies precisely this requirement: a hyper-complex adaptive system evolves in balance within given boundary conditions, but when they change it must adapt. Since it is impossible to return to conditions similar to those of a century ago (it would not even be right, despite the mistakes made, for socio-economic reasons), we do not see other way scientifically and ethically correct.

V. Ingegnoli, *Landscape Bionomics Biological-Integrated Landscape Ecology*,
DOI 10.1007/978-88-470-5226-0_10, © Springer-Verlag Italia 2015

As noted in the previous chapter, in the medical record of the example that we have now presented (Table 9.15), the "therapeutic requirements" are postponed to a second table summarising the criteria for treatment to be prescribed (Table 10.1). As you can see, this table is made so that each set of diagnostic parameters reflects the relative therapeutic criteria [1]. The groups of ecological parameters to be taken into consideration vary in relation to the syndromes encountered in the landscape or in LU examination. In the case of the Venice Lagoon were taken into consideration the following groups of parameters:

(a) *General*: comparison with the characters of higher scales that contain the landscape in question, changes in the type of landscape in the last century.
(b) *Structural*: landscape diversity, grain and contrast, evaluation of the ratio B/T (salt marshes/tidal area), the diversity of LU, hydro-geomorphology.
(c) *Vegetation*: quality (bQ) and BTC and dynamic media in the last century.
(d) *HH apparatus*: HH dynamics in the last century.
(e) *NH apparatus*: NH dynamics in the last century.
(f) *Diagnostic models*: scraps from the models HH/BTC, HH/τ, LM (now updated as g-LM).
(g) *System of residual patches* (i.e. undisturbed patches of strategic interest): chief complaints, blemishes and residual core areas.

In this Table 10.1, therapeutic criteria need to follow an "Executive preparing of criteria for treatment proposed." In fact, it is a different phase from the prescription of criteria, because it must give the inputs to the executive part of intervention, which almost always requires a series of ad hoc projects.

It is noted, then, that for the preparation of executive therapeutic criteria exposed in the above table, the following additional operating steps are necessary:

(1) The preparation of a master plan for setting and controlling the ecological restoration of the lagoon landscape.
(2) The preparation of a master plan of the programming stages of treatment, of the spatial distribution of the elements of the new formation and the preparation of details.
(3) The preparation of a detailed design for excerpts, with control and possible adaptation of the distribution and forms of landscape features newly formed, in general and for LU, after a research on the structural pattern on grain and contrast of ecotopes, on the needed characters of ecotone, etc.

In obvious connection with the above, in the example of Venice the processing of additional criteria of land use planning is to be recommended to the competent authorities for real effectiveness of the Morphological Plan, to be studied with prioritising and incentives (without excluding any constraints). Theoretic criteria have to be inserted in territorial planning.

10.2 Design and Planning Criteria

If we try to define a project, we will see that it is a creative process of intentional organisation of the elements of a complex system, in order to fulfil a set of functions that go beyond the capabilities of individual components (see Sect. 1.3.4, Emerging Properties Principle).

From the definition it follows that it is possible to recognise or infer a design. For example, if from a group of letters scattered randomly on a table (think of one of the many board games to language) comes a phrase, e.g. "Let's go to drink," you can immediately deduce that it is a design, i.e. a configured action, since the occurrence of an order and a purpose. Note, in fact, that there is not a gradual path that leads to a message, given that a single letter does not give you a part of the message and a few extra letters do not give you a little "more than message" (See irreducible complexity).

By studying the landscape it can be proved that it is a coherent system indicating a clear design. "The Earth does not lie," as Aristotle wrote: structure and processes of the landscape reveal, as we have repeatedly stated, a system of

Table 10.1 Landscape of the Venice Lagoon. Therapeutic criteria per diagnostic sets of ecological parameters and suggestions for rehabilitation interventions

Diagnostic parameters and trends of last century		Therapeutic criteria
A- general contest	*Upper spatial scale* • Agr. District L: HH = 67.5; BTC = 1.10; • Water Basin: HH = 75; BTC = 0.90; • Lagoon L (2001): HH = 70; BTC = 0.29 *Upper temporal scale (last century)* • HH 44.8 to 69; SH = 738 m^2/ab • σ = HS/HS* 0.83 to 0.50 • BTC 0.51 to 0.29 Mcal/m^2/year	– Seen ecologic characters, the landscape (L) should be reported to the type "rural-suburban" – HH should reduce about 2 % (or stay) – The average BTC must increase to 0.33 Mcal/m^2/year, in 20 years, tending to 0.40 with planning prescriptions
B-structure	• *Landscape diversity:* ψ = 8.55 to 9.25 τ = 3.83 to 2.95 • *Grain and contrast:* Reduction fractal D; Increase of contrast; • *B/T evaluation* 28.5–11 % • *LU Diversity*: 7 LU, but only 5 open 1 littoral, 3 lagoon, 1 urban; • *Hydro-geomorphology:* Subsidence ≈ 2 mm/year; Erosion; few transport;	– Ecotonal belts are needed to be built in areas of strong contrast – Central urban LU needs semi-natural corridors to connect near LU – The B/T (barren/lagoon) ratio must be inverted to about 18–20 % – New barrens have to be shaped correctly: new exams are needed – Landscape functional diversity (τ) must pass from 2.9 to 3.3 – Submerged prairies have to be widened in the *core areas* – The "Oil Canal" must be isolated with a belt of new barren isles to protect the city of Venice from industrial areas
C-vegetation	• *bQ and BTC:* Woods (50.2; 5.37) Grassland (37.8; 0.55) Arable field (33.5; 0.60) Urban green (27.7; 2.18) *Limonietum* (54.2; 0.46) Shrubs (46.2: 1.30) • *Dynamics 1900–2000* Vegetation tesserae, 47–37.6 % Arable fields, 28–61 %	– Human vegetation must implement their bQ of about 30–40 % and BTC 40–50 % – Natural vegetation have to be protected and shrub belts must regrow – New forest patches and tree corridors have to be planted to reach a correct BTC average – vegetated tesserae have to arrive at least to 40 % of land cover, while arable land have to be reported to 50 %
D- HH apparatuses	*One century dynamics* • RSD, from 0.5 to 4.3 (+) • SBS, from 2.6 to 7.4 (++) • PRD, from 17 to 19.2 (+) • PRT, from 7.5 to 5.6 (–)	– L. apparatuses RSD and SBS can grow not over 12.4 % (today = 11.6) – PRD have to be ecologically enriched – PRT must increase to at least 8–10 %

(continued)

Table 10.1 (continued)

Diagnostic parameters and trends of last century		Therapeutic criteria
E- NH apparatuses	*One century dynamics* • RNT, from 2 to 1.8 (−) • STB, from 14.1 to 8.6 (− −) • ETN, from 7.3 to 4.3 (−) • RSL, from 5.9 to 4.3 (−) • CON, from 3.5 to 2 (−) • EXR, from 3.6 to 4.3 (+) • GEO, from 36 to 39 (+) *Note: recall that the measure of L. apparatuses do not coincide with measures of the types of L. elements*	– RNT must grow over 4–5 % – STB (mainly barrens) must return 10–12 % – ETN have to grow 7–8 %, but it needs detailed exams – CON must double – GEO have to reduce the lagoon water surface through the regrow of new barren isles
A models	*One century dynamics* • HH/BTC scraps −23.5 to −27.5 • HH/τ scraps –4.2 to –2 • LM = 0.85 instead of 1.31	– The landscape was altered since 1900 and it is not completely natural – BTC and τ: see Ingegnoli, 2002
G- system of remnant patches	*Today*: • Area of Disturbances: 53.4 % • Remnant patches, 46.6 % of which ½ buffers belts • *Core areas (inner)* 15.7 %	– Reduce the barrels wetland hunting and especially in the *core areas* – Prohibit illegal fishing and motorboats traffic in some canals

Note **HH** in % of the entire landscape; **BTC** in Mcal/m^2/year; 1 s = period 1900–2000 (±2 anni); **B/T** = barren area/water lagoon surface; b**Q** = bionomic quality of vegetation (LaBiSV) exceptional quality. Even systems established by other natural components may exhibit a design: for instance, every level of biological organisation, e.g. a cell. In fact, the basis of life results in a complex set of functions among which the most typical is the decision-making control, just a project activity.

On the other hand, one should not think a project be an obstacle to a more optimal result: often it is observed that you could do better. So the design does not eliminate the concept of evolution. It deletes however, at least as an omnivalent principle, the interpretations of the evolutionary process as a mere gradually, guided by the case (see Sect. 4.3). It is important to try to understand the organisation (i.e. project) that occurs in a landscape unit (LU) and within its components parts, because to intervene with plans and designs aimed at environmental remediation means having to follow criteria not opposed to such projects, but compatible with it.

It must, however, emphasise that there is no rigid and mechanical method for design, which—on the contrary—is based on all creative and intentional processes. Similarly to what happens for the construction of the models, you need to follow later stages of approximation, often iterative and with Gestalt method. Here the "Method of Painters" mentioned by Konrad Lorenz [2] is to be understood quite literally, since drawing is the heart of projecting and identifies with it (in fact: in Latin to design (deformare or delineare) means to project as in English). Specifically, design demands to proceed from the general to the particular (often by jumps), organising the creation and composition by means of more or less technical drawings gradually enriched with natural or man-made thematic parts, getting to form a system capable of responding to the functional needs of goals for which you have to intervene.

The design of the territory has been developed since the beginning of civilisation, especially after the birth of agriculture (8,000 years B.C.), given that the knowledge about nature and territory were essential for the planning of human settlements, although at the empirical and intuitive level. With the progress of society, already the Roman agronomists had a concept of design of the interventions on the territory not lower to the architects one (see Columella, Vitruvius etc. I–II century [3, 4]), as shown, for example, from "centuriation plans." But only in the eighteenth and nineteen centuries a more comprehensive and scientific approach has started. The most influential teacher for the landscape design was perhaps the American Frederik Olmsted, agronomist and engineer [5], designer of the great Central Park in New York, which at the end of the nineteenth century founded the Graduate Course of Landscape Architecture at Harvard, where Richard Forman [6, 7] still teaches, one of the founding fathers of landscape ecology.

Today the territorial design is carried out in the field of architecture and engineering (territorial and urban plans), and agriculture and forestry (forest and agricultural plans). Only recently, after the European Landscape Convention (ELC) (2000–2006) also naturalists have entered the ranks of the designers of the area [e.g. Ecological Networks, Strategic Environmental Assessment (SEA), Conservation and Protection Plans]. In a form of "base ecologist," a Naturalist should also be formed to read the significant signs of environment and to collaborate with other specialists, in order to then properly design the necessary interventions.

10.2.1 Territorial Planning and Ecological Methodologies

The first modern methods of eco-design, or "Design with Nature" begun in 1968–1969, primarily due to the studies of Ian Mc Harg [8], also cited by Eugen P. Odum in his famous treatise "Fundamentals of Ecology" [9]. These methodologies were divided into four main phases, as follows:

(a) Ecological analysis of the site (*site analysis*) with the use of "overlapping maps."
(b) Professional qualification for the objectives of intervention (*capability*).
(c) Responsiveness better (*suitability*).
(d) Choices feasibility (*feasibility*).

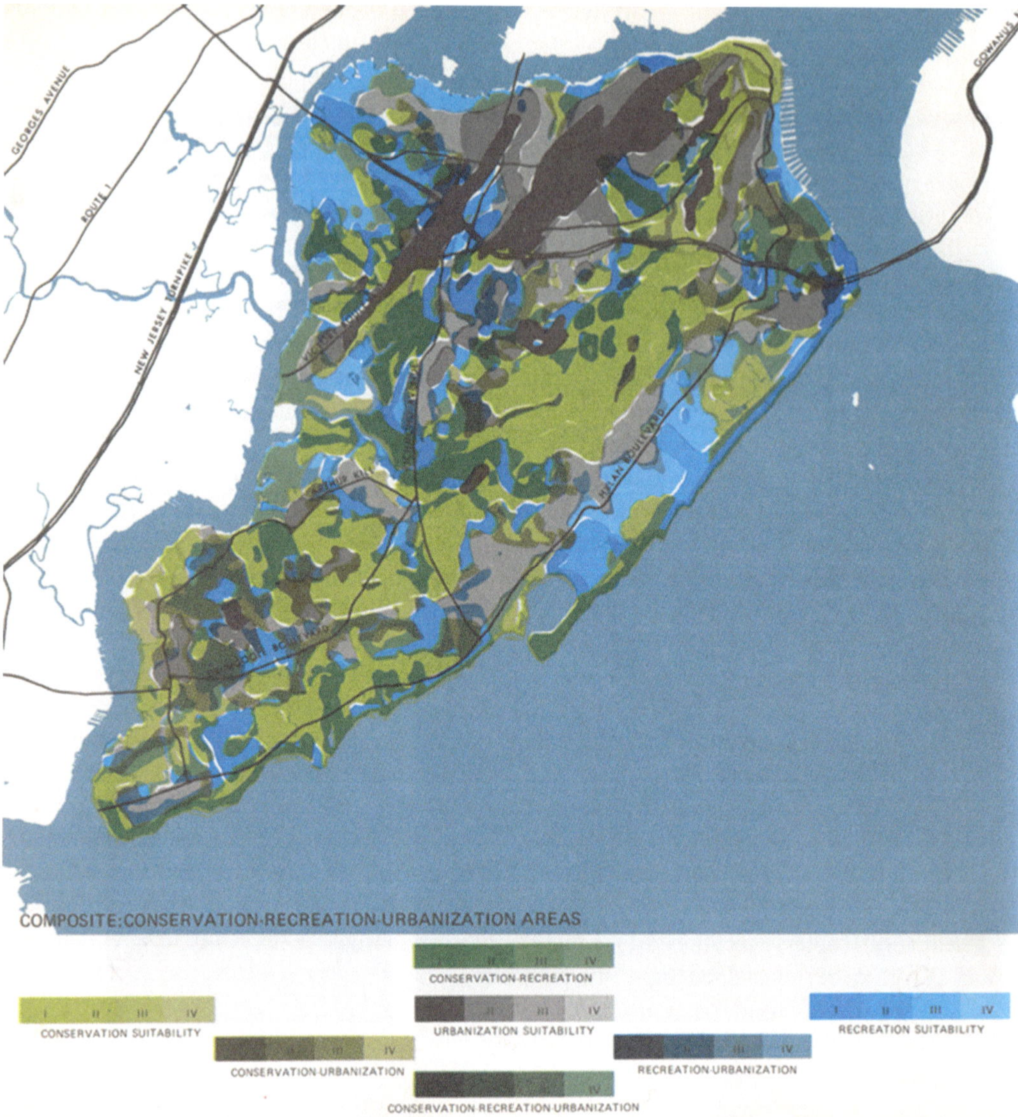

Fig. 10.1 An example of the method of Mc Harg to individuate conservation, recreation and urbanisation areas, near New York. From [8]

Mc Harg demonstrated the possibility of identifying on a relevant map (Fig. 10.1) the areas more responsive to interventions of (1) conservation of nature, (2) recreation and green areas, (3) urbanisation, highlighting with three different colours (each type ranging on 4 tones) following combinations of vocation of the territory:

1. *Conservation suitability*.
2. *Urbanisation suitability*.
3. *Recreation suitability*.

to which were added the four intermediate correspondences:

1–2. *Conservation-urbanisation*

2–3. *Recreation-urbanisation*

1–3. *Conservation-recreation*

1–2–3. *Conservation-recreation-urbanisation*

This method, of course, is not limited to the stages of ecological planning, but continues to the normal development of a draft spatial plan, however, refraining from entering the construction phases.

More recently (2004), as underlined in Sect. 1.1.1, the ELC (European Landscape Convention) has been approved "to promote landscape *protection, management and planning*, and to organise European co-operation on landscape issues." No doubt this is an important goal, but the ELC definition of "Landscape" is not an advanced one: "The "landscape" means an area, *as perceived by people*, whose character is the result of the action and interaction of natural and/or human factors." Consequently, it is very difficult to recommend a territorial planning process based on Landscape Bionomics (LB), as we will see.

At the light of these considerations, it becomes evident that planning and transforming a landscape following the ELC definition is in contrast with the bionomic definition of an "adaptive complex system of biogeocoenosis *which is a specific biological level,*" which requires completely different methodologies and responsibilities. Moreover, being the essential characters of the landscape not limited to spatially dependent processes, but especially to the *intrinsic biological characters* pertaining to this level of life (structural and dynamic), the LB School have to consider:

- New complex integrated functions (e.g. biological and territorial capacity of vegetation; human habitat capacity evaluation, etc.).
- New methods and new applications (e.g. new evaluation of human habitat, new survey of vegetation, etc.).

On the other hand, landscape planning and assessment is a very complex process, which implies relationships among different components, as physical, wildlife, cultural, economic, political dimensions of space. The goals depending from the components may be widely contrasting and have to be accurately evaluated. The process of planning is mainly dependent on the scientific approach, as mentioned before: to manage a complex system a deeply understanding of the system itself is needed. That is why we insisted on the bionomic definition of landscape (LB).

Some general landscape models may help to choose divers approach: for instance, if we define the landscape simply (1) "a perceived geographic area on which interacts natural and human factors" (ELC), we have to build a model with *equivalent* components, because no one is supposed to prevail. On the contrary, if we define the landscape (2) "a specific level of life organisation, with a peculiar behaviour" (LB), the model cannot have equivalent components, because its behaviour, its needs, must be respected even in the presence of man.

For this purpose, it could be interesting to follow the appeal from Linehan and Gross [10] based on a model reported even on a work from the Leuven University (Marcheggiani et al. [11]), named "the landscape goal triangle" (Fig. 10.2).

The first triangle is just the "goals" one: (a) *understanding* matches well the "legibility" and the search for appropriate indicators and metrics; (b) *gaining* refers to actions letting the landscape produce goods and services; (c) *expressing* refers to the way in which our action in the landscape come into reality.

The second triangle is the inner (blue) and it is called the "diagnostic" one: (1) *patterns* refer to structure, spatial configurations, etc.; (2) *functions* refer to provisioning, regulating and socio-cultural activities; (3) *values* refer to significances connected by society to the open space.

The third triangle is the intermediate framework (red) and may be called the "context" one: (1) *spatial context* refers to the relations of the LU with the entire landscape; (2) *political context* refers to the sets of plan of wider areas; (3) *cultural context* refers to the sets of values in a broader area.

This model could be interesting, but without other indications it is supposed to be related with the first (ELC) definition of landscape, in which every component may be equivalent to the other. It could be thought in analogy with the basic components of sustainability, the famous "three E model" (Economy, Ecology, social Equity). In these equal triangles, the concept of ecology and its parameters are considered, but not in the proper way, because this model does not represent a complex living system like the advanced (LB) definition of landscape, together with its intrinsic behaviour.

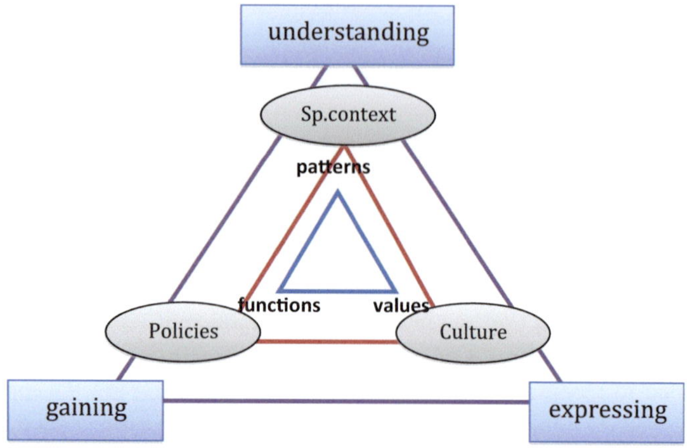

Fig. 10.2 The "landscape goal triangle," composed of three nested triangles accounting for different aspects in order to read, understand and project landscapes

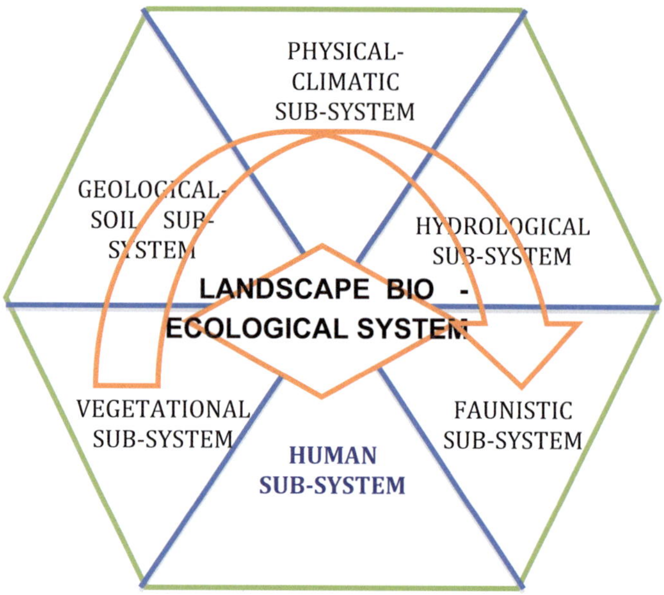

Fig. 10.3 A general model representing the main subsystems which form a complex adaptive landscape system as defined in the more advanced way: the landscape as a bio-ecological (bionomic) system, or living entity. Note that the human sub-system is expressed through a triangle

If we want to plot a general model available to express the landscape as a bio-ecological system, or living entity, we can plot Fig. 10.3. A model like this one, representing the main subsystems which form a complex adaptive landscape system, may be plotted as an hexagon. In fact, its main subsystems have to be at least: (1) physic-climatic, (2) geologic-pedologic, (3) hydrological, (4) vegetational, (5) faunistic, (6) human.

Note that in the hexagon the Human "system" is a subsystem that can be expressed, as any other, through a single triangle: therefore, it is not correct to consider this part as the entire, at least without adding new signs.

In Fig. 10.4, we show an example of the possibility to express the LB triangle model of landscape, adding some signs to the former equal triangle. The orange rhombus around the "spatial

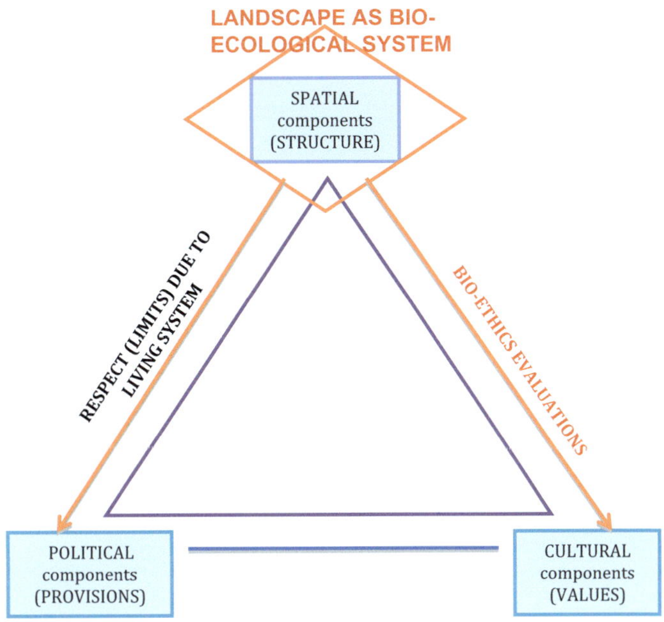

Fig. 10.4 Example of a LB model based on a triangle. The rhombus around the "spatial components" means that these parts are contained into the complex bio-ecological (bionomic) landscape system. Therefore, it is necessary to express at least some constraints towards the other main goals (political and cultural). Note that these limitations are involved with environmental ethics

components" means that these parts are contained into the complex bio-ecological (bionomic) landscape system. But even the human components (political and cultural) are integrated in the complex landscape, so it is necessary to express at least some constraints towards these other main goals. Two arrows, starting from the rhombus, may help for this need.

The first limitation derives from the respect due to a living system: it is possible to derive provisions and socio-cultural activities, but *having care* of the landscape itself. An analogy derives from the note process "host-parasite," in which the dead of the host brings inevitably to the dead of his parasite. But, really, we can express more, because we have to underline that when we recognise a living system, we always find ethical problems. Ethics inquires justice in human actions against the environment. In fact, even the second limitation, towards culture and its values, must be referred to bio-ethics. Remember, again, that we have to study the *landscape pathologies*, but also their *influence on human health*, which may be dangerous *even in absence of pollution*.

An analogous exhortation to landscape bio-ethics should be done regarding the *Strategic Environmental Assessment (SEA)* processes, otherwise dependent from the equal triangle model, which, in its applications, is dominated by economy.

10.3 Landscape Planning: A New Direction

10.3.1 Current Landscape Planning and Assessment

A current planning process generally follows a scheme (or flux diagram) similar to the one shown in Fig. 10.5, linked to the "equal triangle model." The guide discipline is "urban and territorial planning," which can be marginally implemented by economy and ecology. Hence, together with the SEA, this planning is largely practised by architects and engineers, technicians not sufficiently skilled in biological disciplines.

At LU scale, the scheme is quite simple: a set of human and natural components is analysed at present (ex ante), its inputs contribute to the

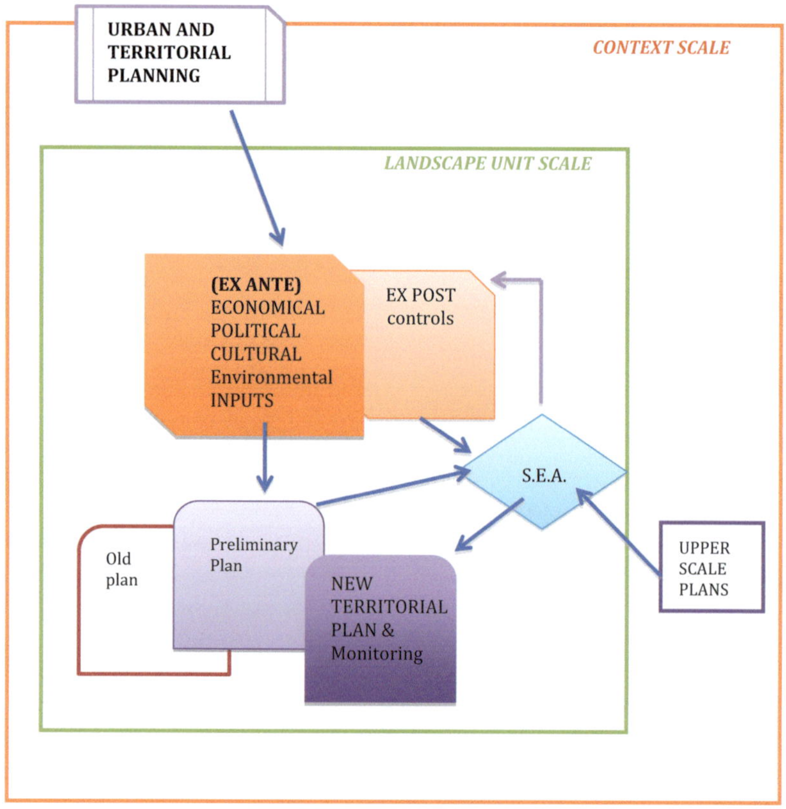

Fig. 10.5 Flux diagram referred to landscape planning processes, as mainly practiced today. Note that environmental aspects are considered, but they are not the leader ones. Moreover, historical-ecological dynamics of the system are not (or insufficiently) evaluated

design of a preliminary plan. This draft plan is controlled through SEA methods, evaluating potential "ex post" impacts and upper scale plan adequacy. SEA results, correcting the draft plan, allow the design of the final plan and its monitoring.

Thus, after the analysis of the existent situation of the landscape (territory), a preliminary plan is composed. As already underlined, the environmental aspects are considered, but they remain limited to pollution levels, recreational services and rare biotope preservation. Functions are first of all intended as socio-economics. No adaptive complex system and no one of its emergent properties are considered. Rarely (or not properly) is even evaluated the historical-ecological dynamics of the landscape system.

Moreover, the LU are normally intended as administrative: so, they very often coincide with the entire municipal territory, falsifying the environmental parameters, e.g. the proportion of urban areas, which should be evaluated following bionomics principles and methods. For instance, a municipality may present only 3.5 % of urbanised area, showing no problem to enhance large built expansions: in reality, applying LB principles, its territory may be composed for examples by two LU, one on the mountain slope, the other in the valley. The first LU is covered by forests and has 0.5 % of urban areas; the second, smaller and mainly cultivated, has 17 % of urban areas. Therefore, the amount of new expansions should be drastically restricted (see the example of Tesero, Sect. 12.2).

The preliminary plane outputs (ex post) have to be controlled through the SEA which is based mainly on the correspondence to the upper scale plans, and the socio-economical implementation: only marginally on the ecological parameters. Here, again, European SEA Directive 42/2001 does not introduce systemic criteria

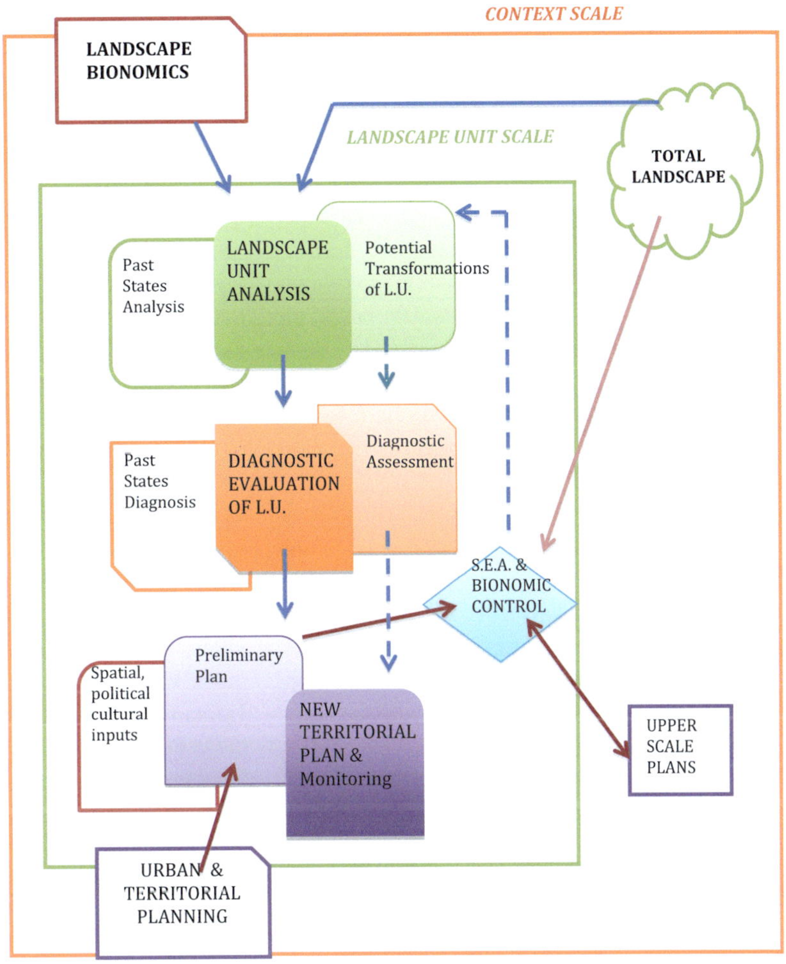

Fig. 10.6 This new flux diagram of landscape planning is very different from the one of Fig. 10.5. Two disciplines are involved, urban and territorial planning and LB. The explicit presence of analytic and diagnostic phases modifies the fluxes diagram with the strategic environmental assessment (*dotted lines*), which is able to change what could be wrong, even in upper scale plans

of LB. The widest information requested in the environmental Report is: "(f) the likely significant effects on the environment, including on issues such as biodiversity, population, human health, fauna, flora, soil, water, air, climatic factors, material assets, cultural heritage including architectural and archaeological heritage, landscape and the interrelationship between the above factors." But we know that *intrinsic biological characters* of the bionomic level cannot be found by interrelating a group of components, because of the Principle of Emergent Properties!

Anyway, the new territorial plan is ready to be approved by the Regional Authorities, after the corrections suggested by SEA Report, which includes compensations and monitoring.

10.3.2 Bionomic Landscape Planning and Assessment

In Fig. 10.6, the flux diagram for a new landscape planning is shown. Even counting sets and linkages in a simple way, we can note that this scheme is at least 2.5 times more complex than the former.

In this diagram, disciplinary and theoretical elaborations and practices derive from (1) Urban and Territorial Planning and (2) from LB. Consequently, at least two professional figures have to cooperate: the "ecoiatra" and the urban planner. Note that the formation of the ecoiatra cannot be the same of the planner (architect or engineer), but it must be, first of all, based on advanced natural sciences, biologically integrated.

The main phases of the process, guided by LB, are supposed to be dynamic, enlarging historical methodologies to the entire landscape system. Analysis and diagnosis are conceptually separated, but in fact they need iterative processes. The LU have to be delimited even in the same municipality, strictly following bionomic criteria. To better understand the planning processes, we need to focalise a more detailed lists of arguments concerning (I) landscape analysis, (II) landscape diagnosis, (III) landscape planning and (IV) SEA.

10.3.2.1 Landscape Unit Analysis (Mainly Directed by an Ecoiatra)

1. Structure (Anatomy).
   (a) LU and their main elements.
   (b) Configurations of elements and fragmentation.
2. Functions (Physiology).
   (a) Human and natural habitats (HH, NH).
   (b) Landscape apparatuses and their bionomic characters: e.g. hydro-geology, forests, prairies, agriculture, residential, urban subsidiary, etc.
   (c) Estimation of the biological territorial capacity of vegetation (BTC).
   (d) Landscape classification and related conditions of normality.
3. Landscape Analysis.
   (a) Basic biological qualities and pollution (water, air, soil).
   (b) Biological conservation and landscape animals.
   (c) Memory conservation (natural and human).
   (d) Presence and types of economical activities.
   (e) Urban structure and functions and mobility network.
   (f) Connectivity and fractal analysis of LU.
   (g) HU/BTC functions and standard BTC classes analysis.
   (h) Complex biodiversity and general metastability of the LU.
   (i) Visual analysis and bionomic semeiotics inferences.
4. Historical Analysis of the LU.
   (a) Past LU reconstruction (possibly 2 periods).
   (b) Past bionomics parameters estimation.
   (c) Remnant historical elements.

As proposed in the above list, the landscape description has to include structural and functional characters, in the sense of "anatomy and physiology" of the landscape. Obviously, the 18 arguments in the list are only a synthesis of the possible elaborations on landscape analysis. The first operation concerns the identification and the delimitation of LU and their main elements, starting from the landscape context and then the reading of configurations of elements and fragmentation of ecotopes. To follow the correct methodology and to understand bionomic terminology see the previous chapters and: "Landscape Ecology, A Widening Foundation" [12], "The impact of the widened Landscape Ecology on Vegetation Science: towards the new paradigm" [13], Ingegnoli and Giglio [14].

To understand the physiology of the integrated area, we have now, for each one of the involved LU, to measure (or estimate) the Human and Natural habitats (HH, NH). From these first data and the dimension of population, it is possible to elaborate the standard habitat (SH) values and the carrying capacity of the LU. Remember that the population density has to be ecological, not geographical. Landscape apparatuses and their bionomic characters (e.g. hydro-geology, forest, prairies, agriculture, residential, urban subsidiary, etc.) need the quite complex elaboration after direct surveys, available to follow bionomic methodologies, especially in vegetation studies. We have to underline also the importance of a correct landscape classification as we will see later.

The evaluation of metastability in vegetation can be estimated through the BTC function, which is the biological territorial capacity. It depends on measures of the degree of the relative metabolic capacity of principal vegetation communities and it measures the degree of the relative antithermic (i.e. order) maintenance of the same main vegetation communities.

The description of basic biological qualities and pollution (water, air, soil) has to be systemic, coherent with LB, thus not treating separate thematic, but cumulative effects. These criteria are available also to understand the biological conservation needs and the presence of landscape animals. Even "memory conservation" must be extended to nature, avoiding to consider only the human signs or monuments.

The presence and types of economical activities may be expressed always in systemic ways, while urban structure and functions and mobility network may start from the work of Forman on urban regions [7], but have to be completed applying HH and SH indexes. Connectivity and fractal analysis are frequently useful to build opportune models. Model analysis is indispensable as in all complex systems, so we indicate the HH/BTC functions, showing one of the topic behaviour of the landscape, the standard BTC classes analysis, the complex biodiversity model and the general metastability of the LU and the concise bionomic state [12, 15].

Visual analysis is not seen as a perceptive-cultural tool, but mainly as bionomic semeiotics inferences. Historical analysis has to be done through the reconstruction of at least two periods (for dynamical needs) to which apply past bionomics parameters estimations. Remnant historical elements (natural and human) have also to be described and interrelated with the complex landscape system.

10.3.2.2 Landscape Unit Diagnosis (Mainly Directed by an Ecolatra)

(1) Undisturbed areas of strategic importance.
(2) Evaluation of ecological networks.
(3) Systemic control of natural and human resources.
(4) Selection of bionomic parameters.
(5) Comparison with normal bionomic parameters.
(6) Limits of sustainability.
(7) LU alterations and pathologies.
(8) Present trends and prognosis.
(9) Therapeutic criteria.
(10) LU vocations.
(11) Rehabilitation projects.
(12) Bionomic control and monitoring.

Being an iterative process, the former analysis begins to put in evidence some evaluation results that can be precious, even if preliminaries. The study and evaluation of "undisturbed areas of strategic importance," following a bionomic methodology able to find correlated patches of "potential core areas of ecological interest," could be very important for a diagnostic frame. This elaboration may be linked to an evaluation of ecological networks and to a systemic control of natural and human resources.

The landscape bionomic theory assures that each landscape type can be classified observing the configurations of their apparatuses [12]. The resultant 16 groups of landscapes and their intermediate combinations, with the specification of their geographic regional characters, allow the classification of hundreds of landscape types. Consequently, the most significant component of the diagnostic process is the selection of bionomic parameters, after a correct landscape classification. These parameters must include indexes available to represent intrinsic landscape behaviour, both in structure and functions. Usually, they may be limited from 15 to 25 attributes, with a peculiar range of normality, able to characterise the landscape type in object, as exposed in the previous chapter.

10.3.2.3 Territorial Planning and Design (Mainly Directed by Territorial Planners)

(1) Choice of strategic criteria and macro-scale planning suggestions.
(2) Socio-economic goals and discussion with territorial management.
(3) Preliminary proposal of expansion zones (human and natural).
(4) Capability and suitability in accordance with LU diagnosis.
(5) Configurations and preliminary design of main subsystems.
(6) Control of structural congruence with bionomic apparatuses.
(7) Equalising criteria for topic patches.
(8) Design of suggested rehabilitation projects.
(9) Design of human and natural networks (road ecology principles).
(10) Preliminary design of the plan.
(11) Draft copy of planning regulations.
(12) Collaboration with SEA procedure.
(13) Correction after SEA responses.

(14) Final drawing of the Territorial Plan.

What previously listed on planning is only approximate. In fact, bionomic principles in planning processes means to design with nature, in the sense stated by Ian Mac Harg [8] and summarised by E.P. Odum [9]. But also a deeper knowledge of the process of design can be useful, because it demonstrates that strong dominance of man in designing choices is an abuse, not compatible with a correct methodology. Let us summarise [12] a design method (omitting its many feedbacks) useful also for planning purposes:

(1) *Preliminary phase.* Illustration of the theme and collection of main data on it; research of the significance of the objectives.

(2) *Setting out.* Main requirements of the project and their implications on the environment; trans-disciplinary linkages; preliminary project criteria and structure typology.

(3) *Investigation.* Project functionality and its compatibility with the environment and local traditions. Evaluation of the behaviour of the project and possible trends.

(4) *Proposition.* Choice of a compatible model and its development; reciprocal relationships with the environment. Times, alternatives and controls.

(5) *Resolution.* Final creative elaboration and executive details; execution of the work and completion of details.

Note that a correct design methodology cannot avoid being confronted with the environment and with local traditions, especially in the first phases. But the main reason for the need to couple this method with nature is dependent on culture. It is essential that a planning methodology was continuously related to basic principles of LB in a synthetic way. Now we also need to add some other principles, more strictly related to planning requirements [16, 12].

1. *Structural congruence.* A good plan must maintain a structural congruence at any scale. This signifies that transformations cannot cancel or alter the spatial relationships among geomorphology and vegetation, hydrology and agriculture, old settlements and their grain and contrast with the surrounding, etc. Instead, design criteria have to follow historical (both natural and human) signs.
2. *Patch shape.* To accomplish several key functions, an ecologically optimum patch shape usually has a large core, with some curvilinear boundaries and narrow lobes, and depends on orientation angle relative to surrounding flows [6].
3. *Aggregates with outliers.* Land containing humans is best arranged ecologically by aggregating land uses, yet maintaining small patches and corridors of nature throughout developed areas, as well as outliers of human activity spatially arranged along major boundaries [6].
4. *Local compensation.* Serious transformations which alter many characters of a LU have to be compensated through an ecologically balanced therapy inside the same unit, not far from the degradation.
5. *Indispensable patterns.* Top-priority patterns for protection, with no known substitute for their ecological benefits, are few large natural vegetation patches, wide vegetated corridors protecting water courses, connectivity for movement of key species among large patches and small patches and corridors providing heterogeneous bits of nature throughout developed areas [6].
6. *Attractors.* In the presence of a landscape element presenting a high attractor potential, it is necessary to insert constraints to avoid the formation of a barrier due to human activities.
7. *Source-sink and boundaries.* Any landscape element presents traits of its boundaries sensitive to the surrounding disposition of source and sink structures. Consequently it is necessary for planning to leave the wider possibility of connections and fluxes regarding those traits.
8. *Complementarity.* The main ecological law (not too much, not too little, just enough) underlines the necessity to avoid any excess; therefore each sub-system of landscape elements needs to have the presence of at least one complementary component.

10.4 Bionomic Criteria for Strategic Environmental Assessment

10.4.1 The SEA Procedure in the European Union

The procedure of SEA, the SEA, is a tool of environmental governance introduced by Dir 2001/42/EC [17]. It is "an important tool for the integration of environmental considerations in the development and adoption of certain plans and programs likely to have significant effects on the environment of the member states." The methods of implementation are as follows:

(1) Screening of the project.
(2) Preparation of the environmental report.
(3) Carrying out of consultations.
(4) Evaluation of the environmental report and the results of the consultations.
(5) Decision.
(6) Information on the decision.
(7) Any monitoring.

As written in the SEA Manual of the European Commission [18], the Environmental Report is a key element of the environmental assessment required by the Directive. Where SEA is required, an Environmental Report must be prepared in which the likely significant environmental effects of implementing the plan or programme, and reasonable alternatives taking into account the objectives and the geographical scope of the plan or programme, are identified, described and evaluated.

The information to be included in the Environmental Report is listed in "Annex I to the Directive" and includes, among other things:

- The environmental protection objectives relevant to the plan or programme.
- The relevant aspects of the current state of the environment (i.e. without implementation of the plan or programme).
- The likely significant effects on the environment, including on issues such as biodiversity, population, human health, fauna, flora, soil, water, air, climatic factors, material assets, cultural heritage, landscape, and the interrelationship between these factors.
- The mitigation measures envisaged; an outline of.
- The reasons for selecting the alternatives dealt with; monitoring measures envisaged.
- A non-technical summary of the information under all the headings in Annex I.

The process of preparing the Environmental Report should start as early as possible and, ideally, at the same time as the preparation of the plan or programme. The preparation of the Environmental Report and the integration of the environmental considerations into the preparation of plans and programmes form an iterative process that contributes to more sustainable solutions in decision making. The Environmental Report should be made available at the same time as any draft plan or programme, as an integral part of the consultation process, and the relationship between the two documents should be clearly indicated.

But the mentioned SEA Manual, page 50–51, shows a table of DPSIR (Driving forces, Pressures, States, Impacts, Responses Model of the EEA), the suggested approach to evaluate the environment effects (Table 10.2). This confirms the deep conflict between the ELC definition of landscape and the landscape bionomic principles. Recalling the mentioned levels of biological organisation (Sect. 1.4.3) nothing in this SEA Manual vision is compatible with the landscape as complex adaptive system of natural and human communities. The environmental components are analysed in separate ways, the landscape is seen as visual object, no systemic behaviour is mentioned.

10.4.2 Discussion on European SEA Criteria

This discussion begins with an example to which the EU SEA brought following its criteria. It is the way pursuing by the Autonomous Province of Trento, that in matters of land use planning is perhaps the most advanced in Italy. The purpose of this discussion is to try to understand the differences among their criteria for SEA and those derived from the principles and methods of LB.

Table 10.2 The DPSIR model approach to evaluate the environmental effects (from SEA Manual, 2005)

Environmental components	Driving forces are underlying factors influencing a variety of relevant variables (D)	Pressure indicators describe the variables which directly cause (or may cause) environmental problems (P)	State indicators show the current condition of the environment (S)	Impact indicators describe the ultimate effects of changes of state (I)	Measures envisaged to prevent, reduce and as fully as possible offset any significant adverse effects on the environment (R)
Biodiversity population Fauna Flora	Transport demand	Deterioration of the living environment (dissection or partial sealing of habitats, soil, water, air, etc.)	Number of (endangered) species, see also the convention on biological diversity 18	Remaining habitat areas capable of carrying the population with the existing biodiversity	Modal shift, emission thresholds, speed limits, access restrictions or pricing, rerouting. Compensatory areas, bridge-tunnels for animals, collection of eluates, noise barriers
Human health	Transport demand	Number and type of vehicles passing by, emission, acoustic quality of the infrastructure	Persons affected by a certain immission level	Changes of the exposure	Modal shift, emission thresholds, speed limits, access restrictions or pricing, rerouting. Noise barriers
Soil	Transport demand	Usage for infrastructure, eluates	Soil usability, geological parameters, number of certain species per m	Changes of soil usability, geological parameters, number of species	Collection and treatment of eluates, minimise soil usage
Water	Transport demand	Usage for infrastructure, eluates	Water quality TOC, COD, concentration of toxics...	Deteriorated water quality	Collection and treatment of eluates
Air	Transport demand	Emissions, infrastructure as wind shield	Immission, exceeding thresholds	Deteriorated air quality	(vegetation, allow air exchange)
Climatic factors	Transport demand	Climate gas emissions, local albedo and heat release	Concentration of green house gases GHG	Global warming, instable climate	Use of Alternative fuels, reduce energy demand
Material assets	Transport demand	Deconstruction, vibration, acid gases	Number and state	Deterioration of number and state of material assets	Decoupling of vibrations
Cultural heritage including architectural, archaeologic	Transport demand	Sealing, deconstruction, vibration, acid gases	Number and state, size of areas where archaeological heritage may be found	Reduction and/ or damage of (potential) archaeological sites	Decoupling of vibrations
Landscape		Build infrastructure	Visual quality	Deteriorated visual quality	Adequate architecture

In response to the solicitation of Europe, the Autonomous Province of Trento has published "Guidelines for Planning and Regional Strategies for the Assessment of the Strategic Plan [19]." The "vision" that is the basis of the Trentino document on SEA shows four main strategies: identity, sustainability, integration, competitiveness, the first two also relating to environment, the other socio-economic. Let us examine them in detail:

- *Identity*, strengthen the recognition of the Trentino territory, enhancing the diversity of landscape, environmental quality and cultural specificity.
- *Sustainability*, directing the use of the territory towards sustainable development, limiting the consumption processes of the soil and of primary resources and promoting urban regeneration and planning.
- *Integration*, consolidate the integration of Trentino in the European context, placing it effectively in major infrastructure networks, environmental, economic and socio-cultural.
- *Competitiveness*, strengthen local capacities for self-organisation and competitiveness and opportunities for sustainable development of the provincial system as a whole.

The information for an understanding of the current conditions of the environment, derived through measurements similar to those shown in Table 10.2, are also to be compared with zoning tabs, summary of the 16 territories in which is divided the province (with obvious criteria of economic geography and urban planning, traditional type). These cards contain:

1. The group of municipalities.
2. The general data on population and economic occupations, tourism and housing.
3. Strengths and opportunities of the territory.
4. Weaknesses and risks.
5. Vocational strategies in key economic-management and attention to the environment.

Finally, the Urban Plan of the Province (PUP) for SEA requires the comparison between the *ex ante* and *ex post* assessments and in some cases even a stage in progress; however, it does not require any assessment fully understood in a historical, evolutionary and clinical-diagnosis. On the other hand, it always requires a comparison with the PUP and other urban plans, that is, with a planning set out to larger scale, but can not necessarily see any critical problems of ecological-landscape, existing or potential, on a local scale that could largely vitiate it.

In addition, consistent with the guidelines of this vision, the description of the types of evaluation of plans is based on the concepts of economy and on the concept of resource management and environmental planners, however, not on ecological concepts. The plan is not seen as a means of therapeutic interventions involving a bio-complex system, but only as a planning instrument. It therefore requires a critical review of the contents of the SEA/PUP, in the light of the most advanced scientific developments in environmental matters. Let me be clear that this criticism cannot be contentious, since it is not a political-administrative opinion, but a scientific demonstration. In Fig. 10.7 are shown the deep differences between the discussed visions.

The strategic lines summarised above are certainly consistent with a "Vision," which also describes an environment "to be considered as excellent," but in obvious dependence on a predominantly socio-economic setting. So even identity and sustainability would be to consider especially within this setting, as shown by the general criteria, in which the main actor is in fact the economy.

As demonstrated by the well-known American economist Herman Daly [20] and after him by several other scholars, it is no longer acceptable that ecology shall depend on economy or urbanism: it should be the opposite! The economy must adapt to the problems of preservation and development of bio-ecology and life, the concept of which cannot be limited, as constantly reiterated by LB, only to organisms but must extend to all levels of biological organisation, including precisely the landscape.

As a result, we do not understand how they can affect the strategic lines, that refer to identity and sustainability, and how other objectives of primary importance for the environment, e.g. history (transformations), health (bio-ecological state

A- MAIN STRATEGIC AIMS OF SEA	
IDENTITY, enhancing the eco-tourism; SUSTAINABILITY, sustainable development of resources; INTEGRATION, in the European context of infrastructure; COMPETITIVENESS, productive development	HEALTH, of landscape systems and of man; MEMORY, natural history and cultural heritage; REHABILITATION, sustainability of the system and therapeutic indications; CONTROL of RESOURCES, systemic, by landscape systems
B- SEA METODOLOGICAL CRITERIA	
URBAN REPORTING Assessment of consistency with the higher-level planning instruments *Analysis*: indicators, economic and environmental planning and simple decision-making models (e.g. SWOT) *Rating*: action to achieve the strategies are evaluated through impacts by thematic area. The assessment is qualitative, for compliance and indeterminacy. *Control*: Planning, with simple methods of descriptive and effectiveness with urban-economic indicators. The following mitigation measures, compensation and monitoring.	BIONOMICS EVALUATION Assessment of the state of health of the landscape units (LU) *Analysis*: indicators of landscape and ecological models bionomic for complex subsystems *Rating*: structure and functions of each LU are valued through a clinical-diagnostic in relation to dynamic states of normal (3 times). Qualitative and quantitative assessment *Control*: trends with and without plan, by means of models of bionomics and ecological indicators of landscape or ecological-economic factors. The following measures of specific therapy and systemic.

Fig. 10.7 Comparison between Trentino Province (*violet*) and Bionomic Principles for SEA (*pale green*)

of the systems under consideration and its implication to human health), etc. are completely absent.

10.4.3 Quality and Quantity of Information

Given the extreme importance and the understandable consequences of application stated herein, it is essential to try to clarify the scope of criticism. To do this it is necessary to refer to the document of the PUP Trentino on SEA, consider the first strategic direction (the "identity") and take the measures suggested referring to the Pilot Study for Mori SEA.

Figure 10.8 shows a first part (white background) containing five ecological parameters estimated for the entire municipality, minimum requirements PUP,[1] while the remainder (yellow) adds another six parameters and extends them even at 4 LU detectable in the municipality of Mori, minimum requirements of LB. We will begin to take a look at the parameters considered:

1. *Areas of pasture and meadow-pasture*: we have measured the existing ones. The PUP adds "restored," but their extent would remain inherently uncertain and, moreover, it would not make sense because the concept of ecological restoration is incorrect in the light of the principles of "order through fluctuations."
2. *Ecologically controlled areas*: this measure is only formal, since often part of these controlled areas have no ecological significance, and sometimes they are not extended enough to preserve the phenomena that may be really of interest.
3. *Protected crops*: the relationship with the total area may be indicative only if you define the LU of competence.
4. *Expansion of the settlement*: land consumption is not significant when measured as m^2/inhab./year. It should be analysed and considered in the perspective of historical evolution.
5. *Reduction of productive agricultural areas*: This measure is fine, but should also be related to the LU.

[1] Data deduced from All.2 Address for Territorial planning strategies and SEA of plans Urban Provincial Plan of Trentino, 2006.

N°	Processes	indicators	units	Mori 3.460 ha	LU-1 1.175 ha	LU-2 602 ha	LU-3 847 ha	LU-4 836 ha
	Average altitude	transects	m (a.s.l.)	602	364	385	795	901
1	Grasslands	Mountain agriculture	% area/ Tot. area	1.7	0.82	0.22	1.74	3.95
2	Preserved areas	Conservation (SIC)	% area/ Tot. area	12.4	18.4	18.6	0	11.9
3	Protected crops	(vines-orchards)	% area/ Tot. area	22.4	30.9	13.3	13.3	10.0
4	Urban expansions (50 yr)	Soil consumption (since 2006)	m^2/ab/yr	2.2	2.0	4.8	1.8	7.1
5	Decrease of crop areas (50 yr)	Crop soil consumption	% area/ Agr. area	24.7	30.2	10.0	16.8	43.9
6	Human Habitat	HH	% LU	40.7	57.9	45.5	30.5	23.3
7	Carrying capacity	SH/SH*	N°	3.35	0.60	5.3	3.04	6.1
8	Biological territorial capacity of vegetation	BTC	Mcal/m^2/yr	3.34	2.33	3.04	3.84	4.47
9	RSD+SBS landscape apparatus	Sum LU apparatuses	% area/ Tot. area	7.5	17.4	3.5	2.44	1.7
10	Trasformation (1860-2007)	Urban/agr. grow	Yr rate Urb/Agr	0.123	0.240	0.045	0.041	0.021
11	Landscape type	classification	HS/HS* x BTC/HU	27,49 AGR-PD	2,41 SUR-RU	34,15 AGR-PT	42,0 AGR-PT	117,8 AGR-FR

LU = Landscape Unit: 1= Mori, 2 = Loppio, 3 = Bassa Val Gresta, 4 = South Mount Biaena.
AGR-PD = agricultural-productive, AGR-PT = agricultural-protective, AGR-FR = forest-agricultural, SUR-RU = rural-suburb an. SH/SH* = ratio local/theoretical.

Fig. 10.8 SEA of Mori (Trento). Evaluation of the strategy "identity": on a *white background* legal requirements; on the total Mori municipality, on a *yellow background* additions according to landscape bionomics. LU = Landscape Unit: 1 = Mori, 2 = Loppio, 3 = Bassa Val Gresta, 4 = South Mount Biaena. SH/SH* = ratio local/theoretical Standard Habitat, (HS/HS*) × (BTC/HU) represents the old HuCE, now updated in LTpE

6. *Human habitat (HH)*: this concept is fundamental to the identity of a LU, as a measure of the general processes of natural and human regulation. The change from just over 20 to almost 60 % of LU gives an idea of the profound difference of ecological subdivision of the territory of Mori, simply impossible to ignore.
7. *Carrying capacity*: recalling it is the relationship between standard habitat (SH) per capita detected and theoretical SH, then provides the degree of autotrophy/heterotrophy of a LU.
8. *Bio-territorial capacity of vegetation*: the BTC is a measure of the level of organisation of a complex adaptive system such as the landscape and is capable of expressing values comparable between components and integrated natural, semi-natural and anthropogenic vegetation. Correlated with the levels of HH you get to form a pattern of normality that effectively contributes to the assessments on the ecological status of a LU.
9. *RSD + SBS landscape apparatus*: These subsystems highlight the functional roles of landscape-ecological context. They are necessarily different from the urban planning concepts although in this case dealing with residential systems and auxiliary (Cf. 2.4.2).
10. *Historical transformation*: in this example it is shown only one of several types of transformation functions of the ecological

system, such as the annual growth rate of urban/agriculture areas in the period 1860–2007. This area is part of the study of the historical evolution of a LU.

11. *Type of landscape*: the landscapes are classified according to indicators capable of binding the carrying capacity with model BTC/HH. As you notice, the municipality of Mori as a whole is an agricultural-productive landscape, but we find in it: a suburban LU, a forest LU and two agricultural-protective LU. Following the directions for the entire territory would lead to make unforgivable mistakes.

We can now carry out the scientific comparison that interests us, going to measure in the two cases, urban-economic criterion vs. landscape bionomic one (U-EC vs. L-B) the information obtained. Note that the measurement of information cannot only take account of the *quantity*, must also consider the *quality*. Quantitatively, the proportion is immediate: instead of 5 parameters we derive 55 (11 times the provincial directions).

For the qualitative part instead we should take into account the differences between *direct* measurements obtained with a single operation (e.g. %) and *systemic-functional* measures, obtainable through more operations. As with the L-B criterion complex measures are 5/11, while the criterion U-EC are 1/5, giving a ratio of processing even only 3–1, we get: 105/7 = 15. This means, therefore, that the L-B criterion provides at least 15 times more ad hoc information. But this disproportion is certainly greater if you apply in full to both policies.

10.4.4 The Bionomic Criterion of Strategic Environmental Assessment

Strategic Environmental Assessment
(Co-directed by an Ecoiatra and a Territorial Planner)

(1) Control of strategic planning criteria.
(2) LU analysis and diagnosis “ex ante” (before the draft Plan).
(3) Structural consistence of the transformation of the LU “ex post” (after the draft Plan).
(4) Bionomic function analysis of the changed LU.
(5) Diagnostic assessment of the changed LU.
(6) Corrections necessaries to the draft Plan.
(7) Consistence with existing macro-scale Plans.
(8) Proposal of correction of macro-scale Plans, when in contrast with point 4.
(9) Final report.

The control of strategic planning criteria must verify the non-prevalence of economical directions and the presence of systemic resource control and health protecting criteria (both for man and landscape).

LU analysis and diagnosis “ex ante” has to be elaborated before the draft Plan following the former list (I). Then we have to measure the structural consistence of the transformation of the LU “ex post” (after the draft Plan) and compare the two cases. Thus, again, the analytical processes and functions of the list (I) must be repeated referred to the new previsions, a potential transformation of the LU. Obviously, the draft plan can be repeated with small variances, constituting two or more scenarios, then tested and compared to choose the best one.

Table 10.3 synthetise in a more complete and congruent way the main sectors of LB theory and applications useful for a complete SEA of a territory. The example of the SEA study of the municipality of Tesero (Sect. 12.2) is linked to this table.

10.4.5 The Screening Pathways for Environmental Assessment

As we may deduce from Table 10.3 and from previous chapters, LB methodologies available to reach the objectives of environmental assessment of a territory are numerous, at least about 35–50. It could be useful to choice the most important of them (about one half) to prepare a screening frame, through which putting in evidence the most opportune path of screening examinations.

Table 10.3 List of the main steps linking theory to application methodologies in the six phases of LU analysis and evaluation

1. D.	1. THEORY	Landscape Bionomics	Application methodologies
2. STRUCTURE	2.1 LANDSCAPE UNIT	LU representative	Surveys
			Delimitation
	2.2 LANDSCAPE ELEMENTS	Types of components	Choice of Typologies
			Measure of Land cover measures (ha and %), grouping per landscape apparatuses
3. L.ANDSCAPE APPARATUSES	3.1 GEOLOGY(GEO)	Geomorphologic system	Descriptive synthesis
			Critical areas
	3.2 HYDROLOGY(EXR)	Main water bodies	Descriptive synthesis (vegetational and faunal)
			Biotic Indexes
	3.3 RESISTANT SYSTEM (RNT)	Forest and wood system	Descriptive synthesis (vegetational and faunal) & Choice of typologies
			LaBiSV sampling T parameters, F parameters, E parameters, U parameters
			BTC, CBSt evaluation, and, if the case, faunal sensitivity
	3.4 RESILIENT SYSTEM (RSL)	Prairies and shrubs	Descriptive synthesis, vegetational (LaBiSV) and faunal
			LaBiSV sampling
	3.5 PROTECTIVE AND CONNECTIVE SYSTEMS (PRT-CON)	Vegetated corridors, remnant wooded patches, urban parks	Descriptive synthesis, vegetational (LaBiSV) and faunal
			Connection evaluation
	3.6 AGRICULTURE (PRD)	Productive system (fields, meadows, orchards, etc.)	Descriptive synthesis, vegetational (LaBiSV) and faunal
			SH-prd evaluation
	3.7 URBANIZED (RSD)	Urbanized system	Descriptive synthesis (LaBiHH)
			SH-rsd evaluation
	3.8 INFRASTRUCTURE AND WORK (SBS)	Industry and traffic	Descriptive synthesis
			SH-sbs evaluation
4. SUSTAINABILITY	4.1 HEALTH	Basic biologic qualities and Cumulative pollutions	Biotic indexes (and epidemiologic data) ecotoxicological data, environmental stresses
	4.2 CONSERVATION	Natural conservation areas (SCI, ZSP)	Descriptive synthesis
			(and bionomic evaluation)
	4.3 VISUAL PERC.	Semantic characters	Semantic analysis
	4.4 ECONOMY	Problems linked to ecology	Possible conflicts
5. L.U. BIONOMIC STATE	5.1 HISTORY	Main transformations	Bionomic indexes, Landscape type, Human Habitat, Average BTC, etc. Research of past maps (min 2 periods), reconstruction of past LU conformations, choice of opportune indexes, Estimation of Historical Data
	5.2 PRESENT	Present characters	Bionomic indexes, Landscape type, Human Habitat, Average BTC, etc. Human habitat (HH), average bTC, photosynthetic area (PhsA), potential of bionomic rehabilitation (PoBR), General metastability of landscape elements (BTC classes), Standard habitat per capita (SH), Carrying Capacity (σ), Landscape type evaluation (LTpE), Evaluation of the undisturbed areas of strategic importance
	5.3 PLANNING	Forecasting planning conditions	Main bionomic indexes updating old plan, new needs per sub-system, priorities etc., transformations
6. DIAGNOSIS	6.1 CHECK-UP	Deviations from the norm	Cumulative EIA
			Bionomic behaviour
			Diagnostic index evaluation of the previous bionomic indexes, balancing of normal parameters. DI evaluation and dynamics, trend and plan suggestions
	6.2 DISFUNCTIONS	Pathological characters	Description of pathologies
	6.3 THERAPEUTIC CRITERIA	Health rehabilitation	Project of interventions
			Urgent priorities

Remember that the aims of this screening paths are the evaluation of the bionomic state of the territory of a municipality, in order to check its pathologic level and to suggest therapeutic interventions. The methodologies of environmental assessment have to be grouped in two sets, depending from the bionomic screening of (A) geographic characters and (B) landscape structure, functions and dynamics. Moreover, we must note that each argument can be articulated in 3–5 level of detail (i.e. scale, commitment, elaboration).

(A) Geographic characters

The set of geographic characters of the municipalities to be examined could be very wide, but the most frequent are:

- A1—the surface of municipal territory [km^2].

Table 10.4 Main geographic characters articulated in synthetic and detailed dimensions

(A)—Geographic characters	Scale 1:25,000			Scale 1:10,000	
A0—Municipalities (EU)	Small	Medium	Wide	Small	Wide
A1—Municipal area [km^2]	<15	15–30	>30	<30	>30
A2—Forest area [ha]	<150	150–500	>500	<500	>500
A3—Landscape Units >1 [n°]	1	2	>2	<2	>2
A4—Altered/normal woods [% A2]	<2	<10	>10	<10	>10
A5—Historical Gardens >1 ha [n°]	0	1	>1	<2	>2
A6—Mountain areas [% A1]	<10	10.1–20	>20	<20	>20
A7—Population size [ab × 10^3]	<5	5.1–10	>10	<5	>5
A8—Natural protection areas [% A1]	<10	<10	>10	<10	>10
A9—Geomorphologic emerg. [% A1]	<10	10.1–20	>20	<10	>10

- A2—the surface of its forested elements [ha].
- A3—the presence of one or more landscape units LU >1 [n°].
- A4—the ratio between altered/normal woods [% A2].
- A5—the presence of wide historical gardens >1 ha [n°].
- A6—the possible presence of mountain areas [% A1].
- A7—the population size, counting even annual tourist %, [ab × 10^3].
- A8—the presence (and type) of areas of natural protection [% A1].
- A9—the possible geomorphologic emergences [% A1].

Each geographic character implies a process of analysis and elaboration of the deduced information. Map requirement, surveys, descriptions, photographic reports, but also historical documentation and the elaboration of population analysis (residents and tourists) and so on (Table 10.4).

(B) Structure, Function and Dynamics Examinations

In analogy with the former point (A), a set of 15 examination methodologies have been listed here, noting that each of them has been exposed in previous chapters:

- B1—Structural frame of the territory: opportune division in LU and ecotopes, and their first characterisation.
- B2—Autotrophic system: basilar analysis to put in evidence the main vegetation formations and the type of local flora.
- B3—Autotrophic system: vegetation surveys following the LaBiSV methodology, in proportion of A2 and A4 data.
- B4—Heterotrophic system: animal presence and main key species of the LU, following the landscape bionomic methodology.
- B5—Landscape Apparatuses: grouping landscape elements, estimation of human habitat (HH) and biologic territorial capacity of vegetation (BTC).
- B6—Heterotrophic system: per capita standard habitat measure (SH) and Carrying Capacity evaluation [SH/SH*].
- B7—Organisation level of LU: evaluation of the BTC classes, Shannon heterogeneity and functional biodiversity (τ) and Landscape Metastability [g-LM].
- B8—Connection and ecologic networks: planar graph design, connectivity and circuitation indexes, networks control and design.
- B9—Disturbances interference: map of man and nature disturbances, buffer belts, evaluation of Undisturbed Areas of Strategic Importance.
- B10—Adjunctive indicators (depending from local needs): river functionality (IFF), lichens survey for air quality, noise and sound level, etc.
- B11—Diagnostic evaluations (DI.EV): parameters choice, landscape type evaluation (LTpE index), normal ranges, deviations from the norm, pathologic levels.

Table 10.5 Main examination processes articulated in synthetic and detailed elaborations

B	(B) Examination processes		Synthetic		Detailed	
B1	Structural frame	General	LU	Ecotopes	LU	Ecotopes
B2	Vegetation analysis	Formations	Flora	Habitat 2000	Flora	Syntassonomy
B3	Vegetation survey [LaBiSV]	<5	5.1–10	10.1–15	17–26	26–34
B4	Animal frame	Main habitats	Key species	Faunal Sensibility	Key species	Faunal Sensibility
B5	L. Apparatus, HH and BTC	L. Apparatus	HH,BTC eval.	HH/BTC mod	HH,BTC eval.	HH/BTC mod
B6	Carrying capacity [SH/SH*]	Equiv. Popul.	SH	SH/SH*	SH	SH/SH*
B7	L. Metastability [g-LM]	BTC classes	g-LM	Model	g-LM	Model
B8	Networks and connections	Planar graph	α, ϒ eval.	Potentiality	α, ϒ eval.	Potentiality
B9	Disturbances interference	Disturbances	Buffers	Evaluation	Buffers	Evaluation
B10	Adjunctive indicators	General	IFF	Lichens	IFF	Lichens
B11	Diagnostic evaluations	Parameters	LTpE	DI.EV	LTpE	DI.EV
B12	Transformation dynamics	Histor. anamnesis	Index hist. eval.	hist. DI.EV	*2 hist. Periods*	*3–4 hist. Per.*
B13	Forcasting planning conditions	Plan analysis	Trend	Model	New criteria	Controls
B14	Therapeutic criteria	Alterations	Suggestions	Discuss. Ther.	Therapies	Controls
B15	Environmental report	Synthetic	Preliminary	Normal	Detailed	Detailed

- B12—Transformation dynamics: historical anamnesis (at least 2 periods), DI.EV to historical structure and functions.
- B13—Forecasting of planning conditions: trend of present complex system, plan proposal consequences for the LU, comparison with bionomic needs.
- B14—Therapeutic criteria: critical analysis of the main LU pathologies, therapeutic suggestions, bionomic rehabilitation criteria.
- B15—Environmental Report: following the landscape bionomic theory with the addition of cumulative pollution impacts.

Even in the case of Table 10.5, the articulation of the 15 examination processes depends from scale (less or more detailed) and territorial areas (Western Europe).

(C) Screening paths matrix

A small matrix (24 rows, 5 columns) can be formed for a synthetic screening proposal (Table 10.6), useful to identify the most opportune paths and the work dimensions.

This matrix put in evidence 120 cells, representing operational steps of the two sets of (A) and (B) examinations into which to choice the most opportune screening path. Two rows are added to the matrix: the first gives the time elaboration coefficient (minimum) in hours per each column; the second registers the total hours per column determined summing the number of cells chosen to form the needed path.

Table 10.7 shows the screening path used in the SEA Pilot Detailed Study of the municipality of Mori (Trentino-S.Tyrol) in 2007–2009. The 24 screening operations summed 7 + 120 + 208 = 335 h of elaboration, to which we have to add the time of trips (by car) from the Office to the locality and the local raids even by foot and ten meetings with the Administrative, about 207 h. It resulted in a work of 547 h very near to the 565 h of the real mentioned work: shorter time preview, near 5 months (real case: 2.5 years for reasons depending by politicians authorities!).

So, after many examples, with an average error of no more than 3–5 %, it is possible to put in evidence a clear work programme for SEA screening. It is also possible to suggest a minimum screening programme for one of the small

Table 10.6 Basic matrix for screening paths of the territory of a municipality

Time elaboration coefficient	*3*	*5*	*7*	12	16
(A)—Geographic characters		**Scale 1:25.000**			**Scale 1:10.000**
A0—Municipalities (EU)	Small	Medium	Wide	Small	Wide
A1—Municipal area [km^2]	<15	15–30	>30	<30	>30
A2—Forest area [ha]	<150	150–500	>500	<500	>500
A3—Landscape Units > 1 [n°]	1	2	>2	<2	>2
A4—Altered/normal woods [% A2]	<2	<10	>10	<10	>10
A5—Historical Gardens >1 ha [n°]	0	1	>1	<2	>2
A6—Mountain areas [% A1]	<10	10–20	>20	<20	>20
A7—Population size [ab × 10^3]	<5	5–10	>10	<5	>5
A8—Natural protection areas [% A1]	<10	<10	>10	<10	>10
A9—Geomorphologic emerg. [% A1]	<10	10–20	>20	<10	>10
(B) Examination processes		**Synthetic**			**Detailed**
B1—Structural frame	General	LU	Ecotopes	LU	Ecotopes
B2—Vegetation analysis	Formations	Flora	Habitat 2000	Flora	Syntassonomy
B3—Vegetation survey [LaBiSV]	<5	5.1–10	10.1–15	17–26	26–34
B4—Animal presence	Main habitats	Kee species	Faunistic Sensibility	Kee species	Faunistic Sensibility
B5—L. Apparatus, HH and BTC	L. Apparatus	HH,BTC eval.	HH/BTC mod	HH,BTC eval.	HH/BTC mod
B6—Carrying capacity [SH/SH*]	Equiv. Popul.	SH	SH/SH*	SH	SH/SH*
B7—L. Metastability [g-LM]	BTC classes	g-LM	Model	g-LM	Model
B8—Networks and connections	Planar graph	α, ϒ eval.	Potentiality	α, ϒ eval.	Potentiality
B9—Disturbances interference	Disturbances	Buffers	Evaluation	Buffers	Evaluation
B10—Adjunctive indicators	General	IFF	Lichens	IFF	Lichens
B11—Diagnostic Evaluations	Parameters	LTpE	DI.EV	LTpE	DI.EV
B12—Transformation dynamics	Histor. anamnesis	index hist. eval.	hist. DI.EV	*2 hist. Periods*	*3–4 hist. Per.*
B13—Forcasting planning conditions	Plan analysis	Trend	Model	New criteria	Controls
B14—Therapeutic criteria	Alterations	Suggestions	discuss. Ther.	Therapies	Controls
B15—Environmental report	Synthetic	Preliminary	Normal	Detailed	Detailed
Max hours per column	***72***	***120***	***168***	***288***	***384***

Time elaboration coefficient in hours

municipalities of the Lombard Plain, as proposed in Table 10.8.

The example of Table 10.8 represents a synthetic and limited work for a small European municipality, in which only 19/24 of the examinations are proposed. But in many cases the European laws need expedite works, if they are completed with impact data on pollution. In this case, the 19 screening operations totalise 18 + 40 + 35 = 93 h of elaborations to which we suppose to add 1/3 of time for car and foot raids, so about 120 h of work, only near 1/5 of the previous case; shorter time preview, even only 1 month.

Table 10.7 Example of the screening path used in the SEA Pilot Study of the territory of the municipality of Mori (Trento)

time elaboration coefficient	3	5	7	12	16
(A)- Geographic characters	**Scale 1:25.000**			**Scale 1:10.000**	
A0- Municipalities (EU)	***small***	***medium***	***wide***	***small***	***wide***
A1- Municipal area [km²]	< 15	15- 30	> 30	< 30	> 30
A2- Forest area [ha]	< 150	150-500	> 500	< 500	> 500
A3- Landscape Units > 1 [n°]	1	2	> 2	< 2	> 2
A4- Altered/normal woods [% A2]	< 2	< 10	> 10	< 10	> 10
A5- Historical Gardens > 1 ha [n°]	0	1	> 2	< 2	> 2
A6- Mountain areas [% A1]	< 10	< 20	> 20	< 20	> 20
A7- Population size [ab x 10³]	< 5	5,1-10	> 10	< 5	> 5
A8- Natural protection areas [% A1]	< 10	< 10	> 10	< 10	> 10
A9- Geomorphologic emerg. [% A1]	< 10	10,1-20	> 20	< 10	> 10
(B) Examination processes	***synthetic***			***detailed***	
B1- Structural frame	general	LU	ecotopes	LU	ecotopes
B2- Vegetation analysis	formations	Flora	Habitat 2000	Flora	syntassonomy
B3- Vegetation survey [LaBiSV]	< 5	5,1-10	10,1-15	17-26	26-34
B4- Animal presence	main habitats	kee species	Faun. Sensibility	kee species	Faun. Sensibility
B5- L. Apparatus, HH & BTC	L. Apparatus	HH,BTC eval.	HH/BTC mod	HH,BTC eval.	HH/BTC mod
B6- Carrying Capacity [SH/SH*]	Equiv. Popul.	SH	SH/SH*	SH	SH/SH*
B7- L. Metastability [g-LM]	BTC classes	g-LM	modello	g-LM	modello
B8- Networks & connections	planar graph	±, Υ eval.	potentiality	±, Υ eval.	potentiality
B9- Disturbances interference	disturbances	buffers	evaluation	buffers	evaluation
B10- Adjunctive indicators	general	IFF	lichens	IFF	lichens
B11- Diagnostic Evaluations	parameters	LTpE	DI.EV	LTpE	DI.EV
B12- Transformation dynamics	histor. anamnesis	index hist.eval.	hist. DI.EV	*2 hist. Periods*	*3-4 hist. Per.*
B13- Forcasting planning conditions	Plan analysis	trend	model	new criteria	controls
B14- Therapeutic criteria	alterazioni	suggerimenti	discuss. Ther.	terapie	controlli
B15- Environmental report	synthetic	preliminar	normal	detailed	detailed
Tot. hours per column	***0***	***0***	***7***	***120***	***208***

10.4.6 Integration of Bionomic Screening with Cumulative Impact Assessment

The bionomic screening matrix linked also with Table 10.3 allows, following LB theory and BTC statistics (e.g. Table 9.2), an integration with cumulative impact assessment, through the Dynamic Computational GIS methodology (DC-GIS) elaborated by G. Magro[2] [21, 22].

[2] G. Magro is the developer of Dynamic Computational G.I.S. (DCGIS) methodology and software, selected as official tool for Environmental screening of waste treatment and disposal plants in Lombardia Region (D.G.R. 11317 Feb. 2010). Magro was also the founder of Q-cumber system, the world wide Geo-Social platform for Environmental participation selected as one of the 100 best projects in the world at the Olympic Start Up Games during Olympic Games of London 2012. Now active system working on Google Maps at www.q-cumber.org

Table 10.8 Proposal of a minimum screening programme for the SEA of a small municipality of Lombard Plain or similar

time elaboration coefficient	3	5	7	12	16
(A)- Geographic characters		**Scale 1:25.000**		**Scale 1:10.000**	
A0- Municipalities (EU)	***small***	***medium***	***wide***	***small***	***wide***
A1- Municipal area [km^2]	< 15	15- 30	> 30	< 30	> 30
A2- Forest area [ha]	< 150	150-500	> 500	< 500	> 500
A3- Landscape Units > 1 [n°]	1	2	> 2	< 2	> 2
A4- Altered/normal woods [% A2]	< 2	< 10	> 10	< 10	> 10
A5- Historical Gardens > 1 ha [n°]	0	1	> 2	< 2	> 2
A6- Mountain areas [% A1]	< 10	< 20	> 20	< 20	> 20
A7- Population size [ab x 10^3]	< 5	5,1-10	> 10	< 5	> 5
A8- Natural protection areas [% A1]	< 10	< 10	> 10	< 10	> 10
A9- Geomorphologic emerg. [% A1]	< 10	10,1-20	> 20	< 10	> 10
(B) Examination processes		***synthetic***		***detailed***	
B1- Structural frame	general	LU	ecotopes	LU	ecotopes
B2- Vegetation analysis	formations	Flora	Habitat 2000	Flora	syntassonomy
B3- Vegetation survey [LaBiSV]	< 5	5,1-10	10,1-15	17-26	26-34
B4- Animal presence	main habitats	kee species	Faun. Sensibility	kee species	Faun. Sensibility
B5- L. Apparatus, HH & BTC	L. Apparatus	HH,BTC eval.	HH/BTC mod	HH,BTC eval.	HH/BTC mod
B6- Carrying Capacity [SH/SH*]	Equiv. Popul.	SH	SH/SH*	SH	SH/SH*
B7- L. Metastability [g-LM]	BTC classes	g-LM	modello	g-LM	modello
B8- Networks & connections	planar graph	±, ϒ eval.	potentiality	±, ϒ eval.	potentiality
B9- Disturbances interference	disturbances	buffers	evaluation	buffers	evaluation
B10- Adjunctive indicators	general	IFF	lichens	IFF	lichens
B11- Diagnostic Evaluations	parameters	LTpE	DI.EV	LTpE	DI.EV
B12- Transformation dynamics	histor. anamnesis	index hist.eval.	hist. DI.EV	*2 hist. Periods*	*3-4 hist. Per.*
B13- Forcasting planning conditions	Plan analysis	trend	model	new criteria	controls
B14- Therapeutic criteria	alterazioni	suggerimenti	discuss. Ther.	terapie	controlli
B15- Environmental report	synthetic	preliminar	normal	detailed	detailed
max hours per column	***18***	***40***	***35***	***0***	***0***

We observe that many bionomic examinations can be elaborated by remote sensing detection, hence rather available to be introduced in a complex informatics programme. The main problem is due to the evaluation of BTC, which depends on vegetation surveys. Moreover, the BTC function represents a component of other important bionomic functions, e.g. the LTpE or the g-LM.

The concept of "transformation deficit" may help in the estimation of the BTC balance in a local LU. The evaluation of the average BTC during the past 130–160 years (Fig. 10.9) is quite surprising, observing that the true

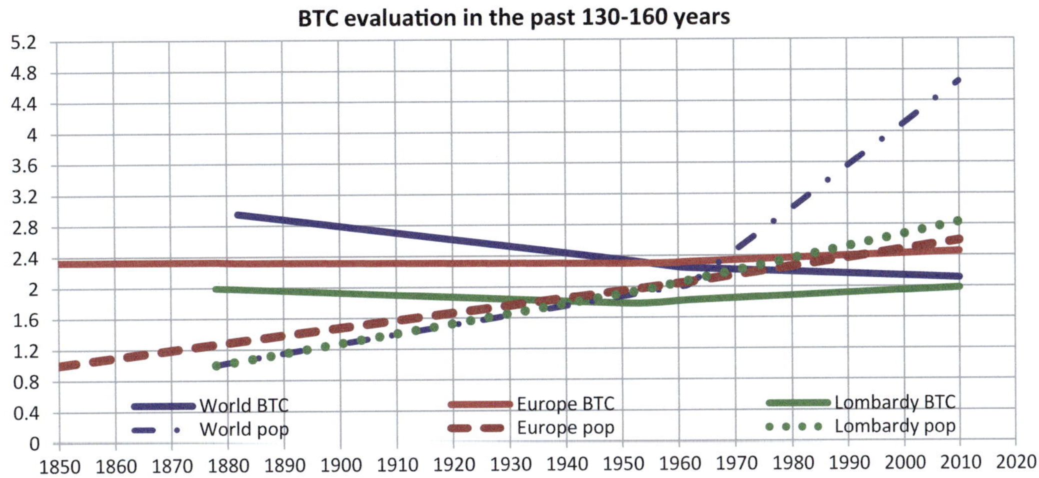

Fig. 10.9 The evaluation of BTC at World (emerged land) scale, European (UE-24) scale and Regional (Lombardy) scale, since the second half of nineteenth century. Note the constant values in Europe

ecological crisis of the World begun in the period of the last World War (1939–1945) and immediately after, when the increase of population grow faster. This is proved by the first books on ecological crisis as "Silent Spring" by R. Carson [23] and "The Science of survival" by B. Commoner [24], both of the first year Sixties.

We can see in Fig. 10.9 the good constancy of the BTC values in European Union (EU-24) with a BTC = 2.35 Mcal/m^2/year in 1850, BTC = 2.30 in 1960, BTC = 2.45 in 2010, notwithstanding the increase of population, from 1 to 2.65 in this period (1850–2010). At regional scale we choose Lombardy, one of the most crowded and urbanised region of Europe; if we exclude the decrease due to the World Wars (−11 %) the BTC is today the same of 1878: BTC = 1.98 Mcal/m^2/year.

The trend lines for European and Lombard BTC with a security coefficient of 25 % bring for the next 20 years (2030) values a bit larger that today: BTC_{EU} = 2.49 and BTC_{LO} = 2.03 Mcal/m^2/year. For the Emerged Land we can forecast, on the contrary, a lower value: BTC_W = 2.07 Mcal/m^2/year.

Anyway, the constancy of BTC permits to consider its trend value as an adequate reference to measure the transformation deficit of a LU, e.g. in Lombardy. A Dynamic Computational GIS can register the distances between each landscape elements transformed by man (resulted from statistic data) and the average BTC of Lombardy, reaching an amount of BTC deficit in the LU under examination. The assessment of all the natural and semi-natural elements should be available

Table 10.9 Differences between surveyed and statistic BTC data in 15 landscape units

Landscape units	Area ha	L. elem.	N° survey	ratio surv/ elem	% total BTC	BTC′ from surveys	BTC′′ from statistics	BTC′′/ BTC′	Difference (%)
PARZANICA, (QUARRY) BG	40.0	4	3	75.0	99.0	5.92	5.49	0.927	7.280
OSIO UDP highway j, BG	58.4	9	6	66.7	95.3	1.35	1.39	1.030	2.963
LAVAZE' pass, TN-BZ	172.6	11	5	45.5	99.5	4.76	5.00	1.050	5.042
MORI, UDP4, TN	836.0	17	11	64.7	99.8	4.47	4.53	1.013	1.342
ROBECCO SN, MI	1,027.2	22	12	54.5	98.7	2.19	2.15	0.982	1.826
CUSAGO, MI	1,151.0	12	6	50.0	91.3	0.86	0.81	0.933	6.721
MORI, UDP1, TN	1,175.2	18	11	61.1	97.5	2.33	2.48	1.064	6.438
BOLLATE, MI	1,312.3	20	6	30.0	79.4	1.03	1.11	1.081	8.050
RONZO-CHIENIS, TN	1,376.0	16	7	43.8	94.7	3.74	3.87	1.035	3.476
TESERO UDP2 TESERO, TN	1,397.5	13	6	46.2	94.5	4.13	4.16	1.007	0.726
TESERO UDP3 LAGORAI, TN	1,562.7	13	4	30.8	91.3	3.77	3.76	0.997	0.265
ALBA-PENIA (CANAZEI) TN	1,584.0	12	3	25.0	92.5	3.54	3.65	1.030	2.965
GALBIATE, Parco BARRO, LC	1,684.1	17	5	29.4	87.2	2.08	2.20	1.058	5.769
TESERO UDP1 STAVA, TN	2,080.2	13	5	38.5	95.8	3.88	3.94	1.015	1.546
ASSE, BELGIUM	5,026.0	16	5	31.3	73.0	1.04	1.14	1.090	9.012
Average	***1,365.5***	***14.2***	***6.3***	***46.2***	***92.6***	***3.01***	***3.04***	***1.02***	***4.23***
Standard deviation	*1,179.2*	*4.5*	*2.8*	*15.6*	*7.6*	*1.56*	*1.52*	*0.05*	*2.85*

to balance that BTC deficit. If not, we can measure the BTC amount needed to this balancing.

The evaluation of natural vegetated elements is possible through Dynamic Computational GIS if the statistical data and vegetation maps of the territory are available; if not this information must be acquired by LU surveys (LaBiSV method). In Table 10.9 we elaborated a group of 15 LU whose assessment was done through vegetational surveys of the field, comparing their results with the use of statistical BTC, derived from about 500 samples in North Italy (90 %) and other Central European surveys.

Note that a survey with BTC analysis on 46.2 % of the landscape elements is sufficient to cover 92.6 % of the total BTC value, because generally few elements express high BTC.

Even if the average error is 4.23 %, we found 1/3 of the results with an error >6.0 %. This is a good result, but we need a wider statistical research, available to better characterise at least 50–60 types of natural and human vegetation.

References

1. Ingegnoli V (2006) Sintesi dell'esame preliminare del paesaggio della laguna di Venezia: cartella clinica e terapie proponibili. VA Valutazione Ambientale n°9:10–18
2. Lorenz K (1981) The foundations of ethology. Springer, Berlin, Wien
3. Columella LJ (1977) (I sec.) De re rustica. Harvard University Press, Cambridge, MA
4. Vitruvius MP (1970) (I sec.) De Architectura. Harvard University Press, Cambridge, MA
5. Olmsted FL (1870) Public parks and the enlargement of towns. Cambridge University Press, Cambridge, MA
6. Forman RTT (1995) Land Mosaics: the ecology of landscapes and regions. Cambridge University Press, Cambridge, MA

7. Forman RTT (2008) Urban regions, ecology and planning beyond the city. Cambridge University Press, Cambridge, UK
8. Mc Harg I (1969) Design with nature. American Museum of Natural History, New York
9. Odum EP (1971) Fundamentals of ecology, 3rd edn. WB. Saunders, Philadelphia
10. Linehan JR, Gross M (1998) Back to the future, back to basics: the social ecology of landscapes and the future of landscape planning. Landsc Urban Plan 42 (2–4):207–223
11. Marcheggiani E, Bomans K, Galli A, Gulinck H (2010) New ways of landscape diagnosis, in Living Landscape: the European Landscape Convention in research perspective, 256–270, Vol1
12. Ingegnoli V (2002) Landscape ecology: a widening foundation. Springer, Berlin
13. Ingegnoli V, Pignatti S (2007) The impact of the widened Landscape Ecology on Vegetation Science: towards the new paradigm. Springer Link: Rendiconti Lincei Scienze Fisiche e Naturali, s.IX, VOL. XVIII:89–122
14. Ingegnoli V, Giglio E (2005) Ecologia del paesaggio: manuale per la conservazione, gestione e pianificazione dell'ambiente naturale ed antropico e delle sue risorse. Esse Libri, Napoli
15. Ingegnoli V (2011) Bionomia del paesaggio. L'ecologia del paesaggio biologico-integrata per la formazione di un "medico" dei sistemi ecologici. Springer-Verlag, Milano
16. Ingegnoli V (1971) Ecologia territoriale e progettazione: significati e metodologia, In: L'ingegnere e la natura. Giornata Internazionale FEANI, Museo della Scienza e della Tecnica, vol. I, pp. 398–400. Coll. Ing. Milano
17. EU Commission (2001) Directive 2001/42/Ec of the European Parliament and of the Council of 27 June 2001 on the assessment of the effects of certain plans and programmes on the environment. Official Journ. of EU Communities
18. EU Commission (2005) The SEA Manual, A Sourcebook on Strategic Environmental Assessment of Transport Infrestructure, Plans and Programmes. EU Commission, DR. Gen. For Energy and Transport
19. Provincia Autonoma di Trento (2006) Piano Urbanistico Provinciale (PUP): 1 allegato 2 – indirizzi per le strategie della pianificazione territoriale e per la valutazione strategica dei piani. Trento
20. Daly HE (1999) Uneconomic growth and the built environment: in theory and in fact. In: Kibert CJ, Wilson A (eds) Reshaping the built environment. Island Press, Washington, DC, pp 73–86
21. Magro G et al (2013) Q-cumber: the environmental-social network for IA. In: Annual conference IAIA2013 "impact assessment: the next generation". Calgary 13-16/05/13, Canada
22. Magro G et al (2012) Multimodeling approach for integrated EIA (EIA&SEA). In: Annual conference IAIA2012 "energy future the role of impact assessment". Porto 27/05/12-01/06/12, Portugal
23. Carson R (1962) Silent spring. H.Mifflin, New York
24. Commoner B (1966) The science of survival. Viking, New York

11 Therapy and Design of the Landscape

11.1 Landscape Design Method, Applications and Sustainability

11.1.1 Main Phases of Design

As stated in the first chapter, in studying the landscape we must proceed from the general to the particular. Therefore, after an exposition on landscape planning and assessment, we may pass to a more detailed, sub-system scale. To understand that eco-design must follow all the stages of design, here we present a general method. The main steps are:

(i) The elaboration of an analysis and a diagnostic assessment of the state of the system on which to intervene
(ii) A proposal to address improvement and sustainable type of treatment (*therapy*) to be followed
(iii) The specification of the design goals to hit and the required minimum sequence of phases to be implemented: preliminary project
(iv) The acquisition of a series of additional knowledge that can facilitate the development of an ecological project
(v) The acquisition of a series of pragmatic knowledge, such as to allow the realisation of the implementation
(vi) The drafting of a plan and a project with the executive processing
(vii) The economic-sustainable control of the construction phases
(viii) The directional ability of the stages of implementation
(ix) The control assessment on the final outcome of the implementation process (testing)

It should be noted that the first two steps include studies of landscape bionomics: analysis, diagnostic evaluation, pathology, therapy (which constitutes the main part of a volume like this). As a result, (i) and (ii) are a large and fundamental part in the activities of a "physician-ecologist" (ecoiatra) who can not avoid to intervene: thus an ecoiatra needs also the other seven steps, to become, so to speak, a "surgeon-ecologist".

(iii) The third point comes more directly in the first settings of Design and requires a creative endeavour not indifferent to the preparation of the preliminary draft or rough. To match the chosen objectives, or the ones agreed with the local authorities, with intervention consistent with the basis of the research and of the clinical diagnostic analysis, which then in practice should lead to formal drawings of application, is not easy. However, there is not an explicit method for this step outside of what we have said earlier, also recalling paragraph on the drawing in the natural sciences (see Sect. 7.3): drawing in the design is of paramount importance both to better understand the existing and to verify the proposed intervention project.

(iv) The fourth issue involves the need for side knowledge. The main subject areas, required for applications of the ecological study of the area, are summarised as follows:

V. Ingegnoli, *Landscape Bionomics Biological-Integrated Landscape Ecology*,
DOI 10.1007/978-88-470-5226-0_11, © Springer-Verlag Italia 2015

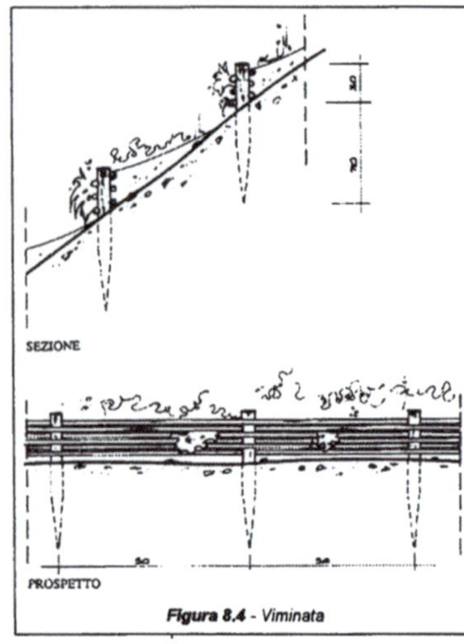

Fig. 11.1 *(To the right)* Viminate used in the restoration environmental design of the extractive quarry of Tavernola Bergamasca on Lake Iseo. In the picture (*left*) some first results after the interventions (Fig. 6.9 shows the state quo ante)

(a) Methods of relief and representation (topography, photography, drawing, etc.).
(b) Methods of estimation and control of disturbing elements (shadows, traffic, noise, water runoff, wind, etc.).
(c) Methods of bioengineering nature (Fig. 11.1) (protection of slopes, green embankments, block of landslides, renaturation of watercourses, etc.).

In Fig. 11.1 we show the details of a "viminata" that were part of in the restoration environmental design of the quarry of Tavernola Bergamasca (Lake Iseo) and the photo of the entire area of intervention of eco-design, developed in the wake of the studies of landscape ecology in the nineties.

(v) The fifth point highlights the need for practical knowledge which mainly concerns the methods of agronomy-forestation. For example, you must know the technical details of hydroseeding, planting of seedlings, planting trees, planting distance, choice of mulch, etc. Another area is that of structural engineering, at least as a principle of knowledge, so you can talk with the engineers (retaining wall, drainage channels, roads work and control, etc.).

(vi) The sixth point is necessary to know the developments of the design according to the technical requirements and approval. This hierarchy consists of:

(a) The final projects = it is the design phase to the next rule, drawn in a more detailed scale and containing reports and descriptive specifications of the materials and the species to be used, required for the approvals from the relevant authorities.
(b) Implementation projects = these developments, sometimes extracts of the final project (if the work is particularly demanding), must be prepared to an even more detailed scale, with construction details (e.g. bioengineering nature) and with useful information for the yard.

(vii) The seventh point is made by the business sector, which is partly anticipated in the final projects, but only briefly, and that is then developed on the fringes of the executive project. The most significant report is the so-called "List of unit prices" which refers to the bulletins and local regulations and market costs of materials, plants, and the workforce. The procurement of the works is part of the economic sector. The main contents of the contract are as follows:

1. The papers of the final design of the works to be carried out

2. The general terms and conditions and tender documents
3. The list of unit prices
4. Security plans
5. The time schedule

(viii) The point eight is of major importance, since it concerns the direction of the work (WD). The implementation of the work requires the institution of the Office of Works management, which consists of the Director of Works and of one or more assistants acting as operational managers or inspectors of the site. In detail, the Director of Works:

1. Cares that the concerned work was carried out in a workmanlike manner and in accordance with the project and the contract
2. Is responsible for the coordination and supervision of the office of the works
3. Interacts exclusively with the contractor for the technical and economic aspects of the contract
4. Has specific responsibility of the acceptance of materials, also on the basis of quality controls of the mechanical properties in relation to the official norms
5. Periodically checks the contractor on the ownership and regularity of the documentation provided for the law, concerning the obligations to its employees

(ix) The ninth point concerns the testing of works. The activity of the realisation of an environmental design (public or private) culminates in the phase of testing that has a dual purpose: on the one hand to verify and certify that the work has been performed in accordance with professional standards and in accordance with the specifications set in the Special Tender and contractual provisions, on the other hand to liquidate the remaining credit of the contractor.

11.1.2 Landscape Bionomics Applications

As you can easily imagine, the (potential) fields of application of the landscape bionomics in Europe, as in all territories, first of all of the West, are of considerable importance. You can, for example, remember the following:

(a) Design and maintenance of parks and nature reserves
(b) Ecological monitoring of a forest
(c) Design and control of ecological networks
(d) Planning for restructuring agricultural landscapes
(e) Environmental recovery projects
(f) Systems projects of urban green areas
(g) Criteria for urban ecology in the planning
(h) Strategic environmental assessment (SEA)
(i) Environmental impact assessment (EIA)
(j) The study of environmental interference (SIA)
(k) Audits of environmental sustainability
(l) Land use plans and landscape
(m) The control environment-human health

Unfortunately, a heavy obstacle to advanced applications of ecology consists of three still very poor conditions, which are present in some European countries (e.g. in Italy):

1. If a technician is well introduced in local government, often he is able to be assigned environmental works even in absence of an effective know-how, at the expense of true experts.
2. The professionals offering more crude, dismissive and even incorrect working methods have always a chance to win, because they may offer time and cost very inferior than serious professionals, these last ones able to propose more scientifically correct methods, but requiring more time and costs. Do not forget that unfair methods give a greater degree of freedom to the most unscrupulous politicians in operating according to its own interests or part, while the most advanced methods provide more guarantees and safer choices, but eliminate almost completely the abuses of freedom of management.
3. The third obstacle is related to the bureaucracy, that is to the technicians of public bodies, especially municipal ones. They are not updated enough and tend to impose a bureaucracy that is almost always short-sighted, non-resilient, not open to scientific breakthroughs,

perhaps because they think that otherwise they would lose ability to control and power.

11.1.3 Environmental Design and Sustainability

All landscape designs need to be elaborated following the principles of sustainability, to guarantee correct relations between ecology and economy.

The true problem is that all the economists sensitive to ecological principles haven't a sufficient ecological background to propose alternative actions of wider cultural perspective. The relationship between ecology and economy is difficult because *both* are generally *misunderstood* and separated [1]. For instance, in economy:

1. Exchange-value is thought to be the measure of richness and to be independent from nature. By contrast, we note that the value of something depends on its role.
2. Our culture rejects any *ethical evaluation* of means and ends. So the end justifies the means. By contrast, ends and means are strictly linked in nature, like a seed to its plant. Therefore, they must not be separated and we have to choose among different ends.
3. Today absolute scarcity is not seen as a limitation, because the low entropy is abundant for our necessities. But these necessities must contain the proper functionality of all the ecological systems.

But also in ecology we have many problems:

1. Man is considered to be outside the ecological processes and in contrast with nature (e.g. E.I.S. environmental *impact* statement). By contrast, human ecosystems co-evolved, co-evolve and bind themselves to all the others.
2. The concept of ecosystem prevails over all the other levels of biological organisation but it is now considered as *ambiguous*. General ecology must consider the entire biological spectrum.

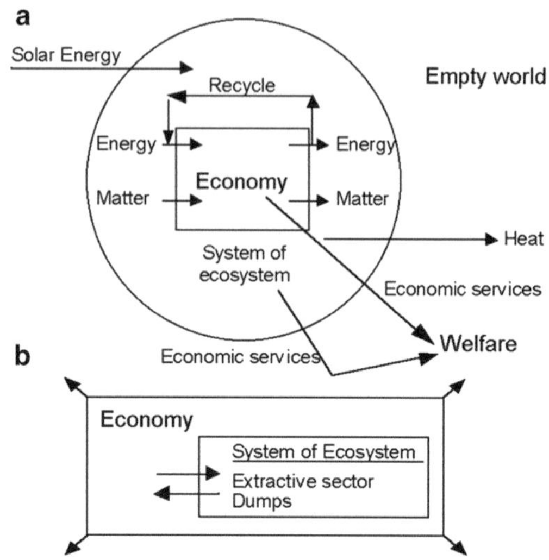

Fig. 11.2 (**a**, **b**) The scheme of the two main paradigms on economy-ecology relationships. (From Daly 1999: redrawn) (**a**) Ecological-Economics (**b**) Neoclassical Economy

3. Pollution problems prevail over all the other environmental questions. By contrast, the degradation and alteration of ecological systems (and of human health) depend first of all on the structural and dynamic dysfunction of systems of ecocoenotopes.

Our capacity to destroy the natural systems of the planet or to choose to save them (and ourselves) makes us a very *special* generation. We are veritably at a *crucial turn* of our history: we hope to change but not only because of disasters and wars. Some scientists claim that economy is impossible to integrate with ecology, essentially because of the presence of the paradigm of growth. This radical position leads to a black pessimism on the man-nature relationships and it could be a dangerous position. Moreover, it seems not to be scientifically correct.

The relations between ecology and economy show at present two major paradigms: the neoclassical one and the ecological-economics one. It is easier to explain their deep differences by referring to Fig. 11.2. The ecological systems, considered as subsystems of the paradigm of economy, postulates that ecology represents merely the extractive and waste disposal sector of economy. Even if these services become scarce, growth may continue without limits, since technology allows us to "grow around" the ecological sector by the substitution of natural capital with human-made capital. It results in the dictates of market prices, if and when prices of natural capital rise. Thus, nature is nothing but a supplier of economy (Fig. 11.2b). The only limit to growth is technology, which is supposed to have no limits.

Since economy is the whole, the growth of the economy is not at the expense of anything else: there is no opportunity cost to growth. Growth does not increase the scarcity of anything, rather it diminishes the scarcity of everything!

The other paradigm, the steady state at optimal scale, is at the opposite end (Fig. 11.2a). In the wide economy-ecology of nature, economy is a subsystem of ecology. Natural living systems are dissipative, not-growing as a whole and materially closed. The flow of solar energy which sustains ecological systems is itself finite and not-growing. The natural environment physically contains and sustains the economy by regenerating the low-entropy inputs that it requires and by absorbing the high-entropy wastes that it cannot avoid generating, as well as by supplying other systemic ecological services.

The two paradigms are per se logical: each is logical within its pre-analytic vision and absurd from the viewpoint of the other. Thus, the true problem is not to integrate ecology and economy, but to overcome the discussed dichotomy.

Herman Daly [2] observed that the faith in economic *growth* can be tempered by the demonstration that growth does not imply only benefits but also costs. Uneconomic growth exists and it is defined as growth that, at the margin, increases environmental and social costs by more than it increases production benefits.

Economy, like ecology, can be divided into macro-, meso- and micro-economy, because the disciplines which have to do with the natural systems are scale-dependent. Thus, some concepts pertain only to a particular level of scale. But some basic concepts have a universal logic: for instance, optimal scale, cost-benefit balance, system metastability and metabolic-antithermic functions. Remember the intrinsic and exportable characters of ecological systems (see Sect. 1.1.2).

Incredibly, in this perspective, macro-economy presents an evident contradiction. Cost-benefit balance is not considered! Let us refer to the GNP (or the GDP): we may think of it as the "built environment"; broadly conceived is the set of all economic activities. Unique among economic magnitudes, GNP is supposed to grow forever. This basic index (GNP) increased about \$ 85–90 per capita per year in the last half century, but the economic benefits are not comparable with its huge growth.

Note that Fig. 11.2a shows two general sources of welfare: (1) services of *human*-made capital and (2) services of *natural* capital. As the economy grows, natural capital is transformed into human-made capital. The law of the diminishing marginal utility of income and the law of the increasing marginal costs tells us that we satisfy our *most* pressing wants first and we sacrifice the *least* important ecological service first. Therefore, growth of economic services takes place at a decreasing rate, while natural services increase at an increasing rate. This process can be expressed in Fig. 11.3.

The MU (marginal utility) curve reflects the diminishing additions of marginal utility to the stock of human-made capital from consuming produced goods and services (Q).

The MDU (marginal sacrifices) curve reflects the increasing marginal cost of growth, as more natural capital is transformed into human-made capital. These costs can be for instance: scarified natural capital services, disutility of labour, disruption of community, sacrifice of leisure, pollution, congestion, environmental destruction and their negative influence on human health.

Two further limits are noted in the diagram: point (*e*), where MU = 0 and further growth is *futile* even at zero cost; and point (*d*), where an ecological *catastrophe* is provoked (MDU goes to infinity). The figure shows that growth out to point (*b*) is economic growth (benefits > costs), while beyond point (*b*) there is uneconomic

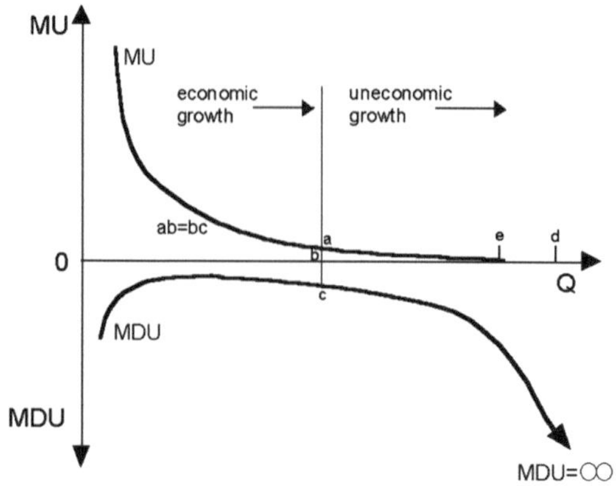

Fig. 11.3 Marginal utility (*MU*) and marginal disutility (*MDU*) vs. quantity of produced goods and services (*Q*) (from Daly 1999, modified). See text for explanation

growth (costs > benefits). Thus, beyond point (*b*), GNP acquires another meaning and its growth does not reflect benefits but costs.

Moreover, we can observe that human-made capital and natural capital are complementary, *not interchangeable*; neoclassical economics is again wrong. This is an important statement. Substitution is reversible, complementation not. Another analogy with the scientific paradigm summarised in Chap. 1: both economy and ecology must follow these new paradigms.

In an effort of the interfaces between economy and ecology, in addition to the concept of sustainability, the "ecological footprint" was proposed, concept introduced in 1996 [3]. Since 1999, WWF periodically updates the calculation of the ecological footprint in its Living Planet Report [4].

The ecological footprint is a statistical index used to measure human demand towards nature. It relates the human consumption of natural resources with the earth's capacity to regenerate them: in other words, it measures the area of biologically productive land and sea needed to regenerate the resources consumed by a human population and to absorb the corresponding waste. Using the ecological footprint, it is possible to estimate how many planet Earth we would take to support humanity if everybody lived according to a certain lifestyle. Comparing the footprint of an individual (or region, or state) with the amount of land available per capita (i.e. the ratio between the total area and population worldwide) you can figure out if the level of consumption of the sample is sustainable or not. To calculate the footprint on a set of consumer spending, relates the quantity of each good consumed (e.g., wheat, rice, corn, cereals, meats, fruits, vegetables, roots and tubers, pulses, etc.) with a constant yield in kg/ha (kilograms per hectare). The result is a surface.

To calculate the impact of energy consumption, it is converted in equivalent tons of carbon dioxide and the calculation is made considering the amount of forested land needed to absorb these tons of CO2. This footprint is however not significant enough, because it does not consider the *integration* of complex systems, neither the concept of landscape as biological organisation, producing data too detached from reality.

From the point of view of landscape bionomics [5], instead, really significant are *changes* in the study of sustainability, the first of them being the overcoming of the concept itself by the principle of "rehabilitation strategy", linked to the redefinition of the landscape as a living complex adaptive system, that is structured thanks to the cooperation and the integration of components into a system following irreversible processes and in which man and nature have co-evolved and co-evolve for mutual selection process of adaptation.

It is not, therefore, to manage the resources in order to postpone as much as possible their depletion, but to understand the relationship between the state of health and diseases of the area under examination, checking the compatibility with the needs and the human health, and to develop criteria of therapeutic rehabilitation with rehabilitative function. Faced with this, the *ethical responsibility* of man (see Chap. 15) is to govern the nature in responsible sense of service, and promoting a more advanced study of its laws allowing us to the clinical-diagnostic methodology and to prepare operators having the courage to follow the therapeutic indications provided by the "environmental physician" (ecoiatra) even when not "politically correct".

It should be noted that, until a few years ago, we distinguished between weak sustainability (where the natural and economic components are considered as interchangeable) and strong sustainability (where the natural components assume a critical priority). But even in the latter case, the sustainable use of ecological systems is often inappropriate, that is insufficient or excessive, since it *does not consider* the *environmental systems as living entities*, nor it works outside of a scope limited to a certain site. For example, the transformation of an area from agricultural to urban one or from industrial to residential one is done being careful to design something of environmentally sustainable, but in the sense of balanced within the territorial element in object! It reiterates that this is insufficient because, as taught by the bionomics of the landscape, what matters are the clinical-diagnostic method and the *context*, that is the system of interacting ecosystems in which the site is inserted and the study of its possible pathologies: namely, rather than sustainability the goal is rehabilitation strategy.

11.2 Applications in Natural and Seminatural Sub-Systems

11.2.1 Intervening in Natural or Semi-Natural Landscapes

The applications of landscape bionomics in natural areas require first of all a clarification of the concept of *naturalness*. In dictionaries the natural (naturalness) is defined as the quality or condition of what is natural, but for conservationists is usually the condition of absence of man and his artefacts. This last consideration, however, denotes a prejudice, that man was not part of nature. As we will see in Chap. 15, the reason is based on an ill concept of freedom: man must never be considered as dependant. It is instead needed a new concept of "*widespread naturalness*" in order to avoid the separation between natural areas under protection and areas exclusively anthropogenic. Moreover, such a separation would be contrary to the concept of ecotissue, (see Sect. 2.3.2), which implies the conservation as naturalness spreading into the human habitat and the preservation of historical and archaeological sites also in natural habitats.

Also keep in mind that nature is not determined once and for all, being open to creative perspectives often unpredictable. It must therefore abandon the Cartesian idea that nature, without understanding history, is mechanistic and therefore may be dominated by men, who become the "engineer" of nature itself. In this new perspective of diffuse naturalness, a cultural landscape, such as agriculture, could also be considered natural. The dividing line is not the presence or absence of man and his activities, but rather the way of management, abuse or without technology, and the contrasts between economy and ecology.

The concept of widespread naturalness, following the principles of landscape bionomics [5], does not admit to base the biological conservation only on parks, even if the parks we want. In the world, the growth of population has increased exponentially in the eighteenth century, with a maximum increase in the twentieth century and a small decrease in the last 30 years. This note sigmoidal curve occurred in a similar way for the growth of protected natural areas, reached about 10^6 km^2 in 1920–1925 to grow nine times as much until 1985 and then increase with a less momentum in recent 27 years (see Fig. 4.1).

In Italy, at the beginning of the third millennium, there are 22 national parks covering a million and a half hectares, or about 5 % of the national territory. The national park completes the protection operated by the regional parks, and vice versa, dealing with very large territories (even several tens of municipalities). In addition to planning and supervision, therefore, the national park must intensify its mission to link tool and enhancement of local realities. The environmental associations had asked for 10 % of the national territory to be protected through parks. This request, however, does not make sense for science, either because the park should not be the only tool of conservation, and because in certain regions 10 % is too little, in others too much. We must avoid, further, that the natural parks and

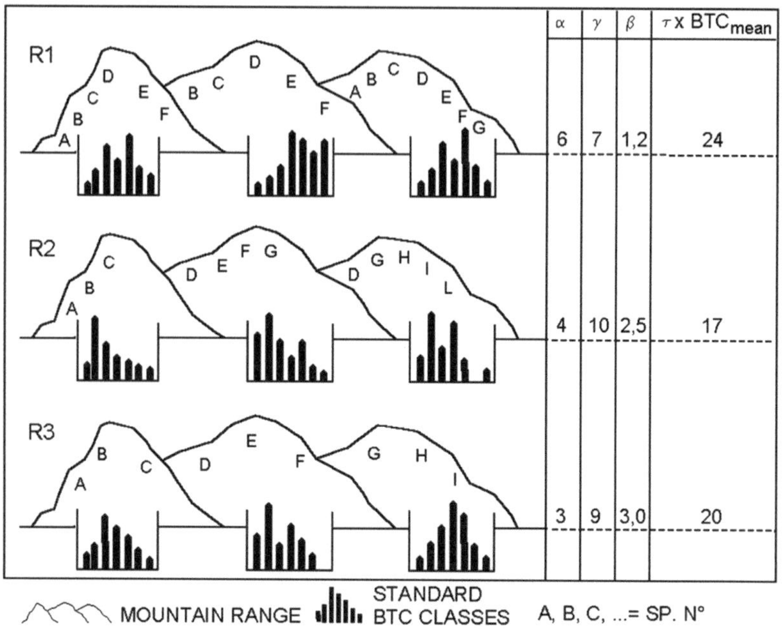

Fig. 11.4 Biodiversity indexes for three landscape units, each with three hills. The first case has the highest α diversity, the second the highest γ, the third the highest β. If we add the distribution of standard BTC values of the components of each ecotope, it is possible to measure also the landscape diversity and add the value of the landscape metastability (LM). The first case has the highest value, thus it seems the best to protect, even if the highest γ index is in the second case

protected areas of Community Interest (SCI, SPA) could become an *alibi* for the surrounding territories toward the destruction of nature.

Nature conservation: in order to better choose between the different areas to be conserved, ecology generally recommends to use biodiversity indicators. We will have to see, as already pointed out, as it dealt only with *specific* biodiversity. So far, only the specific differences were considered:

$$\alpha = \text{species/site};\quad \gamma = \text{species/set of sites};$$
$$\beta = \gamma/\alpha;\quad \eta = \text{species/ecotope}.$$

The assessment of biodiversity can, however, change unexpectedly if we appeal to parameters of landscape ecology (Fig. 11.4). This figure was drawn from Primack [6], so as to refer to an example which has become classic, later modified by Ingegnoli [7] just to demonstrate the need to take account of landscape bionomics. We have three units of landscape (R1, R2, R3) among which to decide the priority of protection. According to the traditional indices will be chosen for R2, in the case of protecting the entire LU. A general index of metastability (LM), however, shows that the best LU is R1, with average LM = 24.

A further aspect of conservation biology that the bionomics of the landscape can definitely improve is the planning of the *boundaries* of protected areas. Too often you notice absurd discrepancies between the landscape units and the boundaries of these areas, whether they are of regional nature reserves, national parks or national or regional SCI (Sites of Community Importance) and SPA (Special Protection Areas). Given the importance of the issue, we will produce a couple of examples of this, taken from a study for the region Friuli-Venezia Giulia (North Italy).

The first example concerns the reserve of Cornino (Fig. 11.5), on the margin of the Carnic Alps, designated to protect the vulture (*Gyps fulvus*), approximately 500 ha [8]. The figure shows the strong difference between the reserve (red line) and the landscape unit (LU). Despite some enlargement towards the Tagliamento river, it remains outside protection the plateau lying behind, which is a site of fundamental importance for birds.

A different, but no less serious case of deficiency in the identification of the boundaries of regional parks is that of Galbiate LU (Lecco). In this LU there is the presence of three parks: the South margin of the Park of Monte Barro, an

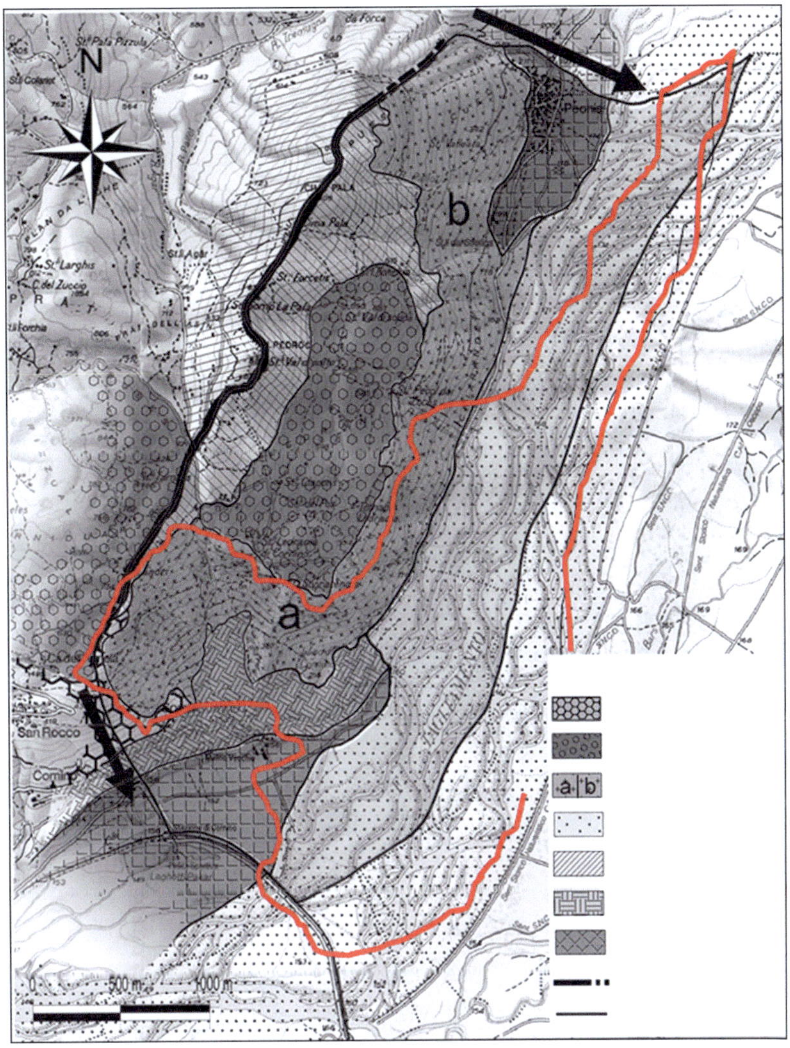

Fig. 11.5 Character of the landscape in which the unit is the Reserve of Lake Cornino, in Friuli. These are the first hills to the west of the Tagliamento river. In *red* (most of) the boundaries of the nature reserve, which do not coincide at all with the LU (from Ingegnoli & Giglio, redrawn). LEGEND (*top down*): steep wooded slopes, karst plateau, rocky habitats range (**a**, naked, **b**, forested), band pertaining to river, mountain slopes and limestone foothills, floodplain, watershed line, LU limit [*arrows* indicate the torrential incisions]. The correct LU delimitation follows the watershed line and the two torrential incisions, while within the river bed it is less clear

offshoot of the Park of St. Genesio and a stretch of the Adda Nord Park (along Lake Garlate) as shown in Fig. 11.6. These parks, however, are not in contact with each other, creating connection problems; in addition, lies there a patch of forest in a good ecological value; however, the best of all the LU is left even without any protection! To propose a project to connect the three parks, including the safeguarding of scrub forest, it had to build a map of the resistances.

Sometimes the map of undisturbed patches of strategic interest (see Sect. 9.2.2) is not enough to set a good project reconnection between protected areas. May then be used the construction of a map of the resistances. It is intended for the whole landscape resistance of the barriers and problems that slow down or prevent a smooth transfer of resources or species (including humans) from the boundary of a LU in question. To build a special map you should consider the value of BTC for each element, the percentage of natural habitats (NH) and four levels of the presence of *barriers* and *disorders* (BD):

1. BD = 0.50 high presence of barriers and disorders
2. BD = 0.33 mean presence
3. BD = 0.25 low presence
4. BD = 0.00 absence of barriers and disorders

The indicator of resistance (RS) is estimated by measuring:

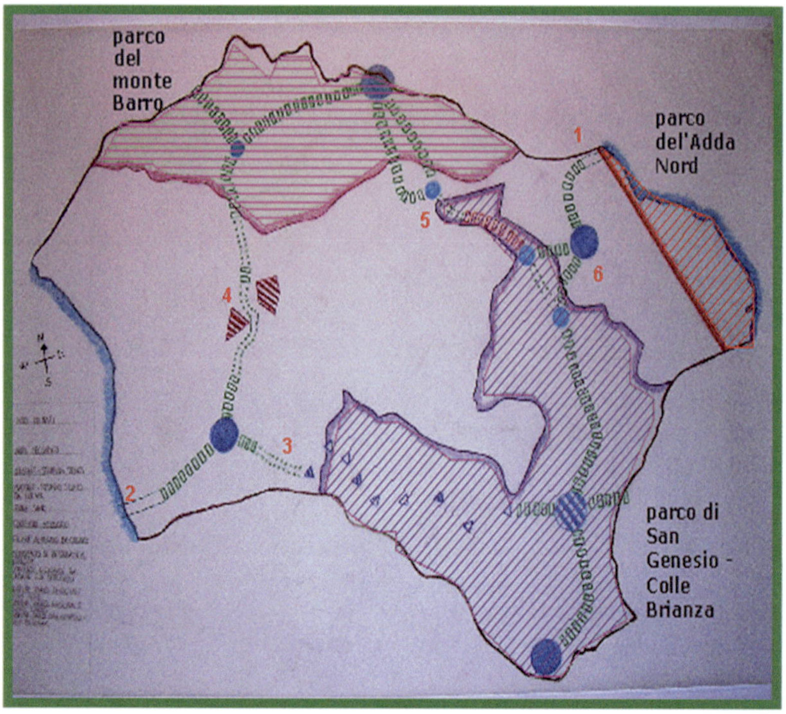

Fig. 11.6 Landscape units of Galbiate (Lecco), in which they scored: at the *top* (*pink*), the area south of the Park of Mount Barro, at the *bottom* (*purple*), the area north of Park St. Genesio, to the east (*red*), a stretch of the Adda Nord Park. Around the *blue stamp* n.6 is the stain of forest with greater BTC across the LU, not included in any park. Note that the three Parks are not connected together

Fig. 11.7 Map of the resistances in the LU of Galbiate (Lecco). The colour intensity is directly proportional to the resistance exerted. See text for details (from Ingegnoli and Giglio, 2005 modified)

$$RS = 10/BTC \times NH\ (\%) \times BD.$$

The landscape resistances values obtained must be divided into four coloured sets, from lightest to darkest (Fig. 11.7):

1. $RS < 1$ for resistance nil or insignificant
2. $1 < RS < 5$ for low resistances
3. $5 < RS < 50$ medium resistance
4. $RS > 50$ for high resistances

From the table of the resistors you can move on to the design and verification of an ecological network.

11.2.2 General Bionomic Prescriptions for Natural Parks and Reserves

Too many times, especially in Europe, conservation areas, as natural parks or reserves or even small biotopes, are under preservation acts, but if we analyse their landscape bionomic status we may register many problems. We already underlined this fact speaking of protective systems (Sect. 4.1). Therefore, it could be methodologically important to synthetise a general frame of bionomic prescriptions related to conservation areas and natural or regional parks, as shown in Table 11.1.

This table exposes nine environmental characters related to parks and, for each attribute, seven prescriptions are synthetised, mainly deduced from landscape bionomics principles, hence 63 items. These points can be present only partially and in this case have to be counted with half value. Suggestions and prescriptions in the table are distributed in this book and do not need any major explanation or discussion.

On the other hand, Table 11.1 gives a good possibility to "check-up" existing parks or to design new conservation areas.

An example may be the control of the Regional Park of Groane (Fig. 11.8), North near Milan, 3,400 ha about 25% of which are wood and forest. Controlling this Park with the prescriptions of Table 11.1, we find only 5 full and 28 partial items, therefore:

$$PkBiCt = (5 \times 1) + (28 \times 0.5) = 19/63 = 0.3016,$$

where PkBiCt = Park Bionomic Characters; Scores: (a) full presence of item = 1.0 (b) partial presence = 0.5

A value of only PkBiCt = 0.30 is very low and, for the moment, is evident that the Groane Park is preserved only to avoid the total destruction of the remnant natural or semi-natural patches in the Metropolitan Area of Milan. But a park like this needs very serious improvements!

11.2.3 Ecological Network Design

One of the first and largest set of ecological protection criteria was elaborated on and proposed by the Society of Conservation Biology (SCB), in the USA. Since its foundation (1987), Michel Soulé, Reed Noss and their collaborators have been working on the "Wildlands Project", with the aim of preserving the entire continental territory of their nation. The modalities of this wide-ranging conservation plan are based on the concept of *"ecological network"* [9], in parallel with a similar concept proposed by Bennett [10] for the European Union.

No doubt that ecological networks present clear *merits*. First of all, their methodological criteria are simple and the comprehensibility of the applied conservation principles gives territorial managers, especially public authorities, the possibility to implement the protection of nature. The territorial surface commitment is generally quite small and the political effects seem to be very much appreciated.

Some aspects of landscape bionomics are transferred even to people with limited knowledge of ecological principles, or to people used to confusing ecology only with pollution. That is why the Habitats and Species Directive of the European Union [11] more or less includes the concept of ecological network, although it leaves much of its realisation to national governments. The development of the "Natura 2000" project included a

Table 11.1 General landscape bionomic prescriptions for natural parks and reserves in temperate climate

Characters	I	II	III	IV	V	VI	VII
Perimeter	Without introflextions or thinning	Control for core areas and ecotopes	Presence of buffer belts	Taking care of the ecological ray of influence	Presence of connection nodes	Segment of porosity for exchange	No barriers against migration
Structure	Patch shape congruent to local characters	Minimum contrast among elements	Preserving ecotonal edges	Controlled fragmentation	Good element heterogeneity	From Agr-prt to For-natural	Including some sink areas
Functions	BTC_{park} > BTC_{region} (10–20 %)	Carrying capacity Min. SH/SH* > 3–4	Human habitat HH < 45–50 (% LU)	General metastability g-LM > 30–35 (bit/Mcal/m^2/year)	Diagnostic index DI > 80–85	Potential core area PCA > 55 %	Presence of higher BTC classes
Forest tesserae	No dominant allocthonous	At least 15–20 % of mature tesserae	Control dangerous insects	Controlled cuttings	Presence of old trees in each Ts	Forest BTC > 7.0–7.5	Forest efficiency CBSt > 36–40
Rivers	IBE cl. I e II	IFF cl. I e II	Fish stair lift if river with bridles	Riparian ecotonal belts	Water intakes with mitigation and compensation	No hydropower > 1 MW	Channelled traits < 10 %
Urbanised and roads	$SH_{urb + sbs}$ <5–7 % HH	Silent area (< 55–60 dB) > 70 %	Max urban green presence	No air pollution Lichen cl. I & II	Percolation BFF > 0.65	Road ecology prescriptions	Local building material & few floors
Sport and recreation Areas	Stop & pic-nick areas with protection	Sport far from biotopes	No sport plants in area of high CBSt	Well designed pathways	No cycle-path in areas of high BTC	Walkways prohibited in certain spots	Marginal and controlled areas for big groups
Crop fields	BTC-prd >1.1–1.2	Prohibited not natural cultivations	No OGM	No pesticide and chemicals	Rural roads not paved	No monocultures	Biologic agriculture > 50 %
Connective elements	Hedgerows without allocthonous	Green network/ roads > 1.1	Controlled connectivity & circuitation	Network efficiency NEf > 1.2	Connections among near parks	Minimum disturbances in every node	Protected animal crossing paths

Fig. 11.8 A map of the Groane Park, North near Milan continuous and dotted red line = bike roads; black dotted lines = railways; grey squares = urban built areas; yellow-green = 'preserved' patches of forest; light green = agricultural fields; other green = woods

network of special areas for conservation (SAC). On the other hand, ecological networks present also many *limitations*. It is necessary to summarise the main questions, because we think that the first positive contribution to nature conservation will be cancelled if we do not change the methodology.

- The scientific basis of the model is excessively simplified: we have the impression that an easy solution to be applied in densely inhabited territories has been transformed into a scientific methodology.
- A landscape is a complex ecological system (see Chap. 1) that cannot be defined simply as a network of patches connected by fluxes of energy, matter and species.
- If the structural model of a landscape is an ecotissue (see Sect. 2.3.2), we have to study its processes, many of which cannot be

reduced to a network! See, for instance, the spread of exotic species which does not follow a network model.

- There is the problem of landscape diversity (τ), which depends on the presence of all the standard classes of territorial capacity of the vegetation, included fields, orchards and gardens.
- The distribution of natural habitat (NH) within a landscape is not distinguishable per landscape elements, but can be imbedded in each element.
- In ecological network design, the core areas cannot be only those having a high ecological quality, because of the source-sink theory.
- The distribution of the minimum resistance of a landscape may indicate a network pattern, but it does not represent the only possible pattern (for instance, new corridors can be designed even in high resistance zones).
- A landscape network can become, in some cases, too rigid with respect to the continuous transformation of many landscape units.
- For certain groups of animals a corridor can be of little or no utility, and even have negative effects: predators, parasites, etc.
- Consequently, landscape ecology risks appearing as an "easy to apply" trans-discipline, therefore allowing many professionals, without an integrated natural background, to plan and design networks.

11.2.3.1 Designing of Networks

Once you are aware of the limits of the concept of network and that you are familiar with the principles and methods of landscape bionomics, then you can think of to address the design of an ecological network. The most important analysis concerns the evaluation of the undisturbed areas of strategic interest (see Sect. 9.2.2) and, if necessary, the evaluation of the resistances. The essential components of the networks are summarised in Fig. 11.9.

However, further investigations are needed. Ecological networks (EN) are formed by the following elements, namely: (a) natural or semi-natural components (margins of forest and

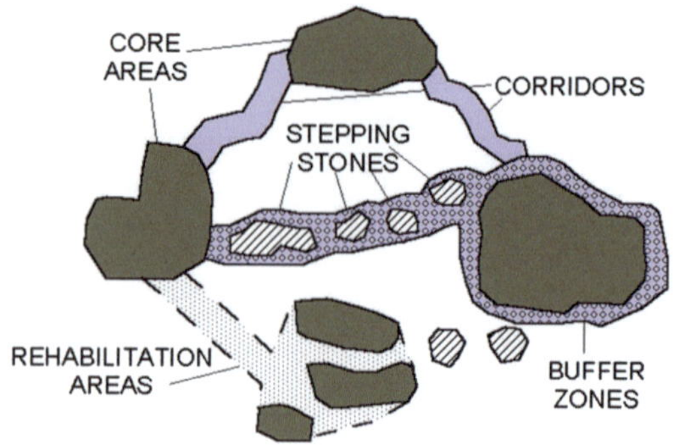

Fig. 11.9 The main structure components of an ecological network. With these components (core areas, corridors, buffer zones, stepping stones, rehabilitation areas) it is possible to put in evidence many aspects of a landscape, but not its entire complex structure

tree-lined avenues, or hydro-graphic), (b) human components (roads, technological networks, perimeters town centres) and (c) area affected by a network (EN area/LU), capable of circumscribing it with a margin of about 80–100 m beyond the perimeter. Also the following parameters are listed:

- *Connection*, (CON) the primary condition for a EN, measured explaining a planar graph of *n* vertices (V) and *m* links (L) as measured by *alpha* and *gamma indexes*, according to Forman and Godron [12]
- *Density*. We distinguish two types of density: (1) of vegetated corridors (DCV), measured in km/km^2; (2) of roads (DST), similar measure
- *Points of intersection* between vegetated corridors and road (PI) and any technological networks, such as high voltage power lines and gas pipelines in the province
- *Network density ratio*, between DCV and DST
- *Network/LU ratio*, including network area and total area of LU (EN/LU)
- *Junctions/links ratio* (PI/L)
- *Interference* (INTF), calculated as (PI/L) × (DCV/DST) × (EN/LU)

The result is a *network efficiency* (NEf) and it can be computed as:

$$NEf = (CON \times DCV) - INTF.$$

Fig. 11.10 Networks in the LU of Loppio (Mori, Trento). In *green* the design proposals. Natural existent elements (corridors or wood edges), Natural elements (design propositions), Perimeters of landscape units (LU), Areas containing the ecological networks, Main technological networks, Fluvial network, Road network, Towns and farms

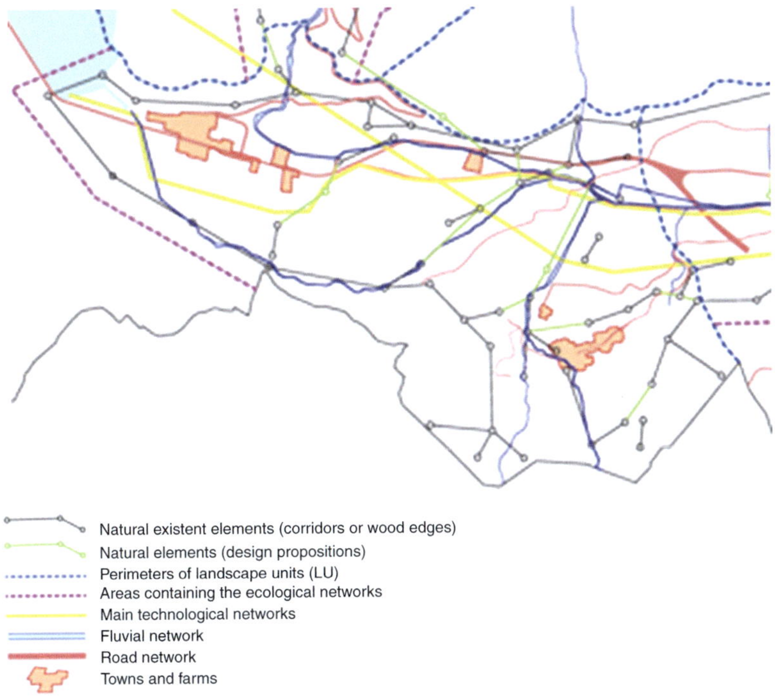

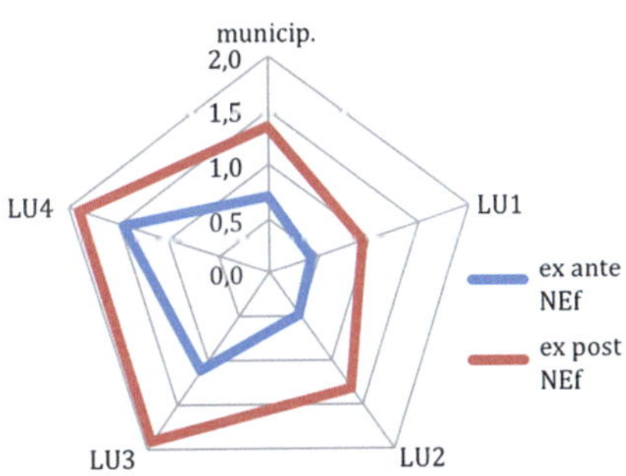

Fig. 11.11 Comparison between the situations ex ante and ex post regarding the overall efficiency of ecological networks of Mori. The greatest improvement is found inLU-2 (Loppio)

An example of design is referred once again to the town of Mori (Fig. 11.11 and Table 11.2). The example concerns the LU-2 Loppio, perhaps the most disastrous in terms of ecological connections.

Table 11.2 Networks efficiency of LU-2 Loppio (6.02 km^2)

Ecological network attributes	Ex ante	Ex post
Vertex V	58	60
Links L	47	62
Circuitation (α index)	−0.090	0.026
Connectivity (γ index)	0.280	0.356
Connection CON (α + γ)	0.190	0.382
Area (a-EN) [km^2]	3.6	3.6
Length [km]	11.230	14.130
Density of corridors [km/km^2] (DCV)	3.119	3.925
Roads [km]	8.720	9.700
Road density [km/km^2] (DST)	2.422	2.694
Intersection points CV-ST, (IP)	6	12
Network density ratio DCV/DST	1.288	1.457
Area EN/LU	0.599	0.599
Interferences (INTF)	0.098	0.169
Network Efficiency (NEf)	*0.493*	*1.332*

EN ecological network, *CV* corridors of vegetation, *CA* core areas, *a-EN* area pertaining to an EN, INTF = (IP/L) * (DCV/DST) * (EN/LU)

The criteria for the project depart from the observation of the shortcomings of the current state (condition ex ante) that are highlighted in a separate table. As a result, complementary interventions are proposed. Let us begin to verify the results of the design choices of the plan for ecological networks by analysing each LU. Many types of project interventions of detail become important in the design of ecological networks. Overcoming barriers committed more than any other issue, especially regarding the presence of road networks. To better understand the comparisons between the situation ex ante and ex post, and to check whether the actions of the project can be judged as sufficient and effective, we present a control plot (Fig. 11.11), which covers the whole territory of Mori, then its four LU: it shows that the project proposals give the most improvement in the network of LU-2 Loppio.

Moreover, considering the relations Core areas/landscape unit (CA/LU) and summing these relations with EN/LU, you can reach values close to 1, given the complementarity of these amounts; the differences to reach unity gives an idea of the dispersion of margin, in these cases fairly contained.

11.3 Applications in Human Infrastructures

11.3.1 Road Ecology

In the design of ecological networks there are different and unavoidable relations with the road networks. Forman et al. [13] have devoted a publication to Road Ecology. The more interested data are undoubtedly: (a) road density (km/km^2), which vary greatly in relation to the distribution of cities and geomorphological emergencies in a territory and (b) road length/1,000 people. You notice large differences in density (km/km^2) between Italy and USA (1.7), the UK (4.0), Germany (4.8), Japan (7.9) and, in analogy, in km/1,000 persons: 7 in UK, 8 in Italy, 8.5 in Germany, 9.5 in Japan, 14 in France, 24 in USA.

Obviously, in landscape bionomics the dynamics of the density over time is also important. Usually, the density increase of roads means a decrease of forest area (induced fragmentation) but, growing agricultural and urbanised area, although beyond a certain level (e.g. about 3 km/km^2 in Massachusetts), forests return to grow. The problem in Italy, however, is at radically different level. Considering the scarcity of arable land in the world (about 14 % of the emerged land) and the Italian situation of drastic decrease of agricultural areas in the last 30 years, the rise of the road surface does not affect much the forests (now mostly located in mountain or hill) but the agricultural areas, which are increasingly thinned and fragmented. In fact, every new road brings with it, in short time, the construction of new industrial and commercial complexes, gradually killing what is the landscape function and breaking up ecotones between urbanised and forest, as well as productive areas [8]. At this rate, soon, agricultural landscapes risk of disappearing, with all their wealth of human and natural culture.

The streets are very powerful corridors of expansion of alien species, so much that, until a few tens of meters from the road edge, in many cases the non-native species can exceed the number of indigenous one. Also the outer bands to the road edge become important for the transit of the animals, and then the arrangement of plants along it must be carefully studied. For a relationship between animal and road density see Fig. 6.8. The range of influence (and noise) of a road from the edge varies greatly in relation to the measured parameters. In our field issues to be evaluated more carefully are:

(a) The microclimatic change compared to the surrounding areas, which can exceed 2 °C in summer, reaches up to about 20–25 m from the edge
(b) Pollution by heavy metals and PCBs, up to almost 200 m
(c) Disturbance to birdlife from the busy road, up to 300–400 m

(d) The same as a very busy highway, up to 700–800 m
(e) Disturbance to large mammals, main road, up to 900–1,000 m
(f) Invasion of non-native plant species, to over 1,000 m
(g) Disorder particulate matter (PM10) from a highway: 40–100 kg/h, up to more than 200–250 m
(h) Disorder NOx and CO from Highway: 100–300 kg/h, up to more than 180–240 m
(i) In the countryside irritating noise (> 60 dB), up to 200 m

Recalling that 1 decibel (dB) is defined as: $10 \times \log_{10} P/P_0$ [where P is the sound pressure] you will notice that the attenuation in dB (respectively 15 and 60 m) from a busy road can be obtained by embankment with trees on the side (about 10–12 dB), while a band of trees alone would give a different and less attenuation, equivalent to 3–5 dB. It should be noted, however, that an attenuation of about 3 dB is equivalent to halving the loudness.

As known, the trend in the design of new roads is to join the network nodes (e.g. cities, towns or industrial areas) possibly in a straight line, for obvious economic reasons, that is, saving time and expense. This often clashes with the requirements of nature protection. We present, therefore, a control method for bionomic assessment of roads design, taking account of ecological parameters, which measure and compare alternative routes (e.g. A and B) on the basis of:

(a) The value of BTC of landscape *elements* crossed: Σ BTC_E
(b) The presence of critical *points* and *barriers* to be overcome with appropriate works: Σ PB
(c) Fragmentation of core area: F_{CA}
(d) Technological works related to the type of road and traffic and related services.

Unless problematic cases, we see that the point (d) is qualitative and can have a weight in the case of minor differences between the alternative routes to examine through. The other three points, however, lend themselves to a diagnostic evaluation, which must proceed by comparing all the measurements of the unit length of the road (Ls):

1. Σ BTC_E [Mcal/m^2/year]/Ls [km] = BTC average unit along the route, which indicates the degree of potential destruction of biological territorial capacity of the vegetation operable from the road in question
2. Σ PB [n °]/Ls [km] = PB average unit, which indicates the degree of technical infrastructure needed to follow that particular route
3. F_{CA} = (mg/in) × L_{CA} (where: mg = margin ratio; in = inside; L_{CA} = length of crossing the core area). Comparing F_{CA}/km = F average unit, which indicates the average fragmentation induced by the route
4. (RI) the road interference on environment/km, that is:

$$RI = BTC/km + PB/km + F_{CA}/km.$$

Consider the example of two alternative paths (I) and (II) as shown by Fig. 11.12: the path (I) is 18 km away, (II) 16.6 km

Path (I)

1. The sum of the BTC of the landscape elements crossed: BTC_E $\Sigma = 20.82$, then BTC_E/km = 1.1567 Mcal/m^2/year.
2. The sum (n °) of barriers and critical points: BP $\Sigma = 3$, then BP/km = 0.1667.
3. The extent fragmented core areas: F_{CA} = 1.5 km, then F_{CA}/km = 0.0833.
4. Interference of the road (RI) total/km: $RI_I = 1.1567 + 0.1667 + 0.0833 = 1.4067$/km.

Path (II)

1. The sum of the BTC of the landscape elements crossed: BTC_E $\Sigma = 44.68$, then BTC_E/km = 2.6916 Mcal/m^2/year.
2. The sum (n °) of barriers and critical points: BP $\Sigma = 2$, then BP/km = 0.1205.
3. The extent fragmented core areas: F_{CA} = 6.4 km, then F_{CA}/km = 0.3855.
4. Interference of the road (RI) total km: RI_{II} = 2.6916 + 0.1205 + 0.3855 = 3.1976/km.

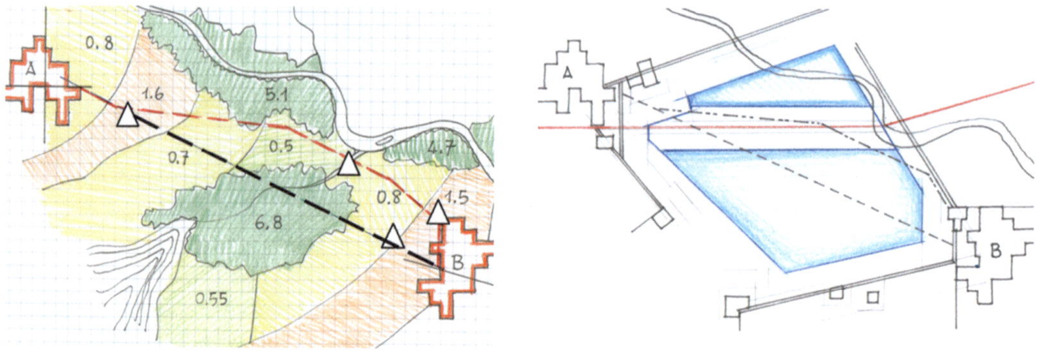

Fig. 11.12 Schematic example of ecological assessment of two alternative routes by road. The path (I) is marked in *red*, is 18 km away and has three critical points; the path (II) is marked in *black*, measuring 16.6 km and has two critical points. As we see on the right (in red a power line) the path (II) alters the potential core area marked on the side. Since we report the values of BTC for each element of the territory concerned, we can apply the indices of RI (interference road environment), from which the path (I) is 56 % less damaging

The comparison between the two alternatives is expressed in the ratio II/I = 3.1976/ 1.4067 = 2.27: accordingly, the path II is 2.27 times worse than the path I.

11.3.2 Design of Degraded Areas

Another topic of considerable interest in environmental design is related to the rehabilitation of degraded areas (e.g. quarries, mines, industrial warehouses, etc.). Therefore, it is necessary to realise the differences between the traditional methods and those that follow the principles of landscape bionomics. We summarise the most significant aspects in the two methods.

A1. *Traditional analysis*. Geological, geotechnical and hydrological if necessary; on local vegetation, with physiognomic method and/or phytosociological. Land use data on mining area and its possible around (not well defined). Human environment: data of the study area (agriculture in particular) in the period before the opening of the activity. Data on visual perception of the extractive industry, from various points of view (considered important).

B1. *Ratings and traditional diagnoses*. Broad assessments, or focused mainly on the perceptive quality. Note: diagnosis is only technical, because the landscape is considered a mere geographic support.

C1. *Traditional therapy*. Concept of "recovery" environmental-based (in more advanced cases): formation of parallel banks, their greening according to "potential vegetation", often with fixed irrigation system, or green painting the walls; re-use of the square base for productive purposes (also in the industrial sense) or ecotechnical (e.g. waste treatment), with provision of Scenes (embankments lined) to mask the residual visual impairment.

A2. *Bionomic analysis*. About traditional data as above. Definition of the unit and the type of landscape. Structural pattern of LU. Landscape apparatuses, habitat and human habitat. Analysis of the vegetation according to the bionomic method. Estimation of the BTC. Analysis of two previous historical periods and dynamics of main parameters.

B2. *Evaluation and bionomic diagnosis*. Assessment of the state of alteration of LU through: the distribution of the standard classes of BTC, processing deficit, abnormal HH and NH, connectivity, etc. Diagnosis by comparison with optimal model of normality.

C2. *Bionomic therapy*. Concept of ecological rehabilitation based on: (1) possible reopening of the quarry for reorganisation with the local geomorphology and reduction of the square base; (2)

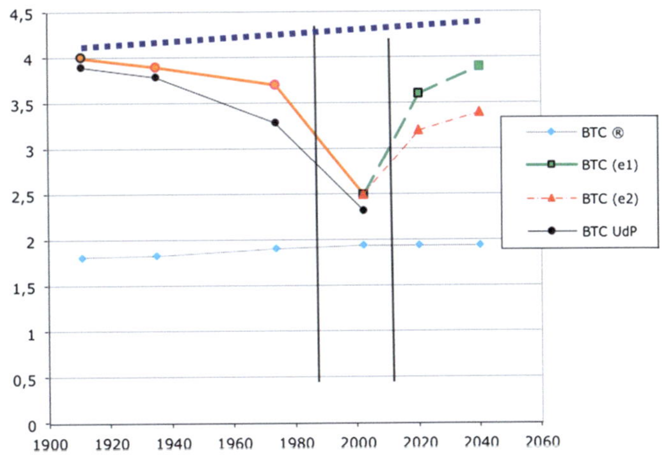

Fig. 11.13 Marble quarry near Brescia. On the ordinate the value of BTC in Mcal/m^2/year. From the top are marked: the trend in the absence of noise (*dotted blue*), the transformation of the quarry (*orange*), the transformation of its UdP (i.e. LU) (*thin black*), the level of BTC_R regional average (*indicators*); the bionomic rehabilitative path (green), the traditional 'recovery' path (dotted red). (from [8] modified)

formation of banks manifold connected, greened with vegetation of suitable bionomic value and diversified in tesserae, with possible temporary irrigation; (3) reforestation of the square residual and/or its agricultural reuse; 4) eventual monitoring of vegetation and invertebrates.

It should be noted that traditional methods can lead to an absurd result, sometimes ridiculous, as can be seen in a trachyte quarry through the hills near Padua, where the "recovery" (diction completely wrong for the bionomics of the landscape) was formed by iron gabions hanging on the rock (first painted grey-green!), with plastic tubes 2 cm in diameter for irrigation and a box with soil to bring some shrubs . . .!

With Fig. 11.13 we want to show the difference in design considered above (A1/A2). The figure represents a comparison between two projects, one traditional and the other bionomic, through the control of the "deficit of transformation" for a marble quarry near Brescia. In the figure, from the top are marked: the trend in the absence of disturbance (blue), the transformation of the quarry (orange), the transformation of UdP = LU which contains the quarry (thin). Starting in 2002, two curves are marked according to the bionomic plan of reorganisation (green broken line) and traditional (red dotetd line). It should be noted that the heavy processing deficit, equal to -1.65 Mcal/m^2/year (2002), is reduced to -0.70 Mcal/m^2/year (2020) when using the principles of landscape bionomics (green line = curve following the bionomic rehabilitative project; red lines = traditional recovery project). So, the bionomic recovery results 83–89 %, vs. a traditional one 74–77 % in the same period (2002–2020).

We now add Fig. 11.14a, b, to demonstrate that the landscape bionomic method must also bring innovations from the *morphological* point of view. In cross-section the same quarry of Brescia highlights the need to act in a different way from the usual in the three parts (1,2,3).

(1) The upper part of the quarry discharged must be reshaped with a slope in which it is possible the planting of vegetation.
(2) The central part can not remain parallel with the steps as usual, but must be reconfigured according to Fig. 11.14b (the details), to allow some effective connection in the quarry.
(3) The basal part is to be filled with what you take away from (1) for a better connection between the lower edge of the quarry and the surrounding vast square, which must also be re-vegetated.

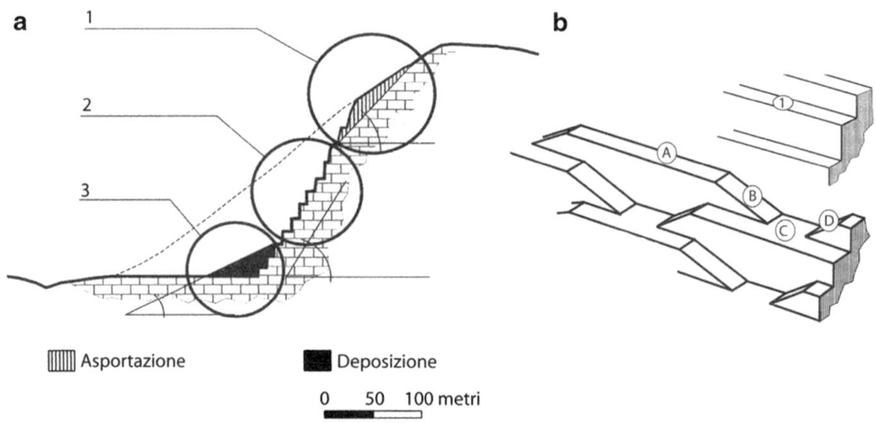

Fig. 11.14 (**a**, **b**) Marble quarry near Brescia. Design criteria for the rehabilitation of the site following landscape bionomic concepts: (**a**) explication in the text; (**b**) details of the central portion to compare a traditional sequence of banks (1) with a more articulated one (A, B, C, D) favourishing spontaneous vegetation recovery too. (from [8])

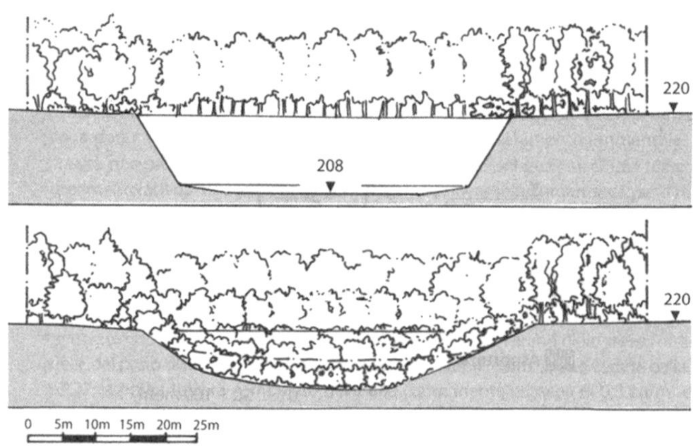

Fig. 11.15 Testing area for the recovery of a quarry margin of the landfill of Gerenzano. It has been demonstrated experimentally (1991) that changing the inclinations of the slopes of the quarry, in just 2–3 years the forest has expanded recovering the entire area (from [8])

Another case in which the morphology becomes important is an experiment done in the margins of environmental remediation project of the discharge of municipal waste in Gerenzano near Milan [14]. In the area of experimentation (Fig. 11.15) for the recovery of a gravel pit on the edge of the landfill, it could be shown that only by changing the inclination of the slopes of the quarry, in just 2–3 years the forest has expanded recovering the entire area. It was a forest of *Quercus robur*, with mantle dominated by *Robinia pseudoacacia* and *Carpinus betulus*.

11.3.3 Urban Green Design and Control

11.3.3.1 Against Formalism in Urban Park Design

Today's dominant ideology (individualism, liberalism and relativism especially) tends to upset the criteria for choosing between "*form*" and "*content*" of a project. Removing the form from content, what you get is just an *empty* formalism. The main cause of this is the renunciation of the pursuit of truth, which leads to a formalistic use of words and concepts (see Chap. 15). This is

Fig. 11.16 A proposal of artificial urban park (no comment!)

noted in the majority of contemporary art, where the only criterion of innovation is the formal technical execution, regardless of content.

As already pointed out, relativism threatens even science, reducing it to scientism. This distortion can be removed, as long as you do not ignore the real essence of phenomena. In our case, designing an urban park can no longer matter today by the scientific demonstration that the landscape is a living entity, that represents a specific level of organisation of life on Earth. Therefore, it is essential to understand the bio-ecological laws and follow a clinical-diagnostic method to act on it. The design criteria of a park must necessarily follow such a theorem.

Even a park in a city (or its sub-urban landscape units) is a portion of a living entity and, for this reason, has a direct impact on the health and human well-being (see Sect. 4.6). If a park design remains anchored in the principles of pure formalism and research of the highest originality, it means "to forget the crucial importance of the quality of human life itself of the inhabitants of the city" (fitting environment). On the contrary, a urban park design has to follow the suggestions of an *ecoiatra*, having to deal with the pathology of landscape units and relationships with human health, well beyond the problems of pollution. A crucial role remains in the design of the form, but the criteria for the formal verification have to change, being no longer reducible only to the visual perception and the caprices of the creative, but rather related to bio-systemic and ecological conditions (i.e. landscape bionomics).

It is necessary to underline that today the whims of a designer can reach hallucinating formalisms: man creates for himself, in total freedom (understood as the absence of constraints) the criteria for his life, and also for its landscape, aided by technology.

For example: vertical walls, "green" lawns of pure geometric shapes (cones, prisms, etc.), trees from opposites habitat held together by technical devices, iron trees (Fig. 11.16), benches of branches of boxwood (but plastic!), coloured lights, abstract art and squares concrete with water jets instead of tree-lined spots (but the surface is always count as a park . . .) and so on! This must be emphasised because, to fully understand an innovative criterion, is not enough to know what to do, you must also know what not to do.

It is important to note, at this point, that in the light of the bio-systemic invoked above, many of the concepts and definitions that are the basis of the applications to the environment and nature are to change profoundly, and this has implications in the definition of sustainability and, perhaps, even more in the environmental design. In the most advanced environmental disciplines, such as landscape bionomics or ecological design, it is now simply impossible not to refer to these innovative definitions, as the true principles of integration to reconstruct in a proper relationship Form and Content.

To understand what should be the real scope of the mentioned principles is necessary to emphasise a cultural change that has disturbing implications in practice. For example, it is no longer possible to design a park that maximises the "foot traffic" or "crossings" and in any case focused only on local utilities.

In fact, if the park must also perform functions at the urban scale, (e.g. UHI, ecological network, etc.) part of it shall not be designed in every component for local users, or even for users as such, otherwise the protection of the health of citizens will be impossible to achieve, as well as other important ecological functions. It must also point out that environmental stresses are detected by the body independently from aesthetics appearance, for obvious reasons of evolution and survival. A "formalistic" drawing in a park, even flashy, is not able to fool a "pattern matching" evolved over thousands of years, in reference to the harmony of the natural bio-systems, where they are lacking.

11.3.3.2 Purposes, Functions and Basic Ecological Criteria

Prerequisite is to understand the role of plant communities in complex systems of ecosystems, even when dealing with urbanised landscapes. In summary, we can say that nature has given to vegetation at least the following tasks:

(a) Regulation of the flows of free energy
(b) Organication of matter
(c) Regulation of the climate
(d) Air purification and oxygenation
(e) Structuring and protection of soil
(f) Structuring of space
(g) Furnishing recreational function (for humans)

As it is known, the vegetation has suffered the greatest destruction in recent decades: the same destruction of wildlife is a consequence. Especially impressive is the destruction within the human habitat: the countryside is depleted of woods, hedges and rows of trees and the cities lack urban green space and wooded habitats.

The primary purpose of an urban park is derived from the broad planning. However, as we will see later (Sect. 13.1), this is not enough to give a valid address and innovative design. Indeed, we know that the modern urban parks were born in 1700 as "places with recreational and ornamental features". However, since the beginning of 1900, due to the gradual destruction of the environment in big cities, the urban park has been assuming prominence in an ecological function, first as:

1. Place to rebalance environmental stresses (responsible for the alteration of the hormonal and immune systems) and place of "reset" of our level of fitness
2. Place of positive influence on urban mesoclimate (See UHI = Urban Heath Isle) and air pollution
3. Sub-regulatory system of the metastability of the local ecological system
4. Ecotope that can contribute to the formation of ecological network, today indispensable to environmental balance at the scale of the urban landscape
5. Area of refuge for flora and fauna

As a result, to the already known functions of urban green, new ones are added, no longer negligible, and new meanings are also added to the concept of sustainability.

We are entering an era where the *majority* of the human population live in urbanised landscapes, no longer in rural or semi-natural environments: consequently, the studies of *urban ecology* should play an increasingly important role. In fact, since a "Green Paper on the Urban Environment" was published by the EEC [15], there has been a crescendo of activity for the greening of the city, following the example of the Pilot Plan for the Ruhr and the system of Green city of Berlin [16–18], which began in the eighties. It must however be noted that, in most cases, it comes to interventions that do not follow the principles and methods of landscape bionomics in the proper sense of the scientific discipline (that is, in a biological sense).

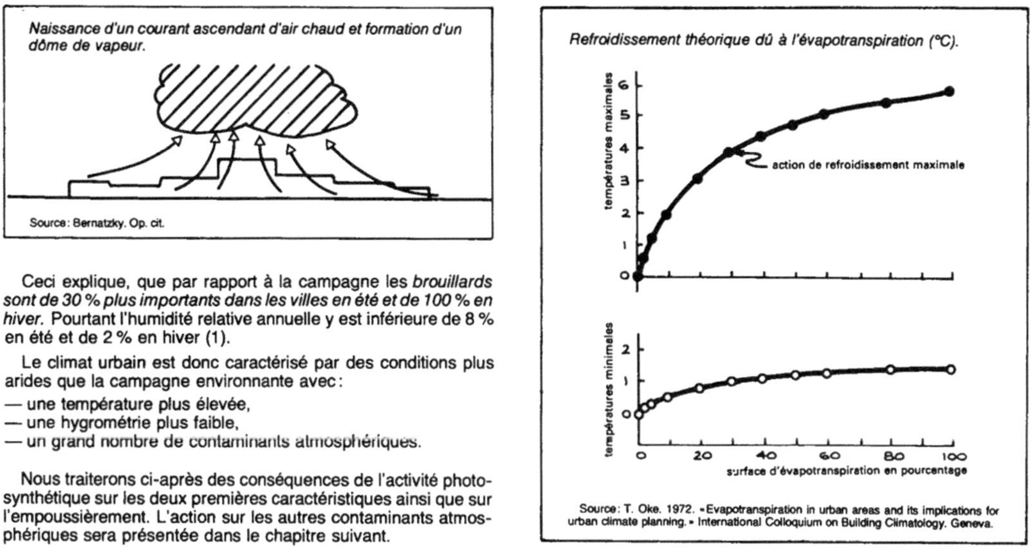

Fig. 11.17 Air pollution is mainly due to upward flows toward the centre. To have a significant temperature drop is necessary to have a vast, complex and multi-layered photosynthetic system. And it is essential to introduce large parks towards the central parts of the city and create true ecological networks. The peripheral green belts worsen the environment

You will notice that, as in Berlin, also in Milan, the urban heat island (UHI) becomes an important issue. The UHI involves, among other things, air pollution in the urban centre, due to convective flows of air (Fig. 11.17). To break this state you need to have parks in areas close to the centre. But the decrease in temperature, due to evapotranspiration, is obtained with adequate potential only if the vegetation is *complex*, that is multi-layered and natural. This becomes an important innovative goal for the design, as a traditional type of urban park is referred only to plant-nursery criteria, not to vegetation one.

If a city, just as Milan, has few green areas in the city centre and draw plans with "green belts" in the suburbs, only formally linked to green areas by internal corridors ("green rays") reduced to bike paths, the result is an inevitable growth in the temperature difference between centre and periphery, due to the increase of the green belt, thus increasing air pollution.

Unfortunately, the urban green is usually considered not clearly defined, by technicians who do not know the principles of ecology, at least in advanced way. The "green" is treated in an undifferentiated way or nearly so. The confusion of terms and concepts leading to the design choices dictated primarily by aesthetic and practical bases, or related to urban standards do not meet ecological reasons, and in any case not very significant. Consider that even today many technicians (agronomists and even planners) evaluate the presence of urban green space per capita as (m^2/inhabitant), which is an ecological blatant nonsense, that does not stand up to the principles of landscape bionomics! In fact, with the same green surface we can have very different levels of ecological organisation (recall Table 9.4).

Note that in cities like Berlin or Vienna, also medium-sized parks (10–15 ha) include parts of complex natural vegetation or, at least, natural shaped, as it can be seen for example in Fig. 11.18 which shows a view of the Turkenshanz Park in Vienna.

If a urban development plan (UDP) prescribes for a city area "urban green areas and public spaces $\geq$ 50 %" this may mean 40 % + 10 % of green spaces or vice versa (!). So it is not clear: (a) what should be the correct proportion, or (b) what

Fig. 11.18 View of a semi-natural wetland in the Turkenshanz Park in Vienna, a park in the city centre of about 15 ha in area, near the University of Agricultural Sciences

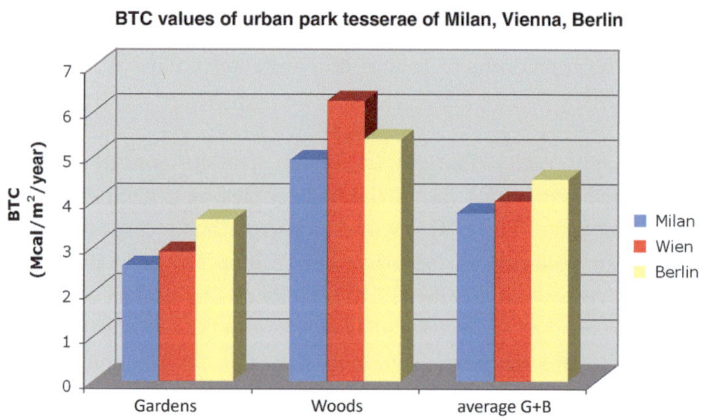

Fig. 11.19 Comparison of the most significant urban parks in Berlin, Vienna and Milan relating only to the biological capacity of the territory. Milan is considerably lower as ecological parameters of its parks, having the lowest average BTC and the increased presence of non-native species

Fig. 11.20 Berlin, Charlottenburg Park, a grey heron along a semi-natural stream

Fig. 11.21 Berlin, Charlottenburg Park, a perspective on the lake and the castle

Fig. 11.22 The tessera 1 within the Park of Monza, formed by a forest of *Quercus rubra* of Canadian origin, with average height of the canopy of about 29 m, but not more than 5.3 BTC

types of urban green would be needed, or (c) at what level of naturalness, or (d) what connections with the surrounding context.

From this example of UDP, we note that, even if you design a neighbourhood with 45 + 15 % green leafy squares, clearly passing out the planning prescriptions, this may be insufficient for the ecological requirements if the green is understood in the floristic not vegetation sense. Only if we define types of urban green and apply appropriate criteria and ecological indices, we can assume credible and effective forecasts, in some cases, even when they are able to stay a few percentage points below the requirements.

Remember the definition of the main types of urban green according to some ecological indices as shown in Table 9.4; it can help to clarify the design choices. We will see that an urban garden has several parameters that do not coincide with those of an urban park, so that the most ecologically significant function, the BTC, is for the garden equal to BTC = 1.3–2.0 while for the park BTC = 2.0–3.0 Mcal/m^2/year.

With regard to non-native species in urban parks, it should be noted that they are rising fast as you get closer to a metropolitan area. Taking into account a background value for the campaign in Central Europe (up Northern Apennines) of approximately 4 % of non-native species for relief (prof. Poldini, personal communication), one can find values up to peaks of nearly 50 % in some parks of Berlin and almost 65 % for Milan, even if the averages are lower (values around the middle), but still equal to about 8–10 times the background value. See also Fig. 11.19, which shows a comparison of the most significant urban parks in Berlin, Vienna and Milan, where you will notice the lower quality of the ecological parks of Milan.

In the most well-designed urban parks, some patches of natural or semi-natural vegetation are structured, in positive contrast with the architectural sections, as you can see in Figs. 11.20 and 11.21, which regards the superb park of Charlottenburg, Berlin. The integration between human culture and nature laws reached in this park a very concrete success.

This question has implications often trivial, based on fashion and the display of originality, however, lead to an obvious uniformity of botanical "garden" compositions. Some Neo-Darwinian scholars also very well-known (e.g. J. Gould [19]) argue that planting exotic trees is not a problem, from the evolutionary point of view. However, these positions do not take into account that it is not correct to talk about evolution only at the species level. It can be shown, in fact, that at the level of ecotope the presence of exotic species can also represent a significant alteration of the structure and ecological functions.

A good example of this fact can be done studying the tessera of American red oak (*Quercus rubra*) highlighted in Fig. 11.22, in the woods of the Historic Park of Monza. This study [20] was conducted on the basis of the LaBiSV method of analysis and evaluation of the vegetation (see Chap. 5). You may notice that the tessera 1 has a high average height (29 m) and great phytomass (649 m^3/ha), but a BTC of only 5.33 Mcal/m^2/year.

This tessera 1 can be compared with an analogous tessera of European oak (*Quercus robur*) always near Milan[1], of about the same age, with an average height of only 24 m, and phytomass volume of 428 m^3/ha, but with a BTC of 7.42 Mcal/m^2/year. The difference in bionomic quality (bQ) changes from 42.3 % for the exotic oaks to 70.4 % for native oaks.

The bionomic limitation of patches of non-native species is, therefore, not a whim of a sentimental naturalist, but an ecological evidence.

References

1. Tavernar E (1999) Lineamenti di economia della conservazione biologica. In: Massa R, Ingegnoli V (eds) Biodiversità, estinzione e conservazione, fondamenti di conservazione biologica. Utet Libreria, Torino

[1] Robecco, Ticino River Regional Park in the "Bosco Fasano".

2. Daly HE (1999) Uneconomic growth and the built environment: in theory and in fact. In: Kibert CJ (ed) Reshaping the built environment. Island Press, Washington, DC, pp 73–86
3. Wackernagel M, Rees W (1996) Our ecological footprint. New Society Press, Gabriola Island
4. WWF (2012) Living planet report, biodiversity, capacity and better choices. WWF International, Gland, Switzerland
5. Ingegnoli V (2011) Bionomia del paesaggio. L'ecologia del paesaggio biologico-integrata per la formazione di un "medico" dei sistemi ecologici. Springer, Milano
6. Primack RB (1998) Essentials of conservation biology. Sinauer Associates Publishers, Sunderland, MA
7. Ingegnoli V (2002) Landscape ecology: a widening foundation. Springer, Berlin
8. Ingegnoli V, Giglio E (2005) Ecologia del paesaggio: manuale per conservare, gestire e pianificare l'ambiente. Simone Edizioni-Esse Libri, Napoli, p 704
9. Noss RF (1992) The Wildlands Project: land conservation strategy. Wild earth (special issue)
10. Bennet MD, Smith JB (1991) Nuclear DNA amounts in angiosperms. Phil Trans R Soc Lond B 334:309–345
11. EEC (1992) Council Directive 92/43/EEC of 21 May 1992 on the conservation of natural habitats and of wild fauna and flora. Official Journal L 206
12. Forman RTT, Godron M (1986) Landscape ecology. Wiley, New York
13. Forman RTT, Spierling D, Bissonette JA, Clevenger AP, Cutshall CD, Dale VH, Fahrig L, France R, Goldman CR, Heanne K, Jones JA, Swanson FJ, Turrentine T, Winter TC (2002) Road ecology: science and solutions. Island Press, Washington, DC
14. Ingegnoli V (1989) Completamento del concetto di funzionalità del bosco, secondo l'ecologia del paesaggio. Ecol Mont Linea Ecologica, XXI 6:38–44
15. EEC (1990) Green paper on the urban environment. COM (90) 218, Bruxelles
16. IBA Emscher Park (1993) An Institution of the State of North-Rhine Westphalia, (ed.) Marion Zerressenabine Radowski
17. MacDonald R (1989) Impressions of urban design in Dortmund. Urban Design Quarterly, 30, pp 18–20
18. Profé b, Renker U, Thierfelder H, Wunnecke A, Haun G (2012) Stadtgrun in Berlin: Raum fur Freizeit und Natureleben. Senatsverwaltung für Stadtentwicklung und Umwelt Kommunikation, Berlin
19. Gould SJ (1994) The evolution of life on Earth. Sc. American CCLXXI:84
20. Ingegnoli V (2006) Criteri di valutazione diagnostica del Parco Reale di Monza secondo l'ecologia del paesaggio. In: LS Pelissetti (ed) Il Parco della Villa Reale di Monza al bicentenario della fondazione. Comune di Monza, pp 29–48

12 Applications in Alpine Landscapes

12.1 Environmental Assessment for the Opening of a Quarry

12.1.1 Introduction: Overcoming a Traditional Method

The municipality of Mori (Trentino-South Tirol, Italy) received a demand (with the required processing) to open a quarry near the village of Talpina-Dos del Gal in 2008–2009, by a company in the extractive sector. Project of cultivation, Technical Report, Environmental Impact Assessment (EIA) and its supplements were presented. The goal of the project was the exploitation of a deposit of prized limestone by opening a quarry of 31,245 m^2, included within a perimeter of 156,325 m^2 authorised by the 4th update of Trentino Province Act due to the presence of an underground quarry, existing since the 1950s, roughly coinciding with this scope.

It must however be noted that this area is now included within the bound of a "Sites of Community Importance" (SCI) IT3120150—Talpina-Brentonico, an important area of environmental protection, especially for the presence of rare species related to traditional agriculture now disappearing, as well as a resting and a reproduction place for migratory birds at long range and a breeding habitat for thermophile species in decline over the Alps; important, finally, to the biodiversity of plant species which are abundant among those thermophiles and those of karst environments. For these reasons, the proposer added to the EIA an Environmental Interference Study (EIS). These studies did not detect any heavy impediment to proceed with regard to the proposed activities, because it was emphasised that the disorders were compatible with the area concerned and that there would have been a "restoration" in three stages, starting after only 2 years to the excavation and ending within 18 years.

Let us examine in detail the submissions. The EIA had examined the emission of dust, noise, visual impacts, impacts from vehicular traffic, the hydrogeological analysis. The EIS that accompanied this project had carried out a survey on "appropriate" evaluation, divided into four phases, as required by the European Commission. These steps can be summarised in the following points: (a) summary information on environmental design, (b) the possible impacts, (c) a threat to the conservation objectives, (d) mitigation measures. The Chart of Habitats of Community Interest showed that the areas in question did not host formations considered important by the EU, but some interaction was allowed for the area in question. Even the dust and traffic could have produced damage to the surrounding vegetated environments, while the destruction of karst microhabitat would have damaged the herpetofauna. Of the 10 factors of potential disturbances, 5 were considered as significant for the interference, as it has been summarised in the EIS in the following way (Table 12.1).

V. Ingegnoli, *Landscape Bionomics Biological-Integrated Landscape Ecology*,
DOI 10.1007/978-88-470-5226-0_12, © Springer-Verlag Italia 2015

Table 12.1 Meaning of the environmental interference due to the opening of the quarry of Talpina (Mori), following the EIS presented by the company

Component	Factor	Interference
Air	Diffuse emissions	Significant (1)
	Traffic	Significant (1)
Water	Water table pollution	Not significant (0)
	Surface water pollution	Not significant (0)
Soil	Geological safety	Not significant (0)
Species	Effects on invertebrates	Not significant (0)
	Effects on birds	Significant (1)
	Effects on habitats	Significant (1)
Noise	Direct noise	Significant (1)
	Reverberation	Not significant (0)
Wisdom	*Environmental restoration*	*Significance: 5/10 (50 %)*

According to the presented EIA, the evaluation of mitigation measures should have been considered in a positive way. Dividing the excavation in three tranches and taking some precautions in the excavation to limit dust would have provided the mitigation. The restoration must have begun after 2 years from the first excavations, leaving 30 % of bare rock, 50 % carry-over of soil (20 cm) seeded with grass (*Festuco-Brometea*) and 20 % of areas to be planted to forest (*Orno-Ostrietum*).

The City Council, being a SCI area, instructed the writer, as an expert of landscape bionomics, of a further environmental assessment [1]. In fact, the environmental studies presented were necessary but not sufficient for a decisive ecological assessment. A lack of systemic considerations and too generic conclusions were not appropriate for an area of nature conservation.

12.1.2 Methodology to Check a Quarry Area

To evaluate the mentioned project in an advanced way, we decide to make reference to the scientific knowledge of landscape bionomics, as exposed in this book (see Table 10.3). In fact, as repeatedly stressed, a landscape represents a specific level of biological organisation, so it is a living entity [2] and we have to approach the study of its ecological state with a clinical diagnostic method, through the comparison with *normal* states of ecological systems under consideration using synthetic ecological indicators, as indices and/or ad hoc functions, already able to integrate, in a hierarchical and transdisciplinary manner, more information and content, simple or complex them to be.

In view of previous studies, especially of the EIS that accompanied the proposed project, we proceeded on the detection and processing of *additional naturalistic* data, completely ignored so far by the proponents, concerning:

(a) The importance of the SCI area also forecasts by Planning and SEA.
(b) The type of limestone substrate of the study area (the possible quarry).
(c) The floristic biodiversity of the area under consideration.
(d) The identification of the ecotope containing the area in question.
(e) The role of the study area in the context of the ecotope to which it belongs.
(f) The evaluation of territorial capacity of vegetation (BTC) at present in the ecotope.
(g) The assessment of the human habitat (HH) in the ecotope.
(h) The historical reconstruction of the ecotope.
(i) The study of the transformation of the ecotope, with projections in the near future, in two alternative cases: (1) natural growth and (2) recover from the quarry.
(j) The study of the transformation of the study area (quarry) with projections in the near

Fig. 12.1 The hump of Talpina-Dos del Gal, about 450 m at the base of Mount Baldo (2,250 m), near Rovereto (Trentino-S. Tirol). *To the right* an example of *Karren* formations

Table 12.2 Wooded tesserae of the ecotope of Talpina, part of the LU-1 of Mori

Forest tesserae	Tree species (%)	SP (n°)	*Hc* (*m*)	*PBv* (*m*^{3}/ *ha*)	*BTC* (Mcal/m^2/ year)	bQ (%)	M-Th	CBSt
Talpina, D-Gal E	*Quercus petraea 21, Carpinus betulus 20, P. nigra 17, Prunus avium 13, Fraxinus ornus 11*	40	12.2	173.1	4.99	50.4	23.04	11.62
Talpina, D-Gal W	*Pinus nigra 43, Ostrya carpin. 25, Quercus pubescens 16, Carpinus betulus 15*	**65**	16.3	140	4.67	47.7	20.62	9.83
Coste di Talpina	*Fraxinus excelsior 26, Frx. ornus 16, P. nigra 13, Ostrya carpin. 11, Larix dec. 10, Fagus sylvat. 8*	50	18.6	201	4.84	45.1	21.89	9.88
Talpina, p. 17°	*Quercus petraea 52, Carpinus betulus 33, Prunus avium 11, Fraxinus ornus 4*	*25*	12.1	126.7	4.52	46.1	19.53	9.01
Talpina, p. 17b	*Fagus sylvatica 55, Prunus avium 24, Ostrya carpinifolia 11, Quercus cerris 9*	*25*	17.2	255.1	6.41	57.4	35.35	20.30
	Average LU-1 Mori Talpina	*41*	*15.3*	*179.2*	*5.09*	*49.32*	*24.08*	*12.13*

future, in the two cases of alternative development and natural recovery after excavation.

(k) The control of these changes in relation to the model HH/BTC.

(l) The diagnostic evaluation of the main parameters involved in the possible ecological transformation.

As you can see, it comes to acquiring a quantity of significant environmental information.

12.1.3 Results of the Study of the Quarry Area

It should be noted that just the area of the proposed excavation assumes considerable importance even for the configuration of the karst substrate, which presents formations with “*Karren*” that are without doubt the most significant of the considered SCI, both for their position on the hump (relevant local structure in the ecotissue) and for deep forms extensively colonised by rock vegetation (e.g., *Asplenietea rupestris*) as well as from different shrubs (Fig. 12.1).

The SCI belongs to LU-1 of Mori and it’s rich in woods: therefore this area is particularly important, because the forests of *Orno-ostrietum* and black pine (*Pinus nigra austriaca*), often mixed, are richer in species of flora compared to similar forests of the entire territory of Mori: 41.0 vs. 31.5 species/tessera. As shown in Table 12.2, the forest tessera of West Doss del Gal presents 65 species!

These woods are young, less than 180 m^3/ha of plant biomass and 14–15 m height, 23–24 % of maturity, thus the concise bionomic state has a CBSt = 12.1 (see Sect. 3.3.2).

The delimitation of the ecotope containing the possible quarry area is shown in Fig. 12.2.

Fig. 12.2 Map of the "Sites of Community Importance" (SCI) IT3120150—Talpina-Brentonico. The *black line* limits the ecotope while the *pink* one the possible quarry area

Its surface measures about 30 ha, 73.4 % covered by woods, while the quarry area is only 3.13 ha. The estimation of the bionomic parameters of the ecotope is exposed in Fig. 12.3. Their values are based on field measures (2009) and on the historical reconstruction of previous values, based on old maps and air-photographic data. Most of the *Pinus nigra* has been planted by foresters about 50 years ago.

The study elaborated a comparison between the two forecast alternative cases of development of the ecotope: (a) in natural conditions and (b) after the quarry opening and restoration (Fig. 12.3). These values show that in the case (b) in 33 years the BTC, after a decrease, reaches the same present value of 4.25. However, this fact does not mean a loss of about 8–9 % respect the natural development (4.25 vs. 4.65), but a "transformation deficit" of about 18–19 %, calculating the area of the figure defined by the two lines of the previous shrubby condition and the quarry development respect to the trapezium delimited by the previous shrubby line and the natural development line.

As it can be seen from Fig. 12.4, the area between the two alternatives (i.e. the historical trend and forecast processing) measures the deficit of transformation. This deficit is clearly detrimental to the ecotope, since that would block the natural development of the same for about

Fig. 12.3 Forecasting of the possible developments of the ecotope of Talpina into the alternative cases: (**a**) natural (*green*) or (**b**) after quarry (*red*)

Elements	ha	%	BTC 1980	BTC 2000	BTC 2009	BTC 2028	BTC 2043	BTC 2028	BTC 2043
Wood	21,80	73,38	4,00	4,60	4,83	5,10	5,30	5,10	5,30
Quarry	3,13	10,52	3,90	4,50	4,67	4,90	5,10	0,60	1,30
Crops	3,50	11,78	1,85	1,85	1,85	1,85	1,85	1,85	1,85
Built	1,28	4,31	0,05	0,05	0,05	0,05	0,05	0,05	0,05
Total	29,70	100,00	**3,57**	**4,07**	**4,26**	**4,48**	**4,65**	**4,03**	**4,25**

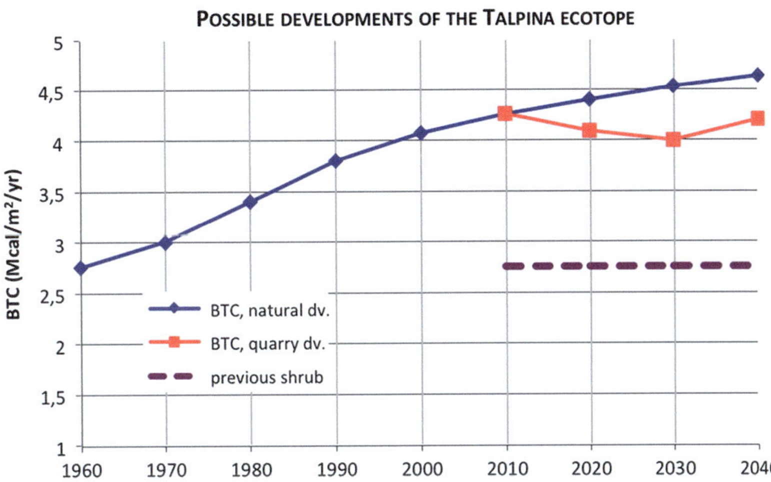

Fig. 12.4 Development of the ecotope of Talpina, from the shrub condition of 1960s and forecast to 2040. The "transformation deficit" is the area between *red* and *blue lines*

70–80 years, well over 18 years provided for by the operation of "excavation and recovery". Moreover, remember that the concept of "recovery" is no more correct, after it was shown that it is impossible to return to the status quo ante, for reasons of non-equilibrium thermodynamics (Sects. 1.3.8 and 1.3.9).

Table 12.3 shows the diagnostic evaluation of the situations ex ante and ex post in the case of the two possible developments of the ecotope. The normal ranges suitable for the type of landscape under examination (remember that it's a semi-natural forest landscape) are compared with the findings of the state ex ante (today) for each of the 12 parameters and evaluated according to the exposed method (Sect. 9.4.2).

In the alternative transformation, that includes the opening of the quarry and the following "recovery" (Table 12.3, columns B), we observe that the same temporal distance of 30 years has a diagnostic index ID = 39.3, instead of 93.9. This result indicates that the ecotope would enter into a pathological stage of "severe dysfunction".

12.1.4 Conclusions

With this synthetic paragraph it is proved that the EIS studies presented together with the project of opening quarry "Talpina-Doss del Gal" were necessary but totally *insufficient* in light of a more advanced ecology, that is systemic and integrative.

In fact, with a general indication of the significance of interference of 50 % you could also think of a mitigation action with a sustainable recovery. Conversely, the results obtained from this study show a diagnostic evaluation of decidedly pathologic level for the future ecotope, therefore unacceptable for a SCI zone.

In the opinion of the writer, therefore, we underline the inadequacy of the methods normally used today for EIS studies as specified in the applicable law.

Table 12.3 Diagnostic index of the possible transformations in the two cases A (natural) and B (artificial)

Parameters	Normal s-nat	Ex ante 2010	Score	A.ex post 2040	Score A	B.ex post 2040	Score B
Human habitat HH	10.0–25.0	22.13	*2*	22.13	*2*	21.29	*2*
Productive apparatus PRD	10.0–30.0	11.78	*2*	11.78	*2*	11.78	*2*
Ecotope BTC (Mcal/m^2/year)	4.5–7.0	4.26	*1.72*	4.65	*2*	4.25	*1.5*
Deviations HH/BTC model	0–20	26.7	*0.55*	20.2	*0.93*	27.1	*0.54*
Forest cover (%)	60–85	83.9	*2*	83.9	*2*	75.48	*2*
Trasf. Deficit (yr)	1.5–25	5	*2*	2	*2*	75	*0*
Natural degree	0.9–1.1	0.9	*2*	1	*2*	0.4	*0.37*
(I) Ecotope			*12.27*		*12.93*		*8.41*
Quarry BTC	5.5–7.5	4.67	*0.99*	5.17	*1.6*	2.1	*0.32*
Deviations HH/BTC model	0–25	19.9	*2*	6	*2*	63.9	*0*
Karst level	0.9–1	0.9	*2*	0.95	*2*	0.2	*0.18*
Trasf. Deficit (yr)	1.5–30	10	*2*	2	*2*	80	*0*
Role in the ecotope	0.8–1	0.8	*2*	0.9	*2*	0.5	*0.52*
(II) Quarry tessera			*8.99*		*9.6*		*1.02*
Diagnostic index (DI)	**85–100**		***88.6***		***93.9***		***39.3***

12.2 Environmental Report of the Territory of Tesero

12.2.1 Introduction and Methodology

The "Reporting of Territorial Plan" of the Province of Trento [3] imposes (a) firstly, the coherence of all local predictions with the evaluation of strategic spatial plans of more general reference (both in spatial and temporal sense) and (b) second, however, it develops specific insights to contextualise the *evaluation with respect to the characteristics* of each area. After the criticism (see Sect. 10.4) towards the insufficient scientific content of traditional planning methods, the second item allows to verify the adequacy of the proposed, more advanced, criterion (see Table 10.3). As in medicine, a complete check-up is not so frequent; anyway, remember that the evaluation of every case of landscape transformation (due to planning purposes or infrastructure building) must elaborate:

(a) *The knowledge of the state ex ante*: summary of the current environmental state of the territory that you have to evaluate, broken down by major subsystems (principles of the discipline, structure, functional systems, sustainability).
(b) *The systemic ex post evaluation*: control and diagnostic evaluation of detail (according to scientific principles) of the sub-examined on the basis of the transformations of the Plan or the building.

We present here a synthesis of the study on the municipality of Tesero (Fiemme Valley) deriving from the SEA (strategic environmental assessment).

12.2.2 Landscape Structure

12.2.2.1 Landscape Units (LU) and Elements

The alpine municipality of Tesero is extended from north to south for about 15 km, covering a total area of approximately 50.4 km^2. The well-known historical centre is situated on the right slope terraces of the Fiemme Valley, in the area where the two tributaries from the valleys of Stava (North) and Lagorai (South) are thrown into the river Avisio.

The division into units of landscape (LU) is readily apparent (remember Fig. 2.8), although the boundaries of these units do not quite match the ecological limits, as almost always happens. We distinguish the following three LU (Table 12.4):

Table 12.4 Differences in the element composition among the LU of Tesero in 2009

	STAVA		TESERO		LAGORAI		Municipality	
L. Elements, 2009	ha	%	ha	%	ha	%	ha	%
	2,080.2	41.26	1,397.6	27.72	1,562.7	31.00	5,040.5	100
Waters	5.6	0.27	16	1.14	23.8	1.52	45.4	0.90
Rocks and gravel	490.5	23.58	45	3.22	516.9	33.08	1,052.0	20.88
GEO + EXR	496.1	23.85	61	4.36	540.7	34.60	1,097.4	21.78
Road system	6.2	0.30	17.2	1.23	3.1	0.20	26.5	0.53
Urbanised	22.5	1.08	135.5	9.70	1.4	0.09	159.4	3.16
RSD + SBS	28.7	1.38	152.7	10.93	4.5	0.29	185,9	3.69
Agricultural	19	0.91	189.7	13.57	9	0.58	217.7	4.32
Crops and trees	3.5	0.17	52.5	3.76	4	0.26	60.0	1.19
PRD (AGR)	22.5	1.08	242.2	17.33	13	0.83	277.7	5.51
Greenland	381	18.32	97	6.94	179.7	11.50	657.7	13.05
Shrubs	50	2.40	9	0.64	56	3.58	115.0	2.28
Pioneer trees	38	1.83	152	10.88	50.2	3.21	240.2	4.77
RSL	88	4.23	161	11.52	106.2	6.80	355.2	7.05
Pinus cembra forest	138	6.63	88	6.30	116	7.42	342.0	6.79
Alpine spruce forest	361.3	17.37	226.1	16.18	343	21.95	930.4	18.46
Fir and spruce forest	335	16.10	271.6	19.43	210.6	13.48	817.2	16.21
Pinus sylvestris forest	229.6	11.04	98	7.01	49	3.14	376.6	7.47
RSN	1,063.9	51.14	683.7	48.92	718.6	45.98	2,466.2	48.93

1. The valley of *Stava*, more than 2,080 ha, only 85 % coincident with its ecological structure: in fact, the upper part of the territory of the hamlet of Pampeago is part of the municipality of Westhofen (127 ha) and the east side of the Cornon should have been part of the municipality of Panchià.
2. The central area of *Tesero* in the Fiemme valley, about 1,397 ha, 95 % coinciding with the ecological zoning.
3. The valley of *Lagorai* of about 1,562 ha, which covers 95 % of the ecological zoning, by excluding the last stretch of the stream Lagorai (administered from Cavalese).

It should be noted that the estimates of the elements in Table 12.5 report precise measures for the entire common while, for the 3 LU in object, data have lower accuracy in any case with an error <5 %. The three LU are very different among them also as *population*. Note that to the 2,827 residents the ecological analysis requires to add the portion of vacationers and tourists, as shown in Table 12.5.

12.2.3 Landscape Apparatuses

12.2.3.1 Geomorphologic and Hydrographical Characters

The LU 1 and 3 (Stava and Lagorai) are extremely diverse as geological formation, so as morphology. As it is well known, summarising and simplifying, we see that the first LU is a limestone-dolomite one (except Zanggen) and, in fact, it descended from the Latemar where, up to Cornon, there are also many volcanic strands-oriented E–W. The Stava creek flows in good part between Permian sedimentary rocks, even older, because of the presence of a fault, which has dislocated them [4]. The LU-2, where the town of Tesero is located, is characterised by the formation of Werfen, whose horizon is called properly basic "Ofioliti of Tesero". It is found in formations raised a 100 m on the wide valley of Avisio, flowing among the loose deposits di Fiemme. The LU-3 (Lagorai), in contrast, is all porphyritic.

Table 12.5 Estimation of the ecologically equivalent population, 2009

Residents and tourists	Municipality	STAVA	TESERO	LAGORAI
Private and agro-tour. (presences)	845	0.5	0.45	0.05
Rooms and BB (presences)	417	0.5	0.45	0.05
Hotel (presences)	933	0.7	0.3	0
Second homes (presences)	1,627	0.5	0.45	0.05
Total	3,822	2,098	1,580	144
Annual permanence	*0.35*			
Seasonal inhabitants and tourists	1,338	734	553	50
Residents (following urbanisation)	2,831	251	2,540	40
Ecologically equivalent inhabitants	**4,168**	**985**	**3,093**	**90**

Table 12.6 Average values of BTC of the main forest types of Tesero

Forest types	Ha	%	BTC (Mcal/m^2/year)
1—*Larix decidua* and *Pinus cembra*	342.0	6.79	6.20
2—*Homogyno-Piceetum*	930.4	18.46	7.53
3—*Picea abies* with *Abies alba*	817.2	16.21	7.78
4—Forests with *Pinus sylvestris*	376.6	7.47	5.74

The hydrogeological risk is still rather limited and does not affect areas of urbanisation. The stretch of river Avisio, which runs through the common of Tesero, is short (not more than 4 km), but important, since the bank vegetation is very rich and includes the rare willow *Myricaria germanica*, a tamarisk tree from Northern Europe. A portion of the river stretch is then protected as SCI (IT 3,120,118). The biological quality of the waters of Avisio is still good, although not as pertaining to the optimal class of IBE or IFF. The Stava creek is a little disturbed stream, while the river was harnessed with concrete walls and rock formations and bridles, after the well-known disaster of 1985. This has greatly diminished the quality of the natural watercourse.

12.2.3.2 Forest System

Given that, for each LU, the largest quantities are the areas of forest (Table 12.4), ranging between 46 and 51 % of their territory, it is necessary to propose an estimate of the formations that characterise these forests, in order to reach ecological indicators with greater credibility. To do this, breakdown is proposed after careful inspections:

1. Formation 1—*Larix decidua* with *Pinus cembra*, mostly *Larici-Cembretae* or larch as well, but also spruce-woods with *Pinus cembra* co-dominant, about 6–8 % of the territory.
2. Formation 2—Subalpine spruce forests, mainly of the type of *Homogyno-Piceetum*, from 16 to 22 % of the territory.
3. Formation 3—Spruce forests with *Abies alba* often co-dominant, as *Carici albae-Abietetum* (altitude >1,200 m), between 13 and 20 % of the territory.
4. Formation 4—*Larix decidua* with *Pinus sylvestris*, forests of *Pinus sylvestris* dominant or mixed with spruces and larches, between 3 and 11 % of the territory.

The average BTC values range from 5.74 to 7.78 Mcal/m^2/year (Table 12.6). Many tesserae of these formations present a good bionomic state, especially spruce forests and spruce with fir patches, reaching BTC of about 8–9 Mcal/m^2/year and CBSt from 50 to 65 and an average height of 30–40 m (Fig. 12.5).

12.2.3.3 Resilient Vegetation

The RSL system regards the pioneer shrub formations and pastures. Even in this case we

Fig. 12.5 Examples of the forest vegetation of the territory of Tesero. *Left*: Spruce forest in Lagorai Valley; *right*, shrub formations with *Pinus mugo*, and alpine prairies on the Mount Cornon, *above* Stava

appeal to field analysis for significant sample, with LaBiSV method. The main formations are summarised into three groups:

1. Formation A= Mountain pines/Alder, mainly at high altitudes, between 2 and 3 % of the territory (Fig. 12.5).
2. Formation B= pioneer formations, in general birch, alder and larch woods, often on former pasture areas abandoned by limited consistency, except in LU-2 where it reaches 12–13 % of the territory.
3. Formation C= Alpine pastures: *Arrhenaterethum elatioris, Trisetetum flavescentis, Nardetum strictae*, etc., which together reach 7–18 % of the territory.

12.2.3.4 Protective and Connective Systems

Remember that agricultural and urbanised ecotopes always need the presence of linear elements (tesserae, corridors) with a prominent functional role of protective character and connection. Historically, these functions were performed by a ubiquitous presence of hedges and rows of farmland and orchards, gardens and tree-lined streets in urban areas. However, these elements are strongly decreased in every European landscape, mainly for economic reasons. Even in our case, the limited presence of gardens and urban gardens and the lack of protective elements in agricultural areas can be seen in Tesero today, agricultural areas which, however, are much lower than in the recent past, as it can be seen above the village of Lago. Also, today, the presence of a freeway in the valley, with obvious barrier effect (even for pedestrians), has accentuated the alteration of the system of which we talk about.

12.2.3.5 Productive System

The change from an agricultural-pastoral economy to a touristic and craft one has greatly diminished the areas with productive role which, as noted, can only be due to agriculture, being the industrial sector only capable of processing. The magnitude of this change in the region Trentino-Alto Adige was among the largest in Italy, being the arable crops decreased from 6.1 in 1928 to 0.7 % in 1998. In just 70 years, a very short time in the history of Alpine landscapes, about 88 % of these areas has been abandoned, while the population grew by almost 50 %!

All this has increased the importance of residual agricultural areas, which must be protected. The Val di Fiemme "field abandonment" was a little less than in the rest of the region, given the tradition of the Magnificent Community: we are fortunately far from what has happened in the neighbouring Val di Fassa, with its indiscriminate urbanisation. We must of course be alert. Today, the sale of local products and the development of farm can revive this sector in a new way.

On the other hand, in the municipality of Tesero, now we can talk about a standard productive habitat still sufficient, if not positive. In fact, adding to the approximately 275 ha of agricultural land another 325 ha of pastures used eminently as pastoral, we arrive at:

$$\begin{aligned} SH_{prd} &= 606.55 \times 10{,}000/4{,}168 \\ &= 1{,}455.25\ m^2/inhab. \end{aligned}$$

But today about 95% of these values are only meadows and pastures. Considering at this altitude (about 900–1,100 m) a SH_{prd}* theoretical minimum of 1,550–1,650 m^2/inh, you get a control value of agricultural self-sufficiency equal to: $SH_{prd}/SH_{prd}^* = 0.90$. This result is evaluated as near sufficient, considering a safety factor of 10 %.

12.2.3.6 Urban Residential and Subsidiary

We all know that the urban residential and the subsidiary systems (Fig. 12.6) have risen strongly in the last century in the region of Trentino-Alto Adige, approximately quadrupling in extension. The first one resisted better than the other towns with an ancient historical centre, as in the case of Tesero, where the increase was a bit more contented. Certainly, the expansion of tourism due to winter sports was important in Pampeago, Val di Stava. The urbanisation of these areas today shows that imbalances are certainly addressed. Even Lago, with cross-country skiing and related service sectors, has similar problems.

The holiday houses are in Tesero fairly small and inconspicuous, as the village of Stava, despite the distortion after the disaster of 1985 and the rebuilt hotels highly questionable, on areas that may contain the remains of dead people under the infamous casting of mud: hotels too big, outside the local tradition. No doubt that the historical centre shows interstitial spaces and recoverable buildings: in addition, it needs urban green spaces as well as parking lots.

Other recent expansions are those due to the industrial-artisan and commercial ones. In Tesero you may notice that these expansions have so far been contained in a well-defined design and without interference on the old town. Also, mobility is efficient, save perhaps a link between the industrial area of Lago and the state road of the Dolomites. With regard to infrastructure, expansion of parking areas in sports or commercial areas may be necessary. However, it can be observed that the ratio SH/SH* as regards the entire urbanisation (RSD + SBS) is decidedly very abundant:

$$\begin{aligned} SH_{rsd+sbs} &= 185.9 \times 10{,}000/4,168 \\ &= 446.02\ m^2/inh \end{aligned}$$

Considering a $SH_{rsd+sbs}$* theoretical minimum of about 250 m^2/inh, we can deduce that the comparison with the normal optimal urbanised area per capita is $SH_{rsd+sbs}/SH_{rsd} + SH_{sbs}^* = 1.78$. This value is too much, you cannot think to a tolerance of up to 40–50 %.

12.2.4 Landscape Sustainability

12.2.4.1 Health and Environment

As enhanced in Sect. 4.6, the assessment of the environment in relation to human health is to be seen in three aspects: (a) the biological state of the basic elements of the environment (air, water, soil), namely the control of pollution; (b) the ecological state of the structural system of

Fig. 12.6 View of the main town of Tesero (1,000 m), LU-2. The strip near the river Avisio is industrial. We can see the small lake of Lago, to the *bottom* of the Fiemme Valley

ecosystems that form the landscape (NB: even in the absence of pollution!); (c) the perceptive aspect of the landscape itself, for psychological reasons.

Regarding the water and soil there are not serious issues in place, rather the river Avisio, as already said, is still to be considered in good ecological state. As regards the air, it had to proceed to an analysis of five stations according to the known Biotic Index-lichens [5]. The stations were the following:

(1) Alpe Pampeago, close to parking lots, 1,725 m.
(2) Palanca location, altitude 1,205 m.
(3) Tesero, in the central square, a height of 1,000 m.
(4) Tesero Lago, near the sawmill, altitude 910 m.
(5) Lagorai Valley, close to the first plateau, altitude 1,425 m.

Figure 12.7 shows a summary of the results, which speak for themselves. The Centre of Tesero is altered.

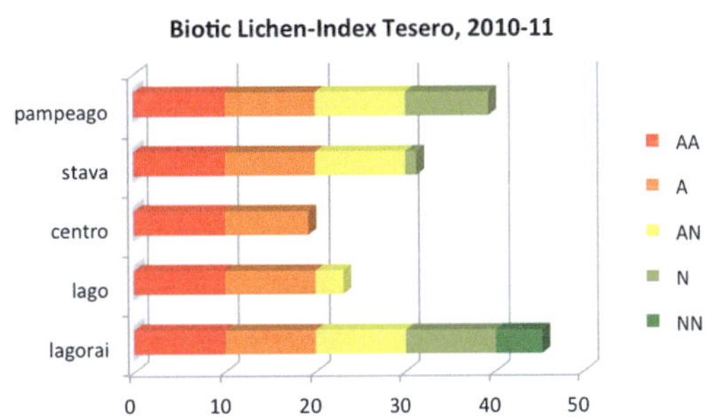

Fig. 12.7 Result of the analysis carried out in the five stations: Pampeago, Stava, Tesero Centre, Lago and Lagorai. Note that in LU-2 (Tesero-Centre) the biological quality of air is altered. Legend: AA = severe impairment, A = alteration, AN = semi-alteration, N = natural low, NN = natural high

12.2.4.2 Areas of Nature Conservation

In the territory of Tesero we can find the following protected zones:

T3120128—Stava, High Valley, example of forest of *Pinus cembra* on dolomitic soil.

It3120097—Chain of Lagorai Mountains. There are considerable floristic rarities. The site is of national interest for the presence and reproduction of animal species in danger of extinction, major glacial relicts.

It3120118—Lago (Fiemme Valley). The interest of the site is linked to the presence of relict *Myricaria germanica*.

It3120160—Lagorai (SPA). The mountain range of Lagorai, located in the central Alps, is one of the less populated areas in the Alps. The most important part of the SPA is made up of massive porphyritic rocks. Lagorai consists of

Fig. 12.8 View of the S–E part of the alpine lake Lagorai (altitude 1,750 m). This lake, with a perimeter of about 1.65 km, is one of the widest of the chain of mount Lagorai. The lake is not included in the SPA

a succession of ridges and peaks, large space rock and altitude grasslands.

These areas depend clearly on European Conservation Lows [6]. After an examination of their perimeters, these protected areas are *insufficient* for the conservation of nature in the territory in question. The reasons are varied and not always immediately apparent without a more complete ecological analysis, which goes beyond this work. However, we can summarise the following observations:

- IT3120118—Lago: The protection belt is totally inadequate, since it does not even include the river floodplain and is limited in length to ¼ of the section crossing the territory of the river Avisio in Tesero.
- IT3120097—It3120160—Lagorai (SPA): the lack of protection of these areas is the most obvious. It's suffice to say that the lake Lagorai (Fig. 12.8) is not protected, and that the U-shaped glacial valley upstream of the lake (Fig. 2.8) is bisected by the SPA. Not to mention the ecotopes around the bog forest at an altitude of about 1,700–1,800 m!

12.2.4.3 Visual Perception

From the point of view of road network, which is significant as today a landscape is perceived especially from roads (!), the entry in the municipality of Tesero is not always happy. In fact, coming from Cavalese along SS.48 (Dolomiti State road) we meet a shopping and craft area, which is very diverse and architecturally poor, while along the highway on the valley floor you can still see the industrial areas and the complex of Lago, not too bright. The same central square (C. Battisti) is still not resolved to the perception of the historical centre. This centre has a wide range of high-profile perceptive paths to the characteristics of courtyards and passages between old houses, which can still be exploited.

It should be emphasised that the issues of perception depend in the first instance by the bio-ecological state of the territory and its LU. Similar to the case of men, it is undoubtedly useful "dress well and have hair care", but it is even more important to stay in good health, an issue essential to have an "attractive look".

12.2.4.4 Signs of Economic Aspects

In Tesero, the need for socio-economic growth does not seem to be too much, for luck. You can easily find today the following questions: (a) residential demand, typically related to demographic trends and housing problems; (b) requests for so-called work activities, namely trade, business, industry, services; (c) demand for tourism development, even beyond winter sports; (d) any need for development and agro-tourism hotel.

However, you should not forget the ever-increasing need for true natural areas, that is high naturalness, which may become in the coming years a highlight of all respect, given the increasing need for rebalancing of too many citizens, forced to live in environments now more or less degraded in large metropolitan areas.

The addresses given by the Territorial Plan of the Province of Trent remains valid abiding by the principles of territorial balance, containment of land consumption, sustainable development.

12.2.5 Bionomic State of LU

12.2.5.1 Historical Aspects

Try to understand the historical transformations of the municipality of Tesero, in the Fiemme Valley, has been very difficult for two reasons: the change of nationality (from Austria-Hungary to Italy, 1919) and the lack of statistics from the province related with ecological data.

The analysis of the iconography of one century ago shows the evident wider presence of agriculture and a relative minor cover of forests, in all the Fiemme Valley, so even in Tesero as we can see from Fig. 12.9. The stronger decrease in agriculture is related with arable fields, passing from 733 to 54 ha (−92 %) in exactly one century (Table 12.7). The same table shows for Tesero a decrease of −97.2 %.

On the contrary, tourism increased continuously from the opening of the road of Dolomites in 1905–1906, and today the ecological population in Fiemme is 15 % more of the residential: 19,578 + 2,932 = 22,510 people (Fig. 12.10).

An estimation of the main landscape changes in the Fiemme Community has been elaborated, based on data since 1910 and 1847 for population [7] as plotted in Fig. 12.10. We can see the increase of forests, especially after the last World war, about 12 % in one century, and the drastic decrease of cropland, mentioned before. The transformations of the municipality of Tesero were proportional to those of its Valley of Fiemme, and this fact can be used to value the landscape conditions of Tesero at the beginning of the past century.

Reduced the forest cover of 12 % and urbanised areas of 49 % (proportional to population), the agricultural sub-system in 1909–1910 should pass from 5.51 to 13.19 % (comprise pastures). Consequently, supposing the forest BTC unvaried (for constant managing reasons), we estimate with a good approximation an average BTC = 3.51 $Mcal/m^2/year$ and an average HH = 22.2 % LU.

12.2.5.2 Present Bionomic State

The measures of the most important bionomic indexes have been calculated for each landscape unit and for the entire territory. In Table 12.8 you can see, as an example, the evaluation of human habitat (HH) and biologic territorial capacity of vegetation (BTC) related to the LU-2 of Tesero, the weighted average of HH resulting equal to

Table 12.7 Drastic decrease of crop fields in the last century both in Fiemme Valley and in Tesero, one of its main municipalities. Note the lack of detailed information for more recent periods (e.g. 2010)

	V. Fiemme	1910	2010	Tesero	1917	2010
Arable fields	ha	%	ha	ha	%	ha
Wheat	131.00	17.87		5.75	10.82	
Rye	318.00	43.38		4.96	9.33	
Barley	194.00	26.47		20.46	38.49	
Oats	1.00	0.14		0.00	0.00	
Corn	89.00	12.14		21.98	41.35	
Total	**733.00**	100.00	**54.3**	**53.15**	100.00	**1.48**

Fig. 12.9 View of Tesero in the years 1900 (*above*) and 2013 (*below*). In the last century, the main transformations of the landscape concerned the passage from an agricultural to a touristic economy. Comparing these figures it is evident the increase of forests and of urbanisation and the decrease of cropland

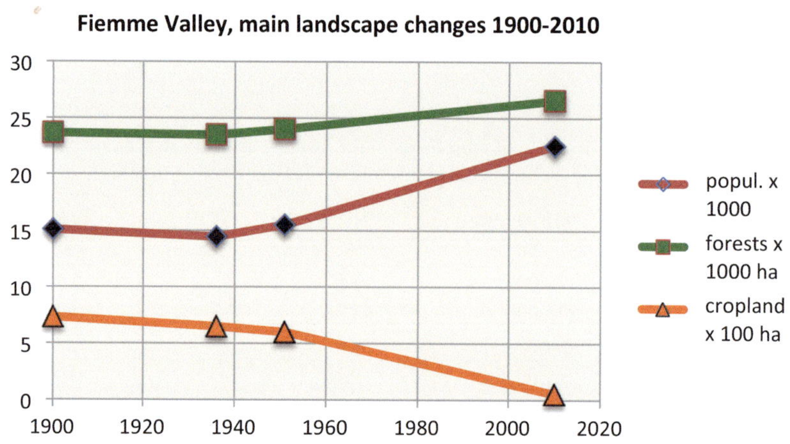

Fig. 12.10 Main landscape changes during the last century within the Fiemme Valley

Table 12.8 Measures of human habitat and BTC in the LU-2 (Tesero), 2009

L. elements	ha	%	HU′	HU 09	BTC′	BTC 09
Water	16.0	1.14	0.10	0.11	0.06	0.001
Rocks	45.0	3.22	0.03	0.10	0.12	0.004
Roads	17.2	1.23	1.00	1.23	0.03	0.000
Urbanised	135.5	9.70	1.00	9.70	0.18	0.017
Agricultural	189.7	13.57	0.90	12.22	0.68	0.092
Wood-prairie	52.5	2.80	0.60	2.25	2.50	0.063
Grassland	97.0	6.94	0.50	3.47	0.65	0.045
(a) Forest with *P. cembra*	88.0	6.30	0.03	0.19	6.20	0.390
(b) Spruce forests	226.1	16.18	0.04	0.65	7.53	1.218
(c) Spruce mixed with fir	271.6	19.43	0.04	0.78	7.78	1.512
(d) Forest with *P. sylvestris*	98.0	8.01	0.03	0.21	5.74	0.460
Tall shrubs	9.0	0.60	0.02	0.01	2.50	0.015
Pioneer formations	152.0	10.88	0.05	0.54	3.80	0.413
Total	1,397.6	100.00				
Weighted average				**31.46**		**4.24**

Table 12.9 Main bionomic parameters of the LU of Tesero, year 2009

Bionomic parameters	STAVA	TESERO	LAGORAI	Municipality
Surface (ha)	2,080.2	1,397.6	1,562.5	5,040.3
Barren (water, rocks, etc.)	19.08	3.50	27.68	17.42
Urbanised and roads (%)	1.38	10.93	0.29	3.69
Human habitat (HH, %)	14.21	31.46	9.82	17.47
BTC (Mcal/m^2/year)	3.90	4.24	3.68	3.91
Population (ecological, n°)	985	3.093	90	4.168
HH surface (ha)	295.57	439.65	153.46	879.97
SH m^2/inhabitants	3,000.7	1,421.4	17,050.8	2,111.3
SH/SH* (SH* = 1,550–2,000)	1.50	0.92	8.53	1.28
LTpE (10 NHd/HH + BTC_{A+F})	80.74	47.93	109.06	68.56
Landscape type	For-Tour	Agr-F-Tour	For-S-Natur	For-Tour

31.46 % of the LU surface and the weighted average of BTC equal to 4.27 Mcal/m^2/year.

These values are normally compatible with the model HH/BTC (see Fig. 7.9). To these values we need also to add the measures of standard habitat per capita (SH), the index of carrying capacity (SH/SH*), the landscape type evaluation (LTpE) and, therefore, the individuation of the landscape types. Table 12.9 summarises these parameters in the present situation.

We may observe that SH registers a poor value in the LU-2 (Tesero) where SH < SH* (about 1,650 m^2/inhabitant at this altitude), consequently the carrying capacity of SH/SH* = 0.92 is near the limit of heterotrophy. The LTpE parameters indicate that the 3 LU pertain to different landscape types: LU-1 forest-touristic, LU-2 agro-forest-touristic, LU-3 forest-seminatural. These results indicate a difficulty to consider the bionomic conditions of the entire municipality as representative of the bionomic state of its LU.

12.2.5.3 Forecasting Planning Conditions

The Territorial Planning proposed in 2010 by the charged technician (prof. Enzo Siligardi) is quite moderate, as exposed in Table 12.10: only

Table 12.10 Territorial Plan proposals per L. apparatus

TESERO municipality	2009		2020		(%)	
L. elements/apparatus	ha	%	ha	%	(−)	(+)
GEO + EXR	1,097.8	21.78	1,093.4	21.69	0.09	
RSD + SBS	185.9	3.69	214.6	4.26		0.57
PRD (AGR)	277.7	5.51	266.2	5.28	0.23	
GREENLAND	657.7	13.05	657.9	13.05		
RSL	355.2	7.05	355.2	7.05		
RSN	2,466.2	48.93	2,453.2	48.67	0.26	
Population, residents	*2,831*		*3,050*			*219*
Tourist permanence	*1,337*		*1,460*			*123*
Total inhabitants	*4,168*		*4,510*			*342*

Table 12.11 Main bionomic parameters of the LU of Tesero, esteem year 2020

Bionomic parameters	STAVA	TESERO	LAGORAI	Municipality
Surface (ha)	2,080.2	1,397.6	1,562.5	5.040.3
Barren (water, rocks, etc.)	19.06	3.5	27.48	17.35
Urbanised and roads (%)	1.66	12.57	0.3	4,27
Human habitat (HH, %)	14.35	32.59	9.82	17.850
BTC ($Mcal/m^2/year$)	3.88	4.10	3.78	3.89
Population (ecological, n°)	1.077	3.343	95	4.515
HH surface (ha)	298.48	455.45	153.46	899.64
SH m^2/inhab.	2,771.4	1,362.4	16,153.7	1,992.6
SH/SH* (SH* = 1,550–2,000)	1.39	0.88	8.08	1.14
LTpE (10 NHd/HH + BTC_{A+F})	79.85	44.81	109.71	67.00
Landscape type	For-Tour	Agr-F-Tour	For-S-Natur	For-Tour

0.57 % of the land should be used to enlarge human activities, about 29 ha only and plus 342 inhabitants in next 10 years.

Table 12.11 is the update of Table 12.9 with the inputs of the plan. In fact, the changes are limited, even if the carrying capacity of LU-2- reached now the limit of heterotrophy. The differences among LU increase of about 6.2 %. These results demonstrate the *impossibility* to consider the bionomic conditions of the entire municipality as representative of the bionomic state of its LU, notwithstanding the legal rules of Province Administration.

12.2.6 Landscape Diagnosis

12.2.6.1 Checking the Deviations from the Norm

Recalling Sect. 9.4, we show Table 12.12, in which 16 ecological parameters were elaborated to control the diagnostic index (DI) in 2009 and with forecast to 2020 and 2030 related to the LU-2 of Tesero. The resulting DI puts in evidence a small increase of DI 2020 and a decrease for 2030. For LU-2 the DI control demonstrates an insufficient action of the new territorial plan. This fact does not compare because the province control is referred only to the entire municipality, where DI passes from 87 to 91. Anyway, it represents a limiting situation when we measure the trend in future planning. Remembering the concept of diagnostic index (DI) (Sect. 9.4.2), it could be useful to show the most important control contained in the SEA Report prepared for the municipality of Tesero, an area of 50 km^2 near the Dolomites of Trento (Fig. 12.11).

Even if the local laws consider only the municipality (not the LU) let us elaborate a control of the general metastability (Fig. 12.12). Considering both information and energy (Sect. 3.7.5) this control is full of significance.

Table 12.12 Diagnostic index (DI) of the LU-2 (Tesero), forecast

Bionomic parameters	Normality	2009	DI-09	2020	DI-20	*DI-30*
RNT (forest cover)	25–65	48.92	2	47.84	2	*2*
BTC For. (Mcal/m^2/year)	6.50–7.5	7.2	2	7.3	2	*2*
Allochtonous formations	0–1	0.3	2	0.3	2	*2*
CON (α + ϒ)	0.65–0.95	0.5	1	0.5	1	*1*
PRD (% cropland and meadows)	15–40	20.8	2	20.2	2	*2*
RSD + SBS (urbanised areas)	2–10 %	10.93	2	12.57	1	*1*
SH_{prd}/SH_{prd}* productive	> 1.1	0.90	1	0.77	1	*0.5*
$SH_{rsd+sbs}/SH_{rsd+sbs}$*	0.9–1.5	1.76	1	1.88	1	*1*
HH (human habitat)	20–40	31.46	2	32.58	2	*2*
SH/SH*(capacità portante)	1.2–1.8	0.92	1	0.88	1	*1*
BTC of LU-2 (Mcal/m^2/year)	1.8–4.5	4.21	2	4.13	2	*2*
$BTC_{LU/}BTC_{mod}$	> 0.95	0.95	2	0.92	2	*1*
LTpE	34–55	47.93	2	44.81	2	*2*
Lichen air quality (IBL)	> 30	21.13	1	24	1	*1*
River functionality (IFF)	> 180	140	1	140	1	*1*
Preserved/needed conservation	> 0.95	0.27	0	0.9	2	*2*
Landscape type	agr-for-tour					
Diagnostic index	85–100		**75.00**		**78.13**	***73.44***

Distance (%) evaluation scores: 0–10 = 2, 10–30 = 1, 30–60 = 0.5, >60 = 0

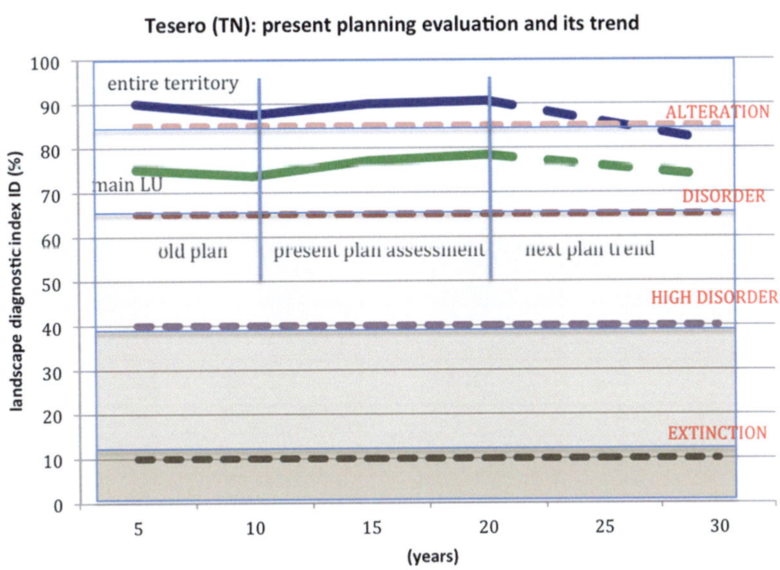

Fig. 12.11 An example of present planning assessment in the territory of the municipality of Tesero (Trento). Note the difference between the diagnostic indexes curve relating to the entire territory (*blue*) or to the main LU, that is LU-2 city of Tesero (*green*). The next plan trend shows a deterioration of the landscape health

Each LU has to refer to an appropriate curve, depending on the amount of soil inaccessibility. We can see that the distances between the present states (coloured dots) and their curves remains of about 75–80 % of the optimum. In the case of LU-2 (Tesero) the red-brown segment indicates a movement of the system passing from 87 to 84 % of the optimum: even considering 10–12 % of security coefficient, we are going towards 16 %, indeed a first alteration of the bionomic state. The municipality shows a very small movement of its environmental complex system (dark blue),

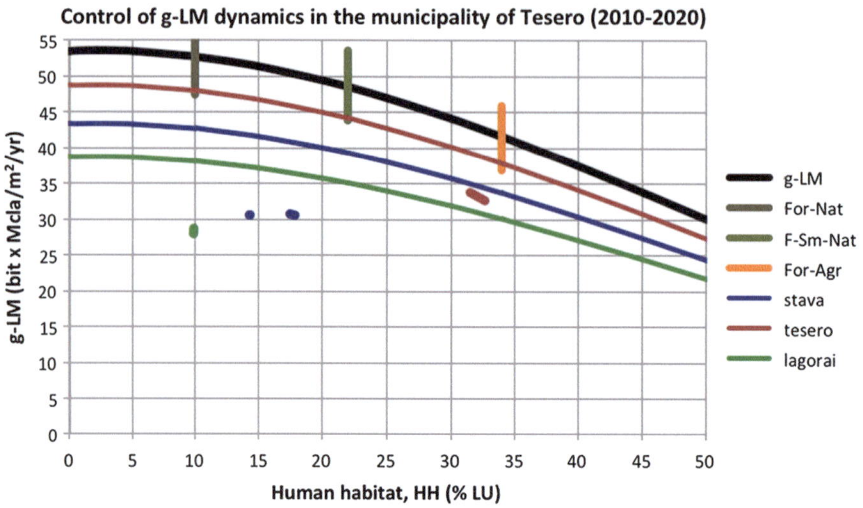

Fig. 12.12 This control of the general metastability (see Sect. 3.7.5) of the municipality of Tesero represents the forecast of the period of validity of the territorial plan. The model condition for the LU-1 (Stava) is the same for the entire municipality

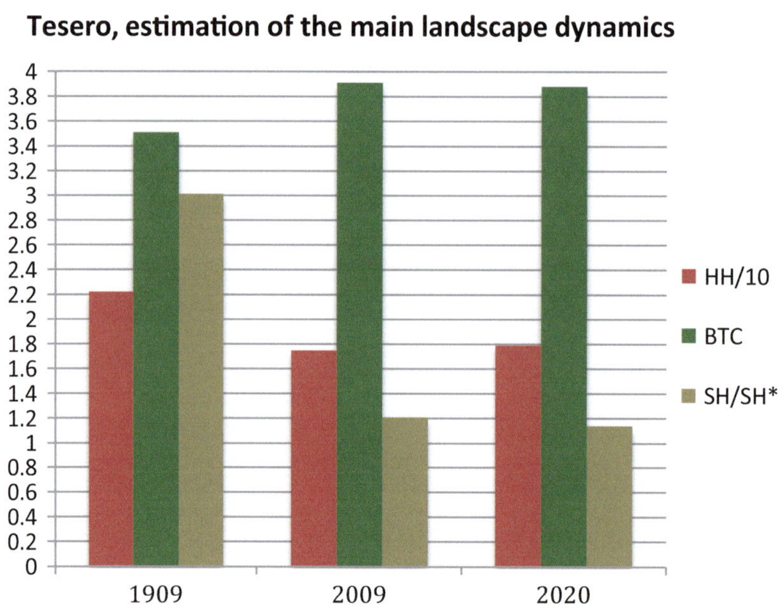

Fig. 12.13 A synthetic control of the main landscape dynamics is plotted in this graph. In the last century the type of landscape is changed, even considering the entire municipality. The Plan forecast in 2020 operates very few changes since 2009

related to the blue curve, but the Plan is not able to reduce the distance to the optimum, which remains quite large: −26.8 %.

In summary, an estimation of the main landscape changes is shown in Fig. 12.13. In the last century the change of the landscape type is

Fig. 12.14 Wide Alpine landscape systems are mainly characterised by spruce forests. View from the Dolomites of Latemar towards the glaciers of Adamello and Presanella (3,539 and 3,559 m a.s.l.)

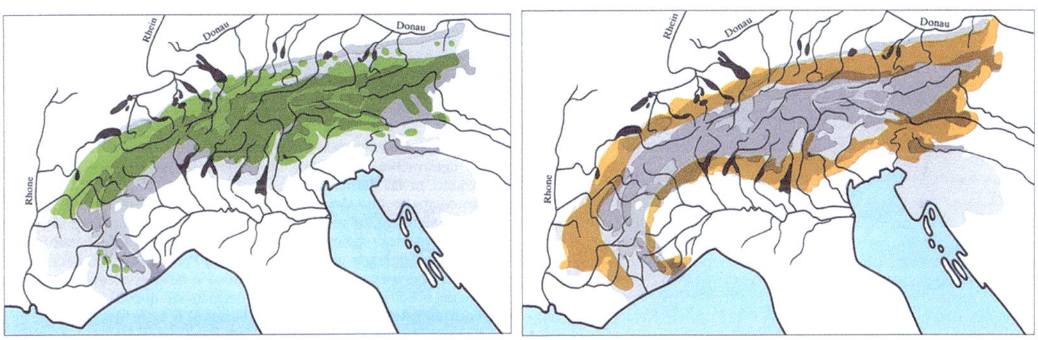

Fig. 12.15 Distribution area of spruce (*Picea abies*) and beech (*Fagus sylvatica*). Note the marginal locations of beech (*orange*) limited mainly to the Pre-Alpine valleys (modified from [8])

evident even at municipality scale. The carrying capacity seems to become the most important problem in the near future, as expressed comparing Tables 12.9 and 12.11. Apart from that, the forecast of the planning actions produces very few changes since the present state: not enough for the local Administration, which added some changes outside the plan, deepening damages and losing the opportunity to start a rehabilitative course!

12.3 Preliminary Research on Alpine Spruce Forests

12.3.1 The Importance of Spruce Forests within Alpine Landscapes

The example of the hilly plateau (1,300–1,500 m) of the Sanctuary of Pietralba-Weissenstein (Fig. 12.14) presents a landscape

Table 12.13 The main European plant associations related with spruce (*Picea abies*) sensu S. Pignatti

Vegetational associations	*Pinus cembra*	*Larix decidua*	*Picea abies*	*Abies alba*	*Fagus sylvatica*	*Pinus sylvestris*	*Acer pseudo-platanus*	Altitude (m)
Larici-Pinetum cembrae	100	81	47					1,700–2,200
Homogyno-Piceetum			100					1,200–1,900
Laricetum deciduae		100	74					1,000–1,850
Veronico urticifoliae-Piceetum		42	96					930–1,700
Oxalidi-Abietetum		50	88	100	100			1,100–1,500
Calamagrostidi-Abietetum			100	73				1,000–1,550
Carici albae-Abietetum			65	100				1,050–1,570
Cardamini trifoliae-Fagetum			80	100	95			880–1,500
Erico-Pinetum sylvestris			60			100		800–1,400
Cardamini pentaphyl.-Fagetum			56	44	100		64	750–1,420
Luzulo albidae-Fagetum			80		100		60	550–1,500

Numbers indicate the frequency of presence in each association

unit characterised by wide spruce forests. No doubt that these forests are the most typical of the Alpine landscape systems.

We can see in Fig. 12.15 the distribution of spruce in the Alps, compared with the areal of beech, which remains marginal. The most part of the inner Alps is covered by spruce forests. Moreover, in Europe, the spruce forests are distributed from Norway to the Alps in a near continuous range, via Carpathian mountains.

The "European forest types categories and types for sustainable forest management reporting and policy", in the EEA Technical Record [9], inserts the spruce forests both in Boreal and Alpine biogeographic regions. This report précises that "Alpine coniferous forest" is a category that grows in climatic conditions similar to those of boreal zone, except for the light regime and length of the day. Cold and harsh climate (short growing seasons) characterises the high altitudes of the Alpine region of Europe; this determines similar altitudinal vegetation belts, though at differing altitudes, on all alpine mountain ranges. Forest tree species composition varies with the vegetation belts (mountainous/subalpine) and site ecological conditions. In addition to strictly boreal conifers (*Picea abies* and *Pinus sylvestris*) we can find *Abies alba*, *Larix decidua*, *Pinus cembra* and *Pinus mugo* as naturally dominant species.

Variation in regeneration patterns and horizontal clustering is also related to vegetation belts. Traditional pastoral farming practices, the mainstay of the mountain economy for centuries, have modified the natural distribution of subalpine forests; pasturing, however, is now rapidly disappearing under the combined pressure of land abandonment and intensification. The management of even-aged stands predominates in the Alpine region; selection cutting management is practised only in small areas of productive forest characterised by mixed forest spruce, fir and beech composition.

Fig. 12.16 Groups of *Picea abies* in forest formations in Val Brembana (Lombardy) mixed with beech at the altitude of about 500–600 m

Table 12.14 Forest cover related to spruce in Central-Eastern Regions of the Alps (ha). Note that about ½ of total amount (10,700 km^2) of forest cover are formed by an absolute dominance of *Picea abies*

Alpine regions	Mountain district ha	Silver fir and spruce	Spruce only	Stone pine + larch and spruce	Beech + spruce and fir	Total (ha)
Piemonte	1,099,813	14,141	18,585	81,569	23,100	137,395
Valle d'Aosta	323,059	1,156	18,230	44,528	231	64,145
Lombardia	968,832	8,005	88,374	47,678	13,136	157,193
Alto Adige	725,197	2,647	175,065	91,693	756	270,161
Trentino	608,276	15,856	137,203	63,038	12,449	228,546
Veneto	535,185	7,097	96,703	40,876	13,439	158,115
Friuli V.G.	334,686	1,858	44,963	11,891	17,762	76,474
Alpine Italy	*4,595,047*	*50,760*	*579,123*	*381,273*	*80,873*	1,092.029
Nothern Alps	792,360	66,900	39,400	3,400	125,000	234,700
W-Central Alps	474,026	12,800	30,800	19,100	1,400	64,100
E-Central Alps	571,144	16,700	52,700	18,000	3,200	90,600
Southern Alps	348,650	15,700	20,100	28,500	15,300	79,600
Alpine Swiss	*2,186,180*	*112,100*	*151,500*	*69,000*	*144,900*	*477,500*
Land Tirol (Austria)	*1,264,771*	*340,000*			*36,000*	*376,000*

More interesting the forest classification of Sandro Pignatti [10], summarised in Table 12.13, with few modifications, in which the species *Picea abies* is dominant or co-dominant in at least 11 vegetational associations. The spruce is full dominating in three associations (*Homogyno-Piceetum*, *Veronico urticifoliae-Piceetum*, *Calamagrostidi-Abietetum*), but is co-dominant in the other eight. Note that also the *Erico-Pinetum sylvestris*, dominated by Scots pine, presents lesser amounts of *Picea abies*, *Larix decidua*, and in low altitudes *Quercus petraea* and *Sorbus aria* [11].

Picea abies is present in a wide range of vegetational belts, from about 500–2,000 m [12, 10]; even in the Southern Pre-Alps it can be seen from 500 to 600 m of altitude, as in Val Brembana (Fig. 12.16). Consequently, the amount of forest cover in the Alpine landscape systems is very wide, as exposed in Table 12.14:

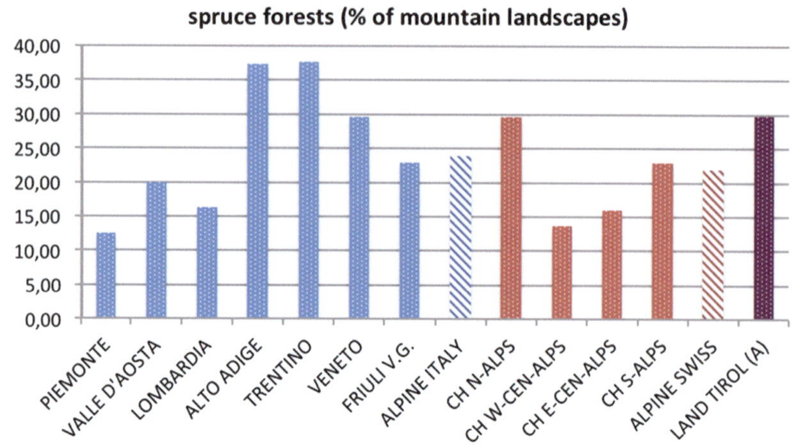

Fig. 12.17 Percentage of spruce forests (pure and mixed) in mountain landscapes of Central-Eastern Alps, from Valais/Val d'Aosta to Friuli/Land Tirol. The region of Trentino-South Tirol presents the higher %

10,920 km^2 of spruce forest in the Italian regions (administrative regions) of which about ½ dominated by pure *Picea*, 4,775 km^2 in Swiss regions (biogeographic regions), 1/3 of which of pure *Picea*, and 3,760 km^2 of pure and mixed spruce in Land Tirol, Austria [13–15].

Figure 12.17 shows the percentages of spruce forests cover in Alpine Regions, from West to East (i.e., from Piedmont to Land Tirol).

12.3.2 Preliminary Research

Given the exceptional importance of the spruce forests in the Alpine landscape systems, it should be interesting to propose a research based on landscape bionomic method of vegetation survey, the LaBiSV (see Chap. 5): in that chapter we underlined that, following traditional geobotany, some basic questions in the study of the landscape remained without answer. It is worthwhile to mention some of these issues: (1) how to consider the contribution of a vegetated tessera to the general metastability (g-LM) of a landscape unit; (2) how to compare ecological data of a forested patch with those of another vegetation; (3) how to use the ecological characters of the different types of vegetation to reach a diagnostic assessment related to a certain landscape unit; (4) how to integrate other ecological parameters of a landscape unit (HH, SH, etc.) with those relating to the vegetation; (5) how to relate ecological forest characters with CBSt; (6) how to evaluate ecotonal belts between associations; etc.

Obviously, a research like this needs a good organisation and adequate financial supports. We could suggest about 18 surveys per region, therefore 180 more surveys, after this preliminary study, based on 38 cases, as reported in Table 12.15.

- These surveys on spruce forests were done principally in the region of Trentino (21) because of some recent study and consulting; we than added 6 surveys in Lombardy, 5 in Land Tirol (Austria), 3 in South Tyrol, 2 in Engadin (Switzerland), 1 in Friuli.
- The vegetational belts are quite well represented, as we can see in Fig. 12.18: 1,000 m, 4 cases; 1,000–1,200 m, 8 cases; 1,200–1,400 m, 7 cases; 1,400–1,600 m, 9 cases; 1,600–1,900 m, 6 cases; 1,900 m, 3 cases.
- Tree height: <26 m, 7 cases; 26–29 m, 17 cases; 29–32 m, 9 cases; >32 m 5 cases.
- *Picea abies* dominance (phytomass volume %): 100 %, 7 cases; 90–100 %, 13 cases; 70–90 %, 12 cases; 50–70 %, 6 cases.
- Volume of plant biomass: <500 m^3/ha, 6 cases; 500–1,000 m^3/ha, 23 cases; >1,000 m^3/ha, 9 cases.
- Species (n°): <22, 7 cases; 22–32, 17 cases; >32, 14 cases.

Table 12.15 Main bionomic data (BTC, bQ, CBSt) from 38 spruce forest tesserae in Alpine landscape systems

Sites of surveys	Altit.	Tree species (%)	SP (n°)	Hc (m)	VpBm (m^3/ha)	BTC (Mcal/ m^2/year)	bQ (%)	CBSt
Tarvisio (UD) Conecofor (2003)	820	*Picea 95, Fagus 5*	*36*	30.5	827	7.70	70.2	39.65
Mezzoldo (BG), Ts 2 (2013)	960	*Picea 54, Abies 44, Fagus 2*	*31*	29.8	730	7.24	67.9	34.23
Andalo (TN) Ts1 (2013)	1,010	*Picea 80, Fag. 19, Pin-syl 1*	49	28	560	7.78	76.7	44.12
Andalo (TN) Ts2 (2013)	1,030	*Picea 59, Abies 25, Fag. 16*	32	33.8	734	8.22	78.3	49.84
Mori (TN) Selva p31 (2007)	1,100	*Picea 92, Fag.7, Larix 1*	*52*	27.5	542	6.91	67.8	31.41
R-Chienis (TN) Piazze (2010)	1,115	*Picea 63, Fag.21, Larix 9, P-syl 7*	*39*	28	574	7.21	70.4	35.24
Munstertal (BZ) Rivaira (2013)	1,160	*Picea 91, Larix 8*	18	23.4	617	6.28	59.7	23.27
R-Chienis (TN) Giazzera (2012)	1,175	*Picea 74, Larix 21, Fag. 1*	54	25.6	477	7.80	78.6	45.43
Val Masino (SO) Conecofor (2004)	1,190	*Picea 53, Abies 27, Larix 11, Fagus 5*	28	22,8	482	6.84	68.1	30.99
Tesero (TN) sotto Cornon (2010)	1,200	*Picea 58, Larix 31, Pin-syl 10*	25	27.1	739	7.20	67.3	33.60
R-Chienis (TN) Est Giazz. (2012)	1,210	*Picea 77, Larix 13, Fag. 7,5*	43	24.2	522	7.41	73.6	38.73
R-Chienis (TN) Fae/ Biaena (2011)	1,270	*Picea 72, Fag. 25, Larix 2*	*36*	27.7	707	8.76	84.6	60.54
Mezzoldo (BG), Ts1 (2013)	1,360	*Picea 54, Abies 44, Fagus 2*	*28*	30.4	804	7.54	69.7	37.87
Scuol (CH) Vulpera (2013)	1,370	*Picea 90, Larix 2, Pin-syl 7*	28	26.2	485	6.31	62.4	24.53
Cavalese (TN) sopra Masi (2013)	1,410	*Picea 75, Abies 21, Larix 4*	30	38.2	1,081	9.82	89.0	78.80
Adamello (BS) V. Avio (2013)	1,420	*Picea 97, Larix 3*	24	29.7	980	8.04	71.9	43.94
Adamello (BS) V. Paghera 1 (2013)	1,430	*Picea 90, Larix 10*	20	26.8	635	7.64	73.9	41.10
Rauth-Novale(BZ) (2013)	1,440	*Picea 97, Abies a 3*	33	39	1,162	8.71	75.7	53.63
Nauders (A), Ts3 (2013)	1,450	*Picea 99, Larix 1*	34	26.5	667	6.89	65.3	30.09
Adamello (BS) V. Paghera 2 (2013)	1,500	*Picea 96, Betula 2*	25	24.9	582	6.89	66.9	30.81
Paneveggio (TN) Mar. Jr. (2013)	1,510	*Picea 100*	20	27.8	743	7.19	67.1	33.44
Paneveggio (TN) Valbona (2013)	1,530	*Picea 100*	43	41.6	1,641	9.53	75.7	63.39
Alba/Penia (TN) Ts2 (2012)	1,560	*Picea 100*	38	26	811	7.26	66.6	33.76
Zermez (CH) (2013)	1,580	*Picea 72, Larix 26, Pin-syl 2*	30	26.2	688	7.53	71.7	38.85
Galtur (A), TS1 (2012)	1,620	*Picea 100*	25	26.1	838	8.31	77.3	50.20
Cauriol (TN) Ts2 (2013)	1,660	*Picea 89, Larix 11*	34	29.1	911	8.82	81.4	59.04
Nauders (A), Ts2 (2013)	1,670	*Picea 91, Pin-syl 9*	21	24.6	504	5.23	50.6	14.23

(continued)

Table 12.15 (continued)

Sites of surveys	Altit.	Tree species (%)	SP (n°)	Hc (m)	VpBm (m^3/ha)	BTC (Mcal/ m^2/year)	bQ (%)	CBSt
Paneveggio (TN) Mar. alto (2013)	1,680	*Picea 100*	26	31.2	1,024	8.38	74.7	49.28
Galtur (A), Ts3 (2012)	1,700	*Picea 100*	20	31.3	1,010	7.35	64.0	33.18
Renon (BZ) Conecofor (2004)	1,740	*Picea 85, Larix 10, Pin-cem 5*	21	27.1	497	6.83	67.9	30.78
Lavazé (TN) Conecofor Z2 (2013)	1,780	*Picea 98, Pin-cem 2*	44	31.6	1,052	9.36	84.6	68.47
Lavazé (TN) Conecofor F (2004)	1,800	*Picea 97; Pin-cem 3*	*28*	32	1,086	8.95	79.6	59.28
Lavazé (TN) Conecofor B (2004)	1,805	*Picea 100*	*20*	26.1	320	6.15	63.7	23.90
Alba/Penia (TN) Ts5 (2012)	1,810	*Picea 98, Larix 2*	22	31.4	1,124	8.77	77.0	55.24
Colbricon (TN) malga (2013)	1,820	*Picea 82, Larix d. 16 Pin-cem 2*	25	27.1	1,020	9.03	81.7	61.89
Colbricon (TN) Ts2 (2005)	1,900	*Picea 86, Larix 7, Pin-. cem 6*	*28*	26.4	332	7.17	74.3	36.84
Alba/Penia (TN) Ts6 (2012)	1,960	*Picea 80, Larix 19, Pin-cem 1*	28	24.7	658	7.65	73.6	41.05
Nauders (A), Ts1 (2013)	1,960	*Picea 82, Larix 16, Pin-cem 2*	24	27.9	878	8.08	74.1	45.70

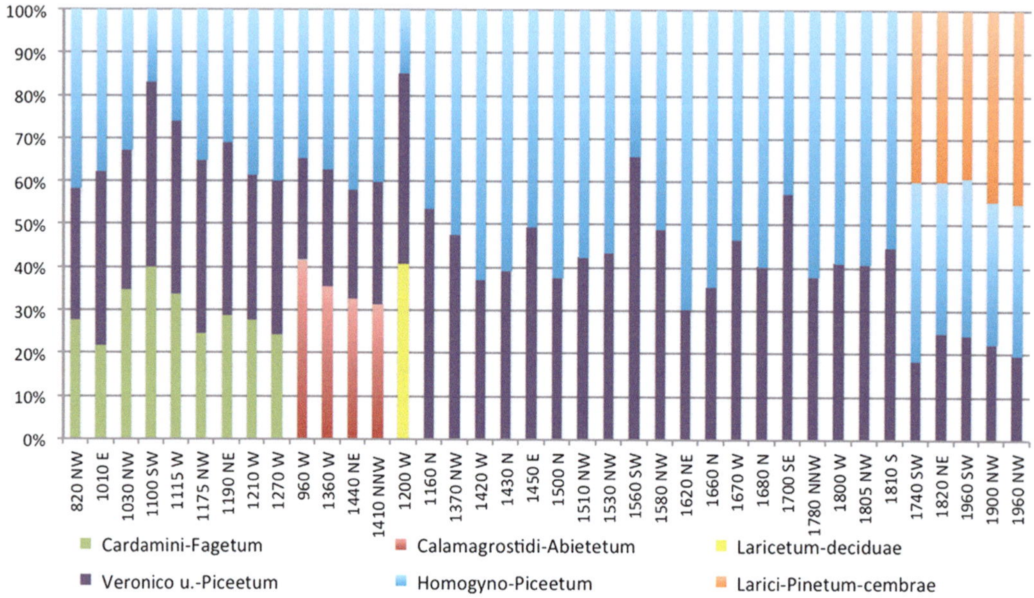

Fig. 12.18 Preliminary surveys of 38 spruce forests: altitudes, exposition and % of key species

Table 12.16 First results of the preliminary study on Alpine spruce forests

Spruce forests: 38 surveys	BA (m^2/ha)	SP (n°)	Hc (m)	VpBm (m^3/ha)	BTC (Mcal/m^2/year)	bQ (%)	CBSt
Averages	**50.34**	**30.58**	**28.64**	**764.33**	**7.70**	**72.20**	**42.27**
Min	22.70	18.00	22.80	320.00	5.23	50.60	14.23
Max	77.00	54.00	41.60	1,641.00	9.82	89.00	78.80

Fig. 12.19 Comparison between a natural and a cultivated spruce forest. The Lavazé Conecofor TS Z2 (*left*) is natural, VpBm = 1,052 m^3/ha, CBSt = 68.47. The Paneveggio Valbona is partially cultivated, VpBm = 1,641 m^3/ha, CBSt = 63.39

12.3.3 First Results

The first results of this preliminary study seem to be interesting. In Table 12.16 we put the averages data per each ecological parameter deduced from these 38 surveys. The number of species (SP n°) per forest tessera is compatible with the data of Pignatti [10] (from 26 to 35 species). The average basal area (BA) is normal. Height (Hc) and plant biomass volume (VpBm) vary from young-adult to mature phases. The BTC and CBSt average values result only 3.5 and 8 % higher than the result of a previous research reported in Table 9.1.

The differences of bionomic state between cultivated and natural forests appear clearly comparing three different cases:

- A high plant biomass volume (VpBm) case "Paneveggio, Valbona" (within Paneveggio-Pale di San Martino Nature Park), cultivated by Austrian foresters at the end of nineteenth century, presenting a very high biomass volume 1,641 m^3/ha, height of the canopy 41.6 m, CBSt = 63.39.
- The Cavalese above Masi case, semi-natural, presenting a lower biomass volume VpBm = 1,081 m^3/ha, height of the canopy 38.2 m, CBSt = 78.8 or even.
- Lavazé Conecofor Ts Z2, 1,052 m^3/ha, height of the canopy 31.6 m, CBST = 68.47 (Fig. 12.19): note that at the North border of the Lavazé tessera it was recently opened a ski cross country skiing [16].

Applying the Ellenberg ecological indexes [17] on light (L), temperature (T), humidity (U) and soil nutrients (N), the non-correlation between altitude and light or humidity is confirmed. Temperature related to altitude leads the change of other ecological parameters as soil nutrients and number of species, as exposed in

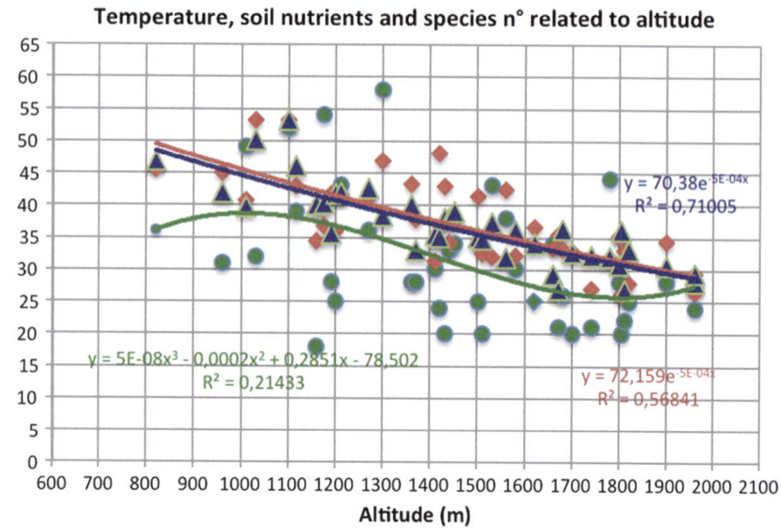

Fig. 12.20 Temperature and soil nutrients following Ellenberg ecological indexes T (*blue*) and N (*red*) in function of the altitude, compared with species number per survey (*green*). Note the differences in their correlations R^2 values

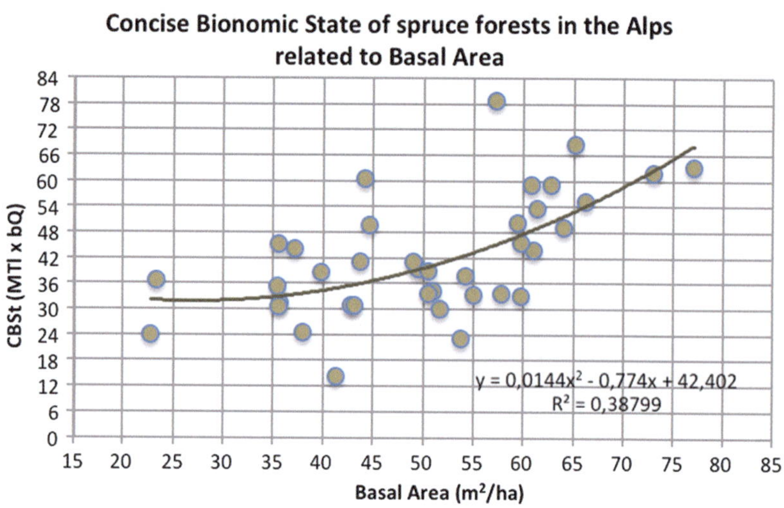

Fig. 12.21 This relation CBSt/BA presents limited changes in the interval BA = 22–45 (about CBSt = +112 %), while with BA from 50 to 75 CBSt increases of 169 %

Fig. 12.20. The good correlation between T/altitude ($R^2 = 0.710$) is lower for N/altitude ($R^2 = 0.568$) and scarce for the relation species/altitude ($R^2 = 0.214$).

The relation between the bionomic efficiency and the basal area, CBSt/BA (Fig. 12.21), seems to indicate little changes of CBSt when the basal area range is comprised between 22 and 45 m^2/ha: in this interval CBSt passes from 32 to 36 (+112.5 %), while when BA vary from 50 to 75 m^2/ha, it changes from 39 to 66 (+169.2 %).

Recalling Fig. 12.18, we observe that key species from different associations are present in each survey. The proportion of their presence varies in relation with the altitude. The two most typical associations (*Veronico urticifoliae-Piceetum* and *Homogyno-Piceetum*) are present in every case study; in low mountain sites key species of beech forests also appear, followed by fir species; in high altitudes, the key species of *Larici deciduae-Pinetum cembrae* appear. Only in 6 surveys (15.8 %) the dominance of a typical association may be fully recognised (key species >60 % vs. 40 % of other key SP), that is one for *Veronico urticifoliae-Piceetum* and 5 for *Homogyno-Piceetum*. These most characterised cases are in the altitude interval from 1,420 to 1,660 m.

Fig. 12.22 Different key species of vegetational associations characterise each survey in a sort of continuum. Anyway, in vegetation belts the main associations present a parabolic optimum

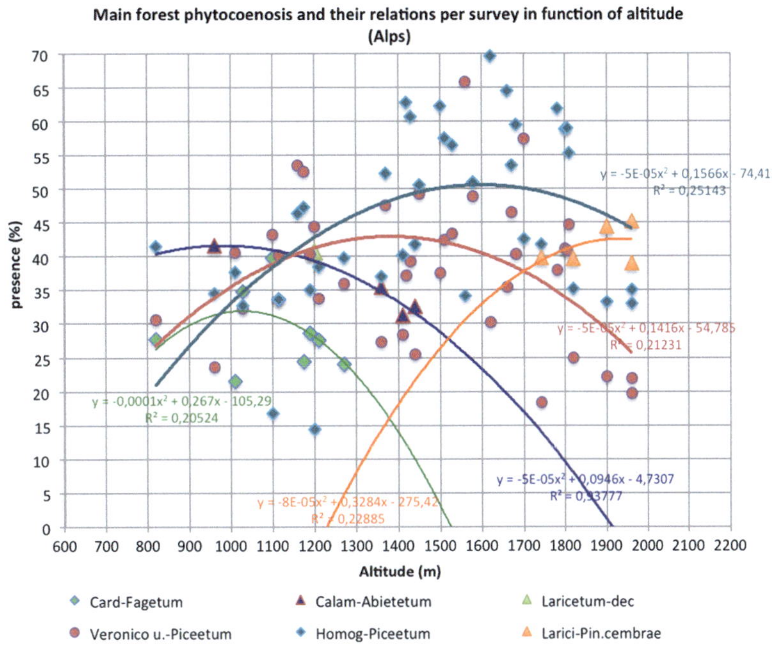

If we try to put in evidence the correlations between the percentage of key species and altitude per each survey (Fig. 12.22), we can see five parabolic curves, representing the main associations. The green curve, having its maximum in the altitude interval 1,000–1,100 m, represents the beech forest association of *Cardamini-Fagetum*. The blue curve, having its maximum in the interval 900–1,100 m, represents the fir association *Calamagrostidi-Abietetum*. The red-ochre curve, having its max in the interval 1,300–1,500 m, represents the *Veronico urticifoliae-Piceetum*. The green-blue curve, having its max in the interval 1,500–1,700 m, represents the *Homogyno-Piceetum*. The orange curve, having its max in the interval 1,800–2,000 m, represents the *Larici deciduae-Pinetum cembrae*.

Note that only the *Homogyno-Piceetum* curve goes above 50 % of key species presence, the other remaining from 32 to 43 %. Let us observe that, considering the interval 900–1,300 m, in the range 25–45 % of key species presence, we find four curves, therefore the most frequent condition is a mixed forest without true dominant association. Even in the interval 1,700–2,000 m, in the range 27–47 % of key species, we find three curves: another difficult situation for the emerging of a dominant association. Only in the central interval 1,300–1,700 m, in the range of key SP presence 40–53 %, we find two curves, with a possible appearance of a dominant association. That is why the mentioned six surveys presenting a clear dominance of a typical association are included in the interval 1,300–1,700 m.

The hypothesis suggested by Pignatti to be tested in a study like this (personal request) is that the presence of a typical association is today difficult to be found because of human disturbances altering the ecological state of the forests: it's a good challenge. In fact, as we have seen from Fig. 12.18, we found many intermediate formations, due to environmental, biological and/or human disturbances. Thus, it is possible to list these different spruce forests in proportion of the dominance of the formation types.

These observations, resulting in a distance from "potential vegetation", lead to stress a diffuse ecological alteration of alpine forests. Anyway, remembering the concept on the "fittest

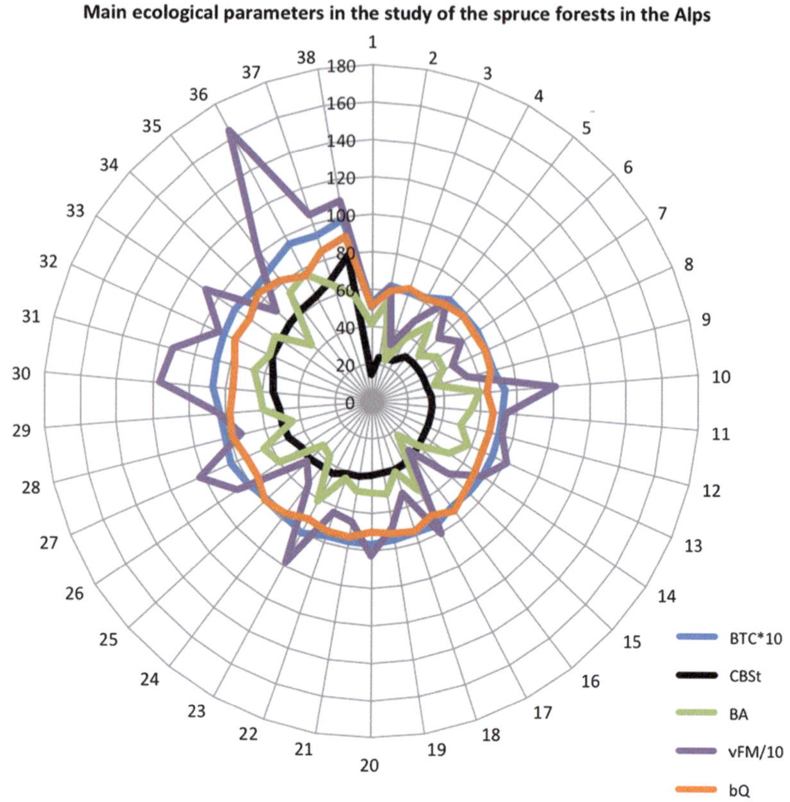

Fig. 12.23 Ecological parameters of the preliminary study on spruce forests, ordered from min to max CBSt

vegetation for" (Sect. 3.3.3), landscape bionomic principles do not accept this assertion. A research like this on spruce forests in Central-Eastern Alps is confirming the poor and fragile basis of the concept of "potential vegetation".

Measuring the CBSt levels in function of the relative dominance of a typical association per each survey (Fig. 3.6) we can see that to a low "purity" of spruce formations does not correspond a low CBSt. On the other hand, mixed spruce forests are not composed randomly. The presence of a clear gradient is evident as the role of vegetation belts with the altitude.

No doubt that we need more data to better understand the relations between structure and function of the spruce forests. For the moment, if we try to plot together the main ecological parameters as basal area (BA), plant biomass volume (VpBm), biologic-territorial capacity (BTC), bionomic quality (bQ) and concise bionomic state (CBSt), a sort of functional capability to reduce even great structural differences emerges. In Fig. 12.23, for instance, we can see the jagged curve of structural parameters BA and VpBm (i.e. vFM in the Figure) contrasting with the linear curve of CBSt.

References

1. Ingegnoli V (2011) Criteri di verifica dell'incidenza ambientale dell'apertura di una cava. In: Ingegnoli V (ed) Bionomia del paesaggio. Springer, Milano, pp 283–290
2. Ingegnoli V (2002) Landscape ecology: a widening foundation. Springer, Berlin
3. Provincia Autonoma di Trento (2009) Indicazioni metodologiche per la Rendicontazione Urbanistica dei PRG e dei Piani dei Parchi Naturali Provinciali (All.III, d.P.P. 24 Nov 2009, n.29-31/Leg.) Trento

4. Dellantonio E (1995) Geologia delle valli di Fiemme e di Fassa. Museo Geologico, Ed. Comune di Predazzo
5. Nimis PL, Castello M (1990) L'uso dei licheni nel biomonitoraggio dell'inquinamento atmosferico. Biol Ambient 14:5–25
6. EUR-LEX (1992) Council Directive 92/43/EEC of 21 May 1992 on the conservation of natural habitats and of wild fauna and flora
7. Pierini A (1852) Statistica del Trentino. Ed. F.lli Perini, Trento
8. Reisigl H, Keller R (1995) Guida al bosco di montagna. Zanichelli, Bologna
9. EEA (2006) European forest types categories and types for sustainable forest management reporting and policy.
10. Pignatti S (1998) I boschi d'Italia: sinecologia e biodiversità. Utet, Torino
11. Kelly DL, Connolly A (2000) A review of the plant communities associated with scots pine (*Pinus sylvestris*) in Europe and an evaluation of putative indicator/specialist species. Invest. Agr Recur For Fuera de serie n°1
12. Ozenda P (1985) La végétation de la chaine alpine. Masson, Paris
13. Speich S et al (2011) Third national forest inventory - result tables on the internet. Second severely extended edition [Published online 31.03.2011] Available from World Wide Web http://www.lfi.ch/resultate/. Birmensdorf, Eidg. Forschungsanstalt WSL.NFI (2011) Results of Switzerland National Forest Inventory
14. Foster K (2003) Tirol atlas. University of Innsbruck, Innsbruck
15. CFS (2007) Inventario nazionale delle Foreste e dei Serbatoi Forestali di Carbonio. Le Stime di Superficie. CRA, Ist Sperim Ass For Agric, Trento
16. Ingegnoli V, Giglio E (2008) Landscape biodiversity changes in forest vegetation and the case study of the Lavazé Pass (Trentino, Italy). Annal di Botanica NS, (Roma) 8:21–29
17. Pignatti S (2005) Valori di bioindicazione delle piante vascolari della flora italiana. Braun-Blanquetia 39:1–97

13 Applications in Plain Landscapes

13.1 Relation Form–Function in the Design of a New Urban Quarter

13.1.1 Synthesis of the Ecological State of Milan

In wide metropolitan areas, sensu Forman [1], the presence of a consistent ecological network is indispensable. The bionomic efficiency of these networks depends on the presence of some "green corridors", linking the central town with rural landscapes, better with a natural park area. In the case of Milan, the only corridor available to give this function is a rural belt from the west side of the city towards the Ticino river (about 20 km wide) and its large Regional Park.

As we can see in Fig. 13.1, creating an ecological network between the central park of Milan (Simplon Park) and the Regional Park of Ticino should be very important. In Fig. 13.2 we expose a transect along this direction: the urban heath isle (UHI) and the insufficient ecological role of the present urban parks are evident, having a biological territorial capacity BTC < 1.95–2.00 Mcal/m^2/year, the average regional value, consequently too scarce possibilities to balance the urban desert. The comparison with the remnant forest of Cusago (BTC $= 7.67$ Mcal/m^2/year) is symptomatic.

To activate the mentioned network, we should have need to restore the vegetation of the urban parks and to insert a new element (e.g. an urban district/quarter) with a good BTC level. The occasion should have started in 2004, with the architectural competition for the building of a new quarter on the place of the old "Fiera di Milano" (35 ha), win by an international group coordinated by Daniel Libeskind (New York) for the company "CityLife".

In the mentioned group of Libeskind, Ingegnoli was charged for the ecological aspect of the project [2]. To understand the ecological state of Milan, indispensable frame to the correct design of the quarter, we recall Table 3.4, which compares the metropolitan area with Berlin, very similar to Milan. From the table it can be immediately derived that the most important problem for Milan is the insufficient amount of protective elements and the too low level of its vegetation biological territorial capacity (average BTC $= 0.43$, being the optimum 0.70 Mcal/m^2/year).

13.1.2 Models and Parameters for the Study

The general ecological situation (Table 13.1) confirms the necessity to insert in the project of the new urban quarter a good park area. That is why in 2006–2007 CityLife charged the Department of Biology of the University of Milan of a

V. Ingegnoli, *Landscape Bionomics Biological-Integrated Landscape Ecology*,
DOI 10.1007/978-88-470-5226-0_13, © Springer-Verlag Italia 2015

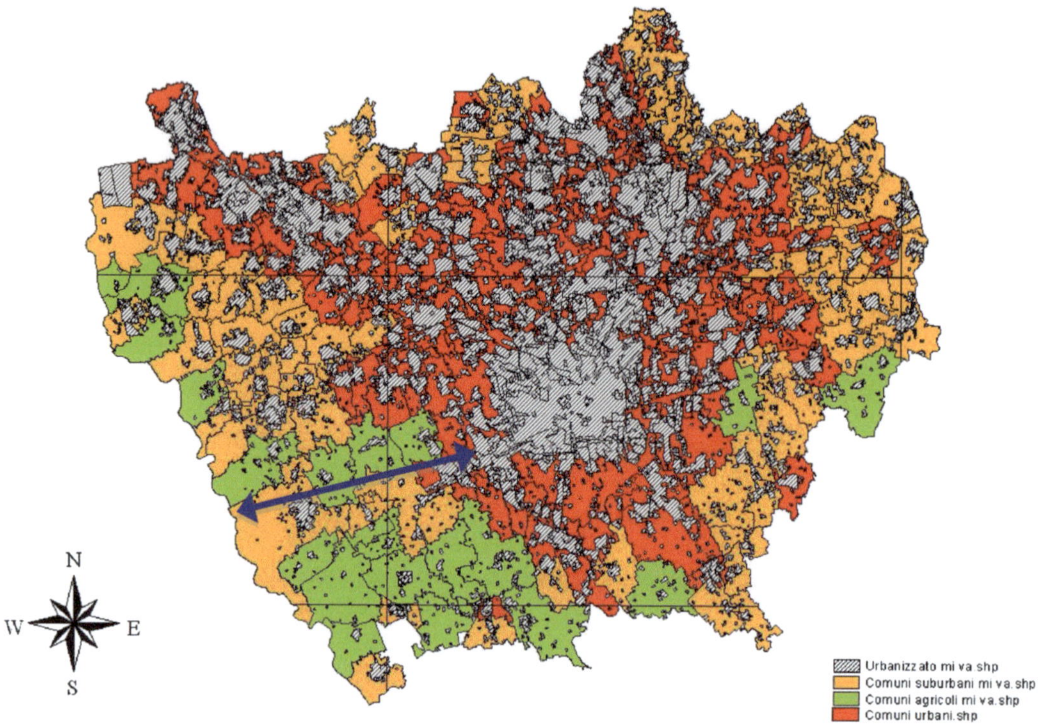

Fig. 13.1 The metropolitan area of Milan (*red* and *grey*) and the green corridor towards the Ticino river (20 km, *blue arrow*) the rural landscape nearest to the centre of the city (*Yellow* = suburban municipalities; *green* = agricultural municipalities)

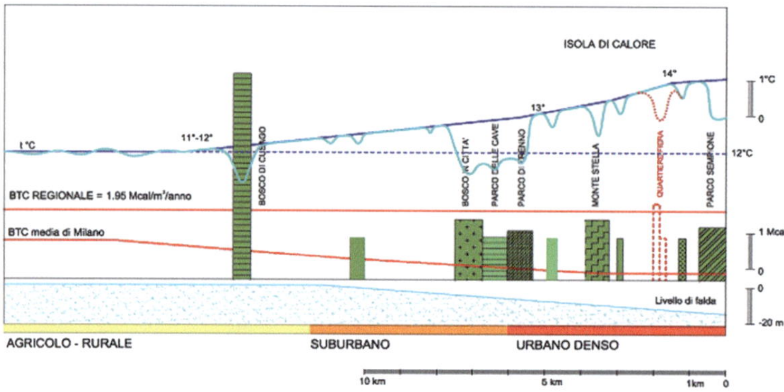

Fig. 13.2 A transect from the regional park of Ticino river to the city of Milan (*right*). Note the UHI, amplifying the annual average temperature from 12 to 14 °C, and the insufficient BTC of the urban parks remaining below the average BTC of the Lombardy Region

research related to the park area of the project. This research was coordinated by the Author and testified that a complex adaptive system (as a landscape unit) is strongly dependent on its configuration even if the components remain of the same extension and with the same function. To evaluate the properties of the form is not simple because the visual analysis is generally insufficient.

The target of the research was the choice of the best model (from an advanced ecological point of view) among diverse master-plan scenarios. Each master-plan model reproduces a

Table 13.1 Urban vegetation types, related to the models

N°	Green types	Composition	%	BTC
1.1	Edge belts [BTC = 1.30]	Meadows or herbaceous formations, both lawn from foot traffic and semi-natural *Arrhenatherion* (about 50 %)	45	0.50
		Patches with trees, only partially exotics, with 50 % of *Quercion robori-petreae* phytocoenosis	25	4.00
		Road green, sport green, etc.	30	0.05
1.2	Core areas [BTC = 2.00]	Meadows or herbaceous formations, both lawn from foot traffic and semi-natural *Arrhenatherion* (about 65 %)	45	0.55
		Patches with trees, only partially exotics with 75 % of *Quercion robori-petreae* phytocoenosis	20	4.20
		Forest biotopes (e.g. *Populion albae* or *Alnion glutinosae* and *Carpinion* or *Quercion robori-petreae* phytocoenosis)	15	6.00
		Tree lined roads, services, playground and/or life paths	20	0.05
2	Square green [BTC = 1.30]	Patches of trees even with exotic plants (not more than 70 %)	35	3.50
		Lawn from foot traffic	40	0.50
		Road green, sport green, etc.	25	0.05
3	Residence green [BTC = 1.40][a]	Traditional garden green, but also about 25 % of *Quercion robori-petreae*	30	4.0
		Lawn from foot traffic	35	0.50
		Road green, sport green, etc.	35	0.05

[a]The residential areas (green = 30 %) will have a BTC = 0.40 Mcal/m^2/year

rectangular area of 25.95 ha, containing: 7.5 ha of residential areas (30 % of which green), 5.2 ha of service buildings, 3.25 ha of squares and free areas, 10 ha of park.

It was a settlement compatible with the Urban Master Plan of Milan, which prescribes for this area "50–50 % built and open spaces-green". In the case of the models (see Fig. 13.7) it was: 48.9 % of built and 51.1 % of open spaces and green (of which 41.6 % closely to green); counting the residential green the percentages became: 40.3 % built and 59.7 % in open spaces and green. The tolerances were considered to be of the order of 1–1.5 %, for obvious reasons of design of the different configurations. The status ex ante or Current State (CS), which summarises a useful synthesis of recent conditions of the study area, was also considered.

The models were first designed in number of ten, to which other five were then added. They included the main possible configurations for the area of the new quarter, according to the criteria drawn in Fig. 13.7. Counting the green residential, the percentage of the green area (without open spaces) resulted equal to 50.2, so the photosynthetic area reached about 13.3 ha. In all models, it was meant that the Green should be articulated in the manner summarised in Table 13.1.

The following types of urban green areas were considered: (i) bands margin, consisting of meadows and wooded spots (with the presence of avenues), for which we can estimate an average of BTC = 1.30 Mcal/m^2/year.; (ii) core areas, formed by more natural meadows, wooded spots with high coverage and a part in forest biotope, wherein the equipment are more contained, for which we can estimate a BTC equal to 2.00; (iii) the green parts of squares, which are similar to the values of the group (i), even if they admit some exotic species; (iv) the green residences, with more parts equipped, and BTC = 1.40 Mcal/m^2/year.

The parameters used for the diagnostic evaluation of the models were obviously divided into two groups on (A) the grounds of the neighbourhood (Qu), (B) the grounds of the urban operative landscape units (LU), defined on the basis of the principles of landscape bionomics. The area of the former Trade Fair, which already historically corresponded to the military exercise area of the late nineteenth century, was considered with all the blocks surrounding it, to the geometric grid of urbanisation next. The normal ranges for each parameter were also set out. The results will be presented in subsequent cards.

A-Quarter

A1-*Fractal dimension of the system of urban green areas* (D): The measurement of this geometric parameter [3] was made according to the method reported (see Sect. 7.1.5). This fractional dimension allows us to check the coverage of the plan which, for a well-distributed vegetation, must normally be D major than 1.5–1.8, because the green must be close to $D = 2$, but not too much (since the grounds must also accommodate other components).

A2-*Ratio area/green perimeter* (A/P%): Green perimeter is constituted here by the sum of park + tree-lined squares. As known, this ratio allows to evaluate the shape: in fact simple and concentrated shapes have A/P ratio greater than lobed or articulated forms. Relatively to the considered cases, the norm was between 30 and 60 % (articulation, but not excessive).

A3-*Core area parks* (%V): In the synthesis afforded by scale models 1/10,000, bands of noise are considered to be 25 m (from buildings, internal squares etc.) and 50 m (from external roads, buildings or tower). The minimum adequate *core areas* in the parks depend on the level of ecological functionality we want to achieve. In our case, to ensure the possibility of arriving at an ecological functionality of the green close to the regional average, we had to consider a normal range between 45 and 75 %.

A4-*Connection of the parts in green* ($\alpha + \gamma$): It is known that the connection of the parts in green is of the utmost importance in the ecology of the landscape [4, 5]. Considering the margin of bands such as green corridors, core areas such as wooded spots, and checking the hypothetical possibility of a link with the outside on each side of the quarter, planar graphs for each model, according to the proposed method, were calculated (see Sect. 7.1.4). The parameters of the circuitry and connectivity $\alpha + \gamma$ were added. This sum should aim at optimal way to unity, but in cases like this we have to keep a tolerance of 15–20 % (experimental data), for which the standard will vary from 0.75 to 1.15.

A5-*Contacts with the outside of the park* (% pV): Measurement of the percentage of the perimeter of the park (and/or tree-lined squares) in contact with the exterior of the land owned by the quarter/neighbourhood. Each neighbourhood park should avoid being completely closed within the inside or completely open to the outside. It is recalled that, for a valid connection with the outside, an opening even lower to 1/3–1/4 of the perimeter is sufficient and that an opening greater than 2/3 is excessive, as it breaks down too ecological benefits to other components of grounds, mainly by transferring them out of it. For our models it seemed reasonable to define normal limits between 20 and 60 %: but it was found that even considering the range 25–65 %, the result of the evaluation doesn't change so much.

A6-*Built area/green perimeter* (m^2/m): This measure is complementary to the previous one and shows the relationship between neighbourhood green spaces and buildings. According to experimental data, which are valid in urban landscapes, this ratio can be considered as normal interval with values around 45–90 m^2/m.

A7-*Structural landscape Diversity* (ψ): The measure of the heterogeneity of the components of a part of a LU was gained through the method given in [6]: the structural diversity of landscape (see Sect. 3.7.1) can be expressed with the formula $\psi = H\,(3 + D)$; remembers that H is the heterogeneity of Shannon and D the dominance, always referred to the number of components of a unit of landscape (LU) or a part thereof. The normal range also takes into account the number of components for model and was considered from 3.6 to 9.1.

A8-*Climatic balance* ((ΔT °C)): In the event of summer days with no wind (worst situation for the Urban Heat Island), experimental data show a decrease in temperature for green area according to the formula: $\Delta T = x^{0.31} - 1$, with x in hectares [7]. The overheating of the built-up areas depends on another experimental formula, deduced from studies of the Institute of the History of Climatology of Milan (Brera), adapted to smaller scales per km, according to the formula: $\Delta T = 0.39 \ln R + 0.00087 R$ (R being the radius of the built in metres). A new settlement should direct its balance very close to 0.00 ° C, with a tolerance of no more than 0.75 °C in the worst conditions.

A9-*Spatial average vegetation bio-potentiality of neighbourhood* ($Mcal/m^2$/year): This quantity, called BTC, is a function of the components of a vegetated LU or a part thereof and is proportional to the ecological metastability of the examined system (see Sect. 3.3.1). The values considered for the element type derive from the literature [6]. In our case, we evaluated the following elements: built up areas with BTC = 0.00; residential areas with green having BTC = 0.40; marginal bands parks and/or corridors lined with BTC = 1.30; core areas of the parks with BTC = 2.00, the normal range depending on the model HH/BTC [6], so for dense cities being from 0.75 to 0.90 [$Mcal/m^2$/year].

A10-*Biopotentiality of the park* ($Mcal/m^2$/year): Using the same values reported above (A9) and taking into account the results on the core areas (A3), it was necessary to estimate the BTC of the parks, which should be forecasted not yet fully (it would be a prediction too far, about 70 years) but halfway (about 30 years) to maturity. For these conditions a range between 1.60 and 1.90 $Mcal/m^2$/year could have been posed as acceptable, being >1.50–1.95 (the average of urban parks of Milan).

B-Landscape Unit (LU)

B1-*Fractal dimension of the system of urban green areas* of LU (D): Referring to what written above (A1), the normal range should have had greater tolerances, given the conditions of low level of ecological LU ex ante, then the following range was considered: $D = 1.45$–1.75.

B2-*Connection LU green* (α + γ): As stated in item A4, here also the normal range should have been reviewed and brought to $(\alpha + \gamma) = 0.5$–0.9, again for reasons of low ecological level of LU ex ante.

B3-*Biopotentiality of LU* ($Mcal/m^2$/year): The same holds of the points A9 and B1 which leaded to the lower normal range around the value BTC = 0.70 and 0.60–0.80 ($Mcal/m^2$/year).

B4-*Radius of ecological influence* (REI, m): This parameter (see Sect. 7.2.1) was measured with the method according to [8], based on four transects perpendicular to the LU: it represents the distance, beyond the grounds of neighbourhood, within which the BTC system of the neighbourhood urban green goes to balance the external shortcomings. The balance value 0.70 $Mcal/m^2$/year derived from the model HH-BTC [6]. Normality must be not less than 110–120 m (up about ½ km).

B5-*Source/sink* (%): Every biological system, being an open system, has input and output resources (organisms, matter, energy, information, etc.). They come from sources and supply the system in question, which could also show resistance to the input, then ejecting the flows to one or more sinks. Please note that these flows follow one of the main laws of ecology: "not too much, not to little, but just enough" [9]. The measurement was performed on transects mentioned in section B4, considering the core areas of the parks as sources and those areas of BTC = 0.00 as sinks. Normality (from experimental data) was 25–75 %.

B6-*Gardens Grain/General Grain*: The average size of the areas of the components of a LU is called grain [10]. This report had to verify that

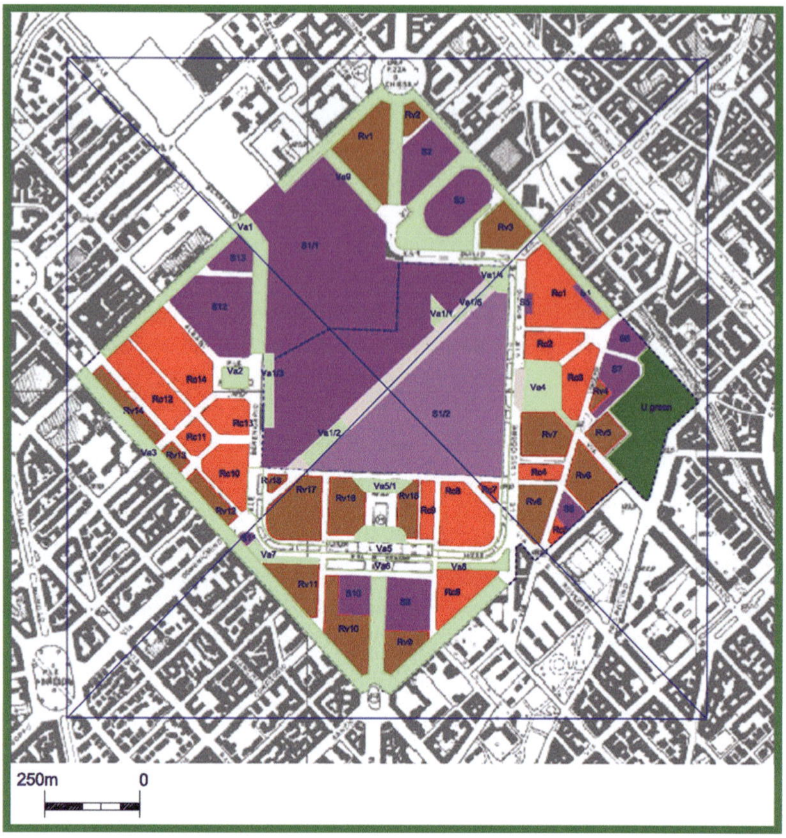

Fig. 13.3 Landscape units (LU) of the former district of "Fiera Milano", the state ex ante (SA). The evaluation of the SA model is obviously very low: equal to 0.162 for the parameters of the quarter and to 0.166 for those of LU. The BTC of LU is approximately 0.35 Mcal/m^2/year (*dark green* = Urban Green; *light green* = tree-lined square; *brown* = residential with gardens; *red* = residential without gardens; *violet* = services and sheds; *grey* = free areas)

the park components were much larger than the average grain size of LU. As is normally good to have park areas at least four times greater than the grain of LU, so the range considered was 4–6.

C-Index of Diagnostic Evaluation and Classification of Results

The method is based on the percentage differences from normality of each parameter found in the different models, as described in Sect. 9.4.2. To get a report that you can legitimately call Index of Diagnostics Evaluation (DI) the values of "score" (Y) must be added and the result must be divided by the theoretical value given by the maximum number of ecological parameters multiplied by 2. In our case we have one DI for the quarter/neighbourhood (QDI), one for the LU (UDI), plus the more general one (DIgen) measured on all the 16 ecological parameters. So: QDI = Y/20; UDI = Y/12; DIgen = Y/32

13.1.3 Results and Evaluation

Let us now review the main case studies, showing maps of the models within their associated LU[1] and tables with the findings of the parameters and evaluations. We'll start with the state of affairs (in 2004) before the opening of the construction site (Fig. 13.3).

[1] More correctly the District of Fiera Milano is an ecotope of a wider LU of North-West. Milan, and the use of the term LU must be intended in an administrative-operative sense. The terms quarter and neighbourhood are used as synonyms to indicate the models in actual fact.

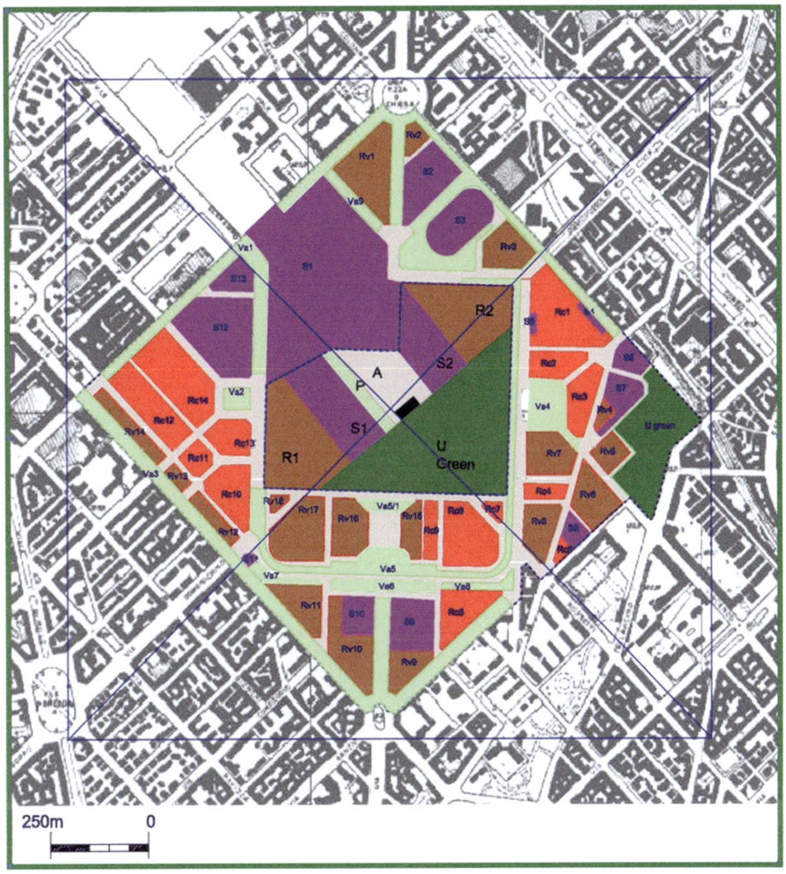

Fig. 13.4 Model M4, included in the minimum unit (LU) of the surrounding cityscape (recognisable as not changing in all the following figures). The neighbourhood park is triangular. The evaluation is equal to 0.71 for the parameters of neighbourhood and 0.50 for those of LU, with average value of 0.634 (*dark green* = Urban Green; *light green* = tree-lined square; *brown* = residential with gardens; *red* = residential without gardens; *violet* = services and sheds; *grey* = free areas)

Table 13.2 Diagnostic evaluation of the model M4

Ecological parameters	Normal	M4	Δ4	Score
Fractal dimension of green (D)	1.5–1.8	1.22	18.80	0.75
Connection of green (α + γ)	0.75–1.15	0.71	5.30	1.65
Core areas (%V)	45–75	43.39	3.60	1.76
Ratio A/P green	30–60	65.76	9.60	1.36
Contacts of green with the outside (%pV)	20–60	53.69	0.00	2.00
Built area/perimeter V (m^2/m)	45–90	131.78	46.40	0.45
Structural diversity (ψ)	3.6–9.1	6.13	0.00	2.00
Climatic balance (ΔT °C)	0–0.75	1.22	62.70	0.31
Neighbourhood BTC ($Mcal/m^2/a$)	0.75–0.9	0.79	0.00	2.00
Park BTC (30 year)	1.6–1.9	1.59	0.60	1.96
Scores			*147.00*	*14.23*
Diagnostic index of the neighbourhood	**0.85–1.0**			**0.711**
Fractal dimension of green (D) LU	1.45–1.75	1.17	22	0.647
Connection of green (α + γ) LU	0.5–0.9	0.31	38	0.514
BTC of LU	0.6–0.8	0.494	17.6	0.826
Radius of ecological influence (min)	110–450	103.46	5.95	1.600
Source/sink (% transects)	25–75	13.85	44.6	0.459
Garden grain/general grain	4.0–6.0	4.11	0	2.00
Scores				*6.046*
Diagnostic index of Landscape Unit	**0.75–0.90**			**0.503**

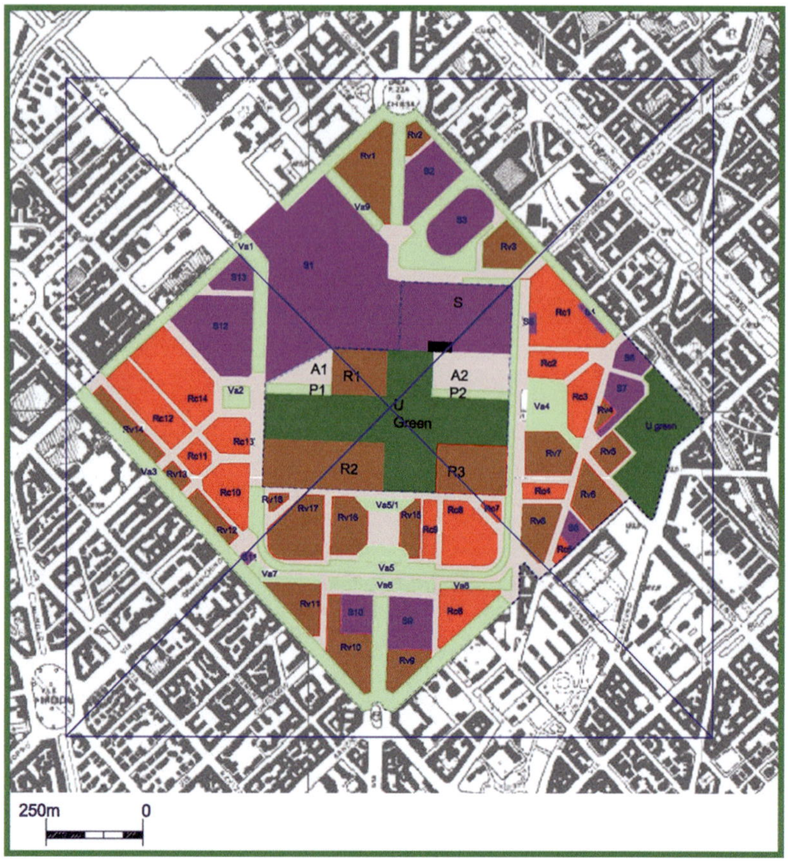

Fig. 13.5 Model M9, placed in the minimum unit of the surrounding cityscape. The neighbourhood park is shaped like a cross. The evaluation is equal to 0.89 for the parameters of neighbourhood and 0.68 for those of LU, with average value of 0.812 (*dark green* = Urban Green; *light green* = tree-lined square; *brown* = residential with gardens; *red* = residential without gardens; *violet* = services and sheds; *grey* = free areas)

Table 13.3 Diagnostic evaluation of the model M9

Ecological parameters	Normal	M9	Δ9	Score
Fractal dimension of green (D)	1.5–1.8	1.41	5.90	1.61
Connection of green ($\alpha + \gamma$)	0.75–1.15	0.875	0.00	2.00
Core areas (%V)	45–75	48.13	0.00	2.00
Ratio A/P green	30–60	52.19	0.00	2.00
Contacts of green with the outside (%pV)	20–60	24.08	0.00	2.00
Built area/perimeter V (m^2/m)	45–90	92.46	2.70	1.82
Structural diversity (ψ)	3.6–9.1	4.96	0.00	2.00
Climatic balance (ΔT °C)	0–0.75	1.10	46.70	0.44
Neighbourhood BTC ($Mcal/m^2/a$)	0.75–0.9	0.79	0.00	2.00
Park BTC (30 year)	1.6–1.9	1.63	0.00	2.00
Scores			*55.30*	*17.87*
Diagnostic index of the neighbourhood	**0.85–1.0**			**0.893**
Fractal dimension of green (D) LU	1.45–1.75	1.209	19.4	0.706
Connection of green ($\alpha + \gamma$) LU	0.5–0.9	0.35	30	0.581
BTC of LU	0.6–0.8	0.494	17.6	0.826
Radius of ecological influence (min)	110–450	295.3	0	2.00
Source/sink (% transects)	25–75	45.6	0	2.00
Garden grain/general grain	4.0–6.0	4.24	0	2.00
Scores				*8.113*
Diagnostic index of Landscape Unit	**0.75–0.90**			**0.676**

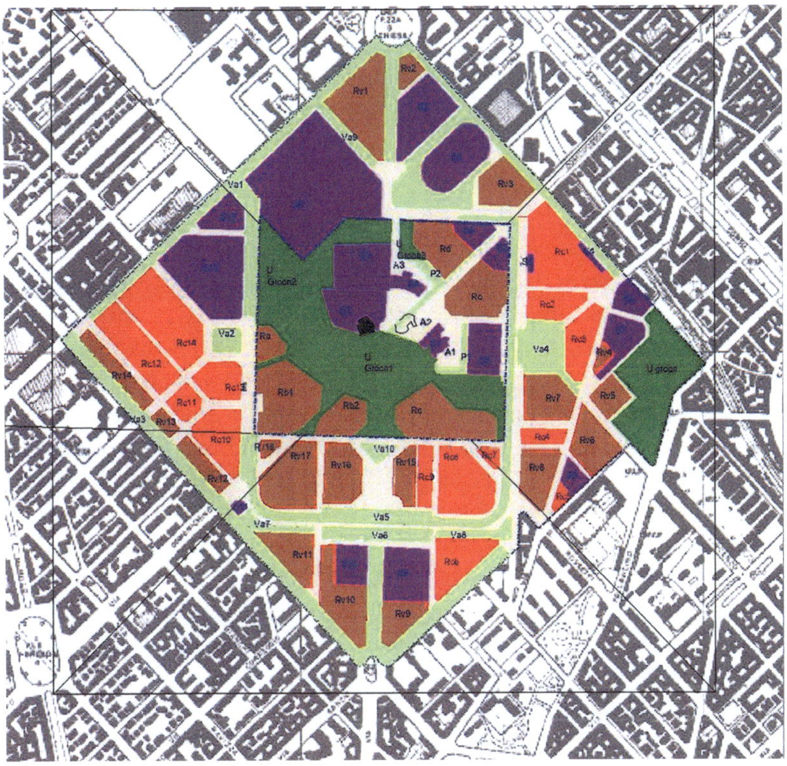

Fig. 13.6 M15, the model more similar to that approved in the PII (Program of intervention). Evaluation is equal to 0.80 for the parameters of neighbourhood and 0.77 for those of LU, with a mean value of 0.786 (*dark green* = Urban Green; *light green* = tree-lined square; *brown* = residential with gardens; *red* = residential without gardens; *violet* = services and sheds; *grey* = free areas)

Table 13.4 Diagnostic evaluation of the model M15

Ecological parameters	Normal	M 15	Δ15	Score
Fractal dimension of green (D)	1.5–1.8	1.69	0	2
Connection of green (α + γ)	0.75–1.15	0.91	0	2
Core areas (%V)	45–75	35.2	21.8	0.65
Ratio A/P green	30–60	26.9	10.3	1.31
Contacts of green with the outside (%pV)	20–60	31	0	2
Built area/perimeter V (m^2/m)	45–90	34.1	24.2	0.63
Structural diversity (ψ)	3.6–9.1	9.5	4.3	1.72
Climatic balance (ΔT °C)	0–0.75	0.67	0	2
Neighbourhood BTC (Mcal/m^2/a)	0.75–0.9	0.75	0	2
Park BTC (30 year)	1.6–1.9	1.53	4.4	1.75
Scores				*16.06*
Diagnostic index of the neighbourhood	**0.85–1.0**			**0.803**
Fractal dimension of green (D) LU	1.45–1.75	1.57	0	2
Connection of green (α + γ) LU	0.5–0.9	0.46	8	1.47
BTC of LU	0.6–0.8	0.49	18.3	0.78
Radius of ecological influence (min)	310–500	194	37.4	0.52
Source/sink (% transects)	25–75	29	0	2
Garden grain/general grain	4.0–6.0	5.6	0	2
Park orientation (% core area)	45–60	55	0	2
Scores				*10.77*
Diagnostic index of Landscape Unit	**0.75–0.90**			**0.769**

As you can see the urban LU is dominated by the sheds of the former Trade Fair and the evaluation of the status ex ante (SA) model is obviously very low: the diagnostic index DI is equal to 0.162 for the parameters of the quarter and to 0.166 for those of LU. The BTC of LU is approximately 0.35 Mcal/m^2/year, lower than the average of Milan city.

We will now show the model M4 (Fig. 13.4), characterised by a triangular park. It should be noted that this model was very well received by the newspapers, however, its ecological assessment (Table 13.2) with diagnostic index is equal to 0.711 for the new quarter, but amounted to only 0.50 for the LU, with average of 0.634, a value certainly not of a good standard.

The model M9 has, instead (Fig. 13.5), a park in the shape of a cross. The ecological evaluation results (Table 13.3) equal to 0.89 for the parameters of quarter and 0.68 for those of LU, with a mean value of 0.812. This assessment is rather high and it was a surprise, since discarded by many designers (but only on the basis of aesthetic perception).

Note that the latest model that is presented in this chapter, M15, with oblong park (Fig. 13.6) and various offshoots, considers one more parameter (the orientation of the park), because we have added the area "pivot" (about 6 ha) to the NW of the old area. The orientation assessment has been carried out with respect to the axis of the ecological network that goes from the park towards the Simplon park to Ticino. The ecological evaluation (Table 13.4) is equal to 0.80 for the quarter parameters and 0.77 for those of LU, with a mean value of 0.786. This is a decent value, although not optimal.

Table 13.5 shows, finally, the ranking of ecological assessments of the design models (Fig. 13.7), with the diagnostic indices of neighbourhood, of landscape units and of the average of the whole. The choice made by the technicians of Citylife was for the M15, one of the best, although none of the studied models reached optimality, which would exceed the value of 0.85. In fact, values of 48.9 % of built and 51.1 % of open spaces and green (of which 41.6 % closely to green) are clearly not sufficient to achieve optimality, but only the sufficiency. Sign that there were some obvious shortcomings in the methods of planning by the City of Milan, which included built too!

Some other considerations, deriving from the setting of landscape bionomics, should be added:

- As a consequence of the Emerging Properties Principle, the insertion of differently designed neighbourhoods within the same exact district triggers a net reaction of diverse connections and relationships, the result of which are strongly diverse and depending on them, as is shown in Table 13.5. Comparing the DI of the quarter, the LU and the general ones, for example within models M1, M2, M8 (and all the proposed models) we can see: M1 QDI = 0.643 and LU DI = 0.439; M2 QDI = 0.812 and LU QI = 0.292; M8 QDI = 0.537 and LU DI = 0.457; so, the DI values of the quarter are always much higher than the DI values of the corresponding unit of the urban landscape. This highlights the fact that considering only the spatial dimension of intervention (although in an environmentally friendly way), without comparing it with the scale of the context, i.e. to evaluate the project without assessing where and how it will be inserted, could lead to wrong choices, even harmful (cfr. models M1 and M2). The more balanced, from this point of view, then the best to be inserted (even if not adequate for the prefixes purposes) resulted to be the model M6.

13.1.4 Conclusions on Form-Function Evaluation

We should emphasise that a search like this shows that it is no longer possible to simply follow the design criteria of the Studies of urban planning and architecture, even if they are of international importance, especially when

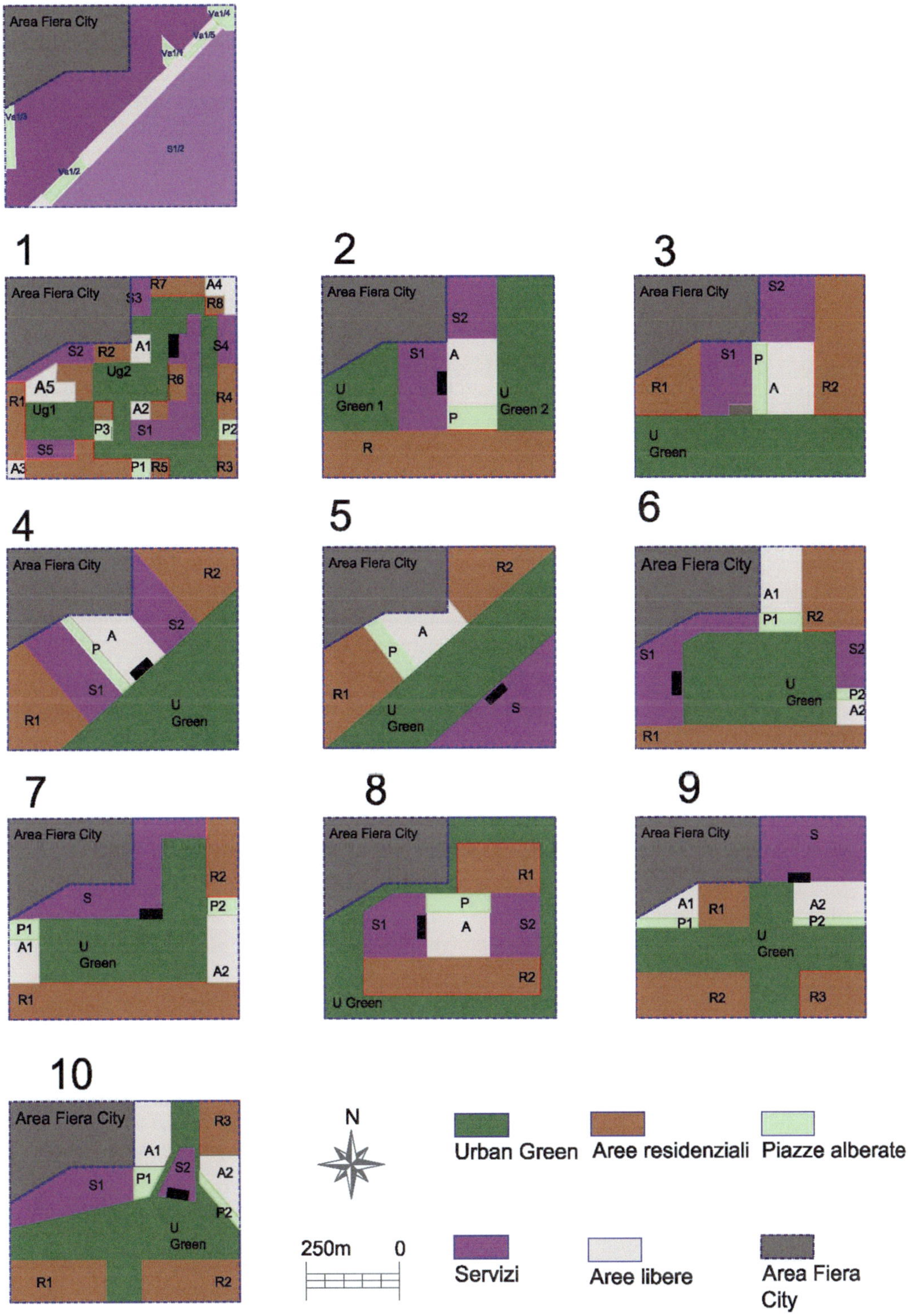

Fig. 13.7 Synthetic diagram of the current state (situation 2004, at the *top left*) and of the better ten models considered in the research on the ecological congruence of the form for the master-plan of the new quarter starting from the ground of the former Trade Fair in Milan (Italy) (*dark green* = Urban Green; *light green* = tree-lined square; *brown* = residential with gardens; *violet* = services and sheds; *grey* = free areas)

Table 13.5 Ranking of the evaluations of the models with diagnostic indices (DI)

Models	Quarter DI	Landscape Unit DI	General DI
SA	**0.162**	**0.166**	**0.164**
M8	0.537	0.457	0.513
M1	0.643	0.439	0.567
M2	0.812	0.292	0.617
M4	**0.711**	**0.503**	**0.634**
M3	0.699	0.580	0.650
M6	0.645	0.669	0.654
M7	0.822	0.671	0.765
M5	0.830	0.658	0.766
M10	0.844	0.646	0.770
M15	**0.803**	**0.769**	**0.786**
M9	**0.893**	**0.676**	**0.812**

the configuration (and the types) of urban green areas is involved as a matter of fundamental importance. Within the landscape units, as within their ecotopes, the functions related to the shape depend mainly on the principles of ecology and are often unpredictable. For example, after an international competition for the former neighbourhood "Fair" of 2004, the triangular shape of the park M4 (proposed by a group of designers arrived among the three winners) was identified as the best by many engineers, not only by newspapers, and yet we have shown how this configuration is rather poor (DI = 0.634), exceeded by seven models and far from best (DI = 0.812).

We should also note that the shortcomings in the comprehension of the urban environment of Milan from an ecological point of view and the lack of knowledge of the principles of landscape bionomics have been such deeper to prevent, in recent decades, to the technicians of the municipal administration, to take optimal decisions for a clear improvement of the environment: for example, a hindrance depends on the limitations of an Urban Plan, being fixed 50 % of green and open spaces and 50 % of built in a neighbourhood inserted in a urban LU, which in the status quo ante had BTC not exceeding 0.35 Mcal/m^2/year, that is lower than the city average BTC (see Table 3.4). In such conditions it is simply futile to try to balance the environment with these simplistic prescriptions (50 + 50 ...), as it is impossible to enter into the optimal range of normality, as demonstrated by the configurations of design models that, as we have seen, however, had tried to overcome these limits: 48.9 % built, and 51.1 % of open spaces and green!

Note, also, that the current planning fails neither to recognise the importance and the need to activate the ecological network Midwest, nor to understand that, for the same air purification, operations such as "Ecopass" or similar would be very more effective if accompanied by an appropriate form of parks, placed in the proper locations of the city and with a vegetation that is not trivial, i.e. parks designed in an environmentally friendly manner. There is only hope that the scientific-technical upgrade offered for an ecological design of the territory, proposed by the writer, could be quickly adopted by the Municipality of the big cities.

13.2 Environmental Impact Assessment of a Large Highway Junction

13.2.1 Introduction and Analysis

The Italian Consortium for Lombard Infrastructure proposed (2009) the final design of the motorway link Dalmine–Como–Varese–Switzerland (Giaggiolo Pass). The last leg (DD) should be in the province of Bergamo and provide a large and complex junction (Fig. 13.8) in the territory of the Municipality of Osio Sotto, in an area south of the European Route E64, between Milan and Bergamo, on the left side of the river Brembo.

Even by a cursory look, it is striking that this area is part of the Local Park of River Brembo (Fig. 13.9) and can be degraded by the impressive project near to be built, with the

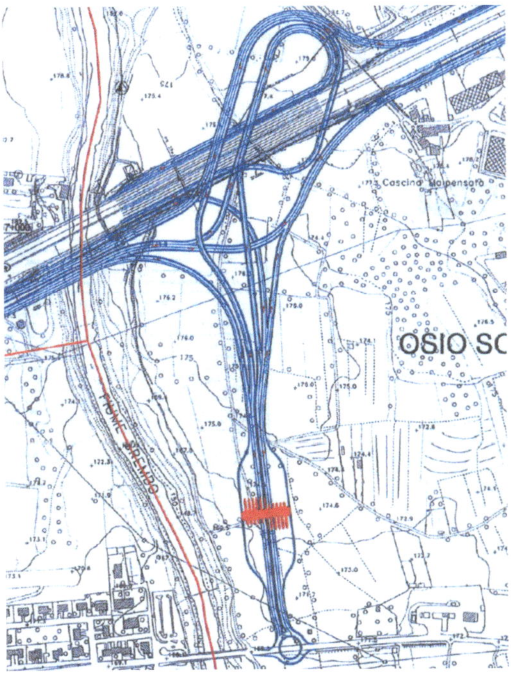

Fig. 13.8 Diagram of the vast amount of territorial development of the project junction of the new "foothills" combining Bergamo to Como, Varese and the Canton Ticino

consequences of serious harm to the area on which it will go to press and on the nucleus of residential houses of the former farmhouse Malpensata still inhabited. The purpose of this Environmental Impact Study consisted, then, in a scientific evaluation in terms of the expected degradation of the area mentioned, so as to highlight both the change in the ecological state of the local system and the repercussions on the residential core.

As can be seen from Fig. 13.10, the limits of this small LU (58.40 ha) are given by the Brembo river to the west, by the E64 highway to the north, by the road to Bergamo to the south and by the industrial area of Osio Sotto to the northeast and, finally, by a separate agricultural-productive not-wooded land in the south-east. Note that this territory is characterised by the presence of forests and tree-lined avenues and urbanised settlements are very limited: the

Table 13.6 Landscape elements and human habitat in 2009

Elements	ha	% LU	%HH	ha HH
Woods, medium BTC	3.10	5.31	5	0.155
Woods, low BTC	4.76	8.20	10	0.479
Tree-lines, high BTC	1.28	2.19	40	0.359
Tree-lines, low BTC	3.20	5.48	60	1.92
Trees plantations	1.37	2.35	60	0.820
Meadows	38.12	65.30	78	29.734
Crops and corn fields	4.20	7.19	90	3.78
Built areas	0.71	1.22	100	1.22
Roads and barrens	1.61	2.76	100	1.61
Total	58.38	100.00	68.65	40.08

residential core of the Malpensata farm in the Northeast and the service area with Restaurant to the South.

The measure, expressed in hectares (ha), of the elements which, in the current state, comprise the landscape units was obtained through a planimeter and the results are shown in Table 13.6. The same table shows the esteems of the percentage of human habitat (HH), measured according to the method prescribed by the bionomics of the landscape. In Table 13.6 we can note, in addition, the main measures of the landscape elements forming of the LU under examination.

13.2.2 Estimated Standard Habitat (SH)

The estimate of the population was made by considering: (1) 6 settlements, next to the industrial area, with a mean of 4 inhab/core; (2) the inhabitants due to the equivalent average presence of users of the park and the restaurant. There was thus obtained a total of 42 inhabitants/year. The result was the following estimate standard habitat $SH = LU \times HH = 58.38 \times 0.6865/42 = 0.954$ m^2/inhab, then a SH/SH^* (σ) = 6.47.

13.2.3 Landscape Apparatuses

In this case it was necessary to measure the following landscape apparatuses: resistant

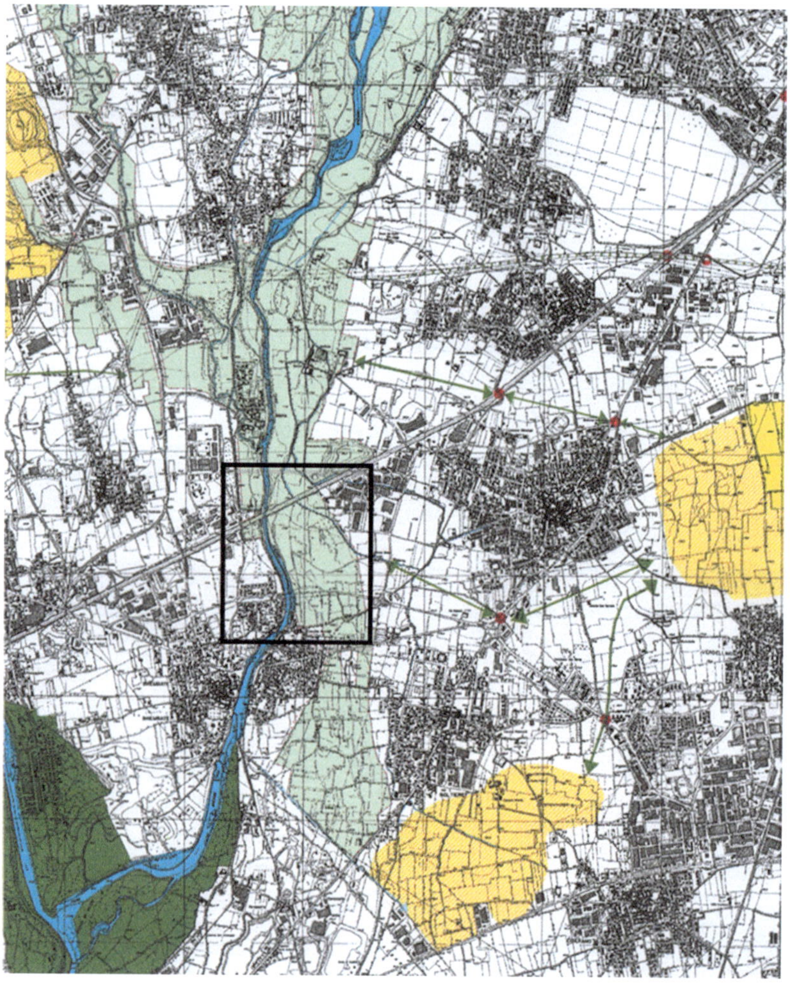

Fig. 13.9 The Local Park of Lower Brembo river, where the *black square* contains the map in the previous figure. It should be noted, however, that the portion to the south of the Park acquires great importance for the presence of the Regional Park of Adda River (*dark green*) (excerpt from the "Charter of territorial framework" of Park Brembo, by the Center Studies on the territory of the University of Bergamo)

(RNT), productive (PRD), protective (PRT), residential (RSD), subsidiary (SBS). Remember that the contents do not coincide with any similar urban or economic terms. The resistant apparatus RNT (15.7 %), formed by patches of forest, wasn't as well developed as it would be expected from an agricultural-protective landscape, but it still remained as the second surface. Instead the protective apparatus, formed by rows of trees, was of excellent level here, as it did being precisely the element distinguishing this type of landscape. These fields are enclosed in a network of often double or triple rows, including channel or the sideliners, what now is very rare even within the Regional Parks.

13.2.4 Vegetation Assessment of the Landscape Unit

It should be noted again that the method LaBiSV deviates from the traditional phytosociological method, since the vegetation that forms a landscape unit cannot be assessed only on the basis of the type of natural species present and placed in relationship with a concept of potential vegetation that has become obsolete. In other words, the assessment of the real state of the vegetation is done not by comparison with a potential abstract existence of another phytocoenosis, generally arboreal, on the indeterminate time scale and with linear development mode, but by

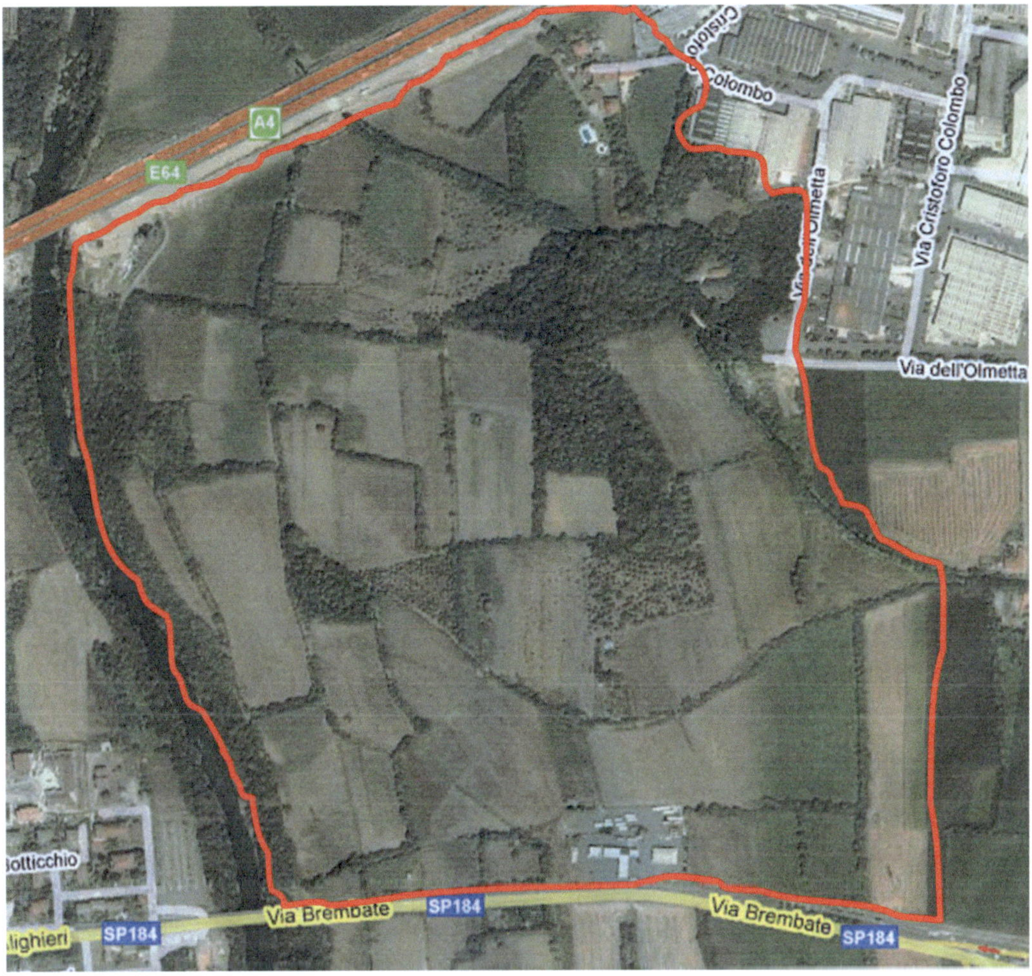

Fig. 13.10 In *red*, the perimeter of the small landscape units (LU) in question, which measures almost 60 ha and is included in the Park of the Lower Brembo river (see Fig. 13.9) for about 90 % (the picture is taken by satellite map by Google)

comparison with *a real potentiality of ecological functioning of the same plant communities under consideration, for defined periods of time and the mode of development of which is not linear* (inflections and bifurcations): that is by comparison with the new concept of "fittest vegetation" (see Sect. 3.3.3).

Therefore, through the new method of parameterisation of vegetation, presented above (see Sect. 5.1.2) and using the standard forms specific for the types of vegetation under examination, it was possible to assess their degree of organisation and consequently their level of quality. In this case, it was patrolled the entire Landscape Unit and a dozen of field surveys were made, about half of wooded tesserae and half of tesserae row of trees.

The results verified in the field were not very different from what it had been found in the Monza Historical Park and non-native plant species weeds were still contained. With some proper interventions in a decade good results could be achieved from these pieces of vegetation which, given the significant presence of *Ulmus minor*, seem mainly to fall into *Querco-Ulmetum minoris* Issler, 1924, the

Table 13.7 Summary of vegetation surveys in the LU

Tessera surface (ha)	BTC Tessera (Mcal/m²/year)	Tessera surface (%)	BTC (Mcal/m²/year)
Wood vegetation tesserae			
1.352	6.00	17.20	1.0321
1.148	5.60	14.61	0.8180
0.600	5.00	7.63	0.3817
1.700	4.70	21.63	1.0166
0.690	4.10	8.78	0.3599
1.004	4.00	12.78	0.5110
0.600	3.50	7.63	0.2672
0.135	3.00	1.72	0.0515
0.630	2.70	8.02	0.2164
= **7.86**	//	100.00	**4.65**
Tree-lines vegetation tesserae			
0.138	3.70	3.08	0.114
0.35	3.60	7.81	0.281
0.36	3.30	8.04	0.265
0.432	3.00	9.64	0.289
0.21	2.80	4.69	0.131
0.460	2.60	10.27	0.267
1.250	2.40	27.90	0.670
1.280	1.80	28.57	0.514
= **4.48**	//	100.00	**2.53**

phytosociological association defined by Pignatti as "...an intermediate stage between woods of *Salix alba* and the *Querco-Carpinetum boreoitalicum* Pignatti 1953, which constitutes the final stage. [...] The few formations of this type that have been preserved, often randomly, are therefore to be considered of high environmental quality and to handle with protection criteria that allow their survival..." [11].

13.2.5 Estimation of Biologic Territorial Capacity of Vegetation (BTC)

The following table shows a summary of the findings on forest vegetation and vegetation of the rows (Table 13.7). The tesserae are divided according to territorial biopotentiality and quantification in hectares was achieved by analogy with the cases detected. The average value of BTC in the woods is modest (4.65 Mcal/m^2/year) but much greater than that of the rows (2.53 Mcal/m^2/year).

After having calculated the partial averages of the woods with mid and low BTC and of the rows with high and low BTC and after having used values from the literature for the remaining components, we took over the landscape elements of the Table 13.7 and calculated, by the weighted average, the values of the territorial biopotentiality of the Landscape Unit, which resulted to be equal to 1.35 Mcal/m^2/year (Table 13.8).

It is definitely a value higher than the average for the Lombard plain, which is around 0.95–1.15 Mcal/m^2/year instead. This justifies the inclusion of the LU into the Park of the Lower Brembo River. However, making a comparison with the model HH/BTC there is an evident failure, as a Human Habitat with value equal to 68.85 % should achieve at least 1.72 BTC Mcal/m^2/year!

Table 13.8 LU elements and evaluation of average BTC. 2009

Elements	Surface (ha)	Surface (% LU)	BTC ($Mcal/m^2/year$)	BTC′ ($Mcal/m^2/year$)
Woods, medium BTC	3.10	5.31	5.66	0.3005
Woods, low BTC	4.76	8.20	4.00	0.3280
Tree-lines, high BTC	1.28	2.19	3.32	0.0727
Tree-lines, low BTC	3.20	5.48	2.22	0.1217
Trees plantations	1.37	2.35	1.60	0.0376
Meadows	38.12	65.30	0.65	0.4245
Crops and corn fields	4.20	7.19	0.85	0.0611
Built areas	0.71	1.22	0.10	0.0012
Roads and barrens	1.61	2.76	0.05	0.0014
Total	58.38	100.00	–	1.35

13.2.6 Landscape Type

As we know, the clinical-diagnostic method requires to highlight the normal ranges for the main ecological parameters, which are obviously different for each type of landscape. From an inspection you can immediately deduce that we are today in the presence of an agricultural landscape, but landscape bionomics distinguish at least 3–4 classes of agricultural landscapes. Helpful in this regard may be the indicator of "Landscape Type Evaluation" LTpE = (10 NHd/HH + BTC_{A+F}), in the present case estimated equal to 16.2. This value excludes both the suburban landscape and the forest-agricultural one; but we cannot attribute the landscape units concerned to an agricultural productive landscape because of the small presence of cultivated land, then the landscape in question can be classified as "agricultural plain protective terraced", a type of landscape of an ecological value generally higher than the simple "productive".

13.2.7 Survey of the Structure: The Method of "Segmented Lines"

Mapping out some "lines" through the most characteristic environmental gradients of the LU in examination and dividing these lines into segments of equal length, we proceeded to the relief of the presence of elements belonging to the local landscape apparatuses, forming a matrix, presenting a structural diagram (see Sect. 7.1.7). The present situation was then compared with the ex post one, as already shown in Fig 7.5. To learn more about the information content for the item H i-th linked to each level of reading, the equation: $H = \log_2 S!/F!\ (S\text{-}F)!$ [bit] has been used, with S equal to the total number of segments that make up the line in question and F to the number of segments in which the i-th element is found (i.e. the absolute frequency of the i-th). Therefore, it was possible to numerically calculate which could have been the resolution able to provide the highest content of global information.

Going back to Fig. 7.5, you find the example of data recorded in the LU in question plotted in the two states ex ante and ex post. It soon becomes evident even to eye the loss of structure! Proceeding in the calculation we get: the total information ex ante H_{exA} equal to 40.449 bit and in the case of ex post H_{exP} = 30.737 bit, with a loss of information equal to 24 %. Extending to all four lines a similar calculation the result is a loss of about 20 %, from 143.36 bits ex ante to 118.59 bits ex post.

13.2.8 Evaluation of the Connection

We used the model of planar graphs (see Sect. 7.1.4), having as purpose the measure of the extent of existing connections between the elements in arboreal vegetation and tall shrubs within an

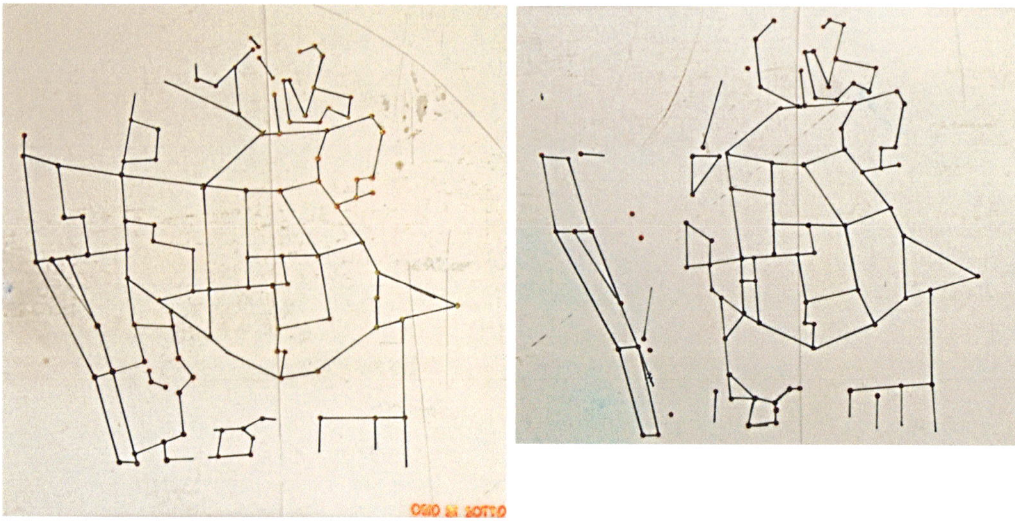

Fig. 13.11 Comparison of planar graphs that represent the nodes and the links connecting the LU of Osio Sotto, in the state ex ante (*left*) and ex post (*right*), then used for the calculation of circuitry and connectivity

ecomosaic, then the verification of the existence of conditions for an ecological network. The application in this case refers to the graph of Fig. 13.11.

In the environmental state ex ante the result was a single graph, with an index $\alpha = 0.1776$ and $\gamma = 0.455$ ($\alpha + \gamma = 0.633$), while in the state ex post we find two separate graphs, that add up to: $\alpha = 0.116$ and $\gamma = 0.414$ ($\alpha + \gamma = 0.530$) with a loss of connection of approximately 16 %, despite the mitigation measures.

13.2.9 Alteration of "Potential Core Areas" (PCA)

As noted in previous chapters, the analysis of the structure of a Landscape Unit cannot stop at the information collected with segmented lines and at the connection level. It is necessary to look what part of the structure was free from disturbances of various kinds, that is to locate the so-called "potential core areas" (sensu [8]). Applying the method of evaluation of the system of residual patches (see Sect. 9.2.2), as shown in Fig. 13.12, it appears that the PCa will contract by more than 60 %, with a change of axis of PCa from E-W to N-S. A real demolition of the area under protection of the Park of Low Brembo!

13.2.10 Further Parameters

Being a Study of Environmental Impact with the highway as the main disorder, we had to verify the change in the relative tranquillity area of the LU under consideration. For this purpose we made many findings in the field, with the sound level meter, to check the current band of noise >60 dB (Leq). A measure of the magnitude of 170–180 m from the E64 Highway and 140–150 m from the provincial road was found. The assumptions for estimating the state ex post have been reported at a shorter distance from the junction, due to the mitigation provided by the great work: 120 m. Despite this caution, the areas that are expected to stay quiet are 57.7 % in the case of ex ante and 27.7 % in the case ex-post, a decrease of approximately 52 %. Since this is a restricted zone, being a park, this will be a very significant impact.

A further ecological parameter is the change of the barrier effect that the North–South axis of the Park meets today respect to the effect in the case of ex post. Today the line passing through the axis of the Brembo park meets two barriers, the E64 motorway and the Provincial, for a total of 56/790 m = 7.09 %. After the work, the barrier effect, enhanced by the presence of the junction, will be

Fig. 13.12 Comparison between the two Potential Core Areas (PCA) of the LU in examination. To the left the current situation; to the right the situation at the end of construction (taking into account mitigation!). In *grey* the primary disorder, in *yellow* the first end of the buffer, in *red* ochre the second buffer strip; in green the PCA. Even at first sight the disruption of the structure of the LU is clear. By measuring the respective PCA a contraction of about 61 % will take place

equal to metres 332/790 = 42.03 %. Also in this case an increase more than remarkable!

13.2.11 Diagnostic Evaluation of the Landscape Unit

As has been pointed out, each biological system—thus also a landscape—over to transform may experience changes in its state and this alteration can become pathological in various conditions. Note the Table 13.9 on the environmental state ex ante (current state). The 15 ecological parameters detected and already recognised in the previous paragraph, their normal ranges, the waste percentage compared to the values found from the norm and the indexes for diagnostic parameter (DI-09) are listed. The normal ranges are based on the type of landscape of LU, which is agricultural-protective and is characterised by agricultural fields with rows and wooded patches, whose values are derived from [6] and subsequent processing (See Sect. 9.4).

The score parametric evaluation is performed by classes of deviations from the norm. You will notice that 9 out of 15 parameters are fully normal, and of the remaining 6 no one has distances greater than 30 %. This means that, despite the E64 has been enlarged, this LU of the Park of the Lower Brembo is quite in good ecological state. To better define of landscape pathology we refer to Table 9.14. From this gradient the state of today's Landscape Unit in question results in diagnostic class II, which denotes a state of mild ecological alteration, not far from normal. An estimate of the ecological state of 10 years ago showed that the Landscape Units was still fully in the standard for a type of agricultural landscape-protective, given the fact that the E64 had not yet been enlarged.

The application of the diagnostic index (DI) to a suitable set of ecological parameters is useful for understanding the present state and the trend from the past, as if to gauge transformation scenarios in the near future. We are therefore able to establish with a high degree of certainty, the diagnosis in the prediction of conversion of the LU in question (a) during the opening of the site and (b) after the construction of the junction of Highway foothills with the E64. This study took place in 2009, so the year 2014 and the year 2024 can be reasonably assumed as the time

Table 13.9 Diagnostic evaluation of the LU of Osio Sotto. Ex ante (2009)

Ecological parameters	Normal AGR. PRT	2009	Distance (from normal) 09	Score DI-09
HH (human habitat)	42–68	68.65	0.9	2
SH/SH*(carrying capacity)	3.0–12	6.47	ok	2
BTC LU (Mcal/m^2/year)	1.90–3.50	1.35	28.9	1
LTpE (10 NHd/HH + BTC_{A+F})	22–40	16.2	26.4	1
BTC Forests (Mcal/m^2/year)	6.50–7.00	4.66	28.3	1
RNT (forest area)	20–55 %	15.7	21.5	1
PRD (cropland)	35–60 %	72.5	ok	2
PRT (hedgerows & plantations)	5–8 %	7.8	ok	2
RSD (urbanised areas)	2–6 %	1.22	ok	2
SBS (industry & traffic areas)	2–4 %	1.61	ok	2
Structure (bit)	140–180	143.4	ok	2
PCA (*potential core areas*)	50–65 %	40.1	19.8	1
Connectivity (α + ϒ)	0.6–0.8	0.63	ok	2
Tranquillity (<60–63 dB) (%)	>60 %	57.7	3.8	2
barriers/axis of Park (%)	0.5–6	7.1	18.3	1
Diagnostic index (DI)	**85–100**			**80.00**
Ecological state	normal	Mild alteration		
Distance (%) Evaluation Scores: 0–10 = 2; 10–30 = 1; 30–60 = 0.5; >60 = 0.				

Table 13.10 Diagnostic evaluation of the LU of Osio Sotto. Ex post (2022)

Ecological parameters	Normal AGR.PRT	10 years after	Distance (from normal) 24	DI-24
HH (human habitat)	42–68	**71.2**	7.9	**2**
SH/SH*(carrying capacity)	3.0–12	**6**	ok	**2**
BTC LU (Mcal/m^2/year)	1.90–3.50	**1.27**	34.2	**0.5**
LTpE (10 NHd/HH + BTC_{A+F})	9.0–55	**10.7**	ok	**2**
BTC Forests (Mcal/m^2/year)	6.50–7.00	**4.56**	29.85	**1**
RNT (forest area)	20–55 %	**17.1**	14.5	**1**
PRD (cropland)	35–60 %	**52.7**	ok	**2**
PRT (hedgerows & plantations)	5–8 %	5.45	ok	**2**
RSD (urbanised areas)	2–6 %	**1.22**	ok	**2**
SBS (industry & traffic areas)	2–4 %	**20.6**	515	0
Structure (bit)	140–180	**118.6**	15.3	**1**
PCA (*potential core areas*)	50–65 %	**15.6**	68.8	**0**
Connectivity (α + ϒ)	0.6–0.8	**0.53**	15.9	**1**
Tranquillity (<60–63 dB) (%)	>60 %	**27.7**	53.8	**0.5**
barriers/axis of Park (%)	0.5–6	42	700	**0**
Diagnostic index (DI)	**85–100**			**56.67**
Ecological state	Normal	Dysfunction		
Distance (%) Evaluation Scores: 0–10 = 2; 10–30 = 1; 30–60 = 0.5; >60 = 0.				

horizon, having to leave at least 2–3 years for the yard and 10–11 years to the growth of the interventions of mitigation and compensation to green in the draft junction (Table 13.10).

By assessing the ecological state in open yard (2011), we can immediately understand that, even if considering a tolerance of 4–5 % in the forecast in favour of a lower environmental impact,

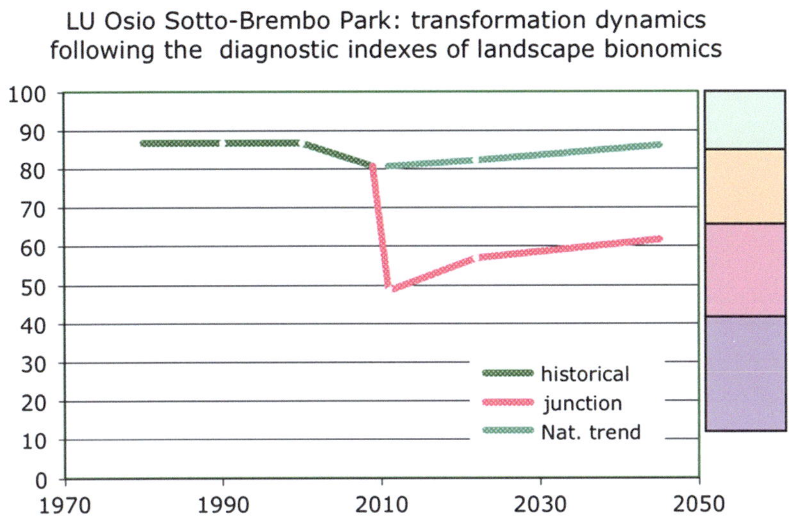

Fig. 13.13 Note that the historical development of LU in question has a natural tendency to return to the margins of normality: however the prospects of transformation due to junction will not succeed in going beyond the sphere of the alteration. The area between the natural tendency and the alteration due to junction measures the processing deficit, unsustainable in a park area. To the right of the pathological levels: *green* = class I (absence of disease), *orange* = class II (low alteration), *pink* = class III (severe alteration)

namely a DI value >48.33 (e.g. 48.3 × 1.04 = 50.2) we would have a change of ecological state in a very short time (2009–2011) that would plunge the diagnostic index from 80.0 to 50.6 (or even 46.4), with a loss of more than 36–42 % of the value DI today (Fig. 13.13).

Wanting to estimate the ecological state 10 years after the closure of the yard (2024), we'll have to consider the works of mitigation and the compensation measures proposed by the project under consideration for release. It is thought that the environmental situation of the LU of Osio Sotto should improve, but not enough, as it's evident processing the necessary calculations to forecast, with the DI reaching values around 56–57 % and no more.

Observing Fig. 13.13 we can evaluate the order of magnitude of the deficit of transformation, which would be the area between the line at DI = 80 → 86 % and below the broken line (ID = 48 → 61.7 %) almost parallel after 2023, then a deficit not only very large but irreversible.

At this point it should be noted that the control of the dynamics following the application of the diagnostic index can be used in cases like this to assume projects less destructive than that proposed by the Italian Consortium for Lombardy Infrastructure. Although beyond the scope required by this environmental impact study, we can guarantee that the noise not incorporated by the LU in question you can be roughly halved, for example by rotating the axis of the junction to the west and moving the connection E64 about 1.2–1.5 km West!

13.3 Bionomic Evaluation of the Project of a Touristic Village in the Littoral of Eraclea (Venice)

13.3.1 Introduction

Preserving the Venice Lagoon during the last millennium has not been easy. The peculiar Authority of the "Magistrato alle Acque" of the "Serenissima Repubblica" was obliged to continue works of hydraulic engineering, arriving to dislocate some river estuaries. It is the case of the river Piave, as exposed in Figs. 13.14 and 13.15.

The river Piave, 220 km, flows in the Adriatic Sea between the municipalities of Jesolo (South) and Eraclea (North). The sea formed littoral dunes forcing the river to go out to sea in an

Fig. 13.14 The dislocation of the estuary of the river Piave, immediately North to the Venice Lagoon. In *red* the new stretch of the river, while the old bed (yellow) was reduced to a canal

Fig. 13.15 The estuarine stretch of the river Piave, about 750 m before flowing in the Adriatic Sea

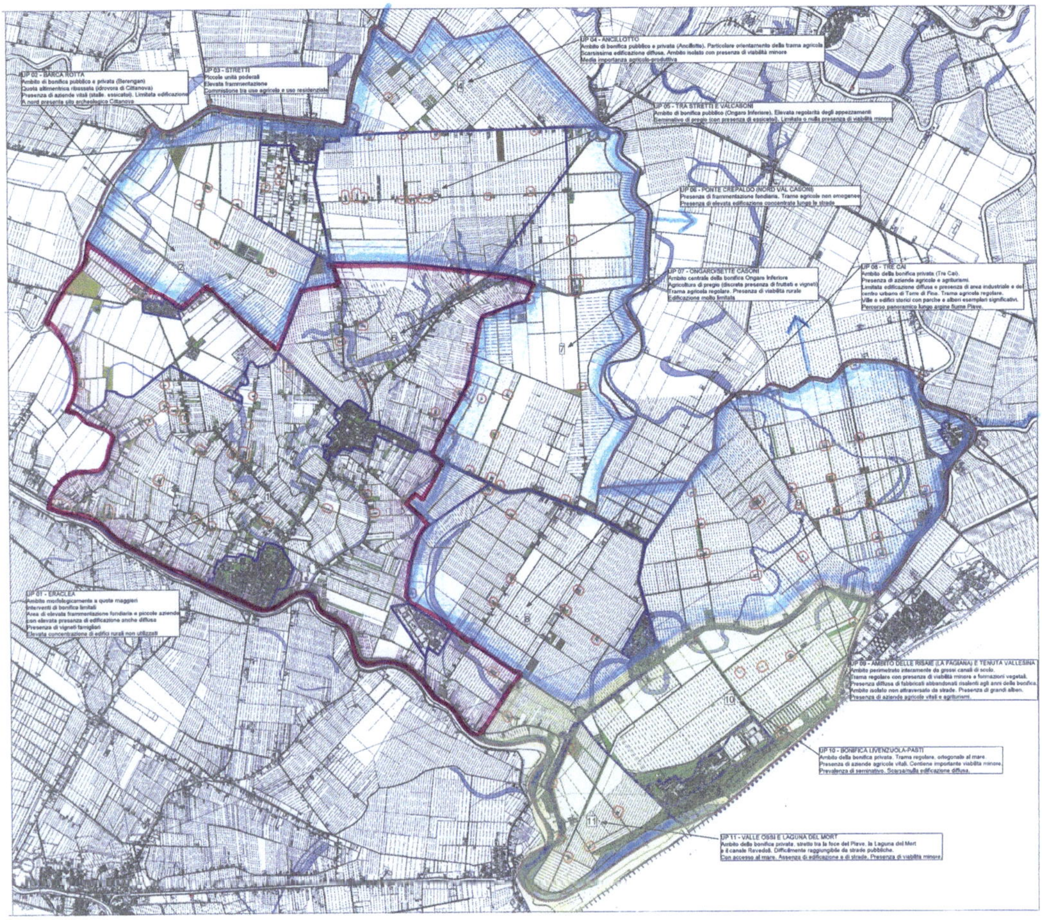

Fig. 13.16 The municipality of Eraclea, about 50 km^2, is composed of three landscape units (LU): the *pale blue* LU-1 was once a lagoon and today this area is below the sea level of 2–3 m. The most part of urbanisation is in the violet LU-2, above the sea for few metres, while Eraclea Mare, the green LU-3, is mainly a touristic village

indirect way still visible in Fig. 3.3. But few decades ago the river opened a direct estuary, reducing the old stretch to a typical small littoral lagoon. Soon the lagoon area was preserved as SAC (Special Area of Conservation) following the EU legislation.

The municipality of Eraclea presents a surface of about 50 km^2. The landscape units (LU) forming this territory are three, as exposed in Fig. 13.16: (1) LU-1 (pale blue) was once a lagoon, and today this area is below the sea level of 2–3 m; (2) the most part of urbanisation is in the LU-2 (violet), above the sea of few metres; (3) LU-3 (green) the littoral Eraclea Mare is mainly a touristic village. In the curved triangle, S-W to Eraclea Mare the project of a new touristic village (2011) was presented to the municipality. Its neighbouring with the SAC area created some contrasts with the public authorities and Virginio Bettini (University of Venice-IUAV), together with the major of Eraclea, asked the opinion of the writer.

13.3.2 Preliminary Analysis and Evaluation

The study of EIA (Environmental Impact Assessment) produced by the company responsible of the project did not found any problem, as is the practice. But the EIA study did not used any principle of landscape bionomics. We want to demonstrate that even a preliminary analysis conducted following our theories is able to find problems.

Table 13.11 Evaluation of BTC and HH of the LU-3 of Eraclea. 2012

LU-3. ERACLEA MARE. 2012	ha	%	BTC′	BTC	HH′	HH
Urbanised	40.4	5.46	0.2	0.011	98	5.355
Roads and squares	7.0	0.95	0	0.000	100	0.947
Beach	15.0	2.03	0	0.000	50	1.014
River and canals	29.7	4.02	0.05	0.002	60	2.410
Lagoon	24.5	3.31	0.1	0.003	30	0.994
Meadows	85.0	11.50	0.55	0.063	75	8.623
Cropland	463.0	62.63	0.7	0.438	89	55.738
Tree-Lines	9.5	1.28	3	0.039	35	0.450
Artificial Woods	8.6	1.16	3.2	0.037	50	0.582
Littoral Pine Wood	40.3	5.45	5.1	0.278	20	1.090
Littoral shrubs and trees	16.3	2.20	3.9	0.086	10	0.220
Tot. LU-3	**739.3**	100.00		**0.96**		**77.42**

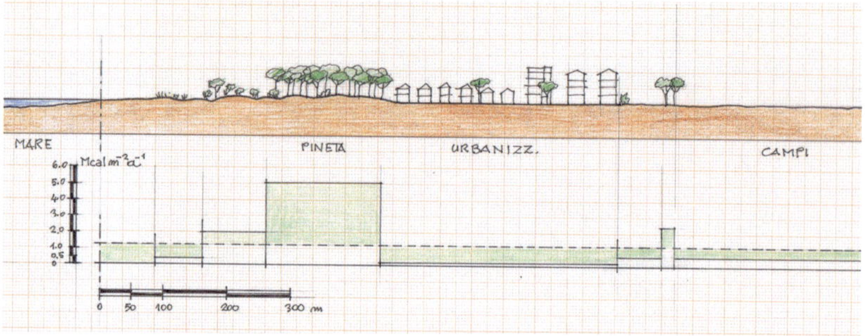

Fig. 13.17 Transect perpendicular to the Sea coast in the LU-3 of Eraclea Mare. Note the positive influence of the local BTC sources

A first synthesis of the bionomic state of the landscape unit LU-3 is summarised in Table 13.11. This landscape is mainly agricultural, but a reforestation of some shrubby dunes (1960) allows the presence of the littoral wood (*Pinus pinaster*), giving major importance to the near SAC area. The average BTC is not so weak as other rural areas in this zone.

A transect (Fig. 13.17) perpendicular to the coast put in evidence the Pine Wood and the position of Eraclea Mare, ½ km from the Sea. A control of the ecological radius of influence (ERI, see 7.2.1) evidenced the good efficiency of vegetation in protecting human settlement and surrounding cropland. Note that the average BTC (dotted line) is BTC = 1.20 Mcal/m^2/year, hence greater than BTC = 0.96, because the LU-3 containing a SAC area must be improved.

The proposed project of touristic village is covered by about 1/3 of urban green, but the types of green are very poor, ecologically speaking. These green components are: (1) in the Northern part a golf course, very technical sport green; (2) in the central part, small house gardens; (3) in the Southern part a belt of green areas for water purification (Fig. 13.18). The BTC values of these technical green are very low, an average from 0.5 to 1.3 Mcal/m^2/year.

The consequences of the building of the new village in the LU-3 can be controlled following the model HH/BTC (see Sect. 7.2.3). Figure 13.19 indicates the movement of the system from the state ex ante (green) to the status ex post (red) and their distance from the curve of normality (blue), which represents a tolerance of about 10–12 %. The failure of the situation is evident. Each principle of biological conservation (remember the SAC area) suggests for this LU-3 an indispensable improving of the present bionomic state. Note that the average BTC of the Veneto Region is 2.06 Mcal/m^2/year. Today the

Fig. 13.18 The project of a new touristic village positioned at the estuary of the old river Piave, strictly near the town of Eraclea Mare

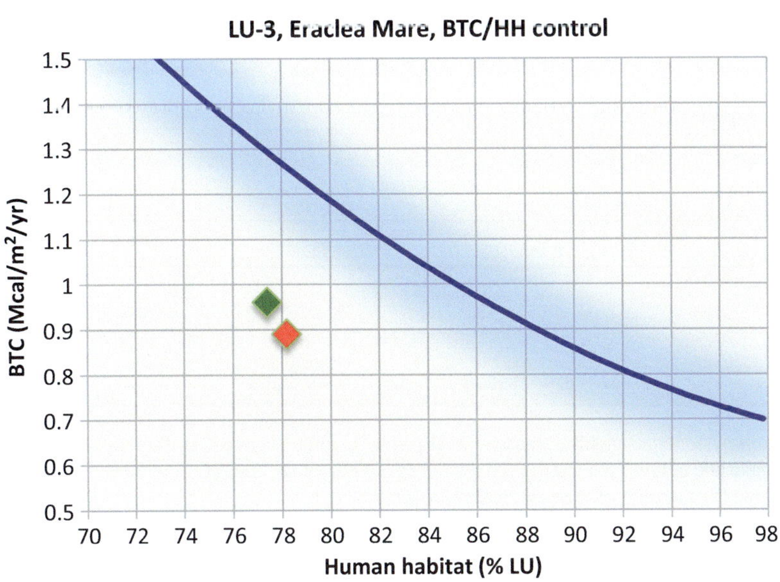

Fig. 13.19 Control of the bionomic state of the LU-3 following the model HH/BTC. The *red square* represents the BTC value ex post the building of the new village

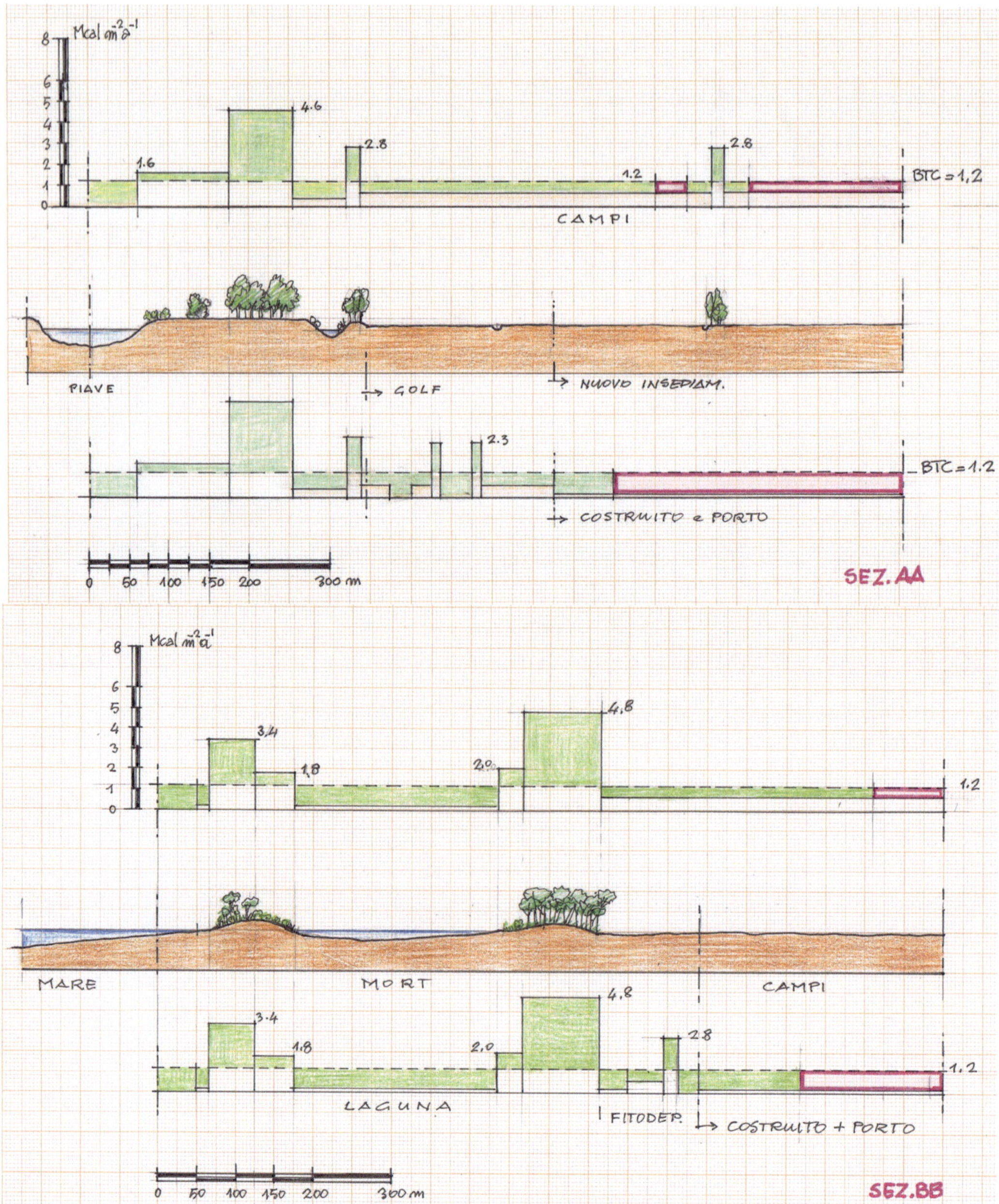

Fig. 13.20 Section AA represents the transect W-E (*above*), while section BB the transect N-S of the area under examination. Above and below each section are plotted the effects of the ecological radius of influence ex ante (*above*) and ex post (*below*) the building of the new touristic village

BTC = 0.96 is only 75 % of the normal and should arrive at least to 85 %; on the contrary, the effect of the new village will decrease the average BTC to 70 %.

Moreover, the control of the ecological radius of influence (REI) of the New Settlement near Eraclea Mare, through the balance of BTC in areas of strong alteration, is not positive. In

Fig. 13.20 the transects above and below each section on the area under examination indicate the deficit of transformation (purple), that is + 280 % for the transect AA and +360 % for BB.

In conclusion, no doubt that we have ecological problems. Even a preliminary analysis of landscape bionomics is able to underline the clear need for revision of the draft resort.

References

1. Forman RTT (2008) Urban regions, ecology and planning beyond the city. Cambridge University Press, Cambridge
2. Ingegnoli V (2004) Criteri di progettazione, valutazione e controllo di un sistema di verde urbano: l'esempio del nuovo quartiere della Fiera di Milano. Valutazione Ambientale 6:5–10
3. Mandelbrot BB (1975) Les objets fractals: forme, hasard et dimension. Flammarion, Paris
4. Ingegnoli V (1993) Fondamenti di Ecologia del paesaggio. CittàStudi, Milano
5. Forman RTT (1995) Land mosaics: the ecology of landscapes and regions. Cambridge University Press, Cambridge
6. Ingegnoli V, Giglio E (2005) Ecologia del Paesaggio: manuale per conservare, gestire e pianificare l'ambiente. Sistemi editoriali SE, Napoli
7. Sukopp H, Hejny S (eds) (1990) Urban ecology. SPB Academic, The Hague
8. Ingegnoli V (2002) Landscape ecology: a widening foundation. Springer, Berlin
9. Van Leeuven CG (1982) Protection of migrating common toad Bufo bufo against car traffic in The Netherlands. Environmental Conservation 9:34–41
10. Forman RTT, Godron M (1986) Landscape ecology. Wiley, New York
11. Pignatti S (1998) I boschi d'Italia: sinecologia e biodiversità. Utet, Torino

14 Comparison Between Two Rural-Suburban Landscapes from Brussels and Milan

Vittorio Ingegnoli, Ernesto Marcheggiani, Hubert Gulinck and Frederik Lerouge

14.1 Belgium and Lombardy, Geographical and Historical Congruences

14.1.1 Belgium and Lombardy, Many Similarities

The idea of the present study derived from a meeting on "landscape interfaces" (see Sect. 4.1.2) in Asbeek/Asse in February 2012, promoted by Hubert Gulink at KU Leuven with Ingegnoli and Marcheggiani. The reference discipline was, obviously, the Landscape Ecology [1 4]. Indeed, many similarities emerge from Belgium and Lombardy and from Brabant and Brianza, both geographically and historically speaking (Figs. 14.1, 14.2 and 14.3).

Although records indicate the city of Milan is about 1,500 years older (as reported by Tito Livio, Historiae, V, 34) than the city of Brussels, founded by the Lotharingians in tenth century (as reported in city of Brussels internet sites), having Milan been founded by the Celts in fifth century B.C., today, as we can see from Fig. 14.4, the two cities are arguably quite similar in many ways:

V. Ingegnoli (✉)
Università degli Studi di Milano, Milan, Italy
e-mail: v.ingegnoli@virgilio.it

E. Marcheggiani
Department of Earth and Environmental Sciences, University of Leuven, Belgium

Polytechnic University of Marche, Ancona, Italy

H. Gulinck • F. Lerouge
Department of Earth and Environmental Sciences, University of Leuven, Belgium

- Bruxelles, area 161.4 km^2, Altitude 13 m; 1,140 × 10^3 inhabitants (50°50′ N; 4°21′ E).
- Milan, area 181.8 km^2, Altitude 120 m; 1,305 × 10^3 inhabitants (45°30′ N; 9°11′ E).

Asse and Bollate also show similarities, being located 15 km N–W of their capital cities, both having about 30–35,000 inhabitants, even though their respective areas are quite different: the territory of Asse (50.08 km^2) is about 3.8 times larger than the territory of Bollate (13.10 km^2). Note that in Fig. 14.4 only the urban area of Bollate is delimitated (about ½ of the whole territory)

Moreover, for both Asse and Bollate a relatively complete sequence of high-quality historical maps is available, from as early as the eighteenth century until today. The aim of this international research is to comprehend more than two centuries of transformation dynamics of landscape structure and functions.

14.2 The Municipality of Asse

14.2.1 General Characters

The location of Asse lies central in the middle of the Brabant Massif, where Cambrian rocks are overlain by Tertiary marine deposits and covered by an aeolian loss deposit. Gently undulating hills characterise this agrarian landscape.

The human colonisation is very old: the Roman road Bavacum—Asse is indicative for the importance of this community during the first century

V. Ingegnoli, *Landscape Bionomics Biological-Integrated Landscape Ecology*,
DOI 10.1007/978-88-470-5226-0_14, © Springer-Verlag Italia 2015

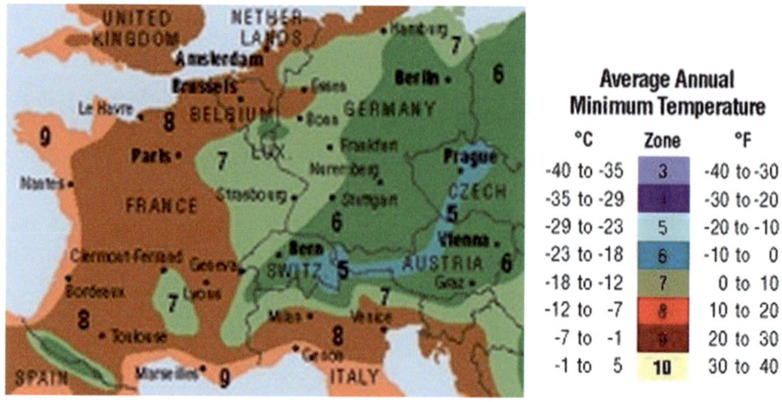

Fig. 14.1 This card of "Plant Hardiness Zones" (USDA) in Central Europe indicates that the Plain and Pre-Alpine areas of Lombardy and the whole of Belgium share the same zones n°7 and 8

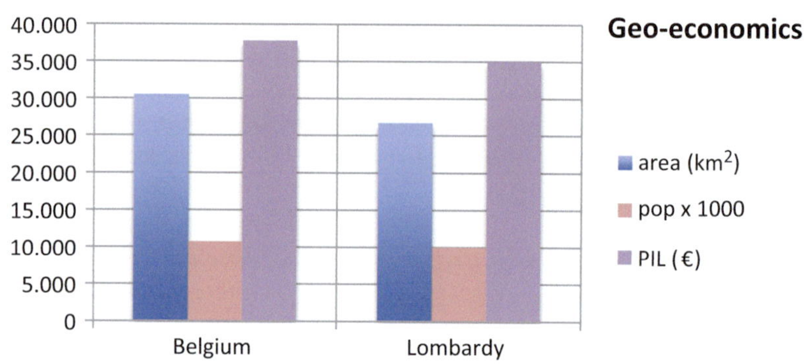

Fig. 14.2 Geographical and historical Lombardy (which comprises Canton Ticino) measures the 87.5 % of Belgium in area, the 93.3 % in GDP and 95 % in population

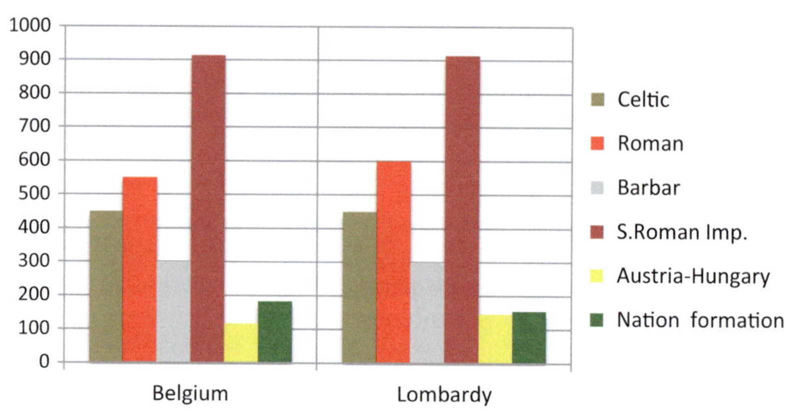

Fig. 14.3 Sequence of principal historic periods (duration in years) in Belgium and Lombardy. The similarity is very high

AC, but Asse was inhabited many centuries before that time (Koninklijke Heemkring Ascania, online, 2014).

The general structure of this landscape is characterised by a mixed presence of agriculture and industry, typically suburban, and by a residential network along the main roads. Remnant forest patches are still present in the landscape, but the fragments are relatively small and their connectivity is weak. Many queries arise on the bionomic state (see Sect. 10.4.4) of this landscape [5].

For instance, a question to be accurately analysed regards the high amount of private gardens, as a "cultural" protective compensation

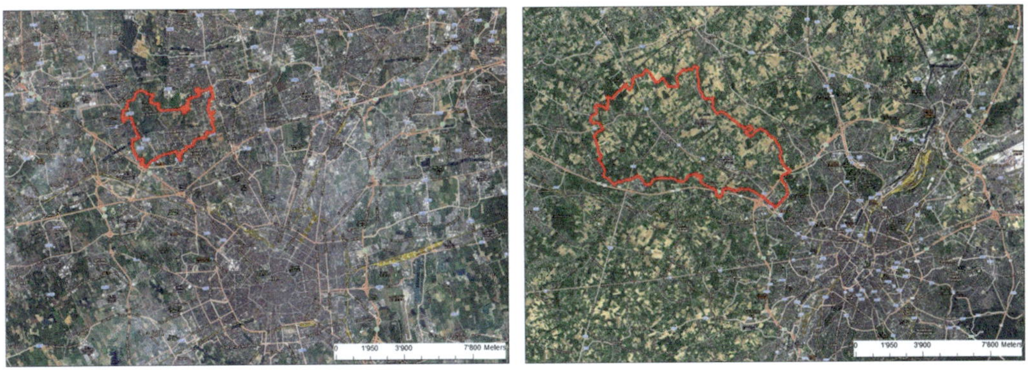

Fig. 14.4 The two cities of Brussels (*right*) and Milan (*left*) and the location of the municipality of Asse and of Bollate. Please note as both municipality are placed N–W of the cities (from Google Maps®). The polygons in *red* represent the demarcation of the administrative boundaries

to the urban expansion. Another question is the bionomic evaluation of forests, and whether there might be protected areas with lower ecological efficiency than non-protected one. A third question is concerned with the landscape type: does it remain in a range of agricultural types or does it reach the rural-suburban structure? Or could it already be in the condition of scattered urban (or suburban-technological) landscapes?

Moreover, we have to check if the present condition of Asse could be similar to the bionomic state of Bollate, or in case they differ, how their transformation dynamics during the past two centuries would have been different.

In Fig. 14.5 two aspects of the residential expansion of Asbeek and Asse, with typical private gardens [6] are shown; in Fig. 14.6 note the presence of horse in agriculture [7], very frequent, and a monumental tree in a forest tessera with only a partial correspondence to the ecology of organisms of that level of maturity.

14.2.2 Vegetation Elements

In parallel with the measure of the land use of the elements of the territory of Asse and their grouping in seven main landscape apparatuses, we surveyed the most significant vegetation types of tessera, following the LaBiSV methodology [4, 17, 18].

Table 14.1 shows the most significant results of the surveys on vegetation elements.[1] Forests present a quite good level of BTC (see Sect. 3.3.1), having an average forest BTC = 6.91 Mcal/m^2/year and reaching a maximum value of BTC = 7.94 Mcal/m^2/year vs. a minimum of BTC = 5.11. It is also possible to find some monumental trees (e.g. of *Fagus sylvatica*) about 35–37 m height, 90–100 cm of diameter, but in a partially altered forest tessera (Fig. 14.6).

We can affirm that in some cases protected forests present BTC and CBSt values lower than the non-protected one, as shown in Table 14.1 and Fig. 14.7. Anyway, the main goal of the vegetation survey is the estimation of the average BTC of the territory, as shown in Table 14.2.

The gardens, very frequent in urban expansions, do not present a high BTC value, being in an early development stage, structurally simple and lacking patches of more natural vegetation. The agricultural fields are cultivated with care, but mainly in traditional way and with a prevalence of mays crops. Their average BTC is not over 0.8 Mcal/m^2/year.

[1] The basic mosaic is generally the vegetational one. The complex structure of a landscape has to integrate diverse components: temporal, spatial, thematic The concept of ecotissue is a broadened complex multidimensional structure represented by a basic mosaic and a hierarchic succession of correlated mosaics and attributes.

Fig. 14.5 The road going from Asbeek to Asse Centre (*left*) and an example of private garden at the margin of the historical Centre of Asse (*right and bottom*)

Table 14.1 The most significant results of the survey on vegetation elements in Asse, LaBiSV methodology

ASSE, Brabant, 2012–2013	SP	Hc	pB	bQ	BTC	M-Th	CBSt
Dominant plant species (%)	n°	m	m^3/ha	%	Mcal/m^2/ year		
FOREST 1—*Quercus rubra 37, Acer pseudoplatanus 20, Populus tremula 17, Quercus robur 14,*	35	29.5	614	56.20	6.53	41.82	23.50
FOREST 2—*Acer pseudoplatanus 30, Prunus avium 22, Fraxinus excelsior 17, Alnus glutinosa 8, Quercus robur4*	36	26.9	584	67.20	7.60	55.11	37.04
FOREST 3—*Fraxinus excelsior 37, Fagus sylvatica 31, Quercus robur 14, Prunus avium 10*	34	29.7	625	67.40	7.70	56.45	38.03
FOREST 4—*Fagus sylvatica 40, Quercus robur 18, Prunus avium 14, Populus nigra 9*	26	22.7	560	58.00	6.60	42.63	24.71
FOREST 5—*Fagus syl. 53, Populus n 19, Quercus rob. 11, Fraxinus excelsior 8, Robinia pseudacacia 7*	23	33.6	585	70.50	7.94	59.72	42.07
FOREST 6—*Salix alba* 100	17	23.2	320	48.10	5.11	27.00	12.98
Average (Forests)	*28.5*	*27.6*	*548*	*61.23*	*6.91*	*47.12*	*29.72*
CORRIDOR 1—*Salix alba* 90, *Quercus robur* 6	20	9.5	73	31.38	2.17	7.48	2.35
GARDEN 1—Garden (private) 1	20	15.7	39	33.16	1.97	7.94	2.63
GARDEN 2—Garden (private) 2	28	14	125	39.60	2.55	12.13	4.80
FIELD 1—Potatoes	9	0.6		40.39	0.71	13.53	5.47
FIELD 2—Mays crop 1	15	2.1		47.76	0.91	23.91	11.42
FIELD 3—Mays crop 2	14	2.2		41.69	0.82	18.68	7.79
Average (other vegetation)	*17.7*	*7.3*		*39.00*	*1.52*	*13.9*	*5.74*

SP species, *Hc* canopy height, *pB* plant biomass volume, *bQ* bionomic quality, *M-Th* maturity threshold, *CBSt* index of concise bionomic state

Fig. 14.6 A view of the agricultural lands (*bottom*). An ecotope with horses near a wood patch (*upper left*) and a monumental tree of *Fagus sylvatica* (about 37 m height, 95 cm diameter) in a partially altered forest tessera

14.2.3 Other Landscape Measures

Recalling the concepts related to landscape apparatuses (see Sect. 2.4), in this Table 14.2 we observe that the productive apparatus (agriculture) is still wide (63.84 %) and the protective one is exceptionally developed (14.16 %), while the built urbanisation remains below 12 %. The main limit comes from the resistant apparatus, because the forest patches are only 4.53 %. At a first view, this landscape does not seem to be suburban.

These seven sets of data are indispensable to evaluate five other landscape functions: human habitat (HH[2]), bio-territorial capacity of vegetation (BTC), concise bionomic state of vegetation (CBSt), photosynthetic area (Phsyn), potential of bionomic rehabilitation (PoBR). The human habitat covers more than three quarters (HH = 78.5 %) of the entire territory, and the bio-territorial capacity (BTC) of the vegetation equals 1.04 Mcal/m^2/year: this includes both suburban-rural and agricultural areas. The other results: CBSt = 7.41; Phsyn = 82.25 % and PoBR = 21.8 %, all evaluated at a territorial

[2] HH, human habitat, the set of areas: (a) *where human population lives, (b) which is managed permanently, (c) in which is added subsidiary energy, limiting the self-regulation capacity of natural systems (NH).*

SH, standard habitat per capita, the vital area per capita [m^2/ab] available for an organism (man or animal), divisible in all its components biological and relational.

SH is the inverse of the ecological density.

SH* is the theoretical minimum standard habitat.

σ *(carrying capacity) = SH/SH**, measures the autotrophy or heterotrophy of a landscape unit (LU).

Table 14.2 The landscape elements of Asse, grouped into 7 L. apparatuses, measured in hectares and %

ASSE, BRABANT, 2010	ha	%	HH′	HH	BTC′	BTC	CBSt′	CBSt	Phsyn	Phsyn	PoBR	PoBR
Built	234.23	4.66	98	4.567	0.15	0.007			0.01	0.047	0.02	0.093
Artificial surfaces	112.49	2.24	100	2.238	0.1	0.002			0.01	0.022	0.03	0.067
RSD + SBS		**6.90**										
Roads	198.93	3.96	100	3.958	0	0.000				0.000	0	0.000
Railroads	7.71	0.15	100	0.153	0.05	0.000				0.000	0	0.000
SBS		**4.11**										
Gardens and parks	711.59	14.16	85	12.034	1.6	0.227	5	0.708	0.75	10.619	0.5	7.079
PRT		**14.16**										
Orchard and vegetable	178.74	3.56	90	3.201	1.2	0.043	4.5	0.160	0.95	3.378	0.15	0.533
Grass	1,234.78	24.57	67	16.460	0.6	0.147	9	2.211	1	24.568	0.2	4.914
Agricultural fields	1,872.24	37.25	87	32.408	0.8	0.298	8.22	3.062	0.96	35.761	0.1	3.725
Other agricultural	101.62	2.02	88	1.779	2.1	0.042	4.8	0.097	0.9	1.820	0.1	0.202
PRD		**63.84**										
Shrubs + abandon. fields	130.18	2.59	40	1.036	1.5	0.039	10	0.259	0.6	1.554	0.25	0.648
Heathland with *Calluna*	0.91	0.02	10	0.002	2	0.000	40	0.007	1	0.018	1	0.018
RSL		**2.61**										
Riparial woods (11.3 %)	25.55	0.51	15	0.076	4.8	0.024	13	0.066	1	0.508	1	0.508
Low dens. woods (81.5 %)	185.24	3.69	12	0.442	5.1	0.188	20	0.737	0.98	3.612	1	3.686
Forest (7.2 %)	16.71	0.33	7	0.023	7.3	0.024	30	0.100	1	0.332	1	0.332
RNT		**4.53**										
Ponds	8.11	0.16	45	0.073	0.15	0.000			0.05	0.008	0.1	0.016
Rivers and canals	6.97	0.14	45	0.062	0.1	0.000			0.05	0.007	0.05	0.007
HGE		**0.30**										
Tot. ASSE	**5,026.0**	**100.0**		**78.51**		**1.04**		**7.41**		**82.25**		**21.83**

The elaborations of other five landscape functions are based on these data: *HH* human habitat, *BTC* bio-territorial capacity of vegetation, *CBSt* concise bionomic state of vegetation, *Phsyn* photosynthetic area, *PoBR* potential of bionomic rehabilitation *Phsyn* photosynthetic areas, *PoBR* potential of bionomic rehabilitation, *RSD* residential, *SBS* subsidiary, *PRT* protective

Fig. 14.7 The first two forest surveys in Asse, dominated by *Acer pseudoplatanus* and *Prunus avium* (*left*) and *Quercus rubra*, *Acer pseudoplatanus* and *Quercus robur* (*right*). Note that the protected area (*right*) expresses a BTC lower than the other: 6.53 vs. 7.60 Mcal/m^2/year

Table 14.3 Calculation of the general metastability

2010	ASSE	
BTC class	P (elem)	P*$\log_2$ P
I	0.1131	−0.356
II	0.6740	−0.384
III	0.1677	−0.432
IV	0.0001	−0.001
V	0.0420	−0.192
VI	0.0033	−0.027
VII	0.0001	−0.001
H	Heterogeneity	−1.393
D	Dominance	0.532
tau	Functional diversity	5.408
BTC	Territorial capacity	1.040
g-LM	**General metastability**	**5.20**

scale, seem quite good values, if compared with the ranges of normality for an agricultural landscape (Table 14.4).

After having divided the BTC of the elements in their seven classes and applied the Shannon equations (using $\log_2$), we register a heterogeneity H = 1.39, a dominance D = 1.41, thus a functional landscape diversity $\tau = 5.408$. As shown in Table 14.3, with an average BTC = 1.04, the general landscape metastability (see Chap. 3) results g-LM = 5.20, a low value indeed.

14.2.4 Historical Characters and Dynamics

Starting from the present state of the examined territory (2010; Fig. 14.8) we looked at three past periods to analyse the transformation dynamics of Asse, depending on the available historical maps: 1950, 1850 and 1777. Here we report the Vandermaelen map (1850) and the Hapsburg maps of Ferraris Figs. 14.9 and 14.10, respectively [8, 9].

Figure 14.11 exposes the most significant transformations of the territory of Asse during 2.3 centuries. These changes were not so strong: we can see a near constant HH, a small decrease of BTC and forest patches (only in the last 50 years), a longer decrease of cropland and a growing of population which induced a lower grow of urbanisation since 1850.

14.2.5 A Synthesis of the Diagnostic Evaluation

Applying the methodology from Sect. 9.4 of this volume, we checked the divergences from the normal state of a landscape like this, typically agriculture-productive (AGR-PRD): in fact, as shown in the following table, the resulting values of the index of landscape typology evaluation (LTpE) are comprised within the range 24–10.2, typical of an AGR-PRD landscape.

Therefore, the choice of the 16 bionomic parameters and their elaborations and measures were compared to the normal AGR-PRD ranges, excluding the population (Table 14.4).

The results of diagnostic index (DI) indicate a good stability during the past two centuries (DI = 85.00), but a drastic change in the recent period (DI = 68.33) that means a decrease of about 20 % in few decades.

This today altered state brings to be at the thresholds of a new type of landscape: the suburban-rural one.

Table 14.4 Diagnostic evaluation of the bionomic state of the LU of Asse, in the period 1777–2010

Bionomic parameters	*AGR-PRD*	1777	Dev	Score	1850	Dev	Score	1950	Dev	Score	2010	Dev	Score
POP/1000		3.7			5.4			16.3			30.0		
LTpE	*10.2–24*	14.52	Ok	2	13.79	Ok	2	13.9	Ok	2	10.58	Ok	2
Urbanised	*8.0–16*	4.4	Ok	2	4.45	Ok	2	7.71	Ok	2	11.01	Ok	2
Gardens	*1.6–12*	3,2	Ok	2	3.24	Ok	2	9.91	Ok	2	14.16	18	1
Agricultural	*45–75*	83	10.7	1	80.25	6.9	2	72.3	Ok	2	67.4	Ok	2
Forest and Heath	*10.0–30*	9.07	9.3	2	8.43	15.7	1	9.91	0.9	2	7.13	28.7	1
Photosyntetic area	*85–95*	88.95	Ok	2	88.91	Ok	2	84.7	Ok	2	82.25	3.2	2
Rehabilit. potential	*25–35*	21.61	13.56	1	21.09	15.64	1	24.2	3.2	2	21.83	12.68	1
HH (%)	*60–82*	75.12	Ok	2	75.56	Ok	2	75.6	Ok	2	78.51	Ok	2
SH	*2,200–9,000*	10272	Ok	2	7087	Ok	2	2332	Ok	2	1315	40.23	0.5
Carrying capacity (σ)	*1.5–9*	3.88	Ok	2	2.95	Ok	2	1.11	26	1	0.89	40.67	0.5
BTC	*1.0–2*	1.19	Ok	2	1.16	Ok	2	1.26	Ok	2	1.04	Ok	2
g-LM	*12–24*	5.00	58.33	0.5	5.50	54.17	0.5	6.30	47.5	0.5	5.20	56.67	0.5
Territorial CBSt	*10.0–24*	9.15	8.5	2	9	10	2	8.8	12	1	7.41	25.9	1
Forest and shrub CBSt	*23–32*	17.63	23.3	1	17.65	23.2	1	17.99	21.8	1	16.38	28.8	1
Potential core areas	*35–50*	62	Ok	2	60	Ok	2	52	Ok	2	46.7	Ok	2
Total score	*27.2–32*			25.50			25.50			25.50			20.50
Diagnostic Index (DI)	*85–100*			**85.00**			**85.00**			**85.00**			**68.33**
Bionomic state evaluation	Normal			Normal			Normal			Normal			Altered

The normal ranges to be compared pertain to an agricultural landscape (AGR-PRD)

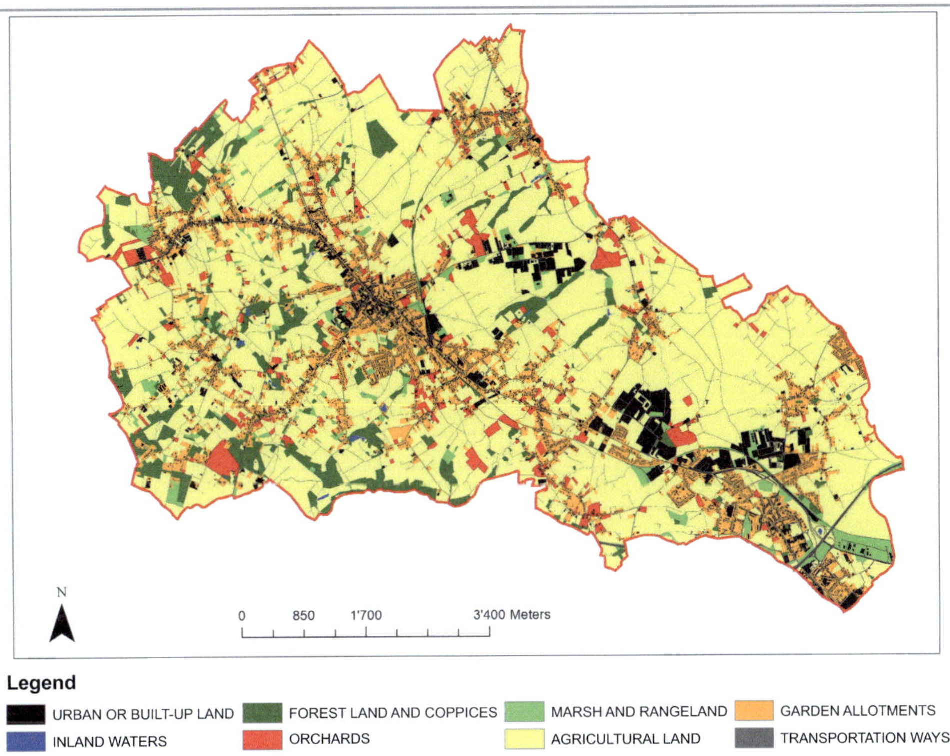

Fig. 14.8 The land use map of Asse, 2010. Note the network shaped expansions outside the historical centres, along the main roads. The industrialisation is most towards Brussels (S–E)

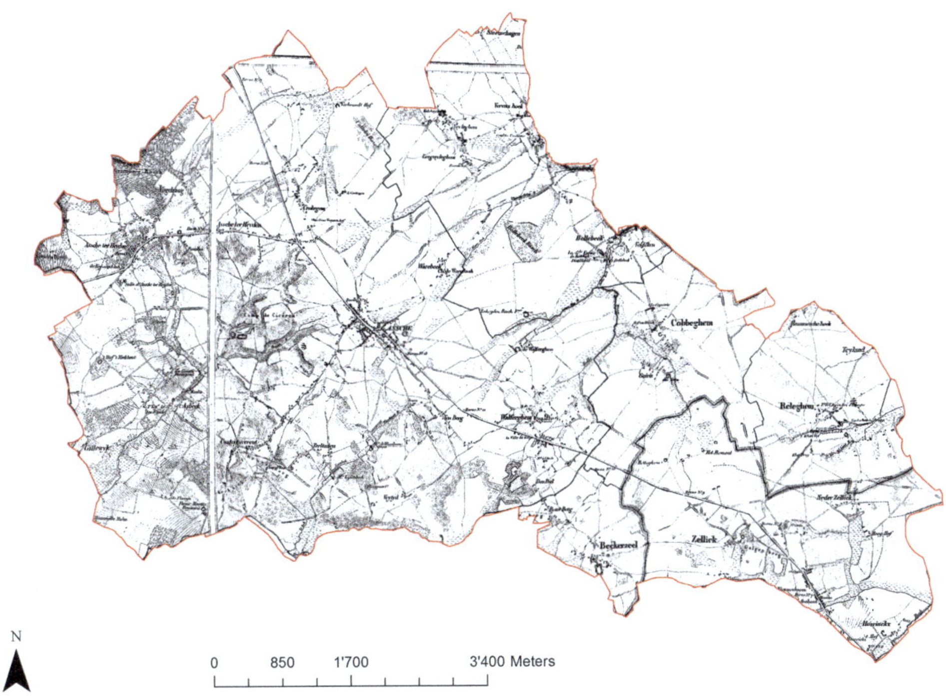

Fig. 14.9 An extract from the historical *Van der Maelen's Kaart* (1850). The map displays the situation during the second half of nineteenth century. Note the overall stability of transportation network (*Source* Royal Library of Belgium, KBR)

Fig. 14.10 The *Ferraris Kaart* (1777) represents an outstanding example of the overall territorial monitoring capacity expressed by the Hapsburg empire in late eighteenth century. In particular the map shows in *colours* the very general structure of the landscape with a wide range of detailed features (*Source* National Geographic Institute, NGI)

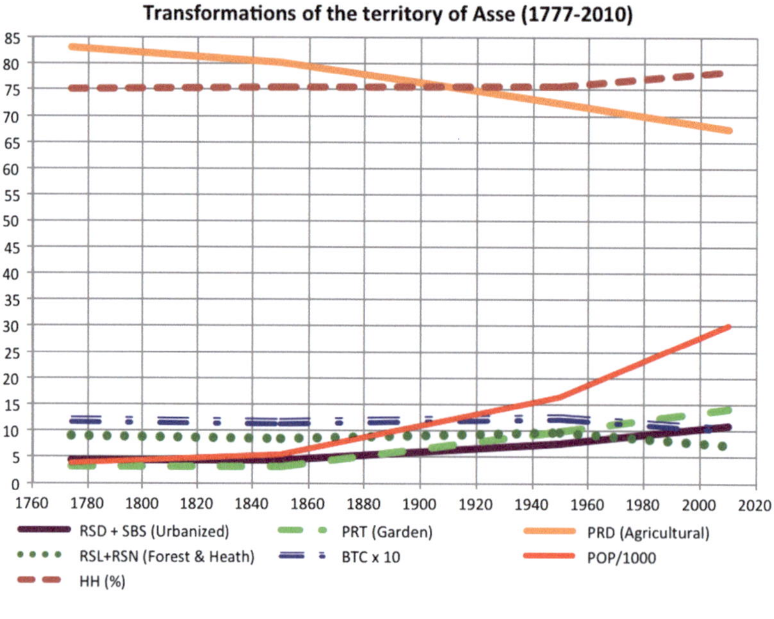

Fig. 14.11 The most important dynamics are characterised by a near constant HH, a small decrease of BTC and forest patches only in the last 50 years, an older decrease of cropland and a growing of population, which induced a lower growth of urbanisation since 1850

14.3 The Municipality of Bollate

14.3.1 General Characters

The location of Bollate is placed at the end of the old glacier tongue of the so-called "Groane," elevated only a few metres (10–20) above the alluvial plain of Milan. The soil is a peculiar ancient clay, rich in iron, and markedly different from the moraine hills of the rest of Brianza [10]. At the Southern part of his tongue, visible in the middle of the Milan metropolitan area in Fig. 14.12 where Bollate is located, we find the first resurgence belt (i.e. natural fountains) of water (between the two blue lines).

The general structure of this landscape, as we can see from Fig. 14.13, is characterised by a strong presence of urban areas around an historical centre, and with many scattered industrial zones and only few patches of cropland. Anyway, in the western part of its territory, the mentioned glacial remnant soil tongue is still visible, with frequent belts of forest and grasses and some remnant patches of heathland. In this green part of Bollate there is a vast and imposing baroque villa with a wide garden (France style): Villa Arcimboldi (1640), even called "Castellazzo di Bollate" (Fig. 14.14), now under restoration. The map of Fig. 14.13 was deduced from the DUSAF (Lombard Region office for territorial analysis) [11] and plotted by dr. Alessandra Rizzi at the University of Milan, an assistant of prof. Ingegnoli for this research.

Fig. 14.12 The location of Bollate in the middle of the Milan Metropolitan Area, at the end of the old glacier tongue of the Groane (*red-brown*), at the beginning of the resurgence *lines* of water (*blue lines*)

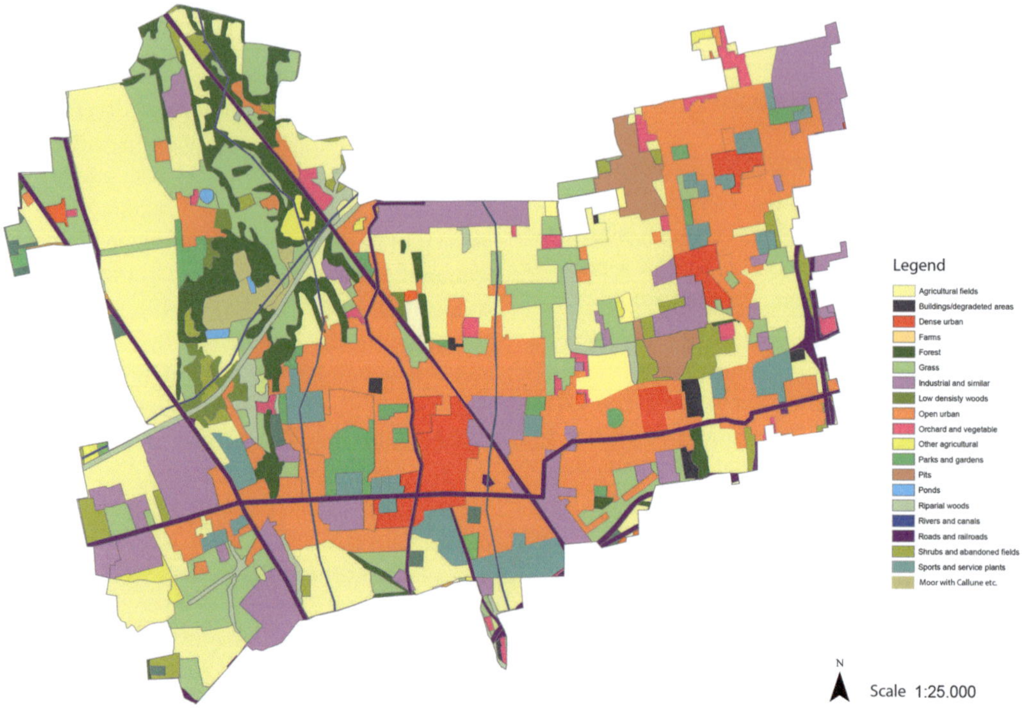

Fig. 14.13 Map of Bollate land use/land cover in 2010, deduced from the satellite surveys of DUSAF, Lombardy Region, and plotted by Dr. A. Rizzi for this research

Fig. 14.14 Bollate, the front of Villa Arcimboldi, called "Castellazzo" (about 1640), with a France style garden, now under restoration

14.3.2 Vegetation Elements

In parallel with the measure of the land use of the elements of this territory of Bollate and their grouping in seven main landscape apparatuses, we surveyed the most significant vegetation types of tessera, following the LaBiSV methodology.

Only few patches of forest are in good natural condition (*Querco-Carpinetum*), as we can see in Table 14.5 and Fig. 14.15. The average BTC of forests is quite modest (BTC = 5.59 Mcal/m^2/year), especially because many of the woods are dominated by *Robinia pseudacacia* with *Prunus serotina*, both exotics coming from the USA (Fig. 14.15).

In excellent condition is the old garden of Villa Arconati, which annexes at least 2 ha of wilderness and forest, as shown in Fig. 14.16. Also the new Urban Park in the centre of Bollate (Fig. 14.16) has some positive remarks. Even the tree corridors present height *Populus nigra* × *deltoids* or *Platanus hybrida* (Fig. 14.17) in two-lines with a lot of smaller plants, especially *Ulmus minor*: they have a BTC = 3.0–3.5 Mcal/m^2/year.

The most interesting vegetation patches are indeed those of heathland, with *Calluna vulgaris* and *Molinia arundinacea*, *Betula pendula*, *Populus tremula* and some *Pinus sylvestris*. These remnants, today about 12 ha but one century ago near 90–100 ha, are probably the most southern heathlands of Northern Europe [12]: an impression is given in Fig. 14.17.

The agricultural fields have decreased much in area and comprise mainly Maize and Wheat, as illustrated in Fig. 14.18.

Table 14.5 The most significant results of the survey on vegetation elements in Bollate, LaBiSV methodology

BOLLATE, Brianza, 2010	SP.	*Hc*	*pB*	bQ	*BTC*	M-Th	CBSt
Dominant plant species (%)	n°	m	m^3/ha	%	Mcal/m^2/year		
FOREST 1—*Quercus robur (+Quercus petraea) 51, Carpinus b. 18, Robinia pseudacacia 21*	16	17	322	50.00	5.29	28.70	14.35
FOREST 2—*Quercus robur 53, Carpinus betulus 31, Robinia pseudacacia 7.5*	19	22.5	580	66.23	7.50	53.79	35.63
FOREST 3—*Quercus rubra 55, Pinus sylvestris 39, Quercus robur 5*	18	20.5	213	51.62	5.26	28.42	14.70
FOREST 4—*Robinia pseudacacia 70, Ulmus minor 22, Prunus avium 6*	21	22.5	384	48.05	5.24	28.23	13.56
FOREST 5—*Robinia pseudacacia 82, Prunus avium 11, Ulmus minor 5*	15	28	342	43.51	4.68	23.16	10.08
AVERAGE (Forests)	*17.8*	*22.1*	*368.2*	*51.88*	*5.59*	*32.46*	*17.66*
HEATHLAND—*Molinia arundinacea 90, Calluna vulgaris 35, Populus tremula 12, Betula pendula 8*	14	1.3	24	72.98	2.12	70.96	51.78
CORRIDOR 1- *Populus nigraxdeltoides 93, Ulmus minor 2, Quercus robur 1, Robinia pseudacacia 1*	35	25.5	253	43.11	3.27	14.50	6.25
CORRIDOR 2—*Platanus hybrida 42, Robinia pseudacacia 17, Corylus avellana 16, Ulmus minor 14*	23	22.4	545	39.78	3.42	15.64	6.22
GARDEN 1- Central Public Garden	38	18.7	51	43.50	2.63	12.78	5.56
GARDEN 2—Villa Arconati Garden	39	25.6	205	60.34	4.01	26.82	16.18
FIELD 1—Mays crop	11	2.3		43.67	0.85	20.31	8.87
FIELD 2—Other crops (average)	10	0.8		39.48	0.77	16.18	6.39
AVERAGE (other vegetation)	*24.3*	*13.8*		*49.00*	2.44	*25.31*	*14.46*

SP species, *Hc* canopy height, *pB* plant biomass volume, *bQ* bionomic quality, *M-Th* maturity threshold

Fig. 14.15 The aspect of two forest patches in Bollate: a *Querco-Carpinetum* (*left*) and a *Robinietum* (*right*)

Fig. 14.16 The aspect of two Garden elements in Bollate: *Quercus robur* of about 35 m from the wilderness part of the Arconati Gardens (*left*), while young mixed woods, with some exotics, characterise the new urban park in the centre of Bollate (*right*)

Fig. 14.17 A large tree-corridor with *Platanus hybrida* (*left*) and a view of the heathland, the most peculiar formation of this territory (*right*)

Fig. 14.18 Two aspects of the cropland, today very reduced (not exceeding 40 %) of Bollate: Mays (*left*) and Weat (*right*)

Table 14.6 The landscape elements of Bollate, grouped into 7 L. apparatuses, measured in hectares and %

BOLLATE 2010	ha	%	HH′	HH	BTC′	BTC	CBSt′	CBSt	Phsyn′	Phsyn	PoBR′	PoBR
Dense Urban	42.38	3.23	100	3.23	0.15	0.005	0.00	0.000	0.01	0.032	0.020	0.065
Open Urban	204.76	15.60	98	15.29	0.2	0.031	2.00	0.312	0.05	0.780	0.025	0.390
Sport and service plants	57.90	4.41	91	4.02	0.8	0.035	3.00	0.132	0.10	0.441	0.040	0.176
RSD		**23.24**										
Roads and Railroads	55.05	4.19	100	4.19								
Industrial and similar	134.92	10.28	99	10.18	0.1	0.010			0.01	0.103	0.020	0.206
Pits	21.63	1.65	97	1.60	0.2	0.003	2.00	0.033	0.05	0.082	0.200	0.330
Building/degraded areas	6.23	0.47	98	0.47	0.05	0.000	1.00	0.005			0.010	0.005
SBS		**16.60**										
Gardens and parks	116.00	8.84	85	7.51	1.62	0.143	7.00	0.619	0.80	7.072	0.500	4.420
PRT		**8.84**										
Orchard and vegetable	18.11	1.38	90	1.24	1.1	0.015	3.00	0.041	0.95	1.311	0.100	0.138
Farms	2.18	0.17	98	0.16	0.1				0.10	0.017	0.040	0.007
Grass	147.12	11.21	67	7.51	0.6	0.067	9.00	1.009	1.00	11.211	0.200	2.242
Agricultural fields	325.91	24.84	87	21.61	0.79	0.196	7.50	1.863	0.96	23.842	0.100	2.484
Other agricultural	8.00	0.61	88	0.54	0.65	0.004	3.50	0.021	0.90	0.549	0.100	0.061
PRD		**36.82**										
Shubs and abandoned fields	32.42	2.47	40	0.99	1.6	0.040	5.00	0.124	0.95	2.347	0.200	0.494
Heathland with *Calluna* etc.	11.70	0.89	10	0.09	2.1	0.019	50.00	0.446	1.00	0.892	1.000	0.892
RSL		**3.36**										
Riparial woods	29.10	2.22	15	0.33	3.9	0.086	8.00	0.177	1.00	2.217	1.000	2.217
Low density woods	11.21	0.85	12	0.10	4.1	0.035	10.00	0.085	0.98	0.837	1.000	0.854
Forest	76.67	5.84	7	0.41	5.8	0.339	17.00	0.993	1.00	5.842	1.000	5.842
RSN		**8.91**										
Ponds	1.88	0.14	45	0.06	0.15	0.000			0.05	0.007	0.050	0.007
Rivers and canals	9.11	0.69	45	0.31	0.1	0.001	2.00	0.014	0.05	0.035	0.050	0.035
HGL		**0.84**										
BOLLATE 2010. Tot.	**1,312.3**	**100.0**		**79.84**		**1.03**		**5.87**		**57.62**		**20.86**

The elaborations of other five landscape functions are based on these data: *HH* human habitat, *BTC* bio-territorial capacity of vegetation, *CBSt* concise bionomic state of vegetation, *Phsyn* photosynthetic area, *PoBR* potential of bionomic rehabilitation

14.3.3 Other Landscape Measures

In Table 14.6 we observe that the productive apparatus (agriculture) is not large (36.82 %) and the protective one is well developed (8.84 %), while the built urbanisation is particularly high 39.84 %: an indisputable bionomic limitation. The resistant apparatus has recovered, mainly due to the establishment of the preservation area of the Groane Regional Park, forest patches cover 8.91 %, and with heathlands included it amounts to 12.27 % of the Bollate area. At a first glance, this landscape seems to be typically a suburban one.

These seven sets of data are indispensable to evaluate other five landscape functions: HH (human habitat), BTC (bio-territorial capacity of vegetation), CBSt (concise bionomic state of vegetation) [13], Phsyn (photosynthetic area), PoBR (potential of bionomic rehabilitation). The human habitat results HH = 79.8 % of the entire territory and the BTC of vegetation BTC = 1.03 $Mcal/m^2/year$, both suburban-rural and agricultural values and (together with PoBR) near the same of Asse. The other results: CBSt = 5.87 and Phsyn = 57.62 %, all evaluated at territorial scale, are on the contrary less good than Asse.

After having divided the BTC of the elements in their seven classes and applied the Shannon equations (using $\log_2$) we register a heterogeneity H = 1.84, a dominance D = 0.97, thus a quite high functional landscape diversity $\tau = 5.714$. As exposed in Table 14.7, with an average BTC = 1.03, the general landscape metastability results g-LM = 5.89, a low value but better than the g-LM of Asse.

14.3.4 Historical Characters and Dynamics

Starting from the present state of the examined territory (2010) we had to find at least three past periods necessary to analyse the transformation dynamics of Bollate. These periods depend on the available historical maps. For the analysis, we opted to work with five time periods: 2000, 1954, 1888, 1836 and 1788.

Here we illustrate 1836 (scale 1:25,000 land use elaboration), from the Hapsburg map of Lombardo-Veneto, shown in Figs. 14.19 and 14.20; moreover an extract of the 1788 map of the Mailand Provinz made by the "Astronomi di Brera" in a scale of about 1:70,000 (Fig. 14.21) [14, 15].

Bollate was the local capital of the III District (West Brianza), with a territory of 92.8 km^2, counting about 17,115 inhabitants (*anno* 1841), growing to 20,971 in only a decade [16]. The population of Bollate in 1841 was about 3,500 inhabitants.

Figure 14.22 shows the most significant transformations of the territory of Bollate during 230 years. These changes are more pronounced as compared to Asse. We can see the growth of HH (from 71.7 to 84.2 % in only 110 years), as well as a relatively recent decrease to 79.8 %, following the huge urbanisation and the establishment of the Groane Regional Park.

A drastic decrease of agriculture resulted from these processes, with a decreasing share from 83.1 to 38.2 % in less than 60 years! As we will see, the landscape was reshaped drastically in this time period.

The small decrease of BTC did not follow the strong decrease of wilderness areas and forest patches: only after 1954 the BTC passed from 1.47 to 1.03, notwithstanding the forest regrowth.

The growth of urbanisation has been following population increase since 1888, more than doubling in 66 years (from 4.25 to 9.95), then arriving to the present 39.8 %.

Table 14.7 Calculation of general L. metastability

	BOLLATE 2010	
BTC class	P (elem)	P*$\log_2$ P
I	0.3644	−0.531
II	0.4245	−0.525
III	0.1220	−0.370
IV	0.0222	−0.122
V	0.0609	−0.246
VI	0.0060	−0.044
VII	0.0001	−0.001
H	Heterogeneity	−1.839
D	Dominance	0.107
tau	Functional Diversity	**5.714**
BTC	Territorial capacity	1.03
g-LM	**gen-L. metastability**	**5.89**

14.3.5 A Synthesis of the Diagnostic Evaluation

Applying the methodology from Sect. 9.4 of this volume, we checked the divergences of the landscape of Bollate from the normal state, in a first period typically agriculture-productive (AGR-PRD), then suburban-rural (SUR-RUR). In fact, the index of landscape typology evaluation (LTpE) indicates a range between 24 and 10.5 for AGR-PRD landscapes and a range between 10.5 and 5.7 for SUR-RUR ones, ranges to be compared with the values shown in Table 14.8.

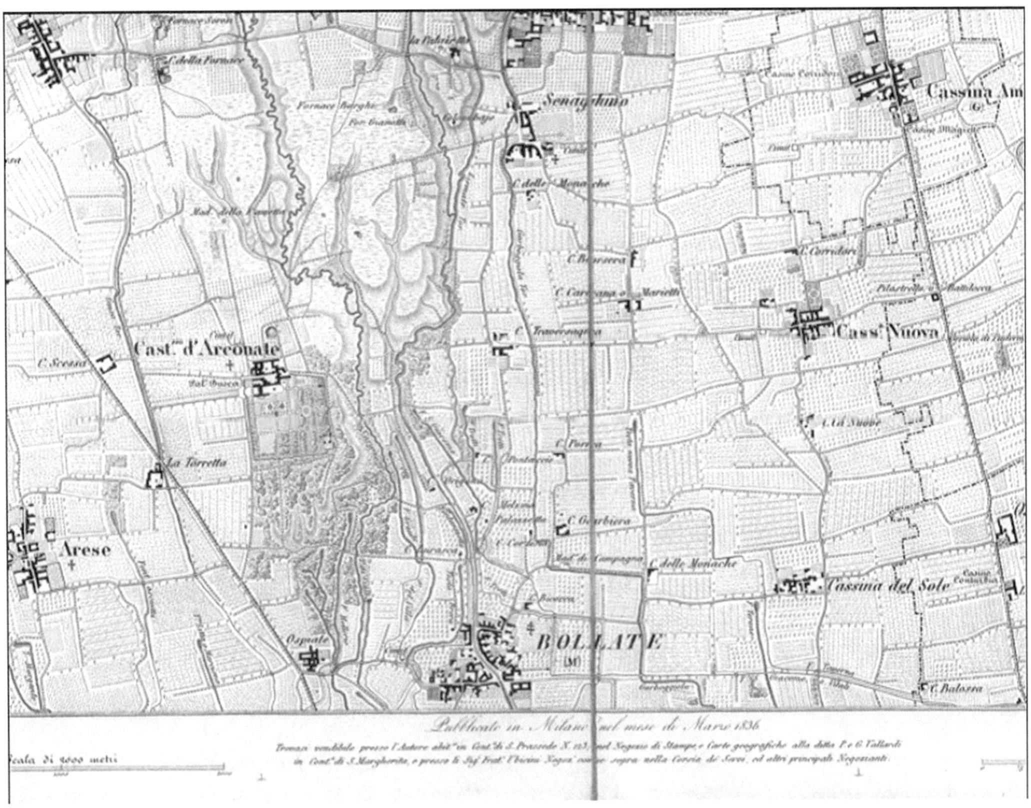

Fig. 14.19 Map of the territory immediately North to Milan scale 1:25,000 edited by Vallardi, 1836. This map is sufficiently precise allowing the translation into land use measures (see Fig. 14.20). Note the wide presence of heathlands, especially towards North, in the Groane area

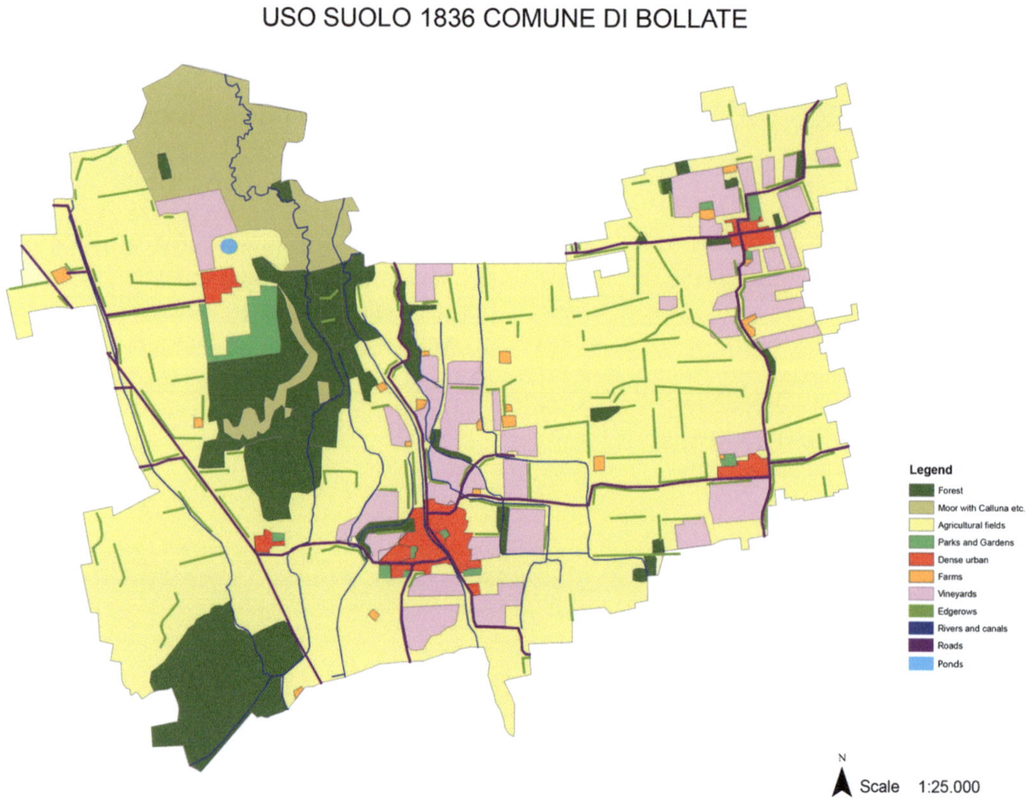

Fig. 14.20 Translation of the foreword map (1836) into land use measures. Note the presence of vineyards

Therefore, we had to divide the diagnostic evaluation in two Tables 14.8 and 14.9.

The choice of the 16 bionomic parameters (the same for Asse) and their elaborations and measures was compared to the normal AGR-PRD ranges, excluding the population (Table 14.8). The results of diagnostic index (DI) indicate a very good stability during the past century (DI = 96.67–88.33) in the period 1788–1888, with a drastic change in the recent period (DI = 66.67 in 1954) that means a decrease of about 25 % in few decades.

The resulting landscape changes breached the threshold of a new type of landscape: the suburban-rural. Table 14.9 shows the new level of the diagnostic index DI when calculated in relationship with the SUR-RUR landscape: 70. This level denotes a state of alteration, being <85.00. Anyway, this value is quite compatible with the current type of landscape, shifting towards a suburban-technologic one: so, it needs interventions but of not very strong rehabilitation therapy.

14.4 Comparison Between the Two Rural Suburban Landscapes

14.4.1 The Present Situation (2010)

Recalling Tables 14.2 and 14.6, we have noted that the built-up lands and the anthropogenic soil sealing of both Asse and Bollate are near the same: 553.36 vs. 522.80 ha. Within the urban area, Bollate has a higher population density, i.e. 63.38 inhabitant/ha, vs. 54.22 inhabitant/ha for Asse (+0.26). This difference is small, because the difference between the territories of Asse and Bollate is quite remarkable: 5,026.0 ha vs. 1,312.3, meaning Asse is +3.8 times larger than Bollate.

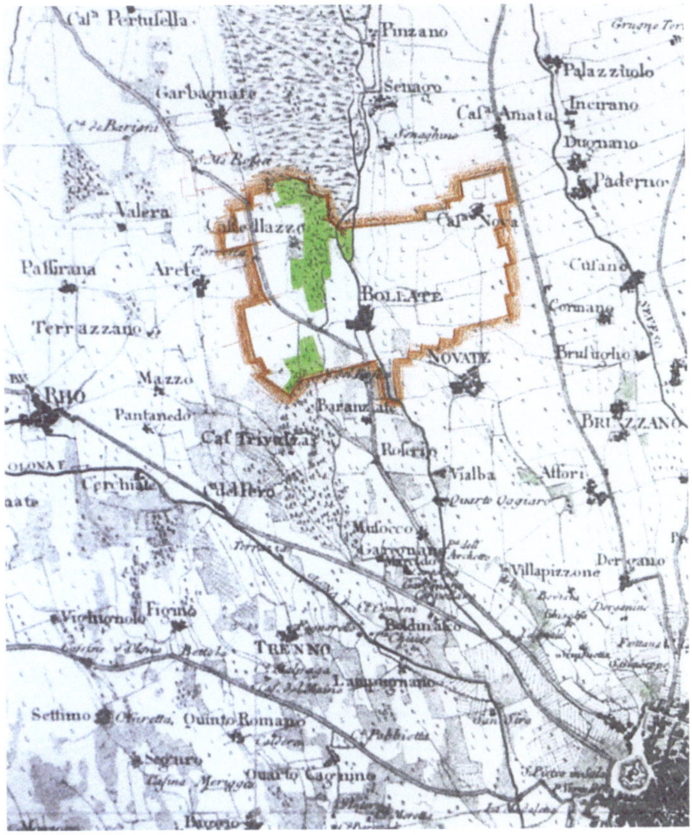

Fig. 14.21 An extract of the Hapsburg map 1788 by "Astronomi di Brera," scale 1:70,000. Note the remnant glacial tongue, covered by forests and heathland, arriving few kilometres North of Milan. The municipality of Bollate is demarcated (*brown*): it was the local capital of the III District of West Brianza

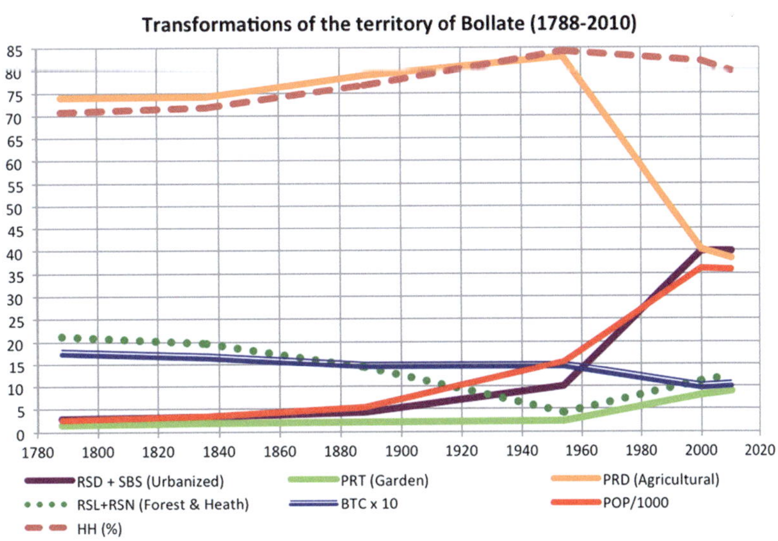

Fig. 14.22 The most important dynamics are characterised by changes in HH, an evident decrease of BTC only in the last 60 years, a strong decrease of cropland after 1954 and a growing of population which induced a huge grow of urbanisation since 1954. Forests regrown after the creation of the preserved area of Groane Park

Table 14.8 Diagnostic evaluation of the bionomic state of the LU of Bollate, in the period 1788–1954

Bionomic parameters	*AGR-PRD*	1788	Dev	Score	1836	Dev	Score	1888	Dev	Score	1954	Dev	Score
POP/1000		2.60			**3.40**			**5.36**			**15.5**		
LTpE	*10.5–24*	21.05	Ok	**2**	19.66	*Ok*	**2**	16.67	Ok	**2**	13.7	Ok	**2**
Urbanised	*8.0–16*	2.82	Ok	**2**	3.27	*Ok*	**2**	4.25	Ok	**2**	9.95	Ok	**2**
Gardens	*1.6–3.5*	1.43	Ok	**2**	1.93	*Ok*	**2**	2.13	Ok	**2**	2.22	Ok	**2**
Agricultural	*45–75*	73.95	Ok	**2**	74.2	*Ok*	**2**	78.85	5.70	**2**	83.1	10.7	**1**
Forest and heath	*10.0–30*	21.33	Ok	**2**	19.59	*Ok*	**2**	14.21	Ok	**2**	4.28	57.2	**1**
Photosyntetic area	*85–95*	92.07	Ok	**2**	90.24	*Ok*	**2**	90.14	Ok	**2**	85.7	Ok	**2**
Rehabilit. potential	*25–35*	30.35	Ok	**2**	26.71	*Ok*	**2**	24.54	1.8	**2**	18.2	27.4	**0.5**
HH (%)	*60–82*	70.73	Ok	**2**	71.74	*Ok*	**2**	76.66	Ok	**2**	84.2	2.70	**2**
SH	*2,200–5,000*	3571.6	Ok	**2**	2768	*Ok*	**2**	1877	14.6	**1**	713	67.6	**0**
Carrying Capacity (σ)	*1.5–9*	1.37	−8.6	**2**	1.11	*Ok*	**2**	0.8	46.6	**1**	0.34	77.3	**0**
BTC	*1.0–2*	1.78	Ok	**2**	1.68	*Ok*	**2**	1.47	Ok	**2**	1.47	Ok	**2**
g-LM	*14–24*	9.80	18.3	**1**	9.2	*23.1*	**1**	7.65	36.3	**0.5**	5.54	53.8	**0.5**
Territorial CBSt	*12.0–24*	11.25	Ok	**2**	10.45	*Ok*	**2**	10.02	Ok	**2**	7.38	26.2	**1**
Forest and shrub CBSt	*23.5–32*	25.79	Ok	**2**	25.05	*Ok*	**2**	25.65	Ok	**2**	23.96	Ok	**2**
Potential core areas	*35–50*	88.5	Ok	**2**	87.2	Ok	**2**	73.6	Ok	**2**	53.8	Ok	**2**
total score	*27.2–32*			29.00			29.00			26.50			20.00
Diagnostic Index (DI)	*85–100*			**96.67**			**96.67**			**88.33**			**66.67**
Bionomic state evaluation	*Normal*			Normal			Normal			Normal			Altered

Table 14.9 Diagnostic evaluation of the bionomic state of the LU of Bollate, in the period 2000–2010

Bionomic parameters	SUR-RUR	2000	Dev	Score	2010	Dev	Score
POP/1000		**36.20**			**35.75**		
LTpE	*5.7–10.5*	5.52	3.16	**2**	5.71	Ok	**2**
Urbanised	*15–27*	39.80	47.4	**0.5**	39.84	47.4	**0.5**
Gardens and sport	*3.5–7*	8.03	Ok	**2**	8.84	Ok	**2**
Agricultural	*30–55*	40.29	Ok	**2**	38.20	Ok	**2**
Forest and heath	*8.0–25*	11.05	Ok	**2**	12.27	Ok	**2**
Photosyntetic area	*60–70*	57.28	4.5	**2**	57.2	4	**2**
Rehabilitation potential	*15–25*	19.73	Ok	**2**	20.66	Ok	**2**
HH (%)	*70–86*	82.02	Ok	**2**	79.84	Ok	**2**
SH	*800–1,650*	297.3	62.9	**0**	293	63.4	**0**
Carrying Capacity (σ)	*0.6–3*	0.21	65	**0**	0.2	65	**0**
BTC	*0.8–1.6*	0.99	Ok	**2**	1.03	Ok	**2**
g-LM	*7.5–15*	5.61	24.6	**1**	5.89	21.5	**1**
Territorial CBSt	*8.0–20*	5.65	29.38	**1**	5.8	27.5	**1**
Forest and shrub CBSt	*22.5–30.5*	14.66	34.8	**0.5**	14.86	34	**0.5**
Potential core areas	*20–35*	22.5	Ok	**2**	21.4	Ok	**2**
total score	*27.2–32*			21			21
Diagnostic Index (DI)	*85–100*			**70.0**			**70.0**
Bionomic state evaluation	*Normal*			Altered			Altered

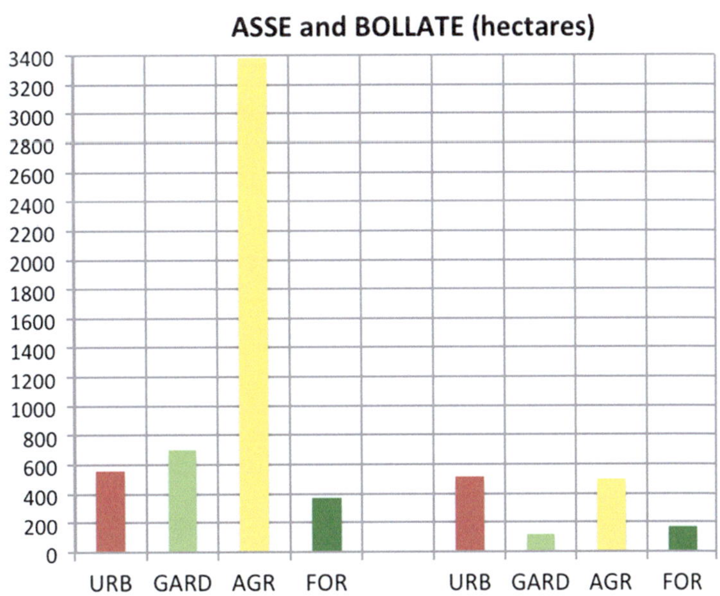

Fig. 14.23 The areas (ha) of the four main subsystems of the territories of Asse (*left*) and Bollate (*right*): URB = built urbanised, GARD = gardens, AGR = agriculture, FOR = forest and shrubs. Note that the difference in size surfaces between the two municipalities are principally linked to cropland

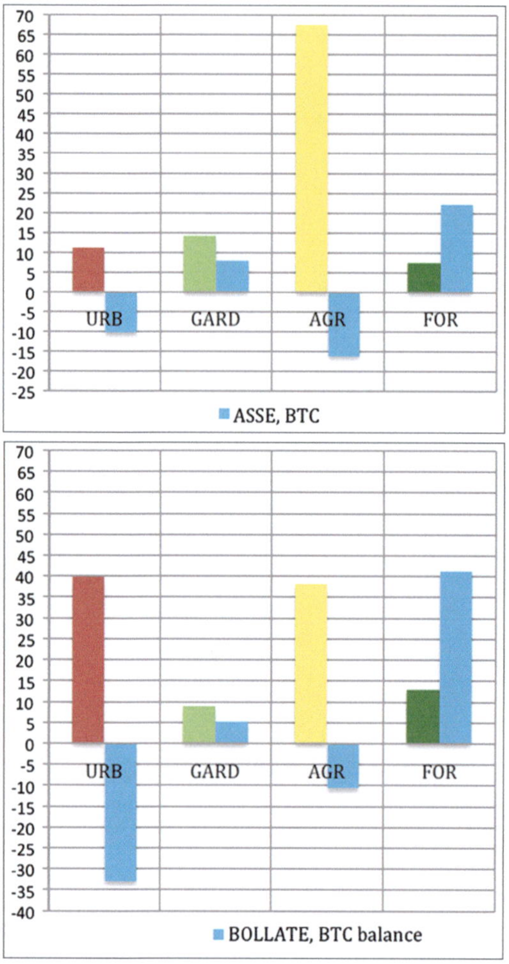

Fig. 14.24 Comparison between the BTC balance of Asse and Bollate, based on the concept of transformation deficit of BTC (sensu Ingegnoli [4])

In Fig. 14.23 we can observe that the difference between these territories is mainly due to agricultural elements, the other subsystems being about of the same order of magnitude.

The relatively high share of gardens in Asse (711.6 ha, 14.16 %), mainly private, is arguably more important for the comparison than the amount of built urban area (553.4 ha, 11.01 %). For Asse, this results in a ratio garden/urban of 128.6 %. The corresponding ratio for Bollate is markedly different, only 22.2 % (5.8 times less). This is probably due to the major surface of Asse, which allows this possibility, given the huge amount of cropland. No doubt that private gardens depend on cultural choices, but we recall the concept of ecological control of human culture, see Sect. 4.2.1.

Even the presence of wilderness (i.e. forest and heathland) is larger in Asse than in Bollate: 358.86 ha, 7.14 % vs. 161.02 ha, 12.27 %, but the ratio forest/agriculture is quite low in Asse 358.86/3,387.5 ha, or only 10.59 % vs. 161/501.3 ha or 32.1 % in Bollate. The main reason for this is probably due to the lower value of the agricultural soil near Bollate, due to the Groane "ferretto."

The concept of transformation deficit of BTC (sensu Ingegnoli [4]) may help to note other interesting observations. Recalling the four main subsystems (each composed of diverse landscape apparatuses): urbanised (URB), garden (GARD), agriculture (AGR), forest and shrub (FOR), their average BTC and the local BTC, we can demonstrate a significant ecological balance (Fig. 14.24). Recall that the average BTC of Asse is very similar to the BTC of Bollate: 1.04 vs. 1.03 Mcal/m^2/year.

On these bases, for Asse it results a local transformation deficit given by AGR + URB $= (-17.5) + (-11.0) = -28.5$ balanced by GARD + FOR $= +8 + 20.5 = +28.5$. Note that this value [% $\times$ Mcal/m^2/year], related with the total surface of 5,026 ha, gives a balance consisting of a huge flux of energy: $5{,}026 \times 0.285 \times 10{,}000 = 14{,}324$ Gcal!

The results for Bollate are: a local transformed deficit given by AGR + URB $= (-11.5) + (-34.0) = -45.5$ balanced by GARD + FOR $= +5 + 40.5 = +45.5$. This value signifies an energy balance of: $1312.3 \times 0.455 \times 10{,}000 = 5{,}971$ Gcal.

The proportion 14,324/5,971 = 2.39 is less than the proportion between the two territories (Sect. 14.1.1), thus Bollate has suffered a stronger transformation.

14.4.2 A Dynamic Perspective (1780–2010)

A transformation like this for Bollate was due to its exceptional increase of population, starting since the beginning of the past century (Fig. 14.25). Having a smaller territory, Bollate counted in the year 1800 about 70 % of the inhabitants of Asse, but in 1920 they had the same number of people, and today Asse counts about 80 % of the population of Bollate.

The minor increase of population in Asse is partly due to its good agriculture condition. Figure 14.25 indicates that, until the year 1990, Asse presented an autotrophic landscape (SH/SH* > 1), while Bollate became heterotrophic since 1860. It is known that population exceeding a normal growth cannot be sustained in a fertile agriculture landscape and emigrates to a city. The low carrying capacity of Bollate since 1780 (SH/SH* = 1.47) attracted the industrial and crafts activities from the near Milan, trigging a feedback, which increased the growth of population.

The natural vocation for agriculture is evident in Asse also analysing the dynamics of its natural areas (forest and shrub) and of its related bionomic functions (BTC and CBSt) and comparing these processes with Bollate.

As exposed in Fig. 14.26, because of the robust exploitation of the fertile soil, in Asse the forest presence was quite low since before 1780, less than one half of Bollate, which descended to the values of Asse only in 1915. The BTC function shows a development near parallel to the forest cover, and also the bionomic efficiency of the territory [13] remained near constant, passing in 230 years from 9.0 to 7.5 (−16.7 %).

On the contrary, in Bollate the BTC function does not follow the forest cover dynamics which, even after the regrowth from the strong fall in the World War II, is today near one half than at the end of eighteenth century. Consequently, the CBSt dynamics changed evidently, passing from 11.5 to 6.0 (−47.8 %).

We reported in Fig. 9.1 the present situation of the wilderness bionomic state of Bollate and Asse. The use of the CBSt index is in this case very symptomatic, because the ratio BTC Asse/BTC Bollate is limited to 129 %, while the ratio CBSt Asse/CBSt Bollate arrives to 185 %. The forest vegetation of Asse is better than the forest of Bollate. In conclusion, the bionomic conditions of Asse should seem to be decidedly better than the one of Bollate.

Anyway, the best synthetic control of the movement of a complex system as a landscape can be checked measuring its general landscape metastability[3] (g-LM), because it is able to integrate energy and information [dimensionally: g-LM = bit × Mcal/m²/year] in function of the human habitat level. The results of these calculations may be surprising:

$$g - LM_{BOLLATE} = 5.89\ [\text{bit} \times \text{Mcal/m}^2/\text{year}]$$

$$g - LM_{ASSE} = 5.19\ [\text{bit} \times \text{Mcal/m}^2/\text{year}]$$

Figure 14.27 highlights, on the plain representing the state of the complex system (HH, g-LM), the sharp differences between the two landscape transformations of Asse (brown) and Bollate (blue) in the last 230 years. The changes in Bollate were 12 times wider than in Asse, but remaining not so far from the optimum (green line). Even today the g-$LM_{BOLLATE}$ = 64.5 % of the optimum, while g-LM_{ASSE} = 52.0 %. No doubt they are both altered,

[3] In complex adaptive self-organised systems the diversity of their components must consider both heterogeneity and information, therefore the proposed landscape diversity index is:

$$H\,(3 + D) = \tau H' = -S_{i=1}^{n} \log_2 P_i [\text{bit}]$$

H Shannon diversity

D the dominance,

τ the synthetic landscape diversity

$\tau \times$ BTC = LM; g-LM = H(3 + D)*BTC [bit × Mcal/m²/year]

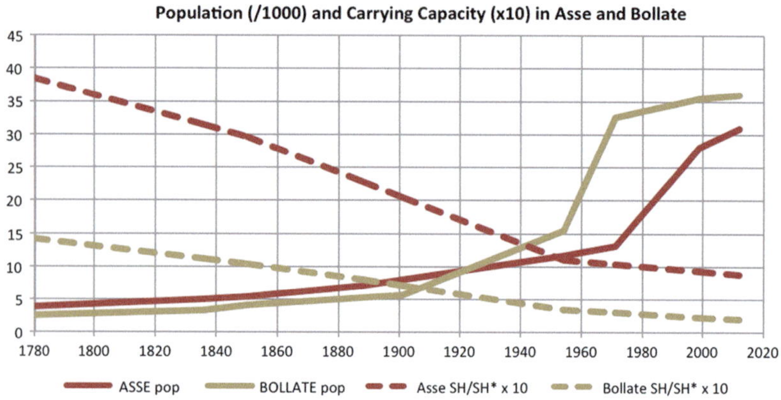

Fig. 14.25 Comparison between the population growth of Asse and Bollate and their Carrying Capacities SH/SH* in the period 1780–2010

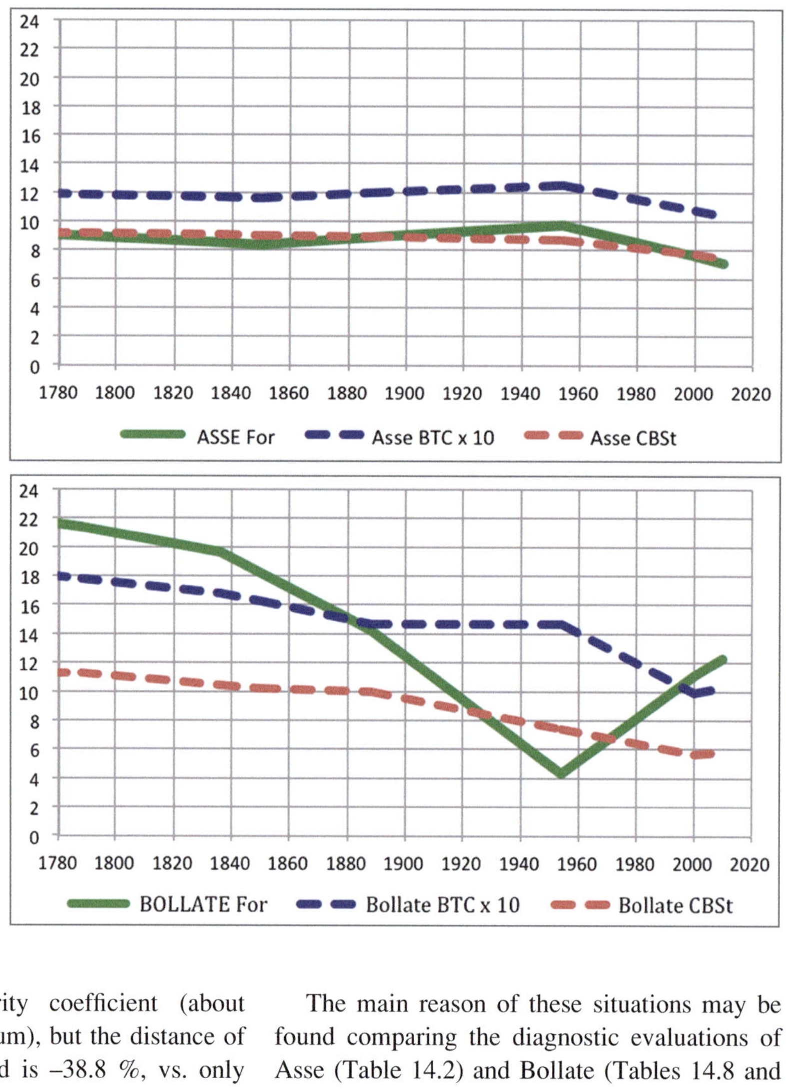

Fig. 14.26 Comparison between the forest and shrub land cover (%) of Asse and Bollate (1780–2010) and their related processes of BTC (Mcal/m^2/year × 10) and CBSt

being outside a security coefficient (about ±10–15 % of the optimum), but the distance of Asse from that threshold is –38.8 %, vs. only –25.5 % of Bollate.

The main reason of these situations may be found comparing the diagnostic evaluations of Asse (Table 14.2) and Bollate (Tables 14.8 and 14.9). The diagnostic index (DI) of Asse resulted

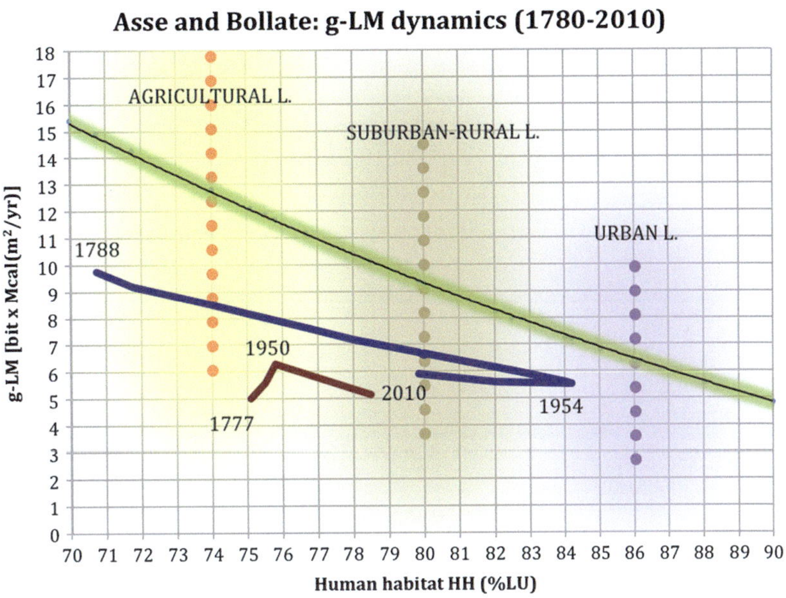

Fig. 14.27 Comparison between the movements of the landscapes of Asse (*brown*) and Bollate (*blue*) registered on the plain representing the state of the complex system (HH, g-LM). Note that today Bollate has a g-LM higher than Asse (5.9 vs. 5.2)

today 68.33 % of the optimum referred to the landscape type "agriculture-productive," very similar to the value of Bollate (same landscape type) in 1954: but, after this period, Bollate overcame the threshold and passed into the other landscape type "suburban-rural," referring to which DI arises now to 70.0, while Asse is now in a belt of instability, that is the threshold between the two types, being not completely structured as suburban-rural but under the influence of Brussels.

Pay attention that only through the reconstruction of the historical dynamics of those two landscapes, that is through the "story of nature" attentive to the process of forming and continuously metamorphosing of individual "local landscapes" (Sect. 8.1.2), we are able to correctly understand the ecological significance of the "position" assumed by the two LU within the plane of the state of the system, going today towards a convergence.

References

1. Forman RTT, Godron M (1986) Landscape ecology. Wiley, New York
2. Naveh Z, Lieberman A (1994) Landscape ecology: theory and application. Springer, New York
3. Zonneveld IS (1995) Land ecology. SPB Academic, Amsterdam
4. Ingegnoli V (2002) Landscape ecology: a widening foundation. Springer, Berlin
5. Marcheggiani E, Bomans K, Gulinck H (2010) New ways of landscape diagnosis, living landscapes: the European landscape convention in research perspective. Bandecchi & Vivaldi, ISBN/ISSN: 9788883414589
6. Bomans K, Duytschaever K, Gulinck H, Van Orshoven J (2010) Tare land in Flemish horticulture. Land Use Policy 27(2):399–406
7. Dewaelheyns V, Bomans K, Gulinck H (eds) (2013) The powerful garden. Emerging views on the garden complex. Aantal Pagina's. Garant, Antwerp, p 237. ISBN 978904412733
8. National Geographic Institute of Belgium NGI-IGN. http://www.ngi.be
9. http://www.geopunt.be/
10. Desio A (1973) La geologia applicata all'ingegneria. Hoepli, Milano
11. Lombardia R (2010) I dati DUSAF. Ersaf, Milano
12. Pignatti S (1998) I boschi d'Italia: sinecologia e biodiversità. Utet, Torino
13. Ingegnoli V (2013) Concise evaluation of the bionomic state of natural and human vegetation elements in a landscape. Rend Fis Acc Lincei. doi:10.1007/s12210-013-0252-2
14. Brenna G (1836) Carta della provincia di Milano, 1:25,000. A. Vallardi, Milano
15. Gabba L (1928) La carta della Lombardia a cura degli Astronomi di Brera. IGM, Firenze

16. Tradati B (1852) Guida Statistica della Provincia di Milano. PL. Pirola, Milano
17. Ingegnoli V, Giglio E (2005) Ecologia del paesaggio: manuale per conservare, gestire e pianificare l'ambiente. Simone Edizioni-Esse Libri, Napoli, p 704
18. Ingegnoli V, Pignatti S (2007) The impact of the widened landscape ecology on vegetation science: towards the new paradigm. Rendiconti Lincei Scienze Fisiche e Naturali, s.IX, XVIII:89–122

15 Landscape Environmental Ethic

15.1 Landscape, Man and Nature

15.1.1 The Nature of Landscape

For the majority of people, talking about landscape means referring to a simple geophysical support to the activities of organisms (plants, animals and humans), presenting an aesthetic structure with historical and cultural connotations. When, more simply, the term landscape is not confused with the term 'scenic view'.

This is a narrow and not biologically correct concept of landscape. For landscape we means (see Chap. 1) a specific level of biological organisation, that is a living entity resulting from the integration of human and natural communities in an appropriate territorial unit [1]. Therefore, the landscape is definable like a hyper-complex bio-ecological system (Fig. 15.1).

The first consequence of this recognition is that the laws that underlie the behaviour of a territory as landscape—the peculiar place for the evolution of man in nature—are largely the same ones that govern the behaviour of every other living entity, although if declined in an opportune manner. To understand this concept you need to consider the concept of life as no longer limited to the individual organism. This is not new to human culture, but a confirmation, from the scientific community, of what our ancestors already sensed through the myths and religions: the Greek-Roman and Celtic sacred woods, the mother Earth, "which sustains and governs us". We consider the role of our Mother Earth as something really special, because it humbly sustains us, nourishes us, in some cases even accept to be robbed, but still governs us (Fig. 15.2).

Note that the claim following which "our Mother Earth is able to sustains and govern us" has been expressed by St. Francis of Assisi, in the well-known Canticle of the Creatures [2]. The reductionist objectivism,[1] still followed by the most conservative scientists, suggests that all the ancient interpretations of nature were wrong, but that is not true. It is, rather, the current concept of nature to be limited: for the ancient Greeks the Physis was not a world of objects but a world of vital processes, as pointed out by Giorgio Israel [3].

We believe essential, therefore, to expose the philosophical basis of the relationship between man and nature and its ethical implications. As you will see, any serious discussion on environmental sustainability and even more on the effects of landscape diseases on human health cannot avoid moral considerations. Similarly, it should be noted that operate on a biological system is deeply different from transforming a mere geophysical substrate. Man's actions towards

[1] Reductionism, which did so much in developing modern science, is outdated in many fields from recent advances in scientific thought, due to the possibility of studying in depth the complex systems, e.g. through the use of computer.

V. Ingegnoli, *Landscape Bionomics Biological-Integrated Landscape Ecology*,
DOI 10.1007/978-88-470-5226-0_15, © Springer-Verlag Italia 2015

Fig. 15.1 The landscape means a specific level of the organisation of life on Earth that is a living entity resulting from the integration of human and natural communities in an appropriate territorial level

living systems should be evaluated before all in ethics. The Environmental Ethics is, moreover, studied in several universities, given that it is the branch of philosophy that asks what actions are right and what are not in the environment.

Surpassing the level of organism, if life is not considered in all its hierarchical structure (from cell to eco-biosphere), we encounter a serious imbalances, as shown by the following two examples, which are located at opposite ends but which are unified by a poorly understood concept of science! On the one hand, doctors who care about the body, with an excess of scientific objectivism, are forgetting the patient as a person; on the other hand, the technicians who deal with environment, land, landscape fail to see the complex nature of the biological systems that claim to heal, showing major weaknesses of the scientific method. This represents a shift from an exaggeration to another. While we plan to transform the traditional medicine in an exact science, we treat the environment in a very rude manner, not wanting to understand the processes that govern it as a hyper-complex system.

To stop these trends a deeper clarification is necessary, which is both epistemological and ethical; however, ethics and epistemology are closely linked in ecological disciplines, since

Fig. 15.2 Drawing "At the Roots of Life", pencil on paper (Vittorio Ingegnoli, 2008), which symbolises the Earth as mother "who sustains and governs us". The design is a real fact: the son of the Author (3.5 years old) took refuge among the roots of the big spruce tree, in the Natural Park of Paneveggio (Trentino-Sud Tirol)

the moral aspects are directly dependent on the address connected to a philosophical vision. In addition, it is essential to clarify what is meant by the concept of misunderstood science. Note the profound revolution of the man-environment relationship that is derived from the above definition of landscape, affecting both the academic world and the civilian and professional one. The traditional general ecology seems to make it hard to understand that landscape ecology is not limited to the study of phenomena at a larger scale, spatial and territorial: in add it studies them deepening their proper biological origin, thus in a biological-integrated way, going to modify various theoretical aspects covering all ecological fields (see Sect. 2.3) and defining itself as landscape bionomics [4, 5].

This revolution involves new aspects, ethical and deontological, for those who practice the profession of 'ecologist' and forces the public and the political class to take a more proper administration of the territory. In addition, this new criterion requires the revision of various provisions of the law on the environment, to be made in the light of principles far more advanced than those followed now. This revolution of the relationship between man and environment requires, finally, the creation of a new profession, that of "doctor of ecological systems" or *ecoiatra*.

15.1.2 Man and Nature Relationships

The majority of people behave towards nature at least superficially, guided by economistic unscrupulous considerations and deceptive ideologies, even if related to environmental sustainability. The reasons for an increasing environmental degradation depend eminently on such behaviour: it is urgent to do everything possible to change these attitudes. Unfortunately, we are still far from what it would be fair to do to achieve a true ethical commitment on the part of ecologists.

To start an argument capable of giving the basic references and incentives for a profound renewal of human behaviour towards nature, it seems not useful to mention the speeches more or less catastrophic about the ecological crisis, both because they are often exaggerated, either because they alarm the 'public opinion', resulting in the opposite effect. Also, if you do not clarify the underlying reasons, capable of leading to a real change, you cannot awaken the sense of responsibility of people. Before that, we need to look for the causes of the current behaviour, in order to show the errors it contains, and fight them.

It can be affirmed that nature has developed an organism, the human being, which is able to summarise the traits of the same creative and cognitive process that generated it. Thus, the controversy over the pre-eminent position of man in relation to any other organism is only specious; not true positions are also those of people who deny the qualities that characterise man as a person: self-awareness, self-reflection, of moral conscience, metaphysical knowledge, ethical commitment, humility, freedom, art, forgiveness, self-sacrifice, holiness. Recall the traditional cultural position of the Western world, summed up in the words which Dante Alighieri attributes to the Homeric hero Ulysses: *Fatti non foste a viver come bruti, ma per seguir virtute e conoscenza* (You were not made to live like brutes, but to follow virtue and knowledge).

Let me be clear, however, that the term "brutes" does not imply denial of the existence of forms of animal intelligence. As written by Rowlands [6], it is ironic that some philosophers continue wondering if some animals have a mind and can even have feelings: "at one time or the other, they should raise their nose from their books and try to train a dog or a wolf", as he did. This also applies to many ethologists, who think they are "more stringent" refraining from interpreting as such actions clearly moved by the sentimental aspects of animal behaviour. The brutes are rather those who don't want to fulfil their duty as men, epistemic and moral duty, regardless of the impotence of the possible victims. The brutes are those who embody the idea of the "banality of evil" introduced by Arendt [7] in his study on the Eichmann trial.

There is no doubt that the position of man in nature involves problematic aspects, as he has to obey the laws of nature that govern complex biological systems and are able to manage many aspects of their own components, including humans. Similarly, man should preside over, and control, the organisation of the natural environment that allows him to live and develop. This position implies a not easy role, because it concerns a domain, even if in obedience.

However, the role of man in nature involves a creative way, but also a clear responsibility. We must recognise that, today, the duty of accountability is not heard, often even forgotten.

There are several reasons, but above all you must track down the root causes in the spread of self-styled progressive ideologies, which, spreading relativism and scientism, greatly reduce the principle of responsibility. We don't want to make a political speech: the opposing sides have both praised (and still praise) the relativistic progressivism and scientism, albeit with different visions. It is not, moreover, an issue "of modern times", even though these out-dated ideologies have suffered a definite strengthening in our days. The epistemological relativism was already present in ancient Greece, when the sophist Protagoras said "...man is the measure of all things...", implying that truth is relative: in the face of these fact Socrates, founder of moral philosophy, was his main antagonist.

Scientism, as exaltation of the scientific method in knowledge even in areas of humanities, can be traced back to the Renaissance, even if it developed with the Enlightenment first, then with positivism and Marxism. It is a materialist vision and affirms that the scientific method, understood as absolute capable of eliminating metaphysics, is the only valid way for knowledge. In the form most prevalent today was founded in the late nineteenth century, by the confluence of the Marxist positivism and historicism.

By placing blind faith in the possibilities of scientific and technological knowledge of man, thus relativising the truth in the name of progress, such a view denies the addiction that is underlying the concept of Creation. The man does not trust in the created world, but in the world to be created, which no longer need trust, but only capacity. The transformation is therefore the fundamental task of man. Progress is dependent on the change in the truth, relative truth, which is essential for 'the good of the cause'. So, today's version of the myth of Icarus leads to see the man as a force opposed to nature, in the name of self-referential freedom.

Nature is no longer a prerequisite able to afford and, at the same time, to limit and to direct human actions, but is an *object* in the hands of man: a man not created, produced by chance and necessity. The only creator is man, the true creator of himself, hence the unfailing faith in science and technology was originated. Adam and Eve ate the fruit of the tree of knowledge and wanted to replace the Creator. This absolute confidence in human capacities devoids nature of practical importance and considers it as a sub-system of the economy. This is evidenced by the fact that, when the scarcity of a resource, or an excessive pollution, threaten to cause radical changes in the environment, perhaps with negative influences on health, science and technology involve with palliative able to avert the danger, maintaining the status quo.

We observe, however, that the greatest scientists reject scientism and wary of preconceptions praising the progress, because, as written by Albert Einstein, "in the laws of nature is revealed a reason so superior that all rationality of thought and of human laws is in comparison a very insignificant reflection".

On the concept of progress unequivocal words were written, to measure the underlying error. Here, by the way, an affirmation of Gandhi, reported by Lanza del Vasto [8], his greatest disciple in Europe: "The progress or the quick make necessary more quick, the redundancy more redundancy, the vanity more vanity, the violence more violence: everything is accelerated, magnifying like an avalanche. The avalanche has to meet a rock where shatter, a village to take away and, at its height, it stops choking all life". Progress as ideology is an attack on nature.

15.2 Knowledge and Responsibilities

15.2.1 Towards a Revolution of Thought

In the face of materialist theories and scientism, which corrupt the society on a global scale, we would need a revolution. The Marxists, of the rest, combined the progress with revolution but, as Bloch wrote [9], that action needs a concrete pole of *hate*: "without the bias of the revolutionary classist point of view only an idealism turned to the past is given, rather than a practice focused on the future". Capitalism has no doubt his guilt, often very serious. However, it is known [10] that the nations most affected by pollution and environmental degradation (the highest in the world) are China and Russia, after decades of "socialist realism"!

In fact, as was pointed out in the previous paragraph, if the dialectical materialism asserts that truth lies only in the change, the truth is that

Fig. 15.3 Photos of Mahatma Gandhi, who wrote, as noted, "The Story of My Experiments With Truth", summarised in the following passage: "Truth is the first thing to look for, after which the Beauty and Goodness will be added by themselves"

which imposes itself: it is, therefore, a scientist and relativist beliefs based in fact on violence. The revolutions in the sign of violence are rebellions that tend to replace the old abuses with new abuses, the ancient tyrannies with new tyrannies, as has undeniably been the Lenin's revolution in Russia. Perhaps is there any alternative to the way of making a revolution?

An example of an alternative there was, we know: the Indian revolution against the British Empire. Mahatma Gandhi (Fig. 15.3), however, said completely other things, as noted by Lanza del Vasto (op. cit.):

> "Look at my spinning wheel: it is called 'revolution' the complete wheel rotation. The stars carry out their revolution in the light and the seasons make their revolution in the flowers and fruits, and human history has to make his revolution in justice and goodness. Those, who want to make fun of me and of my spinning wheel, say: 'You want to go back. You want to delay the clock'. No, my friends, I am the most advanced revolutionary and I just have to let the clock leftovers because it returns itself to the starting point. Revolution is the return to the beginning and the Lord. The ones they stick to the forms of the past and the memory of the dead, and live like the dead, the others fly themselves into new crazy until they fall on deaf ears. But I'll go ahead and I don't lose my way, because I return to the most ancient traditions through the complete revolution which is the total overthrow, but natural, and willed by God, and which happens at the right time".

Despite the success, in the post-war period, of the non-violent revolution in India, it must be assumed that many people consider difficult and unrealistic to think of something similar today. Economic crisis and security measures send into the background the ecological problems, except, in part, those relating to pollution and climate.

This is followed by the increase of agnostics and irreligious persons just after the spread of relativism, scientism and progressive ideas: people who trust only their own reason or who considers science a kind of religion. In these hyper-rationalist lay, we must remember the a-rational character, rather anti-rational, of *Gnostic rationalism*, of which the dialectical materialism is one of the most popular current ideologies.

In the view of environmental ethics, therefore, it is necessary to carry out, on a rational level, the discourse on the "Revolution as a return to the Principle". Then, the true lovers of reason could be pushed to understand, while those who refused it would demonstrate the limits of the bias, i.e. an anti-rational view.

The attempt to find a form of revolution that appeals to a rational behaviour, to get to touch even the heart of men, can benefit not only to the preservation of ecology, but also to scientific research itself, sometimes too compromised with scientism. Therefore, let us ask ourselves if we can find a *principle*, common to reason and heart, at the same time being the basis of knowledge.

15.2.2 Truth and Knowledge

The real search for a responsible knowledge necessarily presupposes the consideration of the boundaries of science, since, for rational thinking, science is the key reference for the human capacity to understand reality.

From the epistemological studies we know that the modern scientific method, i.e. Galilean, deductive, experimental, although not yet been passed, has shown severe limitations. First of all, it is not able to guarantee the certainty [11], because it remains substantially imperfect. We refer, for example, to what demonstrated by Kurt Gödel's theorems of incompleteness [12][2] and Alfred Tarski's one of indefinability [13].

In fact all scientific disciplines have found themselves in crisis, more or less deep, in the course of the past century, which has even shown the flourishing of an exceptional development in all branches of science. Physics has lost his mechanistic confidence, due to the progression of the physics of sub-atomic particles (quantum mechanics) and his faith in an "absolute" time and space, because of the Theory of Relativity; it lost faith in reversibility due to the theory of processes far from equilibrium and, ultimately, lost the certainty in the Laplacian determinism with the development of Deterministic Chaos Theory.

The Mathematical Analysis, in turn, had to abandon its central location (due to the great scientists of the eighteenth and nineteenth centuries) because of a series of events that brought to light the fragility of the building that had developed, perhaps too quickly, and in so many different directions, without an epistemological approach sufficiently robust and thorough.[3]

Scientism, which however often tends to confuse science and technique, believed to avoid the shortcomings of scientific thought resorting to relativism: if the truth is considered relative, you don't need to ensure certainty. This way of thinking, however, presents serious contradictions. It wants to avoid metaphysics, but starts from a preconceived notion of a metaphysical, as in "dialectical materialism". Scientism also does not want to consider the truth as absolute, degrading the ability to know. These important epistemological considerations were reiterated by scholars such as Lorenz and Popper [16] and, more recently, by Israel [15].

Perhaps the biggest limitation of the modern scientific method depends on the fact that the principles adopted by the science of motion (mechanics), starting from Galilee, followed by Newton, Leibnitz and Laplace, became the model for every possible form of science (according to the hierarchy: mathematics, physics, chemistry, biology, psychology and sociology). Even those who took care of absolutely dissimilar spheres of reality, such as the phenomena of life, they had to adapt. The method, called reductionism, which is to reduce all phenomena to the simplest structure below, in the belief that the simplest of all is the physical one, was applied to all fields of science.

This epistemological approach, however, in the world of biology looks very controversial and limited, capable to have deep consequences, well indicated by Lorenz [17] and Israel [3]: the destruction of any notion of subjectivity, planning and teleology, keys of biological thought. Even Jacques Monod [18]—still remembers Israel—was well aware of the epistemological contradiction between scientific objectivism, which implies the rejection of any kind of project, and the teleonomic[4] character of living beings, even if he thought that these profound contradictions could be dissolved. To

[2] Gödel published his most important study at the University of Vienna. It contained the two "incompleteness theorems", according to which: "every axiomatic and consisting system, able to describe the integer arithmetic is equipped with propositions that can neither be proved nor refuted on the basis of the axioms of departure". This means that if a formal system S is consistent (i.e. free of contradictions), then it is possible to construct a formula F syntactically correct but improvable in S: so if a formal system is logically consistent, its non-contradiction cannot be demonstrated staying within the same system.

[3] Hilbert [14] had proposed to arrive to a formulation without contradictions of Mathematics, which was passing through a difficult period after the development of Set Theory. The theorems of Gödel [12] brought to a deep crisis Hilbert's program. It was von Neumann who managed to bring back mathematics to a central role, through an operation of "confidence metaphysics" in a axiomatic approach and through its "pan-mathematics", understood as deductive logic diagram and abstract language of great power [15]. The work of von Neumann led to less emphasis on the differential calculus and more impetus to mathematics timeless; emphasis on algebraic and topological structures, functional analysis, measure theory, the "convex analysis", the fixed-point theorems. It was also von Neumann that led to the birth of the mathematical model, the concept of fundamental importance in applications in all fields.

[4] Teleonomy in biology means the finality inherent to the institutions or living systems.

dissolve that contradiction, however, we must get out of any type of policy prejudice, as the ontological reductionism. We have to proceed trying to integrate science and art, intuition and reason, but on a higher and different epistemological level, able to understand both aspects.

In fact, the only method we have for understanding remains that of "try and error", guided by the principle of approximation to the truth, considered as absolute. It is noted, with Popper, that if the truth is not absolute and objective we could not err, because our mistakes would be just as good as our truth, undermining our ability of understanding. In a broad sense, we can say, too, that without the concept of truth it would be impossible to leave out of consideration the accidental and to abstract the essential, which are the two fundamental bases of perceptual objectification through the well-known process of 'pattern matching'. If the truth is not absolute, it would lead to a dangerous deterioration of ethics because, in the name of the most sacred things, it would lead to allow the most heinous crimes "for the good of the cause" and, as demonstrated by Alexander Solzhenitsyn [19], the lie would have the upper hand. Recall that certain aspects of the Inquisition of the Church in Spain, the wars of religion between Catholics and Protestants, the current Islamic Fedayeen, fall into this serious distortion of reason, as well as the establishment of the Hitler or Stalin concentration camps.

As a consequence, we will not hesitate to affirm that the primary task of all persons gifted by thought is the search for Truth. Truth is the essence of all reality, perceptible or not. The primary role of man, therefore, is the search for Truth. Only in the light of the Truth even seemingly opposite categories, such as science and metaphysic, may have unpredictable and profitable matches,[5] which would be simply wrong to neglect, as noted by Konrad Lorenz [21].

[5] For instance, the recent interest of the medical world for NDE (near death experiences) forced some important neurosurgeon to think possible to demonstrate the existence of our souls, as wrote in 2012 by Eben Alexander of the Harvard Medical School [20].

We now have an answer to the question we posed at the end of the preceding paragraph, because "finding a principle common to reason and heart, at the same time being the basis of knowledge, is possible: it is the absolute Truth"(Gandhi). The heart of the matter lies in the fact that Truth is *ahimsa* in the sense of non-violence, charity, love. Gandhi said that "Truth and Non-violence are two sides of the same coin without effigy and without substance, the first of which is the goal, the second the mean". For many, this finding may seem inconceivable. Still, one can rationally prove that it is not.

Meanwhile, we will say that knowledge implies an act of truth and at the same time emotional involvement, which is love. Above all, it is noted that Truth and Non-violence demonstrate the maximum practicality and rationality of the principle, because if *ahimsa* is not the only way to get closer to the truth, any form of violence could impose the false.

From the foregoing, in practice, the greatest threat to nature, conservation biology, landscape planning, eco-design and sustainable development or even to the medical clinic, is represented by the betrayal of the role of man, expressed as violence against life in all its aspects in the form of pseudo-scientific arrogance. A betrayal claims a principle of justice: it speaks of environmental ethics as a necessary reference in the relationship between man and nature.

As asserted by Gandhi, "Justice is the truth in our acts"; therefore harmony with nature is a question of justice. This has been confirmed ever since the roots of our western civilisation, in the first book of "*De Legibus*", by M.T. Cicero [22]:

> Thus once [man] will have studied heaven, earth, seas and the nature of all the things and will have seen their origin and their possible end, when and how they will perish, what in them is mortal and transient, what divine and eternal, and almost will have touched the Being who governs them, and once [man] will recognise himself as not closed by walls in a limited site, but citizen of the entire world such as a single city: in this magnificence of things, in this spectacle and knowledge of nature, immortal Gods, how much he will have known himself such as the precept of Apollo Pytius! How much he will condemn, how much he will despise and consider as nothing what mass reputes as important. (Chap. XXIII)

The most important ethical issue arises from the fact of translating "Truth in the acts". Also to act upon what science has shown to be necessary usually requires a vision of the causal chains between the laws of nature, pushed to the extreme. For example, there are many cases of construction in Italy (with regular permits!) of homes in the flood plain areas or slopes subject to landslides, despite the notice of geologists, often with tragic consequences. As written by Lorenz [23], this cannot cancel our free will, but can nail us to our responsibilities. Witness to the Truth requires sacrifices, since even extreme. History teaches: Socrates had to drink hemlock, Gandhi was killed by a gunshot, Bishop Luigi Padovese was stabbed to death in Turkey. The martyrs do not count. Christ also, the Son of the Lord, was not spared: history remembers what he said to Pontius Pilate, "I came to testify to the truth". Answer the man of power: "What is truth?"[6]

Yet, the only defence of the Truth is the sacrifice, otherwise you should resort to violence which is its negation, as has already been said. We often forget this and, as a result, we refuse to follow the requirements derived from the consciousness, the laws of nature, environmental ethics, taking an apology, which in fact represent a retreat into a kind of idolatry.

The scientist progressivism is, in fact, the first of the idolatry of the modern world and it is worth to underline how much bad influence in the natural sciences and ecology this idolatry has, as we shall see later. In most cases it resulted in producing pseudo-theories without try to go further, replacing the search for truth with pseudo-truth of convenience.

Even an exasperated reductionism can lead to a similar scientist position, for example when it claims the principle of complete qualitative homogeneity between life and death, from which the controversy on euthanasia and aggressive treatment. That the reductionist epistemology is really dangerous in the study of man and of life in general is provable [3], despite the undoubted achievements of modern medicine. If these ideas become prevaricating facts, man can suffer serious consequences. For example, a scientist medicine is likely to replace the idea of "cure" with that of "repair", reducing the man to machine.

On the contrary, even a theoretical physicist as Albert Einstein [24] claimed to have learned much more from Dostoevsky than from any physical-mathematical. His rejection of the idea that the chance only can be the answer to our questions is not only a matter of faith (Einstein used to say, "God does not play dice"), but of reasoning. As observed by Ilya Prigogine [25] chance and determinism have undoubtedly played a role in the world, but scientists need to avoid prejudice to represent nature relying solely on one of two criteria. These concepts lead to two opposing visions: a world governed by laws that do not give way to the new (determinism) or an absurd world, a-causal, in which you cannot predict or describe anything in universal terms (chance).

As underlined in the first chapter, Einstein wrote in his Autobiography [26] that even prominent scholars (such as Mach) could be hampered by bias in the interpretation of the facts:

> "Prejudice, which still is not gone, is the conviction that the facts can and must be translated into scientific knowledge by itself, without free conceptual construction".

For Einstein [27] independence is determined by the prejudices and philosophical analysis is "the mark of distinction between a mere artisan or specialist and a real seeker of truth".

The epistemology of Einstein (see Fig. 1.6) is based on the experience (E) and proposes a structured set of theoretical assumptions (A, axioms), intended to explain E. The theory identified by A allows to logically derive a series of consequences (S) which will then be compared with E. The result of the comparison will assess the adequacy of theoretical hypotheses A. Note that the only passage from A to S is a logic inference, while in the other two (E→A and S→E) the inference is predominantly intuitive, therefore is a free conceptual construction.

[6] For a believer, the truth is the most beautiful name of God (Saint John Paul II), the ultimate reality of being, and how absolute is the alpha and the omega, the OM, the origin and end of all existence in the universe.

The great scientist anticipated then the epistemological position of Popper, critical rationalism, as indicated by Federico Laudisa [28]. The idea that the conjectural nature of theories can happily live with the aspiration to build true descriptions of the world (interview with Popper, 1989) follows deeply Einstein's thought. This criticism was in perfect harmony with the epistemology of Einstein, who has been repeating, since 1918, that the world's observations undoubtedly determines the theoretical system, nevertheless every direction of thought and logic

> "make observation data to the principles of the theory: this is what Leibniz has so happily called the pre-established harmony".

Even the epistemologist Jacob Bronowski [29], states that what will allow humanity to survive and to continue in scientific discoveries:

> "will not be just or unjust rules of conduct, but more deep illuminations, at the light of which good and evil, means and ends, justice and injustice will be seen in a terrible clarity of boundaries".

These epistemological positions are fully converged with the above writings: these lights will be produced by the light of Truth. The logical process alone is not enough, neither for science nor ethics.

15.3 Acquisition of Consciousness

15.3.1 Idolatry and Distortions in Science

We return once again to what Gandhi said: "The principle common to the reason and the heart, at the same time the base of the power of knowledge, is based on the Truth". Even the first verse of Genesis talks about Principle in the sense of Truth. However, observations derived from the Torah, one could argue, are matters of religion rather than of epistemology and environmental ethics. But as noted by Kurt Gödel, and previously by Galileo Galilei, Thomas Aquinas and Augustine of Hippo, science and faith are not in opposition. The inseparable binomial "Truth & non-violence" is the basis of this statement.

On the contrary, today we often read that those who want to follow a truly scientific conception must be secular, even atheist, "irreducibly atheist to the bone" to avoid having to choose "between faith in God and faith in Darwin" or that to "postulate an agent superior to nature means to stop doing science". The very numerous atheists and secular scientists, often polemical to become acrimonious and sometimes violent towards men of faith, they do not realise that, even scientists who are believers follow an objective and "non-partisan" methodology, as the faith has nothing to do with magic or fanaticism: the same search for truth demands it. The believer has understood what Augustine says in his work, entitled (not surprisingly) "Against the Academics": "Crede ut intelligas, intellige ut credas".

Presumption and contempt, unworthy of men of science, are therefore at the origin of the anathemas against the scientists who are believers: "Dark", "creationists", "reactionaries" whose work must carefully avoid mentioning, although it often turns out later that very similar arguments are sustained from the most intransigent seculars.

Thus, we can return to our discussion on truth as the basis for any epistemological and ethical consideration on nature. The sacrifice that comes from here, that is the sacrifice in bearing witness to the truth and its consequences, is also so linked to this fact that it constitutes the law made at the foundations of civilisation. Civilisation, in the real sense of the term, as suggested by Gandhi, does not consist in the multiplication but in the deliberate and voluntary restriction of needs. In fact, only with self-discipline we can think to accept and follow the requirements derived from the study of the laws of nature, rejecting idolatry and lust. Too often, however, we avoid to understand. Therefore we refuse to follow the requirements derived from the consciousness, the laws of nature, environmental ethics, taking an apology which in fact represents a retreat into idolatry.

It is especially in the name of progress that we have no qualms about destroying nature and its life forms. Repeatedly debated is the exemplary

Fig. 15.4 Historical Park of Monza. Section of the Autodromo which marks a symbolic victory of liberalism against nature and culture. In the background the neo-Gothic building of the "Menagerie". The ancient oak forests, of medieval origin, have been ruined and semi-destroyed by the presumptuous insertion

case of the Historical Park of Monza, which preserves forest relics dating back to the Middle Ages and that was designed by the famous landscape architect Luigi Canonica in 1805, and where in 1922 a large racetrack was built, which has ruined the largest green lung of Milan (Fig. 15.4).

The liberal and scientist relativism, as well as the ontological reductionism, are among the largest idolatry and it is worth remembering how much influence they have in the ecology, as we shall see.

Equally dangerous is the unbridled capitalism, the "growth" economy at all costs: the confidence that it always brings benefits and designs to make more rich and happy all men. From international data reported by Crédit Suisse [30], it appears that, following this ideology, the 8.4 % richest part of humanity has come to use the 83.3 % of the global wealth, leaving to the poorest 68.7 % part only 3.0%! Note that the distribution of resources between adult people with a high wealth (>1,000,000 $) and those with low wealth (<10,000 $) is in the ratio of 32/3,207 = 0.01.

The price of this ideology is, in fact, paid by all the living systems of nature.

Another false idol, which we have already had occasion to mention, is the inexhaustible faith in the ability of human technology, which emerges, for example, in the debate on nuclear power in Italy, dealing with this problem not in a scientific way, but almost like a fairy tale, where the technology would be able to solve everything, as noted with dismay Virginio Bettini [31].

On the other hand, a falsehood also lies in the so-called "green fundamentalism", according to which everything that is man-made is incompatible with nature and the man is considered the same way as animals. Thus, the preservation of the environment too often sustained for political slogans, becomes dangerous for living systems themselves or patently inconsistent and completely free from true scientific considerations. Consider the cases of preconceived opposition to regulate the high water in the Lagoon of Venice or the demonisation of radio waves and electromagnetic fields associated with high-voltage networks.

All the idols of Biblical reference can be symbolised in the 'god Mammon', Aramaic word that has different meanings, but involving money, lie and an explicit reference to original sin. In the painting of Beppe Ingegnoli (1901–1999) the gold Snake symbolising evil generates all violence and destroys the man and the Earth (Fig. 15.5).

Again, we can observe that, when the role of man in nature is guided by the Truth, it produces

Fig. 15.5 *The god Mammon*, painting in tempera and pastel (2×1 m) by Enrico Giuseppe Ingegnoli, known as Beppe Ingegnoli (1901–1999), executed in 1991–1992 to consider the effects of evil which, through the greed that breeds violence, attempts to destroy the man and the Earth

profound ethical implications at all levels of thought and action. It leads mainly a renewal of Science, which will include the natural sciences and ecology, with important consequences in the theories and their applications.

Finally, you need to realise the damage produced by the reductionist scientism or by a distorted view of the truth, in the study of issues affecting the environment and the landscape. We list some of the most significant of these distortions:

- The science of vegetation, when the attempt is to use only the phytosociological method proposed 100 years ago by Braun-Blanquet [32], which showed several limitations, especially at the landscape level (see Sect. 5.1)
- The relationship between human health and environmental alterations, which are limited to the control of pollution, regardless of the pathologies of the landscape (see Sect. 4.5)
- The environmental sustainability, when considering only the geographical territory that support and equalise the "3 E" (environment, economy, equal opportunity) (see Sect. 11.1)
- The evolutionary process, when you pretend to talk about the literal creationism or, on the contrary, insists on Darwinism turning it into an ideology, almost a religion (see Sect. 4.3)
- Climate change, when you want to display at all costs that is solely due to the greenhouse effect and human action
- The loss of biodiversity, when considered eminently specific and caused exclusively by men
- Eco-design, when it remains limited to general ecology which considers the anthropic landscapes as detached from the ecological laws (see Sect. 10.3 and Sect. 11.1)
- The ecological criteria for the systems of urban green space, when, following complacent technocratic and political speeches, when ecological incorrect and backwards choices are accomplished (see Sect. 11.3)

15.3.2 The Hippocratic Oath of the *Ecoiatra*

At the beginning of the chapter, it was pointed out that the landscape is a living entity, the specific level of the organisation of life on Earth (see also Sect. 1.1.1) and, therefore, subject to the laws that govern every other living entity, although declined in an opportune manner. Thus the need for the traditional figure of the doctor, who takes care of body and of the human population, the veterinarian and plant pathologist, to work alongside the new figure of the Physician of environmental systems (*ecoiatra*). So much that in the book "Landscape Ecology: manual to conserve, manage and plan the environment" [33] Giglio and Ingegnoli have included, in analogy with Hippocrates, an "Oath of the Ecoiatra" to strongly indicate the ethical commitment of the ecologist.

Oath of the Ecoiatra

Aware

- That life on Earth is organised in hierarchical levels and that cannot be defined only within each biologic entity but including also each proper environment.
- That the ecologist has to take care of the health of the levels of ecocoenotope and landscape, as well as the doctor is in charge of health state of the organism and population levels.
- About the importance and solemnity of the act that I make and the commitment that I take.

I Swear

- To exert and practise ecology in freedom and independence of judgment and behaviour.
- To pursue the exclusive purposes of the defence of life, the protection of health and the harmony of Nature in all its components (including humans) and the care of Nature itself, according to the clinical-diagnostic method and the Principle of Strategic Rehabilitation.
- To inspire every professional act to responsibility and scientific, cultural and social commitment.
- To never commit acts deliberately causing the destruction of a living system.
- To comply in my business to the ethical principles never using my knowledge against them, in respect of Life.
- To make my work with diligence, skill and prudence in good faith and observing the rules of law which are not in conflict with the purposes of my profession following the most advanced scientific methods.
- To exercise my right to conscience objection when be asked to act in contravention to scientific evidence.
- To oppose actively to all those requests in which the inadequacy of available time or resources may prevent the carrying out of studies and interventions commensurate with the definition and treatment of the health of the ecological system under consideration.
- To entrust my reputation only to my professional skills and study capacities and to my own moral qualities.
- To avoid, even outside of professional practice, every act and conduct prejudicial to the prestige and dignity of the profession.
- To avail of the help of fellow specialists in every situation in which my specific skills are not appropriate.
- To apply myself to all situations, regardless of the spatial-temporal scale and of the level of urbanisation or degradation, with the same commitment.
- To put myself at the disposal of the Competent Authority in case of public calamity.
- To keep secret all that I see or that I have seen, understood or perceived, in the exercise of my profession.

References

1. Ingegnoli V (2002) Landscape ecology: a widening foundation. Springer, Berlin
2. Ingegnoli V (2003) La minorità della sora nostra madre Terra. Implicazioni etiche, ecologiche ed economiche. In: Padovese L (ed) Minores et subditi omnibus. Pontificio Ateneo Antonianum. Laurentianum, Roma
3. Israel G (2010) Per una medicina umanistica. Apologia di una medicina che curi i malati come persone. Lindau, Torino
4. Ingegnoli V (2010) Ecologia del paesaggio: l'ecologia del paesaggio biologico-integrata. In: Gregory T (ed.) XXI Secolo, vol IV. Istituto della Enciclopedia Italiana Treccani. pp 23–33, Roma
5. Ingegnoli V (2011) Bionomia del paesaggio. L'ecologia del paesaggio biologico-integrata per la formazione di un 'medico' dei sistemi ecologici. Springer, Milano

6. Rowlands M (2009) The Philosopher and the Wolf. Tr. It. Il lupo e il filosofo. Lezioni di vita dalla natura selvaggia. Mondadori, Milano
7. Arendt H (1964) Eichmann in Jerusalem. Tr.it. P. Bernardini, La banalità del male: Eichmann a Gerusalemme. Feltrinelli, Milano
8. Lanza Del Vasto G (1954) Vinoba où le nouveau pèlerinage. Tr. It. Vinoba o il nuovo pellegrinaggio. Jaca Book 1980, Milano
9. Bloch E (1954–1959) Il principio speranza. Tr. It. Garzanti 1994, Milano
10. Bradshaw CJA, Xingli G, Navjot S, Sodhi SW (2010). Evaluating the Relative Environmental Impact of Countries. PLoS ONE 5(5):e10440
11. Popper KR (1996) Tutta la vita è risolvere problemi: scritti sulla conoscenza, la storia e la politica. Rusconi, Milano
12. Gödel K (1931) Über formal unentscheidbare Satze der Principia Mathematica und verwandter Systeme. Mh Math Phys 38:173–198
13. Alfred Tarski (1983/1956) In: Corcoran J (ed) Logic, semantics, metamathematics. Oxford University Press, Oxford (Hackett, 1st edition edited and translated by J. H. Woodger)
14. Hilbert D (1922) Die logische Grundlagen der Mathematik. Mathematischen Annalen 88 (12):151–165
15. Israel G (2008) Chi sono i nemici della scienza? Riflessioni su un disastro educativo e culturale e documenti di malascienza. Lindau, Torino
16. Lorenz K, Popper K (1985) Il futuro è aperto. Rusconi, Milano
17. Lorenz K (1978) Vergleichende Verhaltensforschung: Grundlagen der Ethologie. Springer, Berlin, Wien
18. Monod J (1970) Le hazard et la nécessité. Ed. du Seul, Paris
19. Solzenicyn A (1963) Una giornata di Ivan Denisovic. It. Einaudi, 1999. Torino
20. Alexander E (2012) Proof of heaven: a neurosurgeon's journey into the afterlife. Simon & Schuster, Boston
21. Lorenz K (1973) Die Acht Todsünden der Zivilisierten Menshheit. Piper & Co, München
22. Cicero MT (I sec.) De Legibus. I, XXIII. Roma
23. Lorenz K (1981) L'etologia. Bollati Boringhieri, Torino
24. Einstein A (1965) Pensieri degli anni difficili. Bollati Boringhieri, Torino
25. Prigogine I (1996) La fin dès certitudes: temps, chaos et les lois de la nature. Editions odile Jacob, Paris
26. Einstein A (1949) Autobiografisches, tr.it. Autobiografia scientifica, in. A. Einstein Opere scelte, Bollati-Boringhieri, 1988, Torino
27. Einstein A (1944) Einstein Archives, 61–574
28. Laudisa F (2009) Albert Einstein, un atlante filosofico. Ed. Tascabili Bompiani, Milano
29. Bronowski J (1969) Nature and knowledge: the philosophy of contemporary science. Condon lectures, London
30. Credit Suisse (2013) Global wealth report. Berne
31. Bettini V, Nebbia G (eds) (2009) Il nucleare impossibile. UTET Libreria, Torino
32. Braun Blanquet J (1926) Etudes phytosociologiques en Auvergne. G.Mont-Louis, Clermont Ferrand
33. Ingegnoli V, Giglio E (2005) Ecologia del Paesaggio: manuale per conservare, gestire e pianificare l'ambiente. Sistemi editoriali SE, Napoli

Synthetic Glossary

by Elena Giglio

Entries or abbreviations	Field, quality or measure	Definition or short explanation
Adaptive system	Physics	System able to modify itself and to define its proper domain of environmental pertaining perturbation, specifying the admissibility of the different environmental constraints
Alliance	Phytosociology	Taxonomic level superior to association
Allometry	Forestry	The quantitative relation between the size of a part and the whole in a series of related organisms differing in size
Association	Phytosociology	Vegetational— or plant: a plant community presenting a defined floristic composition in relation to specific climatic and edaphic conditions. It represents a vegetation type having a taxonomic name with the Latin suffix—*etum*
Attractor	Mathematics	A geometrical object towards which the trajectory of a dynamic system converges in the course of time
Biological spectrum	Ecology	The whole of the hierarchic levels of life organisation on Earth. Approached by four different points of view, completely and unified by landscape bionomics
Biomass, plant	Ecology	*pB*. The dry weight or the volume above ground of the individuals of a plant species population in a given area
Bionomics	Bionomics	Synonymous of "biological-integrated ecology." The study of the biological-environmental laws of Nature as a complex system of hyper-complex systems
BTC	Bionomics [Mcal/m^2/year]	Biological territorial capacity of vegetation. The flux of energy that a system dissipates to maintain its equilibrium state and its organisational level. Proportional to the state of metastability of the system
BTC Classes	Bionomics (I to IX)	Nine standard ranges of BTC values, corresponding to different sets of vegetation formations, available to indicate the landscape structure and functions
CBSt	Bionomics	Concise bionomic state of vegetation. It relates the maturity level and the bionomic quality to check the bionomic efficiency of phytocoenosis

(continued)

Entries or abbreviations	Field, quality or measure	Definition or short explanation
CLD	Bionomics	Complex landscape diversity. Synthetic index obtained by multiplying the three main biodiversity indexes (structural × functional × apparatus) of the landscape
Complex LU	Bionomics	Interacting disposition of simple, recurring and structuring LU (see), the origin of which pertains to the same geomorphic group of processes
Copying and...	Biology	The first portion of the processes necessary to beget a new living entity, related to hereditary characters for an individual, and to propagule banks and ecological memory for an ecocoenotope
...Coding	Biology	The second portion of the processes necessary to beget a new living entity, essential to transfer significance, not only information: they give the rules of semantics, of reciprocity, etc. to be applied to the copied genetic sequences or some landscape transformation processes to express the life
Deterministic chaos	Mathematics	Non-random systems highly sensitive to initial conditions which lead to diverging outcomes
Determinism	Philosophy	Principle establishing that the dynamic state of a system in a certain time is completely defined by its dynamic state at a given initial moment and by its interaction with the environment. Causality
Dynamics (of a system)	Physics, ecology	System dynamics, or movement of a system. The shift of a system from a state condition A (at time t°) to a state condition B (at time t^{1}) within its proper space of the phases (that is its field of existence)
Disturbance	Ecology	Synonymous with perturbation. A process, or event, able to direct the structuring of an ecological system and to be incorporated by it
Ecocoenotope	Bionomics	Multifunctional entity, representing the elementary unit of integration of the community, the ecosystem and the microchore (spatial contiguity characters) in a definite geographic locality. Part of the territory that is uniform throughout its extent in landform, soil and vegetation. Tessera of the basic mosaic
Ecoiatra	Bionomics	Physician of the living systems from the landscape level up to the whole Earth, expert connoisseur and researcher on their health state and on the Laws of Nature. To be understood in a Hippocratic sense

(continued)

Entries or abbreviations	Field, quality or measure	Definition or short explanation
Ecological control of human culture	Bionomics	Control exerted in many cases—even when the process seems to have other origins, so trough an unconscious transmission of needs— on human culture, on human way of thinking or taking strategic decision (etc.), by natural Laws to balance some strong changes affecting directly or indirectly Nature itself
Ecosystem	Ecology	Part of the ecocoenotope concerning the functional relations between its component communities and the abiotic factors of their environment
Ecotissue	Bionomics	The ecological tissue of the landscape, as the weft and the warp in weaving or the cells in a histologic tissue. Multidimensional conceptual structure representing the hierarchical intertwining, in past, present and future, of the ecological upper and lower biological levels and of their relationships in the landscape
Ecotissue model	Bionomics	The iterative integration of all the simple and complex mosaics (the vegetational one being the basis), of correlated arguments, bonds and information concerning the landscape, at different spatial and temporal scales, guided and hierarchically ordered by the biological laws governing each specific landscape
Ecotope	Bionomics	The smallest unitary landscape element, multidimensional, that owns all the structural and functional characters and properties of the concerned landscape. Minimum system of interdependent ecocoenotopes determined by the topographical recurrence, the geomorphologic origin, the functional configuration and role within the landscape
Emerging property principle	Physics, bionomics	Principle asserting that "...an organic whole/system (a living entity too) is more complex than the sum of its parts..." and that "...the characters of the system are the consequence of the way in which its elements organise themselves". Already cited by EP Odum but not yet understood completely, becoming one of the mainstay of the scientific concept of landscape

(continued)

Entries or abbreviations	Field, quality or measure	Definition or short explanation
Environmental rehabilitation strategy	Bionomics	Updating of the concepts of sustainability (of a *status quo*), mitigation and restoration (of a *status quo ante*) related to an human action in a landscape. It consists of the elaboration of a therapeutic course for a LU (or portion) —through the hierarchical compatibility and integration of landscape health needs and human needs—focused on the rehabilitation of one/more functions
Epistemology	Philosophy	The study of the human knowledge. From the Greek *epistēmē* ("knowledge") and *logos* ("reason"). Along with metaphysics, logic and ethics, it is one of the four main branches of philosophy. In Europe mainly related with the philosophy of science
(the) fittest vegetation (for…)	Bionomics	The most suitable or suited vegetation to the specific climate and geomorphic conditions of a certain, limited, period of time in a certain defined place, for the main range of incorporable disturbances (including man's), in natural and not natural conditions. Patchy in a landscape
HH	Bionomics	*Human habitat*. Area when human populations live and work permanently, limiting the self-regulation capability of natural systems. Limited to human and semi-human LU, ecotopes and apparatuses, may contain a small percentages of natural habitat. Not to be intended and measured as in geographical and urban planning sense
Incompressible information	Mathematics	The term "information" might be misleading, as it depends on the concept of "compressibility." In the Algorithmic Information Theory, the information content of a string is equivalent to the length of the shortest possible "self-contained representation" (i.e. a program) of that string. Following this theory, the complex specified information (CSI) is the "*incompressible*" information, which cannot be synthetised in a more simple form or rule
Integration (hierarchical)	Bionomics	Combination of data, simple and complex elements, models and information related to a living system, ordered on the biological basis of their different degree of importance within a defined contest. For each system it need its history + its scale + its context to be known

(continued)

Entries or abbreviations	Field, quality or measure	Definition or short explanation
LaBiSF	Bionomics	Landscape bionomic survey of fauna. The new bionomic method proposed to study the animal component, currently seeing as two cornerstones: the study of faunal sensitivity and the study for SHf, similar to what is already in depth for human populations
LaBiSHH	Bionomics	The new bionomic method to analyse the main components and characters linked with the concept of human habitat
LaBiSV	Bionomics	Landscape bionomic survey of vegetation. The new bionomic methodology able to determine the state of normality of the ecological parameters of different types of vegetation, to compare the ecological status of natural and man-made vegetated tesserae and to measure the concept of biodiversity at the landscape level (diversity of biological organisation of the context)
Landscape	Bionomics	A proper level of the hierarchy of life organisation between the ecocoenotope level and the ecoregion level, thus a concrete living entity
Landscape apparatus	Bionomics	Complex configuration of multifunctional patches (set of tesserae and/or ecotopes, even not connected) performing a principal dominant function within the ecotissue
Living entity	Biology, bionomics	System presenting the main characters of life. It is synonymous of biological system and it is valid at all scales
Living system	Biology	It includes both biological systems and environmental systems. An ecological system can be considered a hybrid concept in this sense
LTpE	Bionomics	Landscape type evaluation. The bionomic index able to classify the landscape types in relation to HH, NH, Carrying capacity, BTC and agriculture + forestry elements
LU (simple)	Ecology, bionomics	Landscape unit. Eco-bio-noo-geodistrict, or portion of the landscape, owning peculiar structural/functional properties and well-defined (even if sometimes as functional gradient) boundaries, within the entire landscape; not always recognisable and delineable without appropriated structural and/or functional studies. Integrated set of organised ecotopes
Metastability	Chemistry, ecology, bionomy	State of a system oscillating around a central position but susceptible to being diverted to another (metastable) equilibrium state, always far from the zero when concerning living systems

(continued)

Entries or abbreviations	Field, quality or measure	Definition or short explanation
Multidimensionality	Bionomics	Concept used to express the numerousness of dimensions (spatial, temporal, configurative, thematic, thermodynamic, informative, hierarchical, of integration, etc.) involved in a hyper-complex system like a landscape
Multifunctionality	Landscape ecology	Ability of any landscape element to perform more ecological functions, even if with different intensity
NH	Bionomics	Natural habitat. Natural LU, ecotopes and apparatuses dominated by natural components and biological processes, without direct human influence, capable of normal self-regulation: admitting a small share of HH
Non-equilibrium thermodynamics	Physics, chemistry, bionomics	Thermodynamics of not-reversible processes, following a time-oriented arrows, expressing the richness of behaviours within a coherent universe of metastable systems
Operative LU	Bionomics	Portion of one or more LU (see), even belonging to different landscape types, in accordance with the particular problem or function that must be investigated. Merely work-based, often linked to administrative (and not bionomical) boundaries
Order	Physics, mathematics, bionomics	The character of always better organised systems, so the principle permitting the evolution even not in contrast to the second law of thermodynamics. We have to distinguish between energetic and information order
Selection (natural)	Biology, ecology	Often regarded as virtually the sole process of evolutionary change in Nature, in reality only one of the maintenance processes necessary to minimise errors in processes of copying
Self-organising system	Biology	System able to affect or determine its own internal structure, on the basis of information and semeiotics deriving from biological or technological processes, received as matter or energy
SH	Bionomics	Standard Habitat per capita. Reciprocal function of the "specific density"; the real relative proper habitat surface, even of different habitat types, at a specific biological entity's disposal. The theoretic standard habitat, *SH**, represents the minimum optimal habitat surface needed to sustain the individual (human or animal)

(continued)

Entries or abbreviations	Field, quality or measure	Definition or short explanation
Tessera	Landscape ecology	The smallest homogenous unit visible at landscape scale and composing an ecomosaic. In landscape bionomics it corresponds to the ecocoenotope
ψ	Bionomics	Structural diversity of an ecotope/LU, on the basis of the relative dominance of the different ecocoenotope types
τ	Bionomics	Functional diversity of an ecotope or LU, on the basis of the relative dominance of the BTC standard classes
ω	Bionomics	Complex diversity of a LU, on the basis of the relative dominance of the diverse landscape apparatus types
σ	Bionomics	Carrying capacity of a living system, relates the real standard habitat per capita with the theoretical Minimum SH*; equal to the ratio SH/SH*
bQ	Bionomics	Bionomic quality. It can be measured after a survey of a phytocoenosis through a parametric standard form, in proportion to the theoretical optimum; bQ can be expressed in %
CHG, CON, DIS, ETN, EXR, GEO, PRD, PRT, RNT, RPD, RSD, RSL, SBS, SOU, STB,	Bionomics	Abbreviation of different landscape apparatuses, expressing different functions
g-LM	Bionomics	General landscape metastability. This synthetic landscape magnitude considers the functional diversity of distribution of the BTC in a given LU and the amount of information linked to that composition
MtL	Bionomics	Maturity level of a vegetation formation. It can be measured following the LaBiSV methodology

MIX
Papier aus verantwortungsvollen Quellen
Paper from responsible sources
FSC® C105338

If you have any concerns about our products,
you can contact us on
ProductSafety@springernature.com

In case Publisher is established outside the EU,
the EU authorized representative is:
Springer Nature Customer Service Center GmbH
Europaplatz 3, 69115 Heidelberg, Germany

Printed by Libri Plureos GmbH
in Hamburg, Germany